T0214329

Lecture Notes in Computer Science 12746

More information about this subseries at http://www.springer.com/series/7407

Maciej Paszynski · Dieter Kranzlmüller ·
Valeria V. Krzhizhanovskaya ·
Jack J. Dongarra · Peter M. A. Sloot (Eds.)

Computational Science – ICCS 2021

21st International Conference
Krakow, Poland, June 16–18, 2021
Proceedings, Part V

 Springer

Editors
Maciej Paszynski ⓘ
AGH University of Science and Technology
Krakow, Poland

Valeria V. Krzhizhanovskaya ⓘ
University of Amsterdam
Amsterdam, The Netherlands

Peter M. A. Sloot ⓘ
University of Amsterdam
Amsterdam, The Netherlands

ITMO University
St. Petersburg, Russia

Nanyang Technological University
Singapore, Singapore

Dieter Kranzlmüller ⓘ
Ludwig-Maximilians-Universität München
Munich, Germany

Leibniz Supercomputing Center (LRZ)
Garching bei München, Germany

Jack J. Dongarra ⓘ
University of Tennessee at Knoxville
Knoxville, TN, USA

ISSN 0302-9743   ISSN 1611-3349   (electronic)
Lecture Notes in Computer Science
ISBN 978-3-030-77976-4   ISBN 978-3-030-77977-1   (eBook)
https://doi.org/10.1007/978-3-030-77977-1

LNCS Sublibrary: SL1 – Theoretical Computer Science and General Issues

This Springer imprint is published by the registered company Springer Nature Switzerland AG
The registered company address is: Gewerbestrasse 11, 6330 Cham, Switzerland

Preface

Welcome to the proceedings of the 21st annual International Conference on Computational Science (ICCS 2021 - https://www.iccs-meeting.org/iccs2021/).

In preparing this edition, we had high hopes that the ongoing COVID-19 pandemic would fade away and allow us to meet this June in the beautiful city of Kraków, Poland. Unfortunately, this is not yet the case, as the world struggles to adapt to the many profound changes brought about by this crisis. ICCS 2021 has had to adapt too and is thus being held entirely online, for the first time in its history.

These challenges notwithstanding, we have tried our best to keep the ICCS community as dynamic and productive as always. We are proud to present the proceedings you are reading as a result of that.

ICCS 2021 was jointly organized by the AGH University of Science and Technology, the University of Amsterdam, NTU Singapore, and the University of Tennessee.

The International Conference on Computational Science is an annual conference that brings together researchers and scientists from mathematics and computer science as basic computing disciplines, as well as researchers from various application areas who are pioneering computational methods in sciences such as physics, chemistry, life sciences, engineering, arts, and humanitarian fields, to discuss problems and solutions in the area, identify new issues, and shape future directions for research.

Since its inception in 2001, ICCS has attracted an increasing number of attendees and higher quality papers, and this year is not an exception, with over 350 registered participants. The proceedings have become a primary intellectual resource for computational science researchers, defining and advancing the state of the art in this field.

The theme for 2021, **"Computational Science for a Better Future,"** highlights the role of computational science in tackling the current challenges of our fast-changing world. This conference was a unique event focusing on recent developments in scalable scientific algorithms, advanced software tools, computational grids, advanced numerical methods, and novel application areas. These innovative models, algorithms, and tools drive new science through efficient application in physical systems, computational and systems biology, environmental systems, finance, and other areas.

ICCS is well known for its excellent lineup of keynote speakers. The keynotes for 2021 were given by

- **Maciej Besta**, ETH Zürich, Switzerland
- **Marian Bubak**, AGH University of Science and Technology, Poland | Sano Centre for Computational Medicine, Poland
- **Anne Gelb**, Dartmouth College, USA
- **Georgiy Stenchikov**, King Abdullah University of Science and Technology, Saudi Arabia
- **Marco Viceconti**, University of Bologna, Italy

- **Krzysztof Walczak**, Poznan University of Economics and Business, Poland
- **Jessica Zhang**, Carnegie Mellon University, USA

This year we had 635 submissions (156 submissions to the main track and 479 to the thematic tracks). In the main track, 48 full papers were accepted (31%); in the thematic tracks, 212 full papers were accepted (44%). A high acceptance rate in the thematic tracks is explained by the nature of these tracks, where organisers personally invite many experts in a particular field to participate in their sessions.

ICCS relies strongly on our thematic track organizers' vital contributions to attract high-quality papers in many subject areas. We would like to thank all committee members from the main and thematic tracks for their contribution to ensure a high standard for the accepted papers. We would also like to thank *Springer, Elsevier,* and *Intellegibilis* for their support. Finally, we appreciate all the local organizing committee members for their hard work to prepare for this conference.

We are proud to note that ICCS is an A-rank conference in the CORE classification.

We wish you good health in these troubled times and look forward to meeting you at the conference.

June 2021

Maciej Paszynski
Dieter Kranzlmüller
Valeria V. Krzhizhanovskaya
Jack J. Dongarra
Peter M. A. Sloot

Organization

Local Organizing Committee at AGH University of Science and Technology

Chairs

Maciej Paszynski
Aleksander Byrski

Members

Marcin Łos
Maciej Woźniak
Leszek Siwik
Magdalena Suchoń

Thematic Tracks and Organizers

Advances in High-Performance Computational Earth Sciences: Applications and Frameworks – IHPCES

Takashi Shimokawabe
Kohei Fujita
Dominik Bartuschat

Applications of Computational Methods in Artificial Intelligence and Machine Learning – ACMAIML

Kourosh Modarresi
Paul Hofmann
Raja Velu
Peter Woehrmann

Artificial Intelligence and High-Performance Computing for Advanced Simulations – AIHPC4AS

Maciej Paszynski
Robert Schaefer
David Pardo
Victor Calo

Biomedical and Bioinformatics Challenges for Computer Science – BBC

Mario Cannataro
Giuseppe Agapito

Mauro Castelli
Riccardo Dondi
Italo Zoppis

Classifier Learning from Difficult Data – CLD^2

Michał Woźniak
Bartosz Krawczyk

Computational Analysis of Complex Social Systems – CSOC

Debraj Roy

Computational Collective Intelligence – CCI

Marcin Maleszka
Ngoc Thanh Nguyen
Marcin Hernes
Sinh Van Nguyen

Computational Health – CompHealth

Sergey Kovalchuk
Georgiy Bobashev
Stefan Thurner

Computational Methods for Emerging Problems in (dis-)Information Analysis – DisA

Michal Choras
Robert Burduk
Konstantinos Demestichas

Computational Methods in Smart Agriculture – CMSA

Andrew Lewis

Computational Optimization, Modelling, and Simulation – COMS

Xin-She Yang
Leifur Leifsson
Slawomir Koziel

Computational Science in IoT and Smart Systems – IoTSS

Vaidy Sunderam
Dariusz Mrozek

Computer Graphics, Image Processing and Artificial Intelligence – CGIPAI

Andres Iglesias
Lihua You
Alexander Malyshev
Hassan Ugail

Data-Driven Computational Sciences – DDCS

Craig Douglas

Machine Learning and Data Assimilation for Dynamical Systems – MLDADS

Rossella Arcucci

MeshFree Methods and Radial Basis Functions in Computational Sciences – MESHFREE

Vaclav Skala
Marco-Evangelos Biancolini
Samsul Ariffin Abdul Karim
Rongjiang Pan
Fernando-César Meira-Menandro

Multiscale Modelling and Simulation – MMS

Derek Groen
Diana Suleimenova
Stefano Casarin
Bartosz Bosak
Wouter Edeling

Quantum Computing Workshop – QCW

Katarzyna Rycerz
Marian Bubak

Simulations of Flow and Transport: Modeling, Algorithms and Computation – SOFTMAC

Shuyu Sun
Jingfa Li
James Liu

Smart Systems: Bringing Together Computer Vision, Sensor Networks and Machine Learning – SmartSys

Pedro Cardoso
Roberto Lam

João Rodrigues
Jânio Monteiro

Software Engineering for Computational Science – SE4Science

Jeffrey Carver
Neil Chue Hong
Anna-Lena Lamprecht

Solving Problems with Uncertainty – SPU

Vassil Alexandrov
Aneta Karaivanova

Teaching Computational Science – WTCS

Angela Shiflet
Nia Alexandrov
Alfredo Tirado-Ramos

Uncertainty Quantification for Computational Models – UNEQUIvOCAL

Wouter Edeling
Anna Nikishova

Reviewers

Ahmad Abdelfattah
Samsul Ariffin Abdul
   Karim
Tesfamariam Mulugeta
   Abuhay
Giuseppe Agapito
Elisabete Alberdi
Luis Alexandre
Vassil Alexandrov
Nia Alexandrov
Julen Alvarez-Aramberri
Sergey Alyaev
Tomasz Andrysiak
Samuel Aning
Michael Antolovich
Hideo Aochi
Hamid Arabnejad
Rossella Arcucci
Costin Badica
Marina Balakhontceva

Bartosz Balis
Krzysztof Banas
Dariusz Barbucha
Valeria Bartsch
Dominik Bartuschat
Pouria Behnodfaur
Joern Behrens
Adrian Bekasiewicz
Gebrail Bekdas
Mehmet Belen
Stefano Beretta
Benjamin Berkels
Daniel Berrar
Sanjukta Bhowmick
Georgiy Bobashev
Bartosz Bosak
Isabel Sofia Brito
Marc Brittain
Jérémy Buisson
Robert Burduk

Michael Burkhart
Allah Bux
Krisztian Buza
Aleksander Byrski
Cristiano Cabrita
Xing Cai
Barbara Calabrese
Jose Camata
Almudena Campuzano
Mario Cannataro
Alberto Cano
Pedro Cardoso
Alberto Carrassi
Alfonso Carriazo
Jeffrey Carver
Manuel Castañón-Puga
Mauro Castelli
Eduardo Cesar
Nicholas Chancellor
Patrikakis Charalampos

Henri-Pierre Charles
Ehtzaz Chaudhry
Long Chen
Sibo Cheng
Siew Ann Cheong
Lock-Yue Chew
Marta Chinnici
Sung-Bae Cho
Michal Choras
Neil Chue Hong
Svetlana Chuprina
Paola Cinnella
Noélia Correia
Adriano Cortes
Ana Cortes
Enrique
    Costa-Montenegro
David Coster
Carlos Cotta
Helene Coullon
Daan Crommelin
Attila Csikasz-Nagy
Loïc Cudennec
Javier Cuenca
António Cunha
Boguslaw Cyganek
Ireneusz Czarnowski
Pawel Czarnul
Lisandro Dalcin
Bhaskar Dasgupta
Konstantinos Demestichas
Quanling Deng
Tiziana Di Matteo
Eric Dignum
Jamie Diner
Riccardo Dondi
Craig Douglas
Li Douglas
Rafal Drezewski
Vitor Duarte
Thomas Dufaud
Wouter Edeling
Nasir Eisty
Kareem El-Safty
Amgad Elsayed
Nahid Emad

Christian Engelmann
Roberto R. Expósito
Fangxin Fang
Antonino Fiannaca
Christos
    Filelis-Papadopoulos
Martin Frank
Alberto Freitas
Ruy Freitas Reis
Karl Frinkle
Kohei Fujita
Hiroshi Fujiwara
Takeshi Fukaya
Wlodzimierz Funika
Takashi Furumura
Ernst Fusch
David Gal
Teresa Galvão
Akemi Galvez-Tomida
Ford Lumban Gaol
Luis Emilio
    Garcia-Castillo
Frédéric Gava
Piotr Gawron
Alex Gerbessiotis
Agata Gielczyk
Adam Glos
Sergiy Gogolenko
Jorge
    González-Domínguez
Yuriy Gorbachev
Pawel Gorecki
Michael Gowanlock
Ewa Grabska
Manuel Graña
Derek Groen
Joanna Grzyb
Pedro Guerreiro
Tobias Guggemos
Federica Gugole
Bogdan Gulowaty
Shihui Guo
Xiaohu Guo
Manish Gupta
Piotr Gurgul
Filip Guzy

Pietro Hiram Guzzi
Zulfiqar Habib
Panagiotis Hadjidoukas
Susanne Halstead
Feilin Han
Masatoshi Hanai
Habibollah Haron
Ali Hashemian
Carina Haupt
Claire Heaney
Alexander Heinecke
Marcin Hernes
Bogumila Hnatkowska
Maximilian Höb
Jori Hoencamp
Paul Hofmann
Claudio Iacopino
Andres Iglesias
Takeshi Iwashita
Alireza Jahani
Momin Jamil
Peter Janku
Jiri Jaros
Caroline Jay
Fabienne Jezequel
Shalu Jhanwar
Tao Jiang
Chao Jin
Zhong Jin
David Johnson
Guido Juckeland
George Kampis
Aneta Karaivanova
Takahiro Katagiri
Timo Kehrer
Christoph Kessler
Jakub Klikowski
Alexandra Klimova
Harald Koestler
Ivana Kolingerova
Georgy Kopanitsa
Sotiris Kotsiantis
Sergey Kovalchuk
Michal Koziarski
Slawomir Koziel
Rafal Kozik

Bartosz Krawczyk
Dariusz Krol
Valeria Krzhizhanovskaya
Adam Krzyzak
Pawel Ksieniewicz
Marek Kubalcík
Sebastian Kuckuk
Eileen Kuehn
Michael Kuhn
Michal Kulczewski
Julian Martin Kunkel
Krzysztof Kurowski
Marcin Kuta
Bogdan Kwolek
Panagiotis Kyziropoulos
Massimo La Rosa
Roberto Lam
Anna-Lena Lamprecht
Rubin Landau
Johannes Langguth
Shin-Jye Lee
Mike Lees
Leifur Leifsson
Kenneth Leiter
Florin Leon
Vasiliy Leonenko
Roy Lettieri
Jake Lever
Andrew Lewis
Jingfa Li
Hui Liang
James Liu
Yen-Chen Liu
Zhao Liu
Hui Liu
Pengcheng Liu
Hong Liu
Marcelo Lobosco
Robert Lodder
Chu Kiong Loo
Marcin Los
Stephane Louise
Frederic Loulergue
Hatem Ltaief
Paul Lu
Stefan Luding

Laura Lyman
Scott MacLachlan
Lukasz Madej
Lech Madeyski
Luca Magri
Imran Mahmood
Peyman Mahouti
Marcin Maleszka
Alexander Malyshev
Livia Marcellino
Tomas Margalef
Tiziana Margaria
Osni Marques
M. Carmen Márquez
    García
Paula Martins
Jaime Afonso Martins
Pawel Matuszyk
Valerie Maxville
Pedro Medeiros
Fernando-César
    Meira-Menandro
Roderick Melnik
Valentin Melnikov
Ivan Merelli
Marianna Milano
Leandro Minku
Jaroslaw Miszczak
Kourosh Modarresi
Jânio Monteiro
Fernando Monteiro
James Montgomery
Dariusz Mrozek
Peter Mueller
Ignacio Muga
Judit Munoz-Matute
Philip Nadler
Hiromichi Nagao
Jethro Nagawkar
Kengo Nakajima
Grzegorz J. Nalepa
I. Michael Navon
Philipp Neumann
Du Nguyen
Ngoc Thanh Nguyen
Quang-Vu Nguyen

Sinh Van Nguyen
Nancy Nichols
Anna Nikishova
Hitoshi Nishizawa
Algirdas Noreika
Manuel Núñez
Krzysztof Okarma
Pablo Oliveira
Javier Omella
Kenji Ono
Eneko Osaba
Aziz Ouaarab
Raymond Padmos
Marek Palicki
Junjun Pan
Rongjiang Pan
Nikela Papadopoulou
Marcin Paprzycki
David Pardo
Anna Paszynska
Maciej Paszynski
Abani Patra
Dana Petcu
Serge Petiton
Bernhard Pfahringer
Toby Phillips
Frank Phillipson
Juan C. Pichel
Anna
    Pietrenko-Dabrowska
Laércio L. Pilla
Yuri Pirola
Nadia Pisanti
Sabri Pllana
Mihail Popov
Simon Portegies Zwart
Roland Potthast
Malgorzata
    Przybyla-Kasperek
Ela Pustulka-Hunt
Alexander Pyayt
Kun Qian
Yipeng Qin
Rick Quax
Cesar Quilodran Casas
Enrique S. Quintana-Orti

Ewaryst Rafajlowicz
Ajaykumar Rajasekharan
Raul Ramirez
Célia Ramos
Marcus Randall
Lukasz Rauch
Vishal Raul
Robin Richardson
Sophie Robert
João Rodrigues
Daniel Rodriguez
Albert Romkes
Debraj Roy
Jerzy Rozenblit
Konstantin Ryabinin
Katarzyna Rycerz
Khalid Saeed
Ozlem Salehi
Alberto Sanchez
Aysin Sanci
Gabriele Santin
Rodrigo Santos
Robert Schaefer
Karin Schiller
Ulf D. Schiller
Bertil Schmidt
Martin Schreiber
Gabriela Schütz
Christoph Schweimer
Marinella Sciortino
Diego Sevilla
Mostafa Shahriari
Abolfazl
  Shahzadeh-Fazeli
Vivek Sheraton
Angela Shiflet
Takashi Shimokawabe
Alexander Shukhman
Marcin Sieniek
Nazareen
  Sikkandar Basha
Anna Sikora
Diana Sima
Robert Sinkovits
Haozhen Situ
Leszek Siwik

Vaclav Skala
Ewa
  Skubalska-Rafajlowicz
Peter Sloot
Renata Slota
Oskar Slowik
Grazyna Slusarczyk
Sucha Smanchat
Maciej Smolka
Thiago Sobral
Robert Speck
Katarzyna Stapor
Robert Staszewski
Steve Stevenson
Tomasz Stopa
Achim Streit
Barbara Strug
Patricia Suarez Valero
Vishwas Hebbur Venkata
Subba Rao
Bongwon Suh
Diana Suleimenova
Shuyu Sun
Ray Sun
Vaidy Sunderam
Martin Swain
Jerzy Swiatek
Piotr Szczepaniak
Tadeusz Szuba
Ryszard Tadeusiewicz
Daisuke Takahashi
Zaid Tashman
Osamu Tatebe
Carlos Tavares Calafate
Andrei Tchernykh
Kasim Tersic
Jannis Teunissen
Nestor Tiglao
Alfredo Tirado-Ramos
Zainab Titus
Pawel Topa
Mariusz Topolski
Pawel Trajdos
Bogdan Trawinski
Jan Treur
Leonardo Trujillo

Paolo Trunfio
Ka-Wai Tsang
Hassan Ugail
Eirik Valseth
Ben van Werkhoven
Vítor Vasconcelos
Alexandra Vatyan
Raja Velu
Colin Venters
Milana Vuckovic
Jianwu Wang
Meili Wang
Peng Wang
Jaroslaw Watróbski
Holger Wendland
Lars Wienbrandt
Izabela Wierzbowska
Peter Woehrmann
Szymon Wojciechowski
Michal Wozniak
Maciej Wozniak
Dunhui Xiao
Huilin Xing
Wei Xue
Abuzer Yakaryilmaz
Yoshifumi Yamamoto
Xin-She Yang
Dongwei Ye
Hujun Yin
Lihua You
Han Yu
Drago Žagar
Michal Zak
Gabor Závodszky
Yao Zhang
Wenshu Zhang
Wenbin Zhang
Jian-Jun Zhang
Jinghui Zhong
Sotirios Ziavras
Zoltan Zimboras
Italo Zoppis
Chiara Zucco
Pavel Zun
Pawel Zyblewski
Karol Zyczkowski

Contents – Part V

Computer Graphics, Image Processing and Artificial Intelligence

Data-Driven Computational Sciences

Machine Learning and Data Assimilation for Dynamical Systems

MeshFree Methods and Radial Basis Functions in Computational Sciences

Multiscale Modelling and Simulation

Computer Graphics, Image Processing and Artificial Intelligence

Factors Affecting the Sense of Scale in Immersive, Realistic Virtual Reality Space

Jarosław Andrzejczak$^{(\boxtimes)}$ [ID], Wiktoria Kozłowicz, Rafał Szrajber [ID], and Adam Wojciechowski [ID]

Institute of Information Technology, Lodz University of Technology, 215 Wólczańska Street, 90-924 Lodz, Poland
{jaroslaw.andrzejczak,rafal.szrajber}@p.lodz.pl
http://it.p.lodz.pl

Abstract. In this study, we analyze and identify a proper scale value when presenting real world space and everyday objects in immerse VR. We verify the impact of usage of reference points in the form of common objects known to the user such as windows, doors and furniture in the sense of scale in VR. We also analyze user behavior (position, rotation, movement, area of interest and such) in the scale setting task. Finally, we propose optimal scale values for single objects presentation, architectural space with many points of references and a large scale space with less to no points of reference. The experiments were conducted on two groups: the Experts (architects) and Non-experts (common users) to verify the translation of real-world object size analysis skills into the same capacity in the virtual world. Confirmation of the significance of the pre-immersion in VR for a sense of scale accuracy is also described.

Keywords: Sense of scale · Virtual reality · Sense of scale factors

1 Introduction

In recent years, Virtual Reality (VR) has been popularized mainly thanks to the computer games market, but more and more industry companies see the potential in this technology. It opens up huge opportunities for the architecture, allowing exploring the space of the building long before it is built. Still, there are many negative factors including disorientation or dizziness, which prevent correct reception in the virtual world [9]. For VR architecture-related applications, one of the main problems now is the distorted aspect ratio and scale. Allowing the user to fully adapt to a world in which a sense of depth, scale and spatial awareness is mapped, will give VR a huge advantage over traditional forms of information transfer, such as renders, 3D models and animations.

The purpose of the research was to verify the existence of disproportions in the reception of the size of architectural objects in virtual reality in relation to the

© Springer Nature Switzerland AG 2021
M. Paszynski et al. (Eds.): ICCS 2021, LNCS 12746, pp. 3–16, 2021.
https://doi.org/10.1007/978-3-030-77977-1_1

given real dimensions and subjective assessment of the user. It involves examining the impact of various factors (such as pre-immersion[1], the size of the interior and the presence of a reference point in the form of common objects known to the user) on the sense of scale and user's behavior in the VR. The study will be conducted on two groups: Experts (the architects) and Non-experts (common users) to further explore the translation of real-world object size analysis skills into the same capacity in the virtual world.

The contributions to research concerning sense of scale, immersion as well as comfort in Virtual Reality presented in this article are:

- Confirmation of significance of the pre-immersion in VR for sense of scale accuracy.
- Analysis of user's motion (movement and rotation as well as areas of interest - AOI) during the VR scale evaluation task.
- Tests verifying the impact of usage of reference points in the form of windows, doors and furniture on the sense of scale in VR.
- Tests verifying whether the user's professional experience affects the sense of scale in VR (Experts versus Non-experts).
- Tests verifying the impact of view continuity or lack of it in on VR application comfort of use.
- Proposition of optimal scale values for single objects presentation, architectural space with many points of references and a large scale space with less to no points of reference.

We start with a related work overview in the next section. Then the factors affecting the sense of scale detailed description is given. This is followed by an evaluation method along with an information about study group and gathered data overview. Next, test results and their discussion are presented in total for both of the test sessions. Finally, ideas for further development and final conclusions will be given.

2   Related Work

Interesting example of the influence of various factors on the perception of the size of objects in VR is an experiment conducted by Renault to visualize the car's design before the production [3]. The task used in this study was to adjust the interior of the car cockpit. In order to avoid unnatural enlargement of the object in front of the user, a black screen was displayed whenever the scale changed. The study was conducted on four groups made of people who use the selected car model or not and with previous experience with VR or not. Some of them were also allowed to spend a certain time in the cockpit of the real car before the test (pre-immersion). The results of the group with pre-immersion were more

---

[1] Pre-immersion - the user's sense of immersion in digital reality and separating him from the real world extended by the possibility of earlier experiences of a fragment or the whole of the virtual world through its representation in the real world.

consistent, unlike people who could not be inside the car. Also, seeing car before the test had a much greater impact than even everyday use of a given car model. Finally, the authors propose a factor for the correct perception of the scale in the virtual reality environment: for goggles with the head movement tracking 1:0.98, while for devices without head tracking about 1:1.02.

The authors of another study observed not only the disproportion in the perception of real objects, but also the influence of their own body on the perception of the environment [15]. At the start of the study the participants were given a plastic cube for one minute to remember its size. Then they were asked to assess and properly adjust the size of his own hand and a plastic cube in four variants: cube size only (no interaction), the hand only, both with free interaction with the cube, both but sequentially - first the cube, and then the size of the hand. The study showed that users of the VR perceive their own hands as larger than they are in reality, while other objects appear to be smaller. The positive impact of the interaction on the perception of the scale was also proven (third variant).

Another interesting study focuses on the importance of the order and type of environment in which the user is located [7]. The participants were asked to adjust the size of the chair, which they could previously observe and analyze its dimensions in the real world. The subject was introduced to three VR scenes in order: the virtual equivalent of the room in which the experiment was conducted (full pre-immersion), then a futuristic visualization showing the large-size structure and chairs arranged in it and finally the interior of the museum with preserved real proportions. Results showed that it was much easier to determine the scale of object in known space (the worst results were observed in the futuristic interior). By transporting the user from a large-scale world to a much smaller world (or the other way around), his or her perception of the scale of the whole environment changes drastically. For the subjects, the futuristic environment seemed much larger than it actually was.

3 Factors Affecting the Sense of Scale in VR

The sense of scale in Virtual Reality is a sum of many factors. We have to consider physiological and anatomical features of the eye structure as well as elements of psychology of vision. The most important of them are: ability of stereoscopic vision (including IPD - interpupillary distance), parallax effect, Field of View (FOV), Body Base Scaling (BBS), objects known to the user, number of frames per second (FPS), objects out of focus or bad Depth of Field (DOF) and visual aspect. All of those factors have been taken into account by the authors in the experiments described in this article.

Ability to stereoscopic vision is the ability to perceive depth and distance binocularly. Knowing the viewing angle and IPD, the visual cortex calculates the distance to the observed object. However, this only works on objects that are close to the observer (within the convergence phenomenon). At distances greater than about 6 m, the angular differences are too small to detect [4,8].

Parallax effect is a phenomenon based on the incompatibility of images of the same object that is observed from different points. Objects further away seem to move in the same direction as the observer when the objects being closer seem to move contrary to the observer's movement and faster than those that are at distant. Mapping the parallax effect in the virtual world is possible thanks to motion tracking devices.

Field of View (FOV) is a space that a person perceives simultaneously with the fixation point of the pattern. The field of view can differ for each individual due to factors such as the depth of the eye socket, lowered eyebrow or even a drooping eyelid. For VR applications, two separate *FOV* types should be considered: *FOV* of virtual cameras and the second that results from the construction of the VR goggles.

Body Base Scaling (BBS) is the ability to use parts of the body (most often hands) to assess the size of objects in the surrounding environment [12]. Gibson [6] even emphasized that people do not perceive the environment per se, but the relationship between their body and the environment. Additionally, Gedliczka in [5] presents anthropometric measures and their application in functional design.

Object that is known to the user can be used to assess the size of unknown objects. Researchers recognize the role of the module in the always-present architectural elements, such as doors and windows, whose dimensions, dictated usually by a clear functional need, are widely recognized [16].

Frames per second (FPS) or specifically the reduction of the number of displaying frames can affect negatively not only the perception of the scale, but also the entire immersion and cause discomfort. This phenomenon can primarily disturb the parallax effect and the assessment of the distance of objects based on head movements. It can also cause frustration or even malaise and should be assured in value recommended by VR equipment manufacturer [14].

Depth of filed (DOF) represented as blurred objects outside the focus area are widely used in computer graphics and video production. For a VR application that may negatively affect the perception of the scale and create the unwanted impression of movement, thus causing the users to feel unwell. Therefore, setting a photographic DOF effect in VR is not recommended.

Visual aspect understood as providing correctly displayed lights and shadows can significantly improve the perception of space. For objects that are far away, when stereoscopic vision no longer works, the brain recognizes their size and distance thanks to, among others, shadows and perspective.

4   Evaluation

The study was divided into two sessions (named accordingly A and B). The hypotheses in individual sessions were as follows:

First Session (A)

1. Task A1: Displaying a black screen between each change of the scale factor will reduce the impression of unnatural magnification of objects in front of the subjects, making it easier for users to adjust the size of the objects.

2. Task A2: The possibility of experiencing the phenomenon of partial pre-immersion on a real object will make it easier for users to adjust the size of its virtual counterpart.

Second Session (B)

3. Task B1: The impact of pre-immersion as well as user's behavior is the same regardless of the ability to analyze the size of objects in the real world (Experts vs Non-experts).
4. Task B2 and B3: The absolute real size of the room affects its perception in the virtual world.
5. Task B3 to B5: The existence of a reference point in the form of windows, doors and furniture positively affects the sense of scale in the virtual world.
6. Task B1 to B5: There is a translation of real-world object size analysis skills into the same capacity in the virtual world.

4.1 Test Tasks

All respondents were first familiarized with the purpose the study and individual tasks as well as the principles of control (increase/decrease the scale and display of the architectural plan using the controller). Then, for each participant the eye distances from the ground and pupil spacing were measured [2]. After that, the participant was given a specific task regarding the study session.

First Session (A)
Task A1. The aim of the first task was to change the scale of the rectangular cardboard box sized $50 \times 50 \times 100$ cm displayed on the screen with initial scale 1:0.4 (Fig. 1c)) until it reaches a size subjectively corresponding to the dimensions given in the technical drawing (displayed in the virtual space). Two variants were used: one with a black screen appeared for 0.2 s with each object scale change and the other without the black screen. Ten subjects were randomly and evenly assigned to two groups, where each group saw one of these variants first (with or without black screen). We assumed that the use of a black screen with each change of scale forces the user to assess the size of the object with each step again. Thanks to this, the decision on the next step (enlargement or reduction of the object) is more thought-out by the user. In addition, the fact of enlarging the object unnaturally in front of the user may lead to an assessment of its size based on the previous size of the object. The secondary goal of this task was to observe a basic user behaviour during scale assessment task. Scale change, subject position as well as subject rotation was recorded for that purpose.

Task A2. The aim of the task was to determine whether partial pre-immersion will affect the final result of the scale parameter set by users and what impression it will cause for users. The task and participant's distribution to two groups were identical to A1 with the difference with black screen variant was replaced by the possibility of seeing the real box before and during the task (Fig. 1a) and b)).

Fig. 1. The participant of the study during the task consisting in adjusting the scale of the box in a variant with the possibility of experiencing pre-immersion, a) during the first session (task A2); b) during the second session (task B1). c) A1, A2 and B1: box visualization in VR.

The second variant included the same task, but without real object (users had only a drawing available with the dimensions of the tested object).

Second Session (B)
Task B1. The purpose of this task is identical to the A2 task, with the difference that it is carried out by both: ordinary users and experts (Fig. 1b)). The user has a minute to look at the object and remember its subjectively felt size. The user can also at any time remove the glasses and recall the size of the item. In addition, it can display the box's projection with dimensions at any time.

Fig. 2. Scenes used in the second session. a) B2: a large hall with a sloping roof. b) B3: an unfurnished room of standard dimensions, without windows and doors. c) B4: several-room, unfurnished apartment of standard sizes. d) B5: a several-room, furnished apartment of standard sizes.

Tasks B2 to B5. During all further tasks in this session the participants were asked to adjust the size of the room to the dimensions given on the architectural plan. Initially, each room was significantly reduced, relative to its actual size. The examined person has the opportunity to view an architectural plan with selected dimensions at any time and any number of times. The plans were displayed right in front of the user's eyes, while completely obscuring the view of the examined interior in order to avoid the effect of scaling the room relative to the page with the drawn plan. The plans were made in accordance with the basic principles of creating a technical construction drawing. The users were transferred to (Fig. 2):

– Task B2: large hall with initial scale of 1:0.3. The subject will focus only one dimension - height. The 8-m high building with an area of 615 m^2 is characterized by a large number of windows that prevent the feeling of overwhelming.
– Task B3: a rectangular empty room without doors or windows of about 12 m^2 and a height of 2.7 m with initial scale of 1:0.75.
– Task B4: empty apartment with doors and windows and a view of the street and neighboring buildings with initial scale of 1:0.75. The door size was adopted in accordance with standard dimensions as for study participants, adopted because of human anthropometric features [5].
– Task B5: apartment from previous task, but fully furnished with initial scale of 1:0.75. It was extended with standard-size furniture such as kitchen, table, chairs or sofa and a number of accessories (flowers, paintings).

At the end of the test users were asked to complete a short survey regarding the ease and convenience of the tasks performed.

5   Results and Analysis

In total, 40 people of different sexes and ages participated in the study. The number of study participants was selected in accordance with [11,13,17]. Twenty Non-experts (with an average ability to analyze spatial relations) participated in the first study session. The second study session was conducted on two groups, ten people in each: The architects (Experts, characterized by their acquired ability to analyze space in terms of its dimensions and the ability to read architectural plans and compare them with real space) and ordinary users (Non-experts, people unfamiliar with architecture from a professional point of view).

5.1   First Session

Task A1. We did not observe the significant difference between variant without black screen and with it in case of scale value with mean scale values accordingly 0,852 and 0,846 with same Confidence Interval equal 0,1 for p = 0,05^2 (Fig. 3). Participants presented similar behaviour by increasing the scale up to satisfactory value rather than passing this value and decreasing. Also, the plan with box size values was opened only in the first half of this process (mostly at the start of the test). The subjects presented similar movement pattern by moving towards the box and back several times (gray lines on Fig. 4) and looking at the object from different angles from time to time. With no other objects or points of reference in the 3D scene they try to measure object by movement. The black screen variant affects the participant movement being more stationary than in the other variant (less deviation from the center position, Fig. 4). There was no significant difference in subject rotation between both variants.

---

[2] All of the Confidence Intervals (CI) for all of the experiments shown in this study were calculated for $p = 0,05$. Therefore, we will omit that information in a later description writing just CI value.

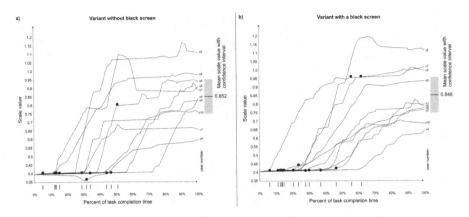

Fig. 3. Scale value and plan opening times for both variants (without black screen on the left, with a black screen on the right). Lines represent scale value change for each participant, with final value on the far right. Also, the mean value with CI is presented. A moment in which the subject displayed architectural plan is marked as dot on a corresponding scale value line as well as vertical line above X-axis. We did not observe the significant difference between those variants in terms of final scale value, scale change process or moments of viewing the plan.

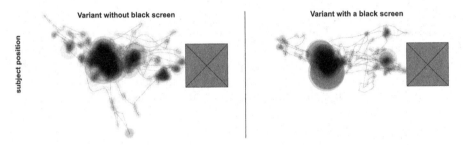

Fig. 4. Cumulative heatmap of the subject's position for both variants (without black screen on the left, with a black screen on the right) gathered five times on the second. The darker the circle, the more participants visits that spot. The bigger the circle, the more time was spent on specific spots. Movement path is presented as connected lines. The position of the box that's been scaled is presented as a grey rectangle with "X" in it. We observed more diverse movement in variant without the black screen.

Even we observe no gain in scale nor the task time (66,9 s ± 20,4 CI for variant without black screen and 54,2 s ± 15,5 CI for black screen variant) all of the participants pointed the black screen variant as more difficult and inconvenient. Seven participants answered in the post-task survey that the black screen *"Made it difficult for me to judge the size of the item"*. Other frequently selected answers were *"It made me feel unwell"*, *"It was irritating"*, *"It made it difficult to focus on the task"* (three people for each answer). Therefore, we decided not to use the black screen in further tasks.

Task A2. Participants have achieved more consistent results with pre-immersion variant (when they can interact with the physical box at the start of the test) with mean final scale value of 0,98 with $CI = 0,057$ rather than 0,885 with $CI = 0,101$ in the variant without physical box (Table 1). Even the Confidence Intervals overlap a bit, the difference is clear with two times lower CI and final scale value much closer to actual object scale in the real world (which was equal 1). Also, without a physical object reference the value of one meter was considered much smaller than it actually was. After pre-immersion in second variant results has improved significantly in most cases, placing close to proper box scale value. On the other hand, the group that starts with pre-immersion achieved much more consistent and closer to real value results with mean difference value at only $0,01 \pm 0,03$ (rather than $0,18 \pm 0,18$ in the other group). In the post-task survey all of the participants found the ability to see the real object easier than the one with the plan with dimensions only (regardless of whether the variant with the pre-immersion was used as the first or as the second). Most often, in the open question responses, users emphasize the possibility of comparing the box with their own body as a feature that made it easier for them to complete the task. They were also more confident of their answers. On the other hand, without pre-immersion it was hard for them to imagine the size of 1 m.

5.2   Second Session

Task B1. We did not observe a significant difference between Experts and Non-experts in case of final scale value (Table 2), scale change process nor behaviour (movement, rotation). At the same time, the benefit of using pre-immersion was once more confirmed both in the results of the survey. The participants pointed this task as one of the easiest from whole study (Fig. 7) and a possibility to see the real object as a helping factor. The mean final scale value oscillates around the actual value of 1 ($1.02 \pm 0,04$ CI for Experts and $1.03 \pm 0,06$ CI for Non-experts).

Task B2. This task showed the least consistent results (highest standard deviation and CI). This scene was often felt by users to be much smaller than it actually was in real world and smaller than apartments in further tasks (Table 2). For example, two of the participants from the group of Non-experts indicated that the correct size of the hall was almost half the size of the actual scale. What is more, this was one of the two tasks (along with an empty room) in which it took the participants the most time to set the final scale value (Table 2). Also, they made the most changes to the scale parameter in this scene with the median value of 86 changes (increase and decrease combined) and CI equals 23 (much higher than tasks B3–B5 with a median value respectively of 38 ± 10 CI in B3, 47 ± 7 CI in B4 and 46 ± 8 CI in B5).

Table 1. Scale value for both variants (with and without pre-immersion). Participants have achieved better results with pre-immersion variant. Also the group that starts with pre-immersion achieved much more consistent and closer to real value results. $FSVpre$ - the final scale value in pre-immersion variant when users can interact with the physical box at the start of the test; $FSVno$ - final scale value for variant without a physical box; $Sdiff$ - final scale value difference between variants; $MSdiff$ - mean scale value difference in each group; $SDdiff$ - standard deviation of the scale value difference in each group; IF group - pre-immersion-first group, users who could see the real box as first variant; NIF group - No-immersion-first group, users who could see the real box as second variant; $Mean$ - mean scale value for all the participants; SD - standard deviation of the scale value; CI - Confidence Interval for the mean scale value.

	User number	$FSVno$	$FSVpre$	$Sdiff$	$MSdiff$	$SDdiff$
NIF group	1	0,68	0,96	+0,28	0,18	0,18
	2	0,83	1,07	+0,24		
	3	1,07	1,01	−0,06		
	4	0,64	1,07	+0,43		
	5	0,8	0,79	−0,01		
IF group	6	0,85	0,91	+0,06	0,01	0,03
	7	0,94	0,95	+0,01		
	8	0,99	0,99	0		
	9	1,05	1,03	−0,02		
	10	1,0	1,02	+0,02		
$Mean$		0,885	0,98			
SD		0,142	0,079			
CI		0,101	0,057			

Table 2. Mean scale value and task time for tasks B1 to B5 and both groups (Experts and Non-experts). B1: cardboard box, B2: large hall, B3: empty room, B4: unfurnished apartment, B5: furnished apartment. For scale, the least consistent case was marked (B2: large hall). For time results, the tasks with the longest task time was marked (B2: large hall and B4: unfurnished apartment). M - mean scale value for all the participants in a particular group; SD - standard deviation of the mean scale/time value; CI - Confidence Interval for the mean scale/time value. The difference between the mean values and the median were in the range 0,0–2,7% (giving 0,0–0,03 in the case of scale value) so they are omitted from the results.

		B1		B2		B3		B4		B5	
		M	CI	M	CI	M	CI	M	CI	M	CI
Scale	Experts	1,03	0,04	**0,96**	0,12	1,02	0,03	1,12	0,06	1,10	0,03
	Non-experts	1,02	0,06	**0,96**	0,22	1,10	0,06	1,18	0,05	1,17	0,05
Time [s]	Experts	61,4	10,5	**87,0**	34,5	59,9	18,3	**89,2**	35,8	45,9	13,5
	Non-experts	57,4	10,2	**59,2**	11,1	56,5	12,2	**64,0**	22,0	44,9	12,7

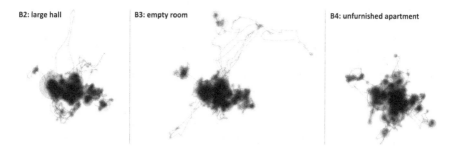

Fig. 5. Subject's position for tasks B2: large hall, B3: empty room and B4: unfurnished apartment gathered five times on the second. Each of visualization presents cumulative heatmap of the subject's position. The darker the circle, the more participants visits that spot, the bigger the circle, the more time was spent on specific spots. Movement path is presented as connected lines. We observed more diverse movement in task with a little or no point of visual reference (B2 and B3) than in a scene with apartment with doors and windows (B4).

We observed more movement in this and following task than in others (Fig. 5). Once again, participants tend to measure the space with continuously moving back and forth from center position. Same type of behavior of even bigger scale can be observed during task B3 (empty room). We conclude that a little or no point of visual reference forced the participants to make an assessment based on movement.

Tasks B3 to B5. In those scenes participants used mainly the doors and windows as referencing points (Fig. 6). That was confirmed in a post-test survey where 16 out of 20 users pointed that out. This scene was also selected as the second easiest in whole test (after cardboard box one, Fig. 7). It is also worth noting that after the appearance of doors and windows in scene four and furniture in scene five, users rated the room as larger than before (Table 2).

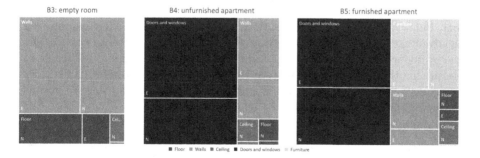

Fig. 6. Areas of interests (AOI, objects being in the center of the view at the time) in tasks with apartment (B3 to B5). Both groups tended to favor (the Experts even more) know objects such as doors and windows over looking up and down on the floor and ceiling. In all three scenes ceiling was observed almost exclusively by Non-experts. *E* - group of Experts, architects; *N* - group of Non-experts, common users.

5.3   Experts vs Non-experts

We observe no significant difference between those groups in case of final scale value (Table 2). However Experts tend to set lower and closer to real world scale value for tasks with architectural space. In the other hand, the common users tend to recognize the space size smaller in VR than it was in real life (scale values higher than 1 and closer to 1,1–1,2 for tasks B3–B5 with apartment setting). Expert results were also more coherent with lover CI than Non-experts in most cases. There was also a difference in the frequency of displaying the plan. The architects remembered the given dimensions faster than the Non-experts group, and they recalled the plan less frequently.

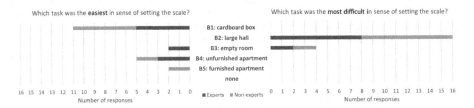

Fig. 7. Posttest survey. On the left: the easiest task; on the right: the most difficult task. Both Experts and Non-experts indicated the task with a cardboard box (B1) and empty apartment, but with doors and windows (B4) as the easiest tasks. Dimensions given to a plan as well as furniture in the scene were considered helpful mostly by the Experts.

Both groups considered a large hall (B2) and a room without doors and windows (B3) to be the most difficult ones (Fig. 7). These opinions are confirmed in the results, as they are the least convergent with each other and frequent changes of scale value were made. Some of the participants tried to obtain information about the surrounding space with their own steps - similar behavior to the one observed in the first test session (we can call it "movement based scaling"). Both groups indicated the task with a cardboard box (B1) and empty apartment, but with doors and windows (B4) as the easiest tasks. They emphasized the advantage of being able to see the actual object and a reference point in the form of doors and windows. At the same time, the overall sense of difficulty in determining the scale differed between the groups (considered more difficult by Non-experts. The mean score of difficulty on a ten-point scale, where 1 means the easiest and 10 the most difficult, for Expert was $4 \pm 3{,}2\ CI$ and $5{,}3 \pm 6{,}1\ CI$ for Non-experts.

6   Conclusion

The aim of this study was to investigate the disproportion in the perception of the size of virtual architectural objects in relation to the given real dimensions and subjective assessment given by the subjects. The study also analyzed the influence of various factors on increasing the spatial awareness of the respondents. The benefit of using pre-immersion in VR was confirmed. The final scale value of the task with the possibility of seeing a real box was more consistent between all users and oscillated around the actual value of 1 (scale equal to 1:1). Surveys showed that users also found the pre-immersion variant easier. We did not observe a significant impact of the presence of a black screen between each change of the scale on the results. At the same time post-test survey showed that the black screen was received negatively by participants and made it difficult to assess the object's size. Although, we did not observe any significant difference between the Experts and Non-experts in case of final scale value, there were noticeable differences in user behavior during the study. Architects viewed a plan with dimensions less often and set lower and closer to real world scale value in tasks with architectural space. Their results were also more coherent with lover CI than Non-experts in most cases. Such factors as doors, windows and furniture causes that the users perceived the space as larger than when it consisted only of the floor, walls and ceiling. The scene with a high hall and a room without doors and windows seemed to be the most difficult for users to assess. The respondents in these tasks tried to measure the room with their own steps. In those scenes, with a little to no known points of reference, we observed behavior that can be called "movement based scaling" (participants tried to measure a space with their steps).

The problem raised in the work has a further, wide field of development. Survey results have shown that it would be a good idea to extend the study to include *BodyBaseScaling* or the human figure as a module helping to determine the size of the object. This element may turn out to be crucial in the case of very high rooms, such as a hall. Another possible development of the study would be to extend it with an Eye-tracker device developed for VR (to track the exact movement of the eyeballs rather than center point AOI) or the automatic image analysis in search of elements affecting the user's sense of scale [10]. We are also considering combining these studies with impression curve study [1] to verify scale impact on computer dame level design process.

To sum up, the study has shown that one can use 1:1 scale when presenting single objects in immerse VR to them being perceived by users as in proper scale. We propose a little bigger scale of 1:1,1 for architectural space with many points of references such as doors, windows and everyday objects. For large scale spaces with less to no points of reference (as empty hall presented in this study) we recommend a bit smaller scale of 1:0,96.

References

1. Andrzejczak, Jarosław., Osowicz, Marta, Szrajber, Rafał: Impression curve as a new tool in the study of visual diversity of computer game levels for individual phases of the design process. In: Krzhizhanovskaya, V.V., et al. (eds.) ICCS 2020. LNCS, vol. 12141, pp. 524–537. Springer, Cham (2020). https://doi.org/10.1007/978-3-030-50426-7_39

2. Brooks, C W., Broish, I.M.: System for Ophthalmic Dispensing. 3rd edn., Elsevier, Philadelphia (2007)
3. Combe, E., Posselt, J., Kemeny A.: 1.1 scale perception in virtual and augmented reality. In: 18th International Conference on Artificial Reality and Telexistence (ICAT), Japan (2008)
4. Dodgson, N.: Variation and extrema of human interpupillary distance. University of Cambridge Computer Laboratory, Cambridge (2004)
5. Gedliczka, A.: Metodyka badań stosowania miar antropometrycznych. Centralny Instytut Ochrony Pracy, Poland (2001). (in Polish)
6. Gibson, J.: The Ecological Approach To Visual Perception. Cornell University, Ithaca (1986)
7. Greenberg, D.: Space Perception in Virtual Reality. Design in Virtual Reality, Cornell University, Ithaca (2016)
8. Gregory, R.L.: Eye and Brain: The Psychology of Seeing. 5th edn., Princeton University Press, Princeton (2015)
9. Jerald, J.: The VR Book: Human-Centered Design for Virtual Reality. Association for Computing Machinery, Morgan and Claypool (2015)
10. Kozłowski, K., Korytkowski, M., Szajerman, D.: Visual analysis of computer game output video stream for gameplay metrics. In: Krzhizhanovskaya, V.V., et al. (eds.) ICCS 2020. LNCS, vol. 12141, pp. 538–552. Springer, Cham (2020). https://doi.org/10.1007/978-3-030-50426-7_40
11. Lazar, J., et al.: Research Methods in Human-Computer Interaction. Wiley Publishing, Indianapolis (2010)
12. Linkenauger, S.A., Witt, J.K., Bakdash, J.Z., Stefanucci, J.K., Proffitt, D.R.: Asymmetrical body perception: a possible role for neural body representations. Psychol. Sci. **20**(1), 1373–1380 (2009)
13. Livatino, S., Hochleitner, C.: Simple guidelines for Testing VR Applications. In: Advances in Human Computer Interaction. In Tech (2008). https://doi.org/10.5772/5925
14. Oculus Developer Website, Guidelines for VR Performance Optimization.https://developer.oculus.com/documentation. Accessed 2 Dec 2020
15. Ogawa, N., Narumi, T., Hirose, M.: Distortion in perceived size and body base scaling in VR. In : Proceedings of the 8th Augmented Human International Conference (AH 2017). Association for Computing Machinery (2017)
16. Sumlet, W.: Skala ludzka w architekturze i przestrzeni mieszkaniowej. Architektura Czasopismo Techniczne Politechnika Krakowska (2012). (in Polish)
17. Tullis, T., Albert, W.: Measuring the User Experience: Collecting, Analyzing, and Presenting Usability Metrics. Morgan Kaufmann, San Francisco (2013)

Capsule Network Versus Convolutional Neural Network in Image Classification
Comparative Analysis

Ewa Juralewicz and Urszula Markowska-Kaczmar[✉] [iD]

Wroclaw University of Science and Technology, Wroclaw, Poland
urszula.markowska-kaczmar@pwr.edu.pl

Abstract. Many concepts behind Capsule Networks cannot be proved due to limited research, performed so far. In the paper, we compare the CapsNet architecture with the most common implementations of convolutional networks (CNNs) for image classification. We also introduced Convolutional CapsNet - a network that mimics the original CapsNet architecture but remains a pure CNN - and compare it against CapsNet. The networks are tested using popular benchmark image data sets and additional test sets, specifically generated for the task. We show that for a group of data sets, usage of CapsNet-specific elements influences the network performance. Moreover, we indicate that the use of Capsule Network and CNN may be highly dependent on the particular data set in image classification.

Keywords: Capsule network · Convolutional network · Image classification · Comparative analysis

1 Introduction

Over past ten years, Deep Learning (DL) has introduced a huge progress in computer vision. This great success is a result of convolutional neural networks (CNNs) application [10]. They make predictions by checking if certain features are present in the image or not. They do not posses ability to check spatial relationship between features. The weakness of CNN is its need for a vast amount of data to train.

A new attractive neural network has been proposed by Hinton [6] in response to the convolutional neural network drawbacks. He indicated the pooling operation used to shrink the size and computation requirements of the network as the main reasons for the poor functionality of CNNs. To solve this problem Capsule Networks and dynamic routing algorithms have been proposed. Capsule Networks are characterized by translational equivariance instead of translational invariance as CNNs. This property enables them to generalize to a higher degree from different viewpoints with less training data. Many successful alternatives have been proposed for the routing process [1, 6, 9] to increase representation

© Springer Nature Switzerland AG 2021
M. Paszynski et al. (Eds.): ICCS 2021, LNCS 12746, pp. 17–30, 2021.
https://doi.org/10.1007/978-3-030-77977-1_2

interpretability and processing time, which proves there are possibilities for Capsule Networks to improve in this area. The experiments conducted in [14] provide an estimate on CapsNet capabilities while taking training time needed to achieve relevant results into account. The experiment setup, however, leaves much space for further improvement. The generalization capability, related to data efficiency, has also been analyzed by training a network to recognize a new category based on a network trained on a subset of initial categories [2]. By concept, capsules hold a more complex entity representation compared to a single neuron and allow for more significant viewpoint variations. This leads to a hypothesis that Capsules need less training data than CNNs to achieve a similar performance quality, which has been confronted in [13]. In the comparison, apart from previously mentioned CapsNet architecture presented in [12] (CapsNet I), a variant with EM (*Expectation-Maximization*) routing algorithm, which was introduced in [6], is compared (CapsNet II). The new version of routing algorithm based on EM is proposed in [3].

Most cited above publications aim to provide an improvement rather than a comparison for the task they consider. This way, it is impossible to directly move observations made in such experiments to contemplate comparative performance, as the experimental setup is not valid. Our work tries to fulfill this gap. In opposite to [8], we focus on a comparison of both specifically designed for image processing neural networks – CNN and Capsule Networks with comparable number of parameters. Neural Networks and Capsule Networks in areas and applications relevant to conceptual differences between the two in a highly comparable environment. Notably, the research is aimed to verify whether Capsule Networks introduce advances in the areas in which they are supposed to be superior to CNNs by intuition and following the core concept differences. In the experiments, we verify the influence of the routing process's utility, increased viewpoint variance coverage, and robustness of Capsule Network to randomly shuffled images about Convolutional Network. The comparison covers the training process of the networks, their features emerging directly from architecture design, and their performance under specified conditions. It allows an understanding of their performance better.

The paper consists of four sections. The next one presents details of compared neural models. Section 3 describes conducted experiments and analysis of obtained results. Conclusions finish the paper.

2   Methodology

In this section, we describe details of models evaluated in the experimental part.

2.1   Convolutional Neural Networks

Convolutional Neural Network (CNN) is especially suitable for image processing because of its structure and the way of information processing. A simple CNN model with one convolutional and one pooling layer is presented in Fig. 1.

It is composed of three different layer types: convolutional, pooling, and fully-connected. Convolutional layers are composed of several feature maps, which are two-dimensional matrices of neurons. Each feature map has its *convolutional filter* applied to the input. Convolutional filters (also called kernels) are applied locally sliding over its input. For each kernel position the convolution operation is performed. Pooling layers typically follow convolutional layers. This operation locally subsamples the output of the preceding convolutional layer. The most commonly used is max-pooling, which is performed by sliding a window of a specified width and height and extracting the current pool's maximum value. The fully connected layers end a processing pipeline in a convolutional network (Fig. 1). The preceding layer output is transmitted to the fully-connected layer after being flattened to form a vector.

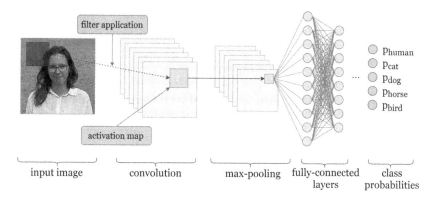

Fig. 1. Simple schema of processing information in CNNs with one convolutional layer, one pooling layer and two fully-connected hidden layers

Training a Convolutional Neural Network is made with the backpropagation algorithm based on minibatch SGD. It iteratively searches the set of weights W that minimizes the *loss function* for data D. For classification problems, the network is trained using the categorical cross-entropy loss function.

It is worth noting that the pooling operation significantly reduces the number of trainable parameters in the following layer allowing for minor input variation while preserving the level of neural activation. In this way, they enrich CNNs with a property referred to as *translational invariance*. It exists only in a limited scope. It makes it necessary for CNNs to be fed with augmented data in terms of rotation, scale, and varying perspective, causing a giant data volume needed for training, which covers as many viewpoints as possible.

There are many convolutional neural networks architectures proposed so far. Currently, the most common CNNs in use are: VGG-16 [14], DenseNet [7] and Residual Neural Network (ResNet) [4].

2.2 Capsule Network

A capsule network is a type of neural network in which the basic low-level node is a *capsule* [5]. Capsule's inputs and output are vectors. Each vector encodes a representation of an entity. The vector's direction indicates the pose of the object, (position and orientation in a specified coordinate system), and the vector's norm indicates the network confidence of this representation. Capsules take vectors $\underline{u}_i$ from a lower capsule layer as input. Then, they multiply them by weight matrices W_{ij} and a scalar *coupling coefficients* c_{ij}, where i is the number of a capsule in the lower layer, and j is the number of a capsule in a successive higher layer, as in Eq. 1.

$$s_j = \sum_{i=1}^{N} W_{ij}\underline{u}_i c_{ij} \tag{1}$$

The multiplication vector result s_j is summed element-wise and normalised using a *squashing function*, producing an output vector v_j, (Eq. 2).

$$v_j = squash(s_j) = \frac{||s_j||^2}{1 + ||s_j||^2} \frac{s_j}{||s_j||} \tag{2}$$

The squashing function reduces the original vector length to the range of $[0, 1]$ while maintaining vector's orientation in space. It also brings a non-linearity to the result. The coupling coefficients are modified proportionally to the level of accordance a_{ij} between prediction vector $\hat{\underline{u}}_{j|i}$ and capsule output v_j. Accordance a_{ij} is simply a scalar value acquired by cosine similarity, which is a product of vectors $\hat{\underline{u}}_{j|i}$ and v_j (Eq. 3).

$$a_{ij} = \hat{\underline{u}}_{j|i} v_j \tag{3}$$

The flow control is obtained by performing *dynamic routing*. Its version called routing-by-agreement is presented in [12]. Routing-by-agreement is an iterative process in which coupling coefficients c_{ij} are calculated. They are interpreted as the probability that a part from the lower-level capsule i contributes to the whole which capsule j should detect.

CapsNet is the first fully-trainable architecture following the ideas behind Capsule Networks. It consists of two parts - an **encoder** and a **decoder**. The encoder is the part that is responsible for building the vector representation of the input. The decoder's role is to achieve the input image's best reconstruction quality based on its representation vector from the encoder.

The encoder is visualised in Fig. 2. It is composed of 3 layers: a purely convolutional layer with ReLU activation function (ReLU Conv1), a PrimaryCaps layer, which is a combination of a convolutional and a capsule layer, and a DigitCaps layer - a pure capsule layer. The PrimaryCaps layer is very similar to the standard convolutional layer with an output extension to a vector and vector-wise application of the activation function. The DigitCaps layer is a capsule layer responsible for producing a vector representation for each of the categories available for the considered task (10 classes in MNIST). The routing operation is applied only between the PrimaryCaps and the DigitCaps layers. The weights in other layers are modified exclusively during the backpropagation algorithm based on minibatch SGD and remain constant in the network's forward pass.

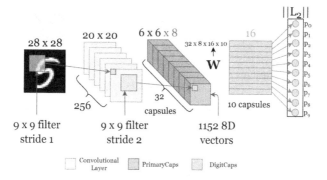

Fig. 2. Architecture of CapsNet encoder with an image from MNIST dataset

The decoder in CapsNet is presented in Fig. 3. The decoder's input is a 16-dimensional vector, which is the output from a capsule from the DigitCaps layer responsible for encoding the true label of the image. The remaining outputs are masked by inserting a value of 0. This masking mechanism and the reconstruction loss incorporation to the entire network's loss function force the separation of label responsibility between capsules in the last layer of the encoder.

The output from the last layer is interpreted as pixel intensities of the reconstruction image. The reconstruction loss is calculated as a distance between the input image and the reconstruction obtained from the capsule representation vector. The loss function for the network training L_c is calculated as in Eq. 4.

$$L_c = L_e + \alpha L_d \tag{4}$$

where L_e is a margin loss function for the classification task and L_d is the loss from the decoder. The classification task loss function is expressed by Eq. 5:

$$L_e = \sum_{k=1}^{K} T_k max(0, m^+ - ||v_k||)^2 + \lambda(1 - T_k)max(0, ||v_k|| - m^-)^2 \tag{5}$$

where: K is the number of considered categories; T_k is a logic value – 1 when the correct label of the sample corresponds with capsule k and 0 – otherwise; λ down-weights the loss function value for categories other than the true label. It has a constant value of 0.5; v_k is the squashed vector representation output by capsule k; m^+ is a margin for positive classification outcome and is equal to 0.9, m^- is a margin for negative classification outcome and is equal to 0.1.

The component L_d in Eq. 4 refers to the decoder training. It is based on the euclidean distance between the target input image and the output from the decoder. It is scaled down by α hyperparameter that is set to 0.0005 as its initial value is much higher than that coming from the margin loss of L_e.

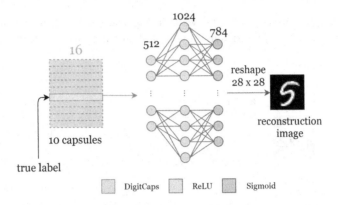

Fig. 3. Architecture of CapsNet decoder with an image from MNIST dataset

2.3 Convolutional CapsNet

Convolutional CapsNet is a self-designed architecture that aims to keep maximum possible similarity to the CapsNet architecture, while – opposite to [15] – remaining a pure Convolutional Neural Network. In reference to Convolutional CapsNet, a term capsule indicates groups of neurons and their similarity to capsules in Capsule Networks. Routing Operation is entirely ignored. By intuition, such maneuver removes the control of preserving the relationships between parts and wholes in situations where they do exist in the image but are spatially disturbed. In the Convolutional CapsNet, the capsule-specific vector-wise operation is replaced with an activation function applied neuron-wise. The activation function can be set as ReLU or Sigmoid.

The differences between CapsNet and Convolutional CapsNet are summarised in Table 1. Any element not mentioned in the Table is exactly the same in Convolutional CapsNet as it is in CapsNet.

Table 1. Characteristics of CapsNet and proposed Convolutional CapsNet architectures

Factor	CapsNet	Convolutional CapsNet
Routing algorithm	Applied	Not applied
Activation funtion after PrimaryCaps and DigitCaps layers	Squashing	Sigmoid or ReLU
Level of activation funtion application after PrimaryCaps and DigitCaps layers	Capsule-wise	Neuron-wise

3   Experiments

Experiments aim to explore the role of capsule network-specific elements on the performance of CapsNet to understand better the mechanisms applied in CapsNet and compare its quality of image classification to the results commonly achieved by CNNs. The compared network architectures need to satisfy the following criteria: maximum architectural similarity, similar number of trainable parameters, comparable performance in terms of classification accuracy.

3.1   Datasets Description

In the experiments, we want to check the robustness of compared networks to input translations and sensibility to spatial relationships between parts and wholes. We prepared two modified datasets based on the original MNIST and CIFAR-10 datasets, to achieve our goal. Both datasets contain images of size 32×32 pixels, and in each case, the training dataset includes 50000, and the test dataset contains 10000 samples.

In the first modification - *VarMNIST* and *VarCIFAR-10* datasets arose as the results of operation similar to those applicable in data augmentation. There are two variants of each test set: *Light VarMNIST* (*Light VarCIFAR*) and *Strong VarMNIST*(*Strong VarCIFAR*). They differ by the extent and amount of input alternations performed, applied with usage of *imgaug* library [11]. Image samples from both data sets for the MNIST dataset are visualized in Fig. 4.

(a) Samples from *Light VarMNIST*   (b) Samples from *Strong VarMNIST*

Fig. 4. Samples from *Light and Strong VarMNIST* test set for each category (by columns)

The second group of test sets is *N-Shattered MNIST* and *N-Shattered CIFAR*sets where $N \in \{2, 4, 8, 16\}$. These datasets are obtained by shattering the original test image sample into N slices and combining them into one image in random order. Samples from N-Shattered CIFAR data sets are visualized in Fig. 5. For VarMNIST and VarCIFAR test sets, the desirable outcome is maximum classification accuracy. For N-Shattered MNIST and N-Shattered CIFAR sets it is its minimum as samples in most cases do not preserve the spatial relationships between parts and wholes.

(a) 2-Shattered CIFAR   (b) 4-Shattered CIFAR

Fig. 5. Visualisations of N-Shattered CIFAR test sets for each category, where *airplane* is the leftmost column and *truck* is the rightmost column.

3.2 Experiment1. Study on Impact of Capsule-Specific Elements on Performance Quality

The experiment aims at checking the role of the routing by agreement procedure and activation function in CapsNet. To perform such comparison, the CapsNet architecture is compared to its sibling network - Convolutional CapsNet. The test is run using *Light Var* and *Strong Var* sets, and finally on N -Shattered test sets for MNIST and CIFAR-10 datasets.

CapsNet and Convolutional CapsNet setup: ReLU Conv1 Layer applies 256 filters of shape 9×9 with a stride 1. Its output consists of 256 feature maps of shape 20×20 with ReLU activation function; PrimaryCaps consists of 32 Capsules. Each Capsule applies 8 filters of size $9 \times 9 \times 256$ with stride of 2. Each filter produces a 6×6 feature map. In total there are 8 such feature maps per each capsule. The dimension of the output vector from a capsule is 8. The weight matrix between the PrimaryCaps and the DigitCaps layer is thus of shape $6 \times 6 \times 32 \times 8 \times 16$, resulting in a single 8×16 matrix between every capsule in the PrimaryCaps and the DigitCaps layers. The encoder network consists of 3 fully-connected layers: 512 ReLU units, 1024 ReLu units and 784 sigmoidal units. Convolutional CapsNet (CCapsNet) has identical architecture.

The hyperparameters value assumed based on tuning CapsNet architecture are presented in Table 2. They were also applied to the Convolutional Cap-sNet (assigned as CCapsNet). Performance of the model in terms of the number of parameters and processing time for the assumed hyperparameter values for MNIST and CIFAR-10 datasets is presented in Table 3. Note that the value of decoder loss is averaged over a number of pixels.

No hyperparameter tuning has been performed in this step. The training process curves of CapsNet and CCapsNet on the MNIST dataset almost overlap. Therefore we skip these charts here. Table 4 shows the results. Using MNIST the CCapsNet network managed to outperform CapsNet in terms of accuracy with $0.9921_{\pm0.0011}$ compared to $0.9917_{\pm0.0007}$ achieved by CapsNet. However, the decoder loss on the validation set is on average higher for CCapsNet $(0.0302_{\pm0.0016})$ than for CapsNet $(0.0244_{\pm0.0012})$. An additional observation is a slight tendency to overfit the training data observed for CCapsNet compared to CapsNet. This tendency is more visible in the case of CIFAR-10 datasets. In

Table 2. Hyperparameters of CapsNet –baseline configuration; PC no – the number of capsules in PrimaryCaps layer (DC for DigitCaps), PC dim – the output vector dimension of each capsule in PrimaryCaps layer (DC in DigitCaps)

Hyperparameter	Value
Optimizer	Adaptive moment estimation (Adam)
Learning rate	0.001
Learning rate decay	0.9
Data batch size	100
Epochs	50
Decoder loss coefficient	0.0005
No. of routings	3
Capsule number and dimension	PC no (32); PC dim (8); DC no (10); DC dim (16)

Table 3. Performance of CapsNet achieved on MNIST and CIFAR-10

Dataset	Train accuracy	Test accuracy	Decoder loss	No of parameters	Train time per epoch
MNIST	0.9996 ± 0.0005	0.9917 ± 0.0007	0.0244 ± 0.0012	8 215 568	208 ± 4.23
CIFAR-10	0.7001 ± 0.0018	0.6629 ± 0.0007	0.0302 ± 0.0016	11 749 120	383 ± 9.13

this case CCapsNet also outperformed CapsNet ($0.7005_{\pm 0.0018}$ vs $0.6629_{\pm 0.0007}$). However, the difference between the training set and test set accuracy and loss value is far greater for CCapsNet. Moreover, overfitting is visible on loss function value curves.

Figure 6 displays charts for CIFAR-10 data set. Each column contains accuracy score, classifier loss function value (margin loss) and decoder loss function value (euclidean distance averaged over pixels) for consecutive epochs. To make the comparison of performance as reliable as possible, the Convolutional CapsNet has been juxtaposed with CapsNet in two variants *CCapsNet* - model with the best achieved accuracy on the test set in 50 epochs - (*best*) , *CCapsNet Matching* - model from the epoch when the test accuracy was the closest to that achieved by CapsNet (*comparable*). Accuracy for all considered test sets and models is shown in Table 4.

Performance Analysis. Contrary to expectations, Convolutional CapsNet achieved better accuracy score in almost all cases. However, training with routing-by-agreement and the squashing function makes the training process more stable than without these components and far less prone to overfitting. The squashing function's application to the decoder's vector representation allows for a slight increase in the accompanying task of image reconstruction. The capsule-specific elements do not make the network react better to test samples modified in terms of translation and noise. It also does not provide considerable improvement in preserving the spatial relationship between capsules, which should be ensured by applying routing-by-agreement between capsules.

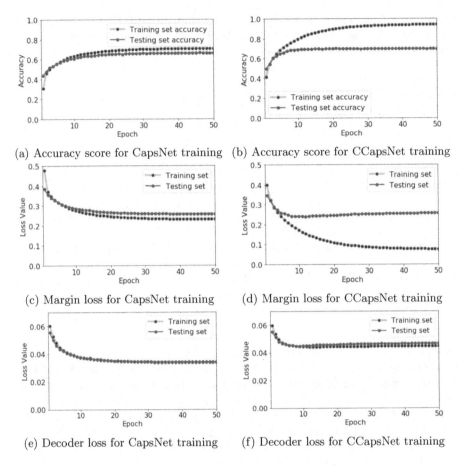

(a) Accuracy score for CapsNet training

(b) Accuracy score for CCapsNet training

(c) Margin loss for CapsNet training

(d) Margin loss for CCapsNet training

(e) Decoder loss for CapsNet training

(f) Decoder loss for CCapsNet training

Fig. 6. Accuracy (a, b), margin loss value for classification (c, d) and decoder loss value (e, f) in training process on CIFAR-10 data set. CapsNet the - left column and Convolutional CapsNet - the right one.

3.3   Experiment2. Comparative Performance of CapsNet and Various CNNs Types on Augmented Test Data

The experiment aims to check whether capsule networks learn a more complex entity representation. The datasets and CapsNet setup are the same as in the Experiment1. The CapsNet model with the best hyperparameter values from the Experiment 1. is used as a baseline. We consider the most popular CNN architectures - VGG, DenseNet, ResNet and its smaller version ResNet II networks. Their characteristics selected based on tuning are defined in Table 5. All architectures used a batch size of 100.

The results for the MNIST data set are shown in Table 6. We can observe that VGG achieved the best scores for all test sets. The situation changes slightly in the case of the comparable models. The best model for *Light VarMNIST* is

Table 4. Accuracy of CapsNet and Convolutional CapsNet (CCapsNet) averaged over 5 runs (best results are bolded)

Data set	MNIST		CIFAR-10		
Architecture	CapsNet	CCapsNet	CapsNet	CCapsNet	CCapsNet Matching
Standard	0.9917	**0.9921**	0.6629	**0.7005**	0.6698
Light Var	0.6443	**0.6894**	**0.5274**	0.5411	0.5211
Strong Var	0.5234	**0.5833**	**0.4841**	0.4828	0.4705
2-Shattered	0.5778	**0.5577**	0.4746	0.4802	**0.4706**
4-Shattered	**0.2512**	0.2553	0.3253	0.2993	**0.2982**
8-Shattered	**0.2481**	0.2661	0.3599	**0.3511**	**0.3501**
16-Shattered	0.1324	**0.1247**	0.2590	**0.2423**	0.2449

Table 5. Characteristics of CNN architectures

Architecture	Depth (layers)	MNIST No. of params	MNIST epoch time [s]	CIFAR No. of params	CIFAR epoch time [s]
CapsNet	6	8 215 568	208	11 749 120	383
VGG	30	8 037 578	34	12 180 170	35
ResNet I	27	8 816 074	170	11 181 642	230
ResNet II	22	272 776	20	237 066	23
DenseNet	144	8 380 090	130	*9 012 202	145

*Comparable parameter space could not be reached due to computer memory limitations

ResNet I and VGG for *Strong VarMNIST* but the difference is extremely low. In the case of CIFAR-10., the best CNN models sometimes surpassed the baseline method, however relation between accuracy on the standard set and transformed sets falls in favor of CapsNet. The observation is further notable in Table 7. We can observe that CapsNet surpasses all CNN models on *Light VarCIFAR* and *Strong VarCIFAR* datasets by far.

Table 6. Accuracy on VarMNIST datasets for CapsNet and CNN models

Comparable	Dataset	CapsNet	VGG	ResNet I	ResNet II	DenseNet
	Standard	**0.9917**	0.9908	**0.9917**	0.9903	0.9903
	Light VarMNIST	0.6443	0.7136	**0.7164**	0.7046	0.6717
	Strong VarMNIST	0.5234	**0.6417**	0.5860	0.6052	0.5738
Best	Standard	0.9917	**0.9927**	0.9923	0.9903	0.9903
	Light VarMNIST	0.6443	**0.7623**	0.6412	0.7046	0.6717
	Strong VarMNIST	0.5234	**0.6923**	0.5380	0.6052	0.5738

Performance Analysis. In the case of the MNIST dataset, it is possible that deep architectures of CNNs managed to learn so many different features that they cover a large part of possible cases. In the case of CIFAR-10, the possible image domain is far greater, thus CNN could not have learnt enough general representations to work well for augmented data. It may also come from higher overfitting of the model for CIFAR-10 than for MNIST.

Table 7. Accuracy for *VarCIFAR* datasets for CapsNet and various CNN models

Comparable	Dataset	CapsNet	VGG	ResNet I	ResNet II	DenseNet
	Standard	0.6629	**0.7019**	0.6602	0.6803	0.6363
	Light VarCIFAR	**0.5274**	0.4559	0.4826	0.4748	0.4731
	Strong VarCIFAR	**0.4841**	0.3862	0.4117	0.3938	0.3994
Best	Standard	0.6629	**0.8519**	0.6602	0.7602	0.6363
	Light VarCIFAR	0.5274	**0.5609**	0.4826	0.5386	0.4731
	Strong VarCIFAR	**0.4841**	0.4608	0.4117	0.4503	0.3994

3.4   Experiment3. Comparison of CapsNet and CNN Models Performance on Randomly Shattered Test Data

This experiment aims to verify the statement that capsules encode the spatial relationship between parts and wholes as opposed to Convolutional Neural Networks. We used N-Shattered vesion of datasets. The experiment setup is the same as in Experiment2.

The results for MNIST dataset are showed in Table 8 and for CIFAR-10 in Table 9. In the case of MNIST, CapsNet always managed to surpass all CNN models. In the case of CIFAR-10, the winner is ResNet.

Table 8. Accuracy for N-Shattered MNIST for CapsNet and CNN models

Comparable	Dataset	CapsNet	VGG	ResNet I	ResNet II	DenseNet
	Standard	**0.9917**	0.9908	**0.9917**	0.9903	0.9903
	2-Shattered MNIST	**0.5778**	0.5974	0.7143	0.7892	0.6480
	4-Shattered MNIST	**0.2512**	0.3341	0.3913	0.4768	0.3685
	8-Shattered MNIST	**0.2481**	0.3220	0.3521	0.3384	0.3349
	16-Shattered MNIST	**0.1324**	0.1759	0.2218	0.1698	0.1882
Best	Standard	0.9917	**0.9927**	0.9923	0.9903	0.9903
	2-Shattered MNIST	**0.5778**	0.5795	0.6839	0.7892	0.6480
	4-Shattered MNIST	**0.2512**	0.2922	0.3688	0.4768	0.3685
	8-Shattered MNIST	**0.2481**	0.3355	0.3457	0.3384	0.3349
	16-Shattered MNIST	**0.1324**	0.1443	0.1976	0.1698	0.1882

Performance Analysis. Similar to previous experiments, observations made on the MNIST data sets are not transferable to CIFAR-10 test results. Ability to encode spatial relationships between parts and wholes by the baseline Capsule Network and compared CNNs seems to be dependent on the level of model under- or overfitting. In the case of the MNIST dataset, where no considerable overfitting is present, CapsNet surpasses all other architectures. There are different behaviours observed for different networks for CIFAR-10. VGG seems to have decreasing performance on N-Shattered test sets along with improvement on standard test set, which is opposite to what is partially observed for ResNet II. We observed that ResNet II highly overfits to training data, which is possibly the reason why it does not recognize shattered images correctly. However,

Table 9. Accuracy scores for N-Shattered CIFAR data sets obtained by CapsNet and different CNN models

Comparable	Architecture	CapsNet	VGG	ResNet I	ResNet II	DenseNet
	Standard	0.6629	**0.7019**	0.6602	0.6803	0.6363
	2-Shattered CIFAR	0.4746	0.5172	**0.4549**	0.5527	0.4640
	4-Shattered CIFAR	0.3253	0.3764	**0.2954**	0.4119	0.3249
	8-Shattered CIFAR	0.3599	0.3635	**0.3241**	0.3886	0.3415
	16-Shattered CIFAR	0.2590	0.2836	**0.2249**	0.3050	0.2529
Best	Standard	0.6629	**0.8519**	0.6602	0.7602	0.6363
	2-Shattered CIFAR	0.4746	0.6202	**0.4549**	0.5943	0.4640
	4-Shattered CIFAR	0.3253	0.4167	**0.2954**	0.4034	0.3249
	8-Shattered CIFAR	0.3599	0.4045	**0.3241**	0.3410	0.3415
	16-Shattered CIFAR	0.2590	0.2701	**0.2249**	0.2229	0.2529

considering only the aspect of spatial relationship encoding, data overfitting is not a disturbing occurrence as what we look for is encoding of parts and wholes and not the input variance covered by the network. Thus a conclusion can be drawn that applying skip connections in the model, like in the case of ResNet and DenseNet, improves network's ability to encode spatial relationships of input data.

4   Conclusions

Capsule Networks provide a framework for shallow architecture with mechanisms addressing common problems present in Convolutional Neural Networks. They have a higher computational complexity but may work better for data sets of complex nature and with little training data. The experiments show that the training process of a CapsNet is very stable and not prone to overfitting, contrary to the tested CNNs. However, the specific cases in which CapsNet may be a better choice as compared to one of the popular deep CNN architectures cannot be clearly specified based on available research as the results and observations highly deviate for different data sets. We can observe that application of routing-by-agreement and the squashing function highly influences performance quality for medical and generated data sets, in their favor.

The experiments presented in this paper could be extended by further tuning of proposed architectures to specified problems. Capsule Networks can be further researched for more complex problems. Application of a Capsule Network to a larger task, like ImageNet classification task or COCO segmentation task would probably give better insight in their characteristics. Moreover, they can be tested in use as embedding builders for other tasks due to their natural way of encoding detected objects in forms of vectors, both for visual and textual data. Based on conducted experiments and available literature, Capsule Networks appear as an alternative model which may outperform other network types for specific problems, but does not show strong advantage over Convolutional Neural Networks, however, due to their consistent design, they remain an area worth exploring.

References

1. Duarte, K., Rawat, Y.S., Shah, M.: Videocapsulenet: a simplified network for action detection. CoRR abs/1805.08162 (2018). http://arxiv.org/abs/1805.08162
2. Gritsevskiy, A., Korablyov, M.: Capsule networks for low-data transfer learning. CoRR abs/1804.10172 (2018). http://arxiv.org/abs/1804.10172
3. Hasani, M., Saravi, A.N., Khotanlou, H.: An efficient approach for using expectation maximization algorithm in capsule networks. In: 2020 International Conference on Machine Vision and Image Processing (MVIP), pp. 1–5 (2020)
4. He, K., Zhang, X., Ren, S., Sun, J.: Deep residual learning for image recognition. In: 2016 IEEE Conference on Computer Vision and Pattern Recognition (CVPR 2016), pp. 770–778. Las Vegas, NV, USA, June 27–30 (2016). https://doi.org/10.1109/CVPR.2016.90, https://www.computer.org/csdl/proceedings-article/cvpr/2016/8851a770/12OmNxvwoXv
5. Hinton, G.E., Krizhevsky, Al, Wang, S.D.: Transforming auto-encoders. In: Honkela, T., Duch, W., Girolami, M., Kaski, S. (eds.) ICANN 2011. LNCS, vol. 6791, pp. 44–51. Springer, Heidelberg (2011). https://doi.org/10.1007/978-3-642-21735-7_6
6. Hinton, G.E., Sabour, S., Frosst, N.: Matrix capsules with EM routing. In: International Conference on Learning Representations (2018). https://openreview.net/forum?id=HJWLfGWRb
7. Huang, G., Liu, Z., Weinberger, K.Q.: Densely connected convolutional networks. CoRR abs/1608.06993 (2016). http://arxiv.org/abs/1608.06993
8. Jiang, X., Wang, Y., Liu, W., Li, S., Liu, J.: CapsNet, CNN, FCN: comparative performance evaluation for image classification. Int. J. Mach. Learn. Comp. textbf9(6), 840–848 (2019). https://doi.org/10.18178/ijmlc.2019.9.6.881
9. Mobiny, A., Nguyen, H.V.: Fast CapsNet for lung cancer screening. CoRR abs/1806.07416 (2018). http://arxiv.org/abs/1806.07416
10. Rawat, W., Wang, Z.: Deep convolutional neural networks for image classification: a comprehensive review. Neural Comput. 29(9), 2352–2449 (2017). https://doi.org/10.1162/neco_a_00990
11. Revision, A.J.: Imgaug: image augmentation for machine learning experiments. https://github.com/aleju/imgaug (2019)
12. Sabour, S., Frosst, N., Hinton, G.E.: Dynamic routing between capsules. CoRR abs/1710.09829 (2017). http://arxiv.org/abs/1710.09829
13. Kenny, S., Neubert Peer, P.P.: Comparison of data efficiency in dynamic routing for capsule networks (2018). https://www.tu-chemnitz.de/etit/proaut/publications/schlegel_2018.pdf
14. Simonyan, K., Zisserman, A.: Very deep convolutional networks for large-scale image recognition. In: International Conference on Learning Representations (2015)
15. Toraman, T., Alakusb, B., Turkoglu, I.: Convolutional CapsNet: a novel artificial neural network approach to detect COVID-19 disease from X-ray images using capsule networks. Chaos Solitons Fractals 140, 110122 (2020). https://doi.org/10.1016/j.chaos.2020.110122

State-of-the-Art in 3D Face Reconstruction from a Single RGB Image

Haibin Fu[1]([✉]) [iD], Shaojun Bian[1,3] [iD], Ehtzaz Chaudhry[1] [iD], Andres Iglesias[2] [iD], Lihua You[1] [iD], and Jian Jun Zhang[1] [iD]

[1] Bournemouth University, Bournemouth BH12 5BB, UK
hfu@bournemouth.ac.uk
[2] University of Cantabria, 39005 Cantabria, Spain
[3] Humain Ltd., Belfast BT1 2LA, UK

Abstract. Since diverse and complex emotions need to be expressed by different facial deformation and appearances, facial animation has become a serious and ongoing challenge for computer animation industry. Face reconstruction techniques based on 3D morphable face model and deep learning provide one effective solution to reuse existing databases and create believable animation of new characters from images or videos in seconds, which greatly reduce heavy manual operations and a lot of time. In this paper, we review the databases and state-of-the-art methods of 3D face reconstruction from a single RGB image. First, we classify 3D reconstruction methods into three categories and review each of them. These three categories are: Shape-from-Shading (SFS), 3D Morphable Face Model (3DMM), and Deep Learning (DL) based 3D face reconstruction. Next, we introduce existing 2D and 3D facial databases. After that, we review 10 methods of deep learning-based 3D face reconstruction and evaluate four representative ones among them. Finally, we draw conclusions of this paper and discuss future research directions.

Keywords: Monocular RGB image · 3D face reconstruction · 3D morphable model · Shape-from-Shading · Deep learning · 3D face database

1 Introduction

Generating 3D face models from 2D images and videos is widely required in industry and academia. It greatly reduces time-consuming and repetitive manual operations of creating 3D shape sequences for every new character. Face reconstruction techniques based on 3D Morphable Face Model and deep learning provide one efficient solution to reuse existing databases and create believable facial animation of new characters just from images or video frames in seconds, which greatly reduce manual operations and loads of time.

3D face reconstruction is widely used in commerce and industry. Traditional 3D reconstruction methods can be divided into two categories: 1) active methods such as Structural light and Laser scanning, and 2) passive methods like Multi-view stereo, Shape-from-shading (SFS) and Data-driven reconstruction. Face reconstruction from a

© Springer Nature Switzerland AG 2021
M. Paszynski et al. (Eds.): ICCS 2021, LNCS 12746, pp. 31–44, 2021.
https://doi.org/10.1007/978-3-030-77977-1_3

single image has been widely investigated in the fields of computer vision and computer graphics. There are many methods tackling this problem. Such as 3DMM methods, SFS, and Deep learning based methods.

Apart from the above static 3D face reconstruction from a single image, reliable detection and recognition of faces has been an on-going research topic for decades. Face detection, face recognition and realistic face tracking remain a challenging task in dynamic environments, such as videos, where the problems of big data and low computational efficiency become much more serious. Reducing the data size but still keeping good realism and details is especially important for 3D face reconstruction in these dynamic environments.

In this paper, we will review the state-of-the-art methods of 3D face reconstruction from a single image. The review will include three categories of 3D reconstruction methods, facial databases, and deep learning-based 3D face reconstruction methods with open source code.

2 Single Image-Based 3D Reconstruction Methods

3D reconstruction is the process of capturing the geometry and appearance of real objects. Face reconstruction from a single image (analysis by synthesis) has been investigated extensively in computer vision and computer graphics communities. To restore the geometry as well as the skin color of the face needs to obtain geometry information and color information from the image. The first step is to choose the geometry representation of a face model. It could be point cloud, depth map/normal map or parametric model. Among them, a parametric model is the most widely used representation. Although faces are different, the geometric mechanism of faces is similar, which lays a foundation for expressing a complex face with few parameters and parameterized expression.

Main 3D face reconstruction methods can be classified into three categories: Shape-From-Shading (SFS), 3D Morphable Face Model (3DMM), and Deep Learning (DL) based 3D face reconstruction. In what follows, we will briefly review them. Since Deep Learning (DL) based 3D face reconstruction is becoming a hot topic, we will give more details about various deep learning based 3D face reconstruction methods with open source code and evaluate four representative ones in Sect. 4, after we make a survey on existing databases for face reconstruction in Sect. 3.

2.1 Shape-from-Shading (SFS) Based 3D Face Reconstruction

Shape-from-Shading [1] is a very basic and classic 3D reconstruction method. The basic principle is to use the brightness information of grayscale images and the principle of brightness generation to obtain the normal vector of each pixel in a 3D space, and finally obtain the depth information according to the normal vector. We can get the geometric structure of an object by analyzing the changes of light and shade on the surface of the object. The reconstruction process is shown in Fig. 1. Shape from shading is known to be an ill-posed problem [2]. Therefore, SFS needs a reliable initial 3D shape as a reference to manually align the input facial image. With this prior information, shape from shading is then used to restore the geometry. Features of facial symmetry have been used by many researchers to constrain this problem [3]. As an efficient and simple low dimensional representation, 3DMM to be reviewed below is widely used as an initial facial shape [4]. SFS can restore fine geometric details by optimizing the process given the light coefficient and reflection parameters. In the recent research study [5], the refinement exploits shading cues in the input image to augment the medium-scale model with fine-scale static and transient surface details. Compared with traditional 3DMM, SFS can get better details of a human face. It is a good way to add details such as acne skin and wrinkles to the 3D face model.

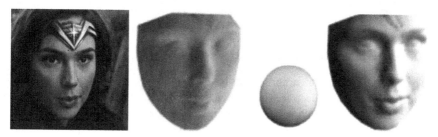

Fig. 1. Shape-from-shading based reconstruction: input image, estimated normal map, and rendered model (from left to right) [6].

2.2   3D Morphable Face Model (3DMM) Based 3D Face Reconstruction

Reconstructing a 3D face shape from a single 2D photograph or from a video frame is an inherently ill-posed problem with many ambiguities. One way of solving some of the ambiguities is to use a 3D face model to aid the task. 3D Morphable Face Models (3DMMs) also called 3D model fitting have been developed for this aim. In 1999, Vetter and Blantz [7] introduced the 3DMM for the synthesis of 3D faces, a principal component analysis (PCA) basis for representing faces. It is one of the most commonly used methods for face reconstruction from a single image [8]. The advantage of using the 3DMM is that the solution space is constrained to represent only likely solutions, so that the problem could be simplified. In 2016, Zhu [9] proposed an automatic reconstruction process for 3DMM. Still, the automated initialization pipelines usually do not produce the same quality of reconstructions when only one image is used. In addition, the 3DMM solutions cannot extract fine details since they are not spanned by the principal components. The common 3DMMs used for facial tracking include Surrey Face Model (SFM), the 3D morphable face model created at the University of Surrey by Huber [10], and Basel Face Model (BFM) created at the University of Basel by Paysan et al. [11]. BFM is a generative 3D shape and texture model. It uses the principal component regression (PCR) model to carry out regression 2D face markers and reconstruct 3D faces with higher shape and texture accuracy (Fig. 2).

Although the 3DMM greatly reduces the data size and raises the computational efficiency, it cannot describe 3D models with high details and good realism due to the reduction of the computational accuracy.

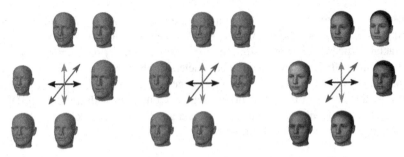

Fig. 2. BFM 2019 [11] shape, expression, and appearance variations.

2.3 Deep Learning (DL) Based 3D Face Reconstruction

In the past few years, deep learning methods have revolutionized computer vision [12,13]. With the development of deep leaning, the classical 3D morphable face modelling and parameter estimation techniques are being replaced by or combined with deep learning [14]. Due to the speed and robustness of deep network, reliable performance is also achieved on large poses and in-the-wild images. Moreover, deep learning methods are adapted at extract fine details like wrinkles. Clearly, deep learning has made some achievements in 3D face reconstruction and face appearance modeling, especially in the following four aspects: frontal face to arbitrary poses, occlusion handling, separation of training and testing, synthetic data and unsupervised real image [15].

Deep learning-based face reconstruction methods normally contain 2 parts: a database and pipeline of 3D face reconstruction process (Fig. 3). One of the most important factors for the success of deep learning-based 3D face reconstruction is the availability of large amounts of training data [16]. We will discuss 2D and 3D face data-bases in Sect. 3.

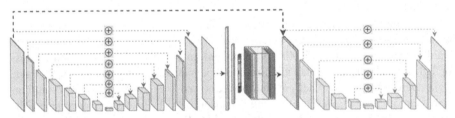

Fig. 3. Pipeline of deep learning based 3D face reconstruction Network architecture for DFDN [17], blue part for coarse model synthesis, and yellow part for fine detail synthesis. (Color figure online)

The pipeline of 3D face reconstruction process can be classified as coarse face reconstruction and fine detail (wrinkles and pores) face reconstruction. For the coarse face reconstruction, face landmarks or 3DMM parameters are treated as supervision signal for training. Dense fine detail reconstruction methods predict dense shape variation rather than low-dimensional parameters.

3 Face Databases

Face databases are fundamental for 3D face reconstruction and evaluation. In this section, we review existing face databases. In general, existing face databases could be divided into two categories: 2D face databases (Table 1), and 3D face databases (Table 2).

Table 1. Comparison of 2D face databases

Dataset	Subjects	Landmarks	Images	Resolution	Male	Year	Ref
PUT	100	30	9,971	2048 × 1536	89%	2008	[18]
Multi-PIE	337	68	750k	3072 × 204	–	2010	[19]
MUCT	276	76	3,375	480 × 640	49%	2010	[20]
SCface	130	21	4.16k	1200 × 1600	88%	2011	[21]
AFLW	–	21	25993	–	41%	2011	[22]
GBU	437	–	6.51k	128 × 128	58%	2012	[23]
Helen	–	194	20330	1200 × 900	–	2012	[24]
300-W	300	68	600	16.2k, 3.3M	–	2013	[25]
FaceScrub	530	–	10.6k	–	50%	2014	[26]
VGG-Face	2,622	–	2.6M	256 × 256	47%	2015	[27]
CelebA	10,177	5	202k	512 × 512	50%	2015	[28]
UMDFaces	8,277	21	8277	–	–	2016	[29]
MegaFace	690,572	49	1 m	100 × 100	–	2016	[30]
LS3D-W	–	68	230k	–	–	2017	[31]
LFW	5,749	–	13k	250 × 250	77%	2018	[32]
FFHQ	–	–	70k	1024 × 1024		2019	[33]
Tufts	113	–	10k	–	34%	2020	[34]

Table 1 above shows some popular 2D face databases since 2008. It provides the information about the numbers of subjects, landmarks, and images as well as resolution, gender of the subjects, year of creation, and the publications for these 2D face databases.

Among the 2D face databases shown in Table 1, the most commonly used 2D face databases are Multi-PIE [19], 300W [25] and CelebA [28]. The CMU Multi-PIE face database contains more than 750,000 images of 337 people with a range of facial expressions [19]. 300W is well known because it was the first facial landmark localization Challenge [25]. CelebFaces Attributes Dataset (CelebA) is a large-scale face attributes dataset with more than 200K celebrity images, each with 40 attribute annotations [28].

Table 2 below shows existing 3D face databases since 2006. The methods used to create 3D databases can be either active or passive. Active methods are Structured Light Systems (SLS), Laser Scanners (LS) and Time-of-Flight (TOF) sensors. Passive approach could be Multi View Stereo (MVS).

Table 2. Comparison of 3D face databases

Dataset	Subjects	Express	Vertex	Source[a]	Male	Texture	Year	Ref
ND2006	888	6	112k	LS	–	640 × 480	2006	[35]
BU-3DFE	100	25	10–20k	SLS	44%	1300 × 900	2006	[36]
BU-4DFE	101	6 ·	30–50k	SLS	43%	1040 × 1329	2008	[37]
UoY	350	10	5–6k	KN	–	360 × 240	2008	[38]
Bosphorus	105	35	35k	SLS	–	1600 × 1200	2009	[39]
BJUT-3D	1200	1–3	200k	LS	50%	478 × 489	2009	[40]
D3DFACS	10	38AU	30k	MVS	40%	1024 × 1280	2011	[41]
Florence	53	1	30–40k	KN	75%	3341 × 2027	2011	[42]
FaceWarehouse	150	20	11k	KN	–	640 × 480	2014	[43]
BP4D-Spontan	41	27AU	37k	MVS	44%	1040 × 1392	2014	[44]
BP4D+	140	7AU	30–50k	MVS	41%	1040 × 1392	2016	[45]
UHDB31	77	1	25k	MVS	69%	2048 × 2448	2017	[46]
4DFAB	180	6	100k	KN	66%	1200 × 1600	2018	[47]
EB+	60	7AU	30–50k	MVS	41%	1040 × 1392	2019	[48]
Facescape	938	20	2 m	MVS	–	4096 × 4096	2020	[49]
BFM	200	–	53490	SLS	50%	768 × 1024	2009	[10]
SFM	169	–	29587	SLS	–	768 × 1024	2016	[11]
LSFM	9663	–	53215		48%		2017	[50]
LYHM	1212	–			50%		2017	[51]

[a]MVS-Multi view stereo, KN-Kinect, SLS-Structure light system, LS-Laser scan.

3D face databases normally contain facial images aligned with their ground truth 3D shapes. These databases are essential for 3D face reconstruction and evaluation. For example, Basel face model (BFM) [11] are commonly used for 3D face reconstruction, and BU-3DFE (Binghamton University 3D Facial Expression), BU-4DFE and BP4D-Spontaneous are normally used for evaluations.

4 3D Face Reconstruction Methods Based on Deep Learning

In this section, we review the recent development of open source 3D face reconstruction methods based on deep learning. First, we identify 10 methods with open source code, which were published between 2018 and 2020. Then, we evaluate four representative ones using their open source code.

The identified 10 methods with open source code are shown in Table 3 and Fig. 4. Among them, UnsupNet [59], PRNet [55], DF2Net [56], and DFDN [17] will be evaluated in Subsects. 4.1, 4.2, 4.3, and 4.4, respectively. In what follows, we briefly review other methods listed in the Table 3.

Table 3. List of 2018 to 2020 open source 3D face reconstruction methods.

Method	Dataset	Vertices	Coarse	Fine	Year	Ref
ExpNet	CK+, EmotiW-17	–	✓		2018	[52]
ITW	300W, BFM	5k	✓		2018	[53]
MGCNet	300W-LP, Multi-PIE	36k	✓	✓	2018	[54]
PRNet	300W-LP, BFM	43,867	✓		2018	[55]
DF2Net	CACD, BU- 3DFE	7–13k		✓	2019	[56]
DFDN	AffectNet	53,149	✓	✓	2019	[15]
Facescape	Facescape	12,483	✓	✓	2020	[49]
Caricature	WebCaricature	6144	✓		2020	[57]
3FabRec	300W	–		✓	2020	[58]
UnsupNet	CelebA, BFM	–		✓	2020	[59]

Fig. 4. An overview of state of the art 3D face reconstruction methods

ExpNet [52] produces expression coefficients, which better discriminate between facial emotions than those obtained using the state-of-the-art facial landmark detection techniques. 3D Face Morphable Models In-the-Wild (ITW) proposed in [53] provide a new fast algorithm for fitting the 3DMM to arbitrary images. The work captures the first 3D facial database with relatively unconstrained conditions. Multi-view Geometry Consistency (MGCNet) [54] first proposes an occlusion-aware view synthesis method to apply multi-view geometry consistency to self-supervised learning. Then, it designs three novel loss functions for multi-view consistency, which are the pixel consistency loss, the depth consistency loss, and the facial landmark-based epipolar loss. Facescape [49] presents a 3D Face Dataset containing 938 subjects and provides a fine 3D face construction method. Based on the constructed dataset and a nonlinear parametric model, Caricature [57] proposes a neural network based method to regress the 3D face shape and orientation from the input 2D caricature image. Fast Few-shot Face alignment by Reconstruction (3FabRec) [57] first trains an adversarial autoencoder to reconstruct faces via a low-dimensional face embedding. Then, it interleaves the decoder with transfer layers to retask the generation of color images to the prediction of landmark heat maps.

Through the investigation, we selected 4 representative methods: UnsupNet, PRNet, DF2Net and DFDN, and run their open source code with our selected image databases. The obtained results are given and discussed in the following subsections.

4.1 UnsupNet

The 3D facial models reconstructed from UnsupNet [59] are shown in Fig. 5. Unsup-Net learns 3D deformable object categories from single RGB images without external supervision. The method is based on an automatic encoder that decomposes each input image into depth, albedo, viewpoint, and illumination. For separating these components without oversight, it works well in symmetric structures.

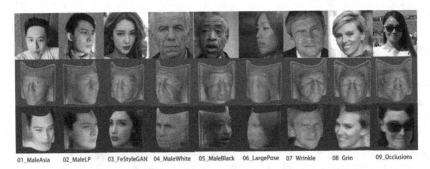

01_MaleAsia   02_MaleLP   03_FeStyleGAN 04_MaleWhite 05_MaleBlack 06_LargePose 07 Wrinkle   08 Grin   09_Occlusions

Fig. 5. The result of UnsupNet

Compared to other methods, UnsupNet could reconstruct the 3D models of human faces, cat faces, cars and other symmetric objects from single RGB images without any supervision or prior shape model such as 3DMM. However, UnsupNet has poor performance in reconstructing 3D faces with large poses and occlusion.

4.2 PRNet

3D face reconstruction and face alignment are closely linked problems. Although a lot of work has been done to achieve both goals by obtaining 3DMM coefficients, it is limited by the quality of the model space. PRNet [55] can simultaneously reconstruct the 3D facial structure and provide dense alignment. A UV position map is created to record 3D position under UV coordinates. By using encoder-Decoder network mode, the joint task of 3D face reconstruction and dense face alignment from single RGB face images can be realized end-to-end. By using the 300W-LP database consisting of 300 indoor and 300 outdoor images downloaded from the web [25] as a training set, it can also get better results for the faces with large variations in poses.

The 3D facial models reconstructed from PRNet [55] are shown in Fig. 6. In general, it can be divided into three parts: 3D face model, UV graph representation of 3D point cloud, and network structure design. The 3D model can be built on the Basel Face Model (BFM) [10] by 3DMM coefficient, while the expression can be built by the FW Model. The sum of the two is then projected to the camera coordinate system through the Pose (attitude Angle, translation and scale) parameters.

The advantages of PRNet [55] are model-free, reference-free, and not Voxel-based. The reconstruction and alignment work fine, including dense alignment of both visible and non-visible points (including 68 key points). The disadvantages are lack of details and low texture resolution. In addition, the distinct stripes could be seen on the model.

01_MaleAsia   02_MaleLP   03_FeStyleGAN 04_MaleWhite 05_MaleBlack 06_LargePose 07_Wrinkle   08_Grin   09_Occlusions

Fig. 6. The result of PRNet

4.3 DF2Net

A deep Dense-Fine-Finer Network (DF2Net) [56] could decompose the reconstruction process into three stages: D-Net, F-Net, and Fr-Net. Training data are generated by 3DMM. D-Net uses U-Net architecture proposed in [60] to map input images to dense depth images. F-Net refines the D-Net output by fusing depth and RGB domain characteristics, and its output is further enhanced by Fr-Net using a novel multi-resolution super column structure. In addition, three types of data: 3D model synthesis data, 2D image reconstruction data, and fine facial images, were introduced to train these networks.

The 3D facial models reconstructed from DF2Net [55] are shown in Fig. 7. The qualitative evaluation shows that DF2Net can effectively reconstruct facial details, such as small crow's feet and wrinkles. It can perform qualitative and quantitative analysis of real world images and BU3DFE datasets [36] with superior or comparable performance to the most advanced methods.

01_MaleAsia   02_MalePortrait 03_FeStyleGAN 04_MaleWhite 05_MaleSmile 06_Glasses   07_Wrinkle   08_Grin   Failure

Fig. 7. The result of DF2Net

Apart from that DF2Net can reconstruct good details, especially for wrinkle and crow's feet, the reduction degree of the image is also very good. The disadvantages of DF2Net are: 1) when the mesh is incomplete, only partially detected face is reconstructed, 2) the eyes are sculptured concave, and 3) large poses, half face, and occlusions are not recognized and could not be reconstructed sometimes. Compared with PRNet, DF2Net is less robust but can create more personalized with individual details.

4.4 DFDN

The Deep Facial Detail Net (DFDN) method [17] is adept at recovering fine geometric details like wrinkles. For facial detail synthesis, it combines with supervised and unsupervised learning. The method is based on Conditional Generative Adversarial Net (CGAN) that adopts both appearance and geometry loss functions. The 3D facial models reconstructed from DFDN [15] are shown in Fig. 8. Compared with other methods, DFDN takes the longest time to reconstruct the 3D models. One of the reasons for this is DFDN conducts emotion prediction to determine a new expression-informed proxy, which is quite useful in animation. On the other side, emotion prediction could also lead to under-regularized global shape and inaccurate prediction for single image input. For example, a neutral face pose with lip drop might lead to open mouth surprise prediction as shown in Fig. 8.

01_MaleAsia   02_MaleLP   03_FeStyleGAN 04_MaleWhite 05_MaleBlack 06_LargePose 07_Wrinkle   08_Grin   09_Occlusions

Fig. 8. The result of DFDN

4.5 Evaluations

This subsection compares the visual results of UnsupNet, PRNet, DF2Net and DFDN shown in Fig. 9. The UnsupNet method shows inaccurate global model and limited details. The global reconstruction by 3DMM and PRNet is normally over-regularized and misses details like winkles. In contrast, DF2Net and DFDN could capture details but the reconstructed 3D model is under-regularized.

All the tests and comparisons were conducted on a desktop PC with an Intel CPU i7-8650U at 2.11 GHz, 16 GB of RAM and NVIDIA GeForce GTX1060. As for the running time for each 3D reconstruction, UnsupNet takes about 12 s to obtain 3D mesh. PRNet takes about 90 ms to obtain both 3D mesh and 68 2D landmarks. The number of vertices of reconstructed mesh is 43,867. DF2Net takes about 20 s to obtain point cloud and takes 420 ms to obtain 3D mesh with 77,767 vertices. DFDN takes about 1 min and 47 s to obtain 3D mesh, texture, and the displacement map with 53,149 vertices.

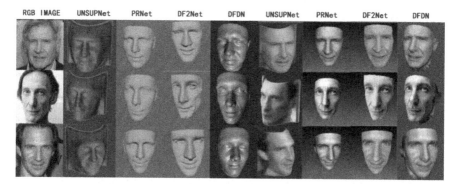

Fig. 9. The mesh and textured results obtained from UnsupNet, PRNet, DF2Net, and DFDN.

5 Conclusions and Future Research Directions

In this paper, we have provided a detailed survey about state-of-the-art methods in 3D face reconstruction from a single RGB image. We reviewed 3D morphable model, shape from shading, and deep learning based 3D face reconstruction methods, introduced commonly used 2D and 3D face databases, identified 10 deep learning-based 3D face reconstruction methods with open source code, and evaluated and compared four representative ones. How to optimize the coarse-to-fine framework and keep the balance of global shape and fine facial details are still unsolved challenges.

In the future, the challenges are how to improve the comparability of 3DMM, find the balance of global shape and fine facial details, add the other parts of parametric models, such as eyes, teeth, hair and skin details, and develop an efficient method to improve the processing time for future scenario use. Moreover, the solution should be robustness and be able to deal with a variety of light conditions; sometimes hard shadows can lead to incorrect normal map or displacement estimates. Furthermore, the solution also needs to take into account the effects of large pose and occlusion (such as hair or glasses). Another limitation of existing techniques is that they cannot handle blurry images and low-resolution images. The geometric detail prediction scheme normally relies heavily on the resolution of pixels. Therefore, the preprocess should be able to sharp the RGB image or correct the perspective to achieve distortion-free wide-angle portrait.

Overall, although existing research has been carried out in 3D face reconstruction, the research in this field is still incomplete and there are still many problems to be studied. Realizing an automatic and real-time system capable of handling unconstrained environmental conditions will open up huge potential for new applications in the gaming, safety and health industries. We hope that this survey will provide a useful guideline for current and future researchers in this field.

Acknowledgements. This research is supported by the PDE-GIR project which has received funding from the European Union Horizon 2020 Research and Innovation Programme under the Marie Skodowska-Curie grant agreement No 778035.

References

1. Zhao, W.-Y., Chellappa, R.: Symmetric shape-from-shading using self-ratio image. Int. J. Comput. Vis. **45**(1), 55–75 (2001)
2. Zhang, R., Tsai, P., Cryer, J.-E., Shah, M.: Shape-from-shading: a survey. IEEE Trans. Pattern Anal. Mach. Intell. **21**(8), 690–706 (1999)
3. Zhu, X., Lei, Z., Yan, J., Yi, D., Li, S.-Z.: High-fidelity pose and expression normalization for face recognition in the wild. In: Proceedings of the IEEE Conference on Computer Vision and Pattern Recognition, pp. 787–796 (2015)
4. Li, Y., Ma, L., Fan, H., Mitchell, K.: Feature-preserving detailed 3D face reconstruction from a single image. In: Proceedings of the 15th ACM SIGGRAPH European Conference on Visual Media Production, pp. 1–9 (2018)
5. Garrido, P., et al.: Reconstruction of personalized 3D face rigs from monocular video. ACM Trans. Graph. (TOG) **35**(3), 1–15 (2016)
6. Sengupta, S., Kanazawa, A., Castillo, C.-D., Jacobs, D.-W.: SfSNet: Learning shape, reflectance and illuminance of faces in the wild. In: Proceedings of the IEEE Conference on Computer Vision and Pattern Recognition, pp. 6296–6305 (2018)
7. Blanz, V., Vetter, T.: A morphable model for the synthesis of 3D faces. In: Proceedings of the 26th Annual Conference on Computer Graphics and Interactive Techniques (1999)
8. Tran, L., Liu, X.: Nonlinear 3D face morphable model. In: Proceedings of the IEEE Conference on Computer Vision and Pattern Recognition, pp. 7346–7355 (2018)
9. Zhu, X., Lei, Z., Liu, X.: Face alignment across large poses: a 3D solution. In: Proceedings of the IEEE Conference on Computer Vision and Pattern Recognition, pp. 146–155 (2016)
10. Huber, P.: Real-time 3D morphable shape model fitting to monocular in-the-wild videos. University of Surrey (2017)
11. Paysan, P., Knothe, R., Amberg, B., Romdhani, S., Vetter, T.: A 3D face model for pose and illumination invariant face recognition. In: 2009 Sixth IEEE International Conference on Advanced Video and Signal Based Surveillance, pp. 296–301. IEEE (2009)
12. Liu, P., Yu, H., Cang, S.: Adaptive neural network tracking control for underactuated systems with matched and mismatched disturbances. Nonlinear Dyn. **98**(2), 1447–1464 (2019)
13. Gerig, T., et al.: Morphable face models-an open framework. In: 2018 13th IEEE International Conference on Automatic Face & Gesture Recognition (FG 2018), pp. 75–82. IEEE (2018)
14. Egger, B., Smith, W., Tewari, A., Wuhrer, A., Zollhoefer, M.: 3D morphable face models—past, present, and future. ACM Trans. Graph. (TOG) **39**(5), 1–38 (2020)
15. Zollhöfer, M., Thies, J., Garrido, P.: State of the art on monocular 3D face reconstruction, tracking, and applications. Comput. Graph. Forum **37**(2), 523–550 (2018)
16. Sun, L., Zhao, C., Yan, Z., Liu, P., Duckett, T., Stolkin, R.: A novel weakly-supervised approach for RGB-D-based nuclear waste object detection. IEEE Sens. J. (2018)
17. Chen, A., Chen, Z., Zhang, G., Mitchell, K., Yu, J.: Photo-realistic facial details synthesis from single image. In: Proceedings of the IEEE International Conference on Computer Vision, pp. 9429–9439 (2019)
18. Kasinski, A., Florek, A., Schmidt, A.: The PUT face database. Image Process. Commun. **13**(3–4), 59–64 (2008)
19. Gross, R., Matthews, I., Cohn, J., Kanade, T., Baker, S.: Multi-pie. Image Vis. Comput. **28**(5), 807–813 (2010)
20. Milborrow, S., Morkel, J., Nicolls, F.: The MUCT landmarked face database. Pattern Recognit. Assoc. S. Afr. **201**(0) (2010)
21. Grgic, M., Delac, K., Grgic, S.: SCface–surveillance cameras face database. Multimedia Tools Appl. **51**(3), 863–879 (2011)

22. Koestinger, M., Wohlhart, P., Roth, P., Bischof, H.: Annotated facial landmarks in the wild: a large-scale, real-world database for facial landmark localization. In: 2011 IEEE International Conference on Computer Vision Workshops, pp. 2144–2151. IEEE (2011)

23. Lui, Y.-M., Bolme, D., Phillips, P.-J., Beveridge, J.-R.: Preliminary studies on the good, the bad, and the ugly face recognition challenge problem. In: 2012 IEEE Computer Society Conference on Computer Vision and Pattern Recognition, pp. 9–16. IEEE (2012)

24. Zhu, X., Ramanan, D.: Face detection, pose estimation, and landmark localization in the wild. In: Conference on Computer Vision and Pattern Recognition. IEEE (2012)

25. Sagonas, C., Antonakos, E., Tzimiropoulos, G., Zafeiriou, S., Pantic, M.: 300 faces in-the-wild challenge: database and results. Image Vis. Comput. **47**, 3–18 (2016)

26. Ng, H.-W., Winkler, S.: A data-driven approach to cleaning large face datasets. In: 2014 IEEE International Conference on Image Processing (ICIP), pp. 343–347. IEEE (2014)

27. Parkhi, O.-M., Vedaldi, A., Zisserman, A.: Deep face recognition (2015)

28. Liu, Z., Luo, P., Wang, X., Tang. X.: Deep learning face attributes in the wild. In: Proceedings of the IEEE International Conference on Computer Vision, pp. 3730–3738 (2015)

29. Bansal, A., Nanduri, A., Castillo, C.-D., Ranjan, R., Chellappa, R.: UMDfaces: an annotated face dataset for training deep networks. In: 2017 IEEE International Joint Conference on Biometrics (IJCB), pp. 464–473. IEEE (2017)

30. Shlizerman, I.-K., Seitz, S.-M., Miller, D., Brossard, E.: The MegaFace benchmark: 1 million faces for recognition at scale. In: Proceedings of the IEEE Conference on Computer Vision and Pattern Recognition, pp. 4873–4882 (2016)

31. Bulat, A., Tzimiropoulos, G.: How far are we from solving the 2D & 3D face alignment problem. In: Proceedings of the IEEE International Conference on Computer Vision, pp. 1021–1030 (2017)

32. Huang, G.-B., Mattar, M., Berg, T., Learned-Miller, E.: Labeled faces in the wild: a database for studying face recognition in unconstrained environments (2008)

33. Karras, T., Laine, S., Aila, T.: A style-based generator architecture for generative adversarial networks. In: Proceedings of the IEEE Conference on Computer Vision and Pattern Recognition, pp. 4401–4410 (2019)

34. Panetta, K., et al.: A comprehensive database for benchmarking imaging systems. IEEE Trans. Pattern Anal. Mach. Intell. **42**(3), 509–520 (2020)

35. Faltemier, T.-C., Bowyer, K.-W., Flynn, P.: Using a multi-instance enrollment representation to improve 3D face recognition. In: 2007 First IEEE International Conference on Biometrics: Theory, Applications, and Systems, pp. 1–6. IEEE (2007)

36. Yin, L., Wei, X., Sun, Y., Wang, J., Rosato, M.-J.: A 3D facial expression database for facial behavior research. In: 7th International Conference on Automatic Face and Gesture Recognition, pp. 211–216. IEEE (2006)

37. Yin, L., Chen, X, Sun, Y., Worm, T., Reale, M.: 3D dynamic facial expression database. In: 8th International Conference on Automatic Face & Gesture Recognition, pp. 1–6. IEEE (2008)

38. Heseltine, T., Pears, N.: Three-dimensional face recognition using combinations of surface feature map subspace components. Image Vis. Comput. **26**(3), 382–396 (2008)

39. Savran, A., et al.: Bosphorus database for 3D face analysis. In: Schouten, B., Juul, N.C., Drygajlo, A., Tistarelli, M. (eds.) BioID 2008. LNCS, vol. 5372, pp. 47–56. Springer, Heidelberg (2008). https://doi.org/10.1007/978-3-540-89991-4_6

40. Yin, B., Sun, Y., Wang, C., Ge, Y.: BJUT-3D large scale 3D face database and information processing. J. Comput. Res. Dev. **46**(6), 1009 (2009)

41. Cosker, D., Krumhuber, E., Hilton, A.: A FACS valid 3D dynamic action unit database with applications to 3D dynamic morphable facial modeling. In: 2011 International Conference on Computer Vision, pp. 2296–2303. IEEE (2011)

42. Bagdanov, A.-D., Bimbo, A.-D.: The florence 2D/3D hybrid face dataset. In: Proceedings of ACM Workshop on Human Gesture and Behavior Understanding, pp. 79–80 (2011)

43. Cao, C., Weng, Y., Zhou, S., Tong, Y., Zhou., K.: FaceWarehouse: a 3D facial expression database for visual computing. IEEE Trans. Vis. Comput. Graph. **20**(3), 413–425 (2013)
44. Zhang, X., Yin, L., Cohn, J.-F.: BP4D-spontaneous: a high-resolution spontaneous 3D dynamic facial expression database. Image Vis. Comput. **32**(10), 1–6 (2014)
45. Zhang, Z., et al.: Multimodal spontaneous emotion corpus for human behavior analysis. In: Proceedings of the IEEE Conference on Computer Vision and Pattern Recognition, pp. 3438–3446 (2016)
46. Le, H.-A., Kakadiaris, I.-A.: UHDB31: a dataset for better understanding face recognition across pose and illumination variation. In: Proceedings of the IEEE International Conference on Computer Vision Workshops, pp. 2555–2563 (2017)
47. Cheng, S., Kotsia, I., Pantic, M., Zafeiriou, S.: 4DFAB: a large scale 4D database for facial expression analysis and biometric applications. In: Proceedings of the IEEE Conference on Computer Vision and Pattern Recognition, pp. 5117–5126 (2018)
48. Ertugrul, I.-O., Cohn, J.-F., Jeni, L.-A., Zhang, A., Yin, L. Ji, Q.: Cross-domain au detection: domains, learning approaches, and measures. In: 2019 14th IEEE International Conference on Automatic Face & Gesture Recognition, pp. 1–8. IEEE (2019)
49. Yang, H., Zhu, H., Wang, Y., Huang, M., Shen, Q.: FaceScape: a large-scale high quality 3D face dataset and detailed riggable 3D face prediction. In: Proceedings of the IEEE/CVF Conference on Computer Vision and Pattern Recognition, pp. 601–610 (2020)
50. Booth, J., Roussos, A., Ponniah, A., Dunaway, D., Zafeiriou, S.: Large scale 3D morphable models. Int. J. Comput. Vis. **126**(2–4), 233–254 (2018)
51. Dai, H., Pears, N., Smith, W.-A., Duncan, C.: A 3D morphable model of craniofacial shape and texture variation. In: Proceedings of the IEEE International Conference on Computer Vision, pp. 3085–3093 (2017)
52. Chang, F., Tran, A.-T., Hassner, T., Masi, I., Nevatia, R., Medioni, G.: ExpNet: landmark-free, deep, 3D facial expressions. In: 2018 13th IEEE International Conference on Automatic Face & Gesture Recognition, pp. 122–129. IEEE (2018)
53. Booth, J., Roussos, A., Ververas, E., Antonakos, E., Ploumpis, S., Panagakis, Y.: 3D reconstruction of "in-the-wild" faces in images and videos. IEEE Trans. Pattern Anal. Mach. Intell. **40**(11), 2638–2652 (2018)
54. Shang, J., Shen, T., Li, S., Zhou, L., Zhen, M.: Self-supervised monocular 3D face reconstruction by occlusion-aware multi-view geometry consistency (2020)
55. Feng, Y., Wu, F., Shao, X., Wang, Y., Zhou, X.: Joint 3D face reconstruction and dense alignment with position map regression network. In: Proceedings of the European Conference on Computer Vision, pp. 534–551 (2018)
56. Zeng, Xi., Peng, X., Qiao, Y.: DF2Net: a dense-fine-finer network for detailed 3D face reconstruction. In: Proceedings of the IEEE International Conference on Computer Vision, pp. 2315–2324 (2019)
57. Zhang, J., Cai, H., Guo, Y., Peng, Z.: Landmark detection and 3D face reconstruction for caricature using a nonlinear parametric model (2020)
58. Browatzki, B., Wallraven, C.: 3FabRec: fast few-shot face alignment by reconstruction. In: Proceedings of the IEEE/CVF Conference on Computer Vision and Pattern Recognition, pp. 6110–6120 (2020)
59. Wu, S., Rupprecht, C., Vedaldi, A.: Unsupervised learning of probably symmetric deformable 3D objects from images in the wild. In: Proceedings of the IEEE/CVF Conference on Computer Vision and Pattern Recognition, pp. 1–10 (2020)
60. Ronneberger, O., Fischer, P., Brox, T.: U-Net: convolutional networks for biomedical image segmentation. In: Navab, N., Hornegger, J., Wells, W., Frangi, A. (eds.) MICCAI 2015. LNCS, vol. 9351, pp. 234–241. Springer, Cham (2015). https://doi.org/10.1007/978-3-319-24574-4_28

Towards Understanding Time Varying Triangle Meshes

Jan Dvořák⬚, Petr Vaněček⬚, and Libor Váša⁽⊠⁾⬚

Department of Computer Science and Engineering, University of West Bohemia,
Pilsen, Czech Republic
{jdvorak,pvanecek,lvasa}@kiv.zcu.cz

Abstract. Time varying meshes are more popular than ever as a representation of deforming shapes, in particular for their versatility and inherent ability to capture both true and spurious topology changes. In contrast with dynamic meshes, however, they do not capture the temporal correspondence, which (among other problems) leads to very high storage and processing costs. Unfortunately, establishing temporal correspondence of surfaces is difficult, because it is generally not bijective: even when the full visible surface is captured in each frame, some parts of the surface may be missing in some frames due to self-contact. We observe that, in contrast with the inherent absence of bijectivity in surface correspondence, volume correspondence is bijective in a wide class of possible input data. We demonstrate that using a proper intitialization and objective function, it is possible to track the volume, even when considering only a pair of subsequent frames at the time. Currently, the process is rather slow, but the results are promising and may lead to a new level of understanding and new algorithms for processing of time varying meshes, including compression, editing, texturing and others.

Keywords: Time varying mesh · Model · Animation · Tracking · Analysis · Surface

1 Introduction

Time Varying Meshes (TVMs) are appearing more commonly in recent years, especially because of the improved methodology (hardware and processing) used for acquiring 3D surface data at the required framerate. Photogrammetry in particular, supported by depth sensing technologies, such as time-of-flight cameras and pattern projection scanning, have enabled capturing deforming shapes, such as articulated human faces or whole bodies, with relative ease. Most of these approaches rely on a tight synchronization of the scanning devices, which allows processing data at each time instant separately, making the TVM a natural output format of the scanning process.

TVMs find application in a variety of fields, such as movie/gaming industry, visualization, telepresence, live sport broadcasting and others, which take

© Springer Nature Switzerland AG 2021
M. Paszynski et al. (Eds.): ICCS 2021, LNCS 12746, pp. 45–58, 2021.
https://doi.org/10.1007/978-3-030-77977-1_4

advantage of the versatility of this representation. On the other hand, their application is limited by the large size of data needed for the representation: a common sequence lasting about a minute with 100k vertices in each frame takes on the order of gigabytes to store/transmit. This is mainly because each frame carries its own connectivity, which in many cases requires a portion of the overall bit budget that is comparable with the geometry. In contrast with dynamic meshes, which share the common connectivity, this is a major disadvantage, topped by the general difficulty of exploiting the temporal coherence of the data exhibited by the TVM representation.

Another hindrance inherent to TVMs is the missing temporal correspondence. This makes certain common tasks, such as attaching an artificial prop to the 3D model, difficult. Similarly, it is difficult to exploit the temporal coherence of the surface albedo for efficient texturing using a shared map and image, since the UV unwrapping (and thus the texture itself) must be different in each frame.

Extracting the temporal correspondence would make a great step towards better understanding of the captured surface deformation. Typically, captured performances frequently involve limbs being connected or merged with each other or with the torso in the reconstruction. While human observers easily identify such occurrences and distinguish them from true merging and/or deformation, for automatic processing algorithms such distinction is difficult and leads to problems when establishing the temporal correspondence.

The main issue is that surface correspondence is inherently not bijective: in some (most) frames, certain parts of the surface are missing, even when the surface is captured from all possible viewpoints, due to the self-contact. Typically, parts of the surface may disappear for several frames and then eventually reappear. It is even possible, that the whole of the surface is not completely visible in any of the input frames.

Ideally, we would like to track the whole surface and add the invisible parts that are in contact to the frames where they are not visible. Such approach could allow for consistent, bijective correspondence tracking. Unfortunately, this is a difficult problem, which could be visualized in a simplified 2D case of tracking silhouettes, as shown in Fig. 1. The data in this case can be interpreted as a 3D (2D + time) object, and the task translates to cutting the shape along certain ridges in order to complete the silhouettes. Formulating the criteria for the cuts seems difficult and such process is necessarily prone to errors.

Our main observation is that there is a different, orthogonal approach to the problem, which eliminates much of the difficulty. Rather than tracking the surface, we propose to track the volume of the captured objects. In many practical scenarios, the overall volume changes negligibly, and it is possible to track it bijectively over the length of the sequence. Splitting the volume into finite size elements, it is much easier to formulate the criteria/priors for correct tracking: the whole of the volume in each frame should be covered, and the volume elements should move coherently. With volume correspondence, it is then much easier to establish surface correspondence and eventually add the missing parts of surface in any frame.

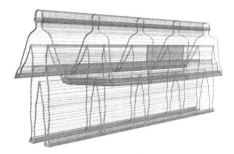

Fig. 1. A Sequence of 2D silhouettes interpreted as a 3D volume. Tracking the surface translates to cutting the 3D object at the red locations. (Color figure online)

Our main contribution is a working volume tracking pipeline, verified on data from multiple sources. The key ingredients are:

1. choice of appropriate representation
2. proper tracking initialization
3. formulation of objective function/energy.

2   Related Work

One could track the evolving surface by propagating certain canonical frame (e.g. first frame) using a non-rigid alignment method. Such methods are usually based on minimizing an energy which evaluates the quality of the alignment. However, only the methods that do not require the correspondences information as an input can be considered, since this information is not known a priori. Myronenko et al. [22,23] proposed to model the unknown correspondences using a Gaussian mixture model, enforced the movement of points to be spatially smooth and used the expectation-maximization algorithm to optimise the energy. Li et al. [18] also treat the correspondences as probability, however, they model the movement of points using deformation graphs. In their subsequent work [2] this method was utilized to track evolving surfaces. Yoshiyasu et al. [31] constrain the underlying deformation to be as-conformal-as-possible, i.e. preserving angles. Such property preserves the structure of the deformed mesh. Utilizing the volume the surface encloses was already considered in the past. Huang et al. [12] proposed to use centroidal voronoi tessellations with connection to signed distance function to model volumetric features used in the alignment process. Tracking of surfaces based on non-rigid alignment has, however, quite a few limitations. Such tracking process unfortunately highly depends on the selection of the propagated frame and might suffer from the error accumulation.

Compression of time-varying geometry, i.e. sequence of triangle meshes or point clouds representing a surface with dynamic topology is a closely related field of research in the sense that its main goal is to find a reduced representation of the data. Quite a few methods utilizing the temporal coherence of the data have been already proposed. The most common approach is to store the surface in a spatial data structure (e.g. grid or octree) and exploit the coherence of occupancy of such structures between frames instead of the coherence of surfaces [7,8,10,16,21,26]. Inspired by video compression methods [5,9,17,20,24,30] use motion compensation, where parts of the surface are predicted by parts of the previous frame. However, such predictions are, in general, not correspondences between frames. Since the video compression is a mature field of research, it is also utilized in so-called geometry-video coding [11,14,25,29], where the geometry of the surface is mapped to a video, which is then compressed using the spatio-temporal coherence in the image domain. The temporal coherence of the surfaces themselves is exploited only in few methods. Yang et al. proposed to match the decimated frames [15]. Another approach is to utilize skeletal information [6,19]. However, both approaches can be used only for data with constant genus.

3   Algorithm Overview

We assume that the input data take the form of TVM: a sequence of triangle meshes captured at certain framerate, usually 25–30 fps. Each frame consists of a set of vertex positions representing the geometry, and a set of integer triplets, representing the triangles forming the connectivity. While there is a certain inter-frame coherence of the geometry, since the subsequent frames represent similar deformations of the same shape, there is no coherence in the connectivity: it is assumed to be completely independent in each frame. For simplicity, we also assume that each input mesh is complete and watertight. Later we will discuss how this assumption can be lifted without modifying much of the proposed pipeline.

We aim at building a compact data structure that captures the temporal development of the shape represented by the input TVM. The structure should provide some form of temporal correspondence, which enables deeper insight into the semantics of the shape, as well as more efficient processing of the input data.

Our proposal is to represent the shape by a fixed number of points (denoted *centers*), each representing a small volume surrounding it. The locations of the centers will vary in time, their positions will be denoted $\mathbf{c}_i^f$, representing the 3D position of the i-th center in frame f.

In contrast with the difficulty of specifying conditions for proper surface data structure, formulating the conditions for the volume elements is comparatively easy. In particular, we wish the centers to

- cover all parts of the input object in each frame,
- be distributed evenly over the volume of the objects in each frame and
- move consistently between frames, i.e. nearby centers should move in similar direction.

Such representation can be used in many applications. It allows reconstructing a surface with naturally changing genus (as described in Sect. 10). It can be used to track the surface and add parts missing due to self-contact. Note that the data structure may also complement the original input instead of replacing it: in particular, it can serve as a reference for compression purposes.

When determining the model parameters (center positions in each frame), we will proceed frame by frame, considering only a pair of subsequent frames at the time. On one hand, this restriction means that the algorithm is not exploiting all the information that is available. On the other hand, such approach eventually allows for online processing of the inputs, given that the size of the data and processing power allows for real-time computations.

First, we will distribute the centers uniformly over each input shape, as will be detailed in Sect. 4. This ensures that every part of every frame will be covered by the model.

Next, we proceed from a previous frame to the next. We determine the likely correspondence using the Kuhn-Munkers ("hungarian") method (Sect. 5), and then we optimize an objective function described in Sects. 6, 7 and 8, which ensures a smooth, coherent inter-frame transition while preserving the sampling uniformity.

4   Uniform Object Sampling

We will be sampling the volume occupied by the object in each frame, denoted V_f, interpreted as an infinite set of points. A constant number of samples (centers) will be placed in each frame. In our experiments, 1000 centers worked well as a compromise between representation accuracy and processing time. If more centers are needed, it is easier to add them in post-processing, after the initial batch has been tracked.

We aim at achieving a uniform distribution of the centers. As with many similar problems, formulating the objective precisely goes a long way towards finding the solution. In our case, a uniform distribution can be identified by looking at the volumes that surround each center in certain sense. We may define a cell associated to i-th center as the set of all points $\mathbf{p} \in V_f$, such that $\|\mathbf{p} - \mathbf{c}_i\| \leq \|\mathbf{p} - \mathbf{c}_j\|$ for every j. Such structure is very similar to the Voronoi cell of i-th center, with the only difference that it is limited to the volume V_f. The sampling uniformity can be directly linked to the standard deviation of the cell sizes.

The objective can be thus reformulated as minimizing the spread of cell sizes. It is well known (and easy to see) that such objective can be achieved using the Lloyd iteration algorithm, where in each iteration the centers are shifted towards the center of mass of their associated voronoi cell, resulting in the so-called Centroidal Voronoi Tesselation (CVT). In order to stop the centers from diverging to infinity, the center of mass must be computed with respect to a certain weight function, in our case it is the indicator function of the volume V_f.

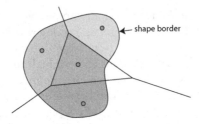

Fig. 2. Cells (shades of blue) associated with centers (green) can be non-convex and even non-continuous, but they are always finite, which contrasts the Voronoi cells, which are always convex, but often infinite. (Color figure online)

Despite the similarity to Voronoi cells, finding the center of mass of the cells associated with centers is not easy, since they are more general, potentially non-convex and even non-continuous (see Fig. 2). Therefore we will opt for computing the centers of mass quickly at the cost of certain approximation error.

The approach we choose is based on sampling the indicator function using a regular volume grid. The indicator function that determines for each voxel whether or not it lies within the volume occupied by the object is evaluated using the ray-shooting technique, essentially evaluating the number of intersections with the surface along a certain ray. Choosing axis aligned rays allows reusing the previously computed intersection points for evaluating a whole column of the regular sampling grid. Note that this is only possible for watertight models. A more general approach could use the generalized winding number [13] instead of the indicator function, possibly generalizing the algorithm to incomplete models, however, at this point, we have not experimented with this possibility.

Centers are initialized randomly into cells of the regular volume grid where the indicator function is positive. Next, we approximate the center of mass of each center associated cell by averaging the positions those voxel centroids, that have positive indicator function and the given center as its closest center. This is done by a single pass over all voxels, searching for the nearest center using a KD-tree. Finally, all centers are shifted to their respective associated centers of mass.

As a result, we obtain a uniform sampling of each frame. Note that the result strongly depends on the initialization [3], because the objective function generally exhibits a broad basin of low values, with many shallow local minima. We will exploit this property in the subsequent steps by regularizing the solution using the smoothness term.

First, however, we need to initialize the inter-frame correspondences of centers, which we detail in the next section.

5   Initial Correspondence Estimation

The task at hand can be formulated as an optimal transport problem, with certain cost function. We wish to find a mapping between the centers of the previous (source) and current (target) frame, such that (i) corresponding points are as close together as possible, (ii) each source point is associated to a unique target point, or an affine combination thereof, and (iii) each target point is associated to a unique source point, or an affine combination thereof.

Apart from spatial proximity, captured by simple length squared cost function, the transport cost should also reflect whether the trajectory connecting the source and target points is located within the object, or passes outside, since the latter often indicates that the correspondence involves center points that belong to separate components or topologically distant locations (see Fig. 3). On the other hand, in case of fast movement, even correct correspondences sometimes pass "outside" of the objects, and therefore such correspondences cannot be ignored by the algorithm completely. Having the input sampled, it is relatively easy to sample every possible trajectory and adjust the cost, so that segments outside of the object are penalized: in our implementation, the penalization is done by multiplying the proportional part of the cost outside of the object by 10, which is a constant that worked well in all our experiments.

The optimal transport task at hand can be solved using the Sinkhorn iterations algorithm [4]. Each source center is associated to a weighted combination of target points, which can be averaged using their weights to obtain a unique corresponding center in the target frame. This approach has the advantage of natural smoothing of the correspondences.

In our experiments, however, we have encountered difficulty with setting the λ parameter of the Sinkhorn iterations algorithm. In essence, the parameter controls the "reach" of each center. When set too low, the algorithm does not converge to a proper solution of the problem at hand (the mapping weights were not affine), while setting it too high leads to a too poor approximation of the solution, where each source center is associated with many target centers and averaging them leads to shrinkage of the centers in the target frame. Unfortunately, in many cases it is impossible to reach a compromise value that eliminates both these effects.

Since the number of centers is constant in every frame, the problem can also be formulated as a special case of the optimal transport: the optimal assignment, where each source center is associated with exactly one target center. An optimal solution can be found efficiently using the Kuhn-Munkers ("hungarian") method. However, in contrast with the Sinkhorn iterations, the result can be a very uneven result (see Fig. 4), because of the lack of weighting and the random character of sampling of each frame. This causes problems in the following steps, because such state represents a local minimum of the objective function we are going to minimize. Fortunately, this issue can be solved using a technique described next.

6 Smoothness Energy, Smoothing

One of our objectives is achieving a smooth movement of centers, i.e. that nearby centers move in a similar direction. It is important to note that we do not wish to penalize the amount of movement itself: it follows from the nature of the input data that there is movement present in the sequence.

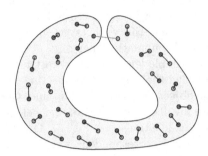

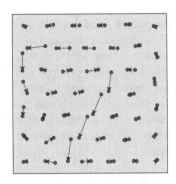

Fig. 3. Red and green dots represent different samplings of the same gray domain. The blue correspondence lies partially outside of the domain, and thus should be penalized. (Color figure online)

Fig. 4. Green and red dots represent stationary points of the Lloyd algorithm over the same square domain. The blue lines represent the correspondences as found by the Kuhn-Munkers method, minimizing the sum of squared distances. (Color figure online)

The model can be interpreted as a vector field $\mathbf{v}$ defined over the domain V_f of the source frame, sampled as $\mathbf{v}(\hat{\mathbf{c}}_i) = \mathbf{c}_i - \hat{\mathbf{c}}_i = \mathbf{v}_i$, where $\hat{\mathbf{c}}_i$ is the position of the i-th center in previous frame. The object of interest is the smoothness of this vector field, which can be captured by its Laplacian $\Delta\mathbf{v}$. Smooth (harmonic) vector fields exhibit zero length Laplacian, and the sum of the squared lengths of $\Delta\mathbf{v}$ at uniform sample points can be interpreted as the amount of deviation from smoothness.

In our case, the vector field is sampled irregularly, which makes evaluating its Laplacian harder, but not impossible. In particular, the discrete Laplacian proposed by Belkin [1] can be used. This allows us to express an energy E_s that captures the smoothness of the vector field as follows:

$$E_s = \sum_{\hat{\mathbf{c}}_i \in C} \|\Delta\mathbf{v}(\hat{\mathbf{c}}_i)\|^2 = \frac{1}{|C|} \sum_{\hat{\mathbf{c}}_i \in C} \| \sum_{\hat{\mathbf{c}}_j \in C} H^t(\hat{\mathbf{c}}_i, \hat{\mathbf{c}}_j)(\mathbf{v}_j - \mathbf{v}_i)\|^2, \quad (1)$$

where $H^t(\mathbf{x}, \mathbf{y}) = \frac{1}{(4\pi t)^{\frac{5}{2}}} e^{-\frac{\|\mathbf{x}-\mathbf{y}\|^2}{4t}}$ is a Gaussian kernel with parameter t. Gradient of such energy consists of following partial derivatives:

$$\frac{\partial E_s}{\partial \mathbf{v}_k} = \frac{\partial E_s}{\partial \mathbf{c}_k} = \frac{8}{|C|^2} \sum_{\hat{\mathbf{c}}_i \in C} (H^t(\hat{\mathbf{c}}_k, \hat{\mathbf{c}}_j))^2 (\mathbf{v}_k - \mathbf{v}_j) \quad (2)$$

$$= \frac{8}{|C| \cdot (8\pi t)^{\frac{5}{2}}} \frac{1}{|C|} \sum_{\hat{\mathbf{c}}_j \in C} H^{\frac{t}{2}}(\hat{\mathbf{c}}_k, \hat{\mathbf{c}}_j)(\mathbf{v}_k - \mathbf{v}_j) \quad (3)$$

$$= -\frac{8}{|C| \cdot (8\pi t)^{\frac{5}{2}}} \hat{\Delta}\mathbf{v}(\hat{\mathbf{c}}_k), \quad (4)$$

where $\hat{\Delta}$ is a discrete Laplacian defined with half the kernel width of Δ.

7   Sampling Energy

Apart from smoothness, we also need to preserve the uniform density of sampling. As discussed before, this property can be quantified as

$$E_u = \frac{1}{2} \sum_{\mathbf{c}_i \in C} \|\mathbf{m}_i - \mathbf{c}_i\|^2, \tag{5}$$

where $\mathbf{m}_i$ is the center of mass of a cell associated with i-th center.

Treating the centers of mass as constants, the gradient consists of following partial derivatives:

$$\frac{\partial E_u}{\partial \mathbf{c}_i} = \mathbf{m}_i - \mathbf{c}_i. \tag{6}$$

The location of each center of mass $\mathbf{m}_i$ can be approximated as described in Sect. 4.

8   Overall Energy

Having the two energy terms that we wish to minimize, the tracking can be formulated as minimization of the overall energy $E = E_s + \alpha E_u$, where α is a weighting constant. Having the gradient of both terms expressed in Eqs. 4 and 6, such minimization can be done using the standard gradient descent technique, i.e. in a series of steps, the centers are shifted in the direction of the sum of the two gradients. The procedure is terminated after a set maximum of iterations has been performed, or when the gradient is sufficiently small for each $\mathbf{c}_i$.

8.1   Pre-smoothing

Note that the initial result of the Kuhn-Munkers method naturally represents a local minimum of E_u, since it maps the centers of the source frame directly to the centers of target frame, which have been optimized for uniformity during initialization. This is somewhat unfortunate, because the gradient of E_s is often not strong enough to exit the local minimum.

This issue could be addressed by a smooth variation of the weighting parameter α, starting with a stronger influence of E_s and then gradually increasing the influence of E_u. Such approach, however, leads to problems with formulating the stopping condition, because it enforces a certain number of iterations in order to reach the final ratio of term influences.

In our experiments, we have therefore used a different empirical approach: we start with a few iterations (50 worked well with our data) that only consider E_s, which lead the optimization out of the local minimum, and then we continue with the full energy E, using a constant value of α. This way, we can stop the iteration whenever the gradient is small enough (smaller than 0.1 mm in each component in our experiments), since the energy expression does not change during the descent.

9   Results

We have tested the proposed algorithm on data from two sources: a commercially available TVM sequence of professional quality (denoted *casual_man*), and a pair of sequences provided for academic purposes (denoted *samba* and *handstand*), consisting of mesh sequences of constant connectivity [28]. Note that even though the second dataset consists in fact of dynamic meshes rather than TVMs, we do not exploit this fact and treat the datasets as if they were TVMs. Table 1 summarizes some basic properties of the used input datasets.

Table 1. Datasets used in experiments.

Dataset	No. of frames	(Average) no. of vertices	Uncompressed size (.obj)[MB]
casual_man	546	36 145	4 086
samba	175	9 971	113
squat	250	10 002	159

Using a volume grid of 512 cells along the longest dimension and a stopping condition of 0.1 mm in each component of the gradient descent direction, we have processed the *casual_man* dataset at 30.7 s per frame on average. This framerate was achieved using a purely CPU implementation written in C#, including all the necessary steps (volume grid sampling, initial point distribution, energy minimization). We have used a common Intel Core-i7 9700 CPU. As for the other datasets, the processing time per frame was only slightly shorter, because most of the processing cost depends on the resolution of the volume grid, which was constant over all the experiments. The results are shown in Fig. 5. The means of quantifying the tracking quality naturally depends on the particular application, some of which we discuss next.

Fig. 5. Tracking results, showing the first, middle and last frame for each sequence. Coloring is done by the first three principal trajectory coefficients, notice the consistency of the coloring throughout the sequence.

10   Applications

It is possible to interpret the resulting model as an approximate representation of the input deformable surface. A mesh can be extracted from the set of centers in each frame by extracting the iso-surface of an appropriately designed function that represents the distance from the centers, ideally blended in some way. One particular choice is using

$$f(\mathbf{x}) = -ln\Big(\sum_{\mathbf{c}_i \in C} exp\big(- k(\|\mathbf{c}_i - \mathbf{x}\| - r)\big)\Big)/k, \qquad (7)$$

where r is a minimum radius of sphere surrounding each center and k is a parameter influencing the blending distance. The function is sampled using a regular grid and its iso-surface can be extracted using a standard technique, such as marching cubes. The resulting triangle mesh can be compared with the input using some tessellation oblivious metric, such as the mean nearest inter-surface distance. Figure 6 shows the resulting Hausdorff distances for the *casual_man* sequence, using 1000, 2000 and 4000 centers.

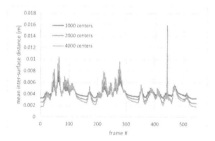
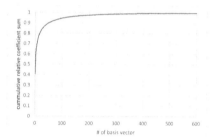

Fig. 6. Mean inter-mesh distances for the *casual_man* dataset.

Fig. 7. Cummulative sums of coefficient magnitudes. 90% of coefficients are captured by the first 50 eigenvectors, 99% are captured by 346 eigenvectors.

We also observe that the trajectories of the centers are located in a small subspace of the full space of trajectories. A basis of the space can be found using PCA, similarly to what has been previously done for vertex trajectories in dynamic meshes [27]. A reduced basis can be used to capture most of the variance present in the data, leading to even more efficient representation. Figure 7 shows the cumulative relative sums of coefficient magnitudes, showing that most of the variance is concentrated with the first few eigenvectors.

The model can also be used as a compressed version of the input data. Storing the complete data needed for reconstruction, i.e. 1000 trajectories, using 32-bit floats for coordinates, amounts to roughly 6 MB of data for the *casual_man* sequence. Projecting onto the most important eigentrajectories, it is possible to reduce the amount data down to about 2 MB without sacrificing the quality. Should a TVM be compressed down to equivalent size, each frame must be stored in no more than 3.7 kB. Current state of the art compression techniques require about 16 bits per vertex, i.e. each frame must be simplified from the original 36k vertices down to 1830, which necessarily drastically degrades the visual quality.

11   Conclusions

We have described a volume tracking algorithm for analyzing Time-Varying meshes. It is based on simple energies, yet it produces feasible results, both perceptually and numerically, despite the fact that the tracking is done using the information from only a pair of subsequent shapes. As far as we know, there is currently no competing method that is able to analyze a sequence of meshes with the generality and precision provided by our method.

This result can find many applications. The resulting model can be used to analyze the deforming bodies: it captures the nature of the deformation, it allows tracking volumes even when they visually merge with others or get occluded or hidden.

In the future, we would like to work on speeding the process up, potentially using a GPU implementation of some of the time consuming steps. Additionally, we would like develop means increasing the number of tracked volumes, yielding a finer representation. We believe that with a proper initialization provided by the proposed method, tracking additional volumes should be manageable. Finally, we would like to better analyze the relations of the centers, building some kind of connectivity that captures which volumes move together and how. Having such structure could in turn help with further improving the tracking.

Acknowledgement. This work was supported by the project 20-02154S of the Czech Science Foundation. Jan Dvořák was also partially supported by the University specific student research project SGS-2019-016 Synthesis and Analysis of Geometric and Computing Models. The authors thank Diego Gadler from AXYZ Design, S.R.L. for providing the test data.

References

1. Belkin, M., Sun, J., Wang, Y.: Discrete Laplace operator on meshed surfaces. In: Proceedings of the Twenty-Fourth Annual Symposium on Computational Geometry, SCG 2008. Association for Computing Machinery, New York (2008)
2. Bojsen-Hansen, M., Li, H., Wojtan, C.: Tracking surfaces with evolving topology. ACM Trans. Graph. 31(4), 53-1 (2012)
3. Celebi, M.E., Kingravi, H.A., Vela, P.A.: A comparative study of efficient initialization methods for the k-means clustering algorithm. Expert Syst. Appl. 40(1), 200–210 (2013)

4. Cuturi, M.: Sinkhorn distances: lightspeed computation of optimal transport. In: Burges, C.J.C., Bottou, L., Welling, M., Ghahramani, Z., Weinberger, K.Q. (eds.) Advances in Neural Information Processing Systems, vol. 26, pp. 2292–2300. Curran Associates, Inc. (2013)

5. de Queiroz, R.L., Chou, P.A.: Motion-compensated compression of dynamic voxelized point clouds. IEEE Trans. Image Process. **26**(8), 3886–3895 (2017). https://doi.org/10.1109/TIP.2017.2707807

6. Doumanoglou, A., Alexiadis, D.S., Zarpalas, D., Daras, P.: Toward real-time and efficient compression of human time-varying meshes. IEEE Trans. Circ. Syst. Video Technol. **24**(12) (2014). https://doi.org/10.1109/TCSVT.2014.2319631

7. Garcia, D.C., de Queiroz, R.L.: Context-based octree coding for point-cloud video. In: 2017 IEEE International Conference on Image Processing (ICIP), pp. 1412–1416 (2017). https://doi.org/10.1109/ICIP.2017.8296514

8. Garcia, D.C., Fonseca, T.A., Ferreira, R.U., de Queiroz, R.L.: Geometry coding for dynamic voxelized point clouds using octrees and multiple contexts. IEEE Trans. Image Process. **29**, 313–322 (2020). https://doi.org/10.1109/TIP.2019.2931466

9. Han, S., Yamasaki, T., Aizawa, K.: Time-varying mesh compression using an extended block matching algorithm. IEEE Trans. Circuits Syst. Video Technol. **17**(11), 1506–1518 (2007). https://doi.org/10.1109/TCSVT.2007.903810

10. Han, S., Yamasaki, T., Aizawa, K.: Geometry compression for time-varying meshes using coarse and fine levels of quantization and run-length encoding. In: 2008 15th IEEE International Conference on Image Processing, pp. 1045–1048 (2008). https://doi.org/10.1109/ICIP.2008.4711937

11. Hou, J., Chau, L., He, Y., Magnenat-Thalmann, N.: A novel compression framework for 3d time-varying meshes. In: 2014 IEEE International Symposium on Circuits and Systems (ISCAS), p. 2161–2164 (2014). https://doi.org/10.1109/ISCAS.2014.6865596

12. Huang, C., Allain, B., Franco, J., Navab, N., Ilic, S., Boyer, E.: Volumetric 3d tracking by detection. In: 2016 IEEE Conference on Computer Vision and Pattern Recognition (CVPR), pp. 3862–3870 (2016). https://doi.org/10.1109/CVPR.2016.419

13. Jacobson, A., Kavan, L., Sorkine, O.: Robust inside-outside segmentation using generalized winding numbers. ACM Trans. Graph. **32**(4), 1–12 (2013)

14. Jang, E.S., et al.: Video-based point-cloud-compression standard in MPEG: from evidence collection to committee draft [standards in a nutshell]. IEEE Signal Process. Mag. **36**(3), 118–123 (2019). https://doi.org/10.1109/MSP.2019.2900721

15. Yang, J.-H., Kim, C.-S., Lee, S.-U.: Semi-regular representation and progressive compression of 3-d dynamic mesh sequences. IEEE Trans. Image Process. **15**(9), 2531–2544 (2006). https://doi.org/10.1109/TIP.2006.877413

16. Kammerl, J., Blodow, N., Rusu, R.B., Gedikli, S., Beetz, M., Steinbach, E.: Real-time compression of point cloud streams. In: 2012 IEEE International Conference on Robotics and Automation, pp. 778–785 (2012). https://doi.org/10.1109/ICRA.2012.6224647

17. Kathariya, B., Li, L., Li, Z., Alvarez, J.R.: Lossless dynamic point cloud geometry compression with inter compensation and traveling salesman prediction. In: 2018 Data Compression Conference, p. 414 (2018). https://doi.org/10.1109/DCC.2018.00067

18. Li, H., Sumner, R.W., Pauly, M.: Global correspondence optimization for non-rigid registration of depth scans. In: Computer Graphics Forum, vol. 27, pp. 1421–1430. Wiley Online Library (2008)

19. Lien, J.M., Kurillo, G., Bajcsy, R.: Multi-camera tele-immersion system with real-time model driven data compression. Vis. Comput. **26**(1), 3 (2010). https://doi.org/10.1007/s00371-009-0367-8

20. Mekuria, R., Blom, K., Cesar, P.: Design, implementation, and evaluation of a point cloud codec for tele-immersive video. IEEE Trans. Circuits Syst. Video Technol. **27**(4), 828–842 (2017). https://doi.org/10.1109/TCSVT.2016.2543039

21. Milani, S., Polo, E., Limuti, S.: A transform coding strategy for dynamic point clouds. IEEE Trans. Image Process. **29**, 8213–8225 (2020). https://doi.org/10.1109/TIP.2020.3011811

22. Myronenko, A., Song, X.: Point set registration: coherent point drift. IEEE Trans. Pattern Anal. Mach. Intell. **32**(12), 2262–2275 (2010)

23. Myronenko, A., Song, X., Carreira-Perpinán, M.A.: Non-rigid point set registration: coherent point drift. In: Advances in Neural Information Processing Systems, pp. 1009–1016 (2007)

24. Santos, C.F., Lopes, F., Pinheiro, A., da Silva Cruz, L.A.: A sub-partitioning method for point cloud inter-prediction coding. In: 2018 IEEE Visual Communications and Image Processing (VCIP), pp. 1–4 (2018). https://doi.org/10.1109/VCIP.2018.8698661

25. Schwarz, S., Sheikhipour, N., Fakour Sevom, V., Hannuksela, M.M.: Video coding of dynamic 3d point cloud data. APSIPA Trans. Signal Inf. Process. **8**, e31 (2019). https://doi.org/10.1017/ATSIP.2019.24

26. Thanou, D., Chou, P.A., Frossard, P.: Graph-based compression of dynamic 3d point cloud sequences. IEEE Trans. Image Process. **25**(4), 1765–1778 (2016). https://doi.org/10.1109/TIP.2016.2529506

27. Váša, L., Skala, V.: CODDYAC: connectivity driven dynamic mesh compression. In: 3DTV Conference Proceedings (2007)

28. Vlasic, D., Baran, I., Matusik, W., Popović, J.: Articulated mesh animation from multi-view silhouettes. ACM Trans. Graph. **27**(3), 1–9 (2008)

29. Xu, Y., Zhu, W., Xu, Y., Li, Z.: Dynamic point cloud geometry compression via patch-wise polynomial fitting. In: ICASSP 2019–2019 IEEE International Conference on Acoustics, Speech and Signal Processing (ICASSP), pp. 2287–2291 (2019). https://doi.org/10.1109/ICASSP.2019.8682413

30. Yamasaki, T., Aizawa, K.: Patch-based compression for time-varying meshes. In: 2010 IEEE International Conference on Image Processing, pp. 3433–3436 (2010). https://doi.org/10.1109/ICIP.2010.5652911

31. Yoshiyasu, Y., Ma, W.C., Yoshida, E., Kanehiro, F.: As-conformal-as-possible surface registration. Comput. Graph. Forum **33**(5), 257–267 (2014). https://doi.org/10.1111/cgf.12451

Semantic Similarity Metric Learning for Sketch-Based 3D Shape Retrieval

Yu Xia, Shuangbu Wang[✉], Lihua You, and Jianjun Zhang

National Centre for Computer Animation, Bournemouth University, Poole, UK
swang1@bournemouth.ac.uk

Abstract. Since the development of the touch screen technology makes sketches simple to draw and obtain, sketch-based 3D shape retrieval has received increasing attention in the community of computer vision and graphics in recent years. The main challenge is the big domain discrepancy between 2D sketches and 3D shapes. Most existing works tried to simultaneously map sketches and 3D shapes into a joint feature embedding space, which has a low efficiency and high computational cost. In this paper, we propose a novel semantic similarity metric learning method based on a teacher-student strategy for sketch-based 3D shape retrieval. We first extract the pre-learned semantic features of 3D shapes from the teacher network and then use them to guide the feature learning of 2D sketches in the student network. The experiment results show that our method has a better retrieval performance.

Keywords: Sketch · 3D shape · Retrieval · Metric learning · Semantic feature

1 Introduction

The virtual 3D shape plays an increasingly important role in our daily lives due to the rapid development of digitalization techniques, such as visual effects, medical imaging and 3D printing. How to retrieve a desired 3D shape among a great number of 3D shapes is a popular research topic in many years [1–4]. Compared to using texts and 3D shapes as queries, sketches can easily describe the detailed information of complex 3D shapes, and are also more intuitive and convenient for humans to use. Therefore, sketch-based 3D shape retrieval has attracted considerable attention in the community of computer vision and graphics [5,6].

The main challenge for sketch-based 3D shape retrieval is the big domain discrepancies [7]. First, sketches are represented in a 2D space while 3D shapes are embodied in a 3D space, so their heterogenous data structures make it extremely difficult to directly retrieve 3D shapes from a query sketch. Second, sketches are abstract free-hand drawings, which usually consist of several simple lines and contain very limited information. Conversely, 3D shapes are realistic geometric objects and have many details of their shape characteristics. Third, sketches are presented with only one view of 3D shapes, and it is very hard to find the best

© Springer Nature Switzerland AG 2021
M. Paszynski et al. (Eds.): ICCS 2021, LNCS 12746, pp. 59–69, 2021.
https://doi.org/10.1007/978-3-030-77977-1_5

or most similar view of 3D shapes according to query sketches. Figure 1 gives some examples of sketches and corresponding 3D shapes from the same class, and shows the large domain gap between them.

Fig. 1. Some examples of sketches and corresponding 3D shapes

In order to tackle the aforementioned challenge of sketch-based 3D shape retrieval, a variety of research efforts have been dedicated to this task, and their main purpose is to improve the retrieval accuracy. There are mainly two ways to achieve the accuracy improvement: 1) learning robust features representations for both sketches and 3D shapes [8–10], and 2) developing effective ranking or distance metrics between sketches and 3D shapes [7,11,12]. Due to the great success of deep convolutional neural networks (CNNs) applied in the image feature extraction in recent years, all state-of-the-art methods have used deep metric learning for sketch-based 3D shape retrieval and achieved a better retrieval accuracy compared with traditional methods [13]. However, these studies have two weaknesses. First, they address the domain discrepancy problem by mapping sketches and 3D shapes into a joint feature embedding space, where the similarity is measured using a shared loss function. It is difficult to effectively reduce the domain discrepancy because sketches and 3D shapes cannot be aligned perfectly within the same embedding space. Second, they have two different network structures to extract features of sketches and 3D shapes, respectively, and the parameters of the two networks are unshared and updated simultaneously during the training process, which leads to a high computational cost.

In this paper, we propose a novel semantic similarity metric learning to overcome the above-mentioned disadvantages of recent studies. Note that the aim of sketch-based 3D shape retrieval is to find 3D shapes belonging to the class labels of query sketches, so their label spaces are shared and can be used as a semantic embedding space. In such a semantic space, sketches and 3D shapes are aligned perfectly [7]. Inspired by the knowledge distillation technique, which uses a large teacher network to guide a small student network [14], we adopt a teacher-student strategy to obtain efficient networks for learning semantic similarity between sketches and 3D shapes. It can not only reduce the computational burden but also make the semantic features alignment easier. In our method, our proposed metric learning network consists of a teacher network and a student network, as shown in Fig. 2. The teacher network is a pre-trained classification network based on MVCNN [15] to extract the semantic features of 3D shapes and the student network is a transfer network based on ResNet-50 [16] to learn the semantic features of sketches. We train the transfer network by the guide of a new similarity loss for optimizing the semantic feature distance between sketches and 3D shapes. The main contributions of our work are listed as follows:

- A metric learning network using the teacher-student strategy is proposed to conduct sketch-based 3D shape retrieval in a joint semantic embedding space.
- A similarity loss function is developed to optimize the semantic feature distance between sketches and 3D shapes.
- Several experiments are carried out on a large benchmark dataset of sketch-based 3D shape retrieval and show that our method outperforms other state-of-the-art methods.

The remaining parts of this paper are organized as follows. The related works on sketch-based 3D shape retrieval and the teacher-student strategy in metric learning are briefly reviewed in Sec. 2. Our proposed method is described in Sec. 3 and the experimental results and analysis are presented in Sec. 4, and finally the conclusion is drawn in Sec. 5.

2   Related Works

Our proposed method is related to sketch-based 3D shape retrieval and the teacher-student strategy in metric learning. In this section, we briefly review the most related work in the two fields.

2.1   Sketch-Based 3D Shape Retrieval

In the early stage, most sketch-based 3D shape retrieval methods relied on the handcrafted features for describing sketches and 3D shapes [4,5]. With the rapid growth of CNNs, learning-based methods have developed in recent years. Wang et al. [11] used two projection views to characterize 3D shapes and applied a Siamese network to learn a joint embedding space for sketches and 3D shapes. Zhu et al. [17] developed pyramid cross-domain neural networks to reduce the cross-domain discrepancies between sketches and 3D shapes. To address the same problem, Chen et al. [8] proposed a cross-modality adaptation model using an importance-aware metric learning method. Unlike these projection-based methods, Dai et al. [12] presented a deep correlated metric learning method to mitigate the discrepancy by directly extracting the feature of 3D shapes, and Qi et al. [7] used the PointNet network to extract 3D shape features and developed a deep cross-domain semantic embedding model. Chen et al. [13] developed a deep sketch-shape hashing framework for sketch-based 3D shape retrieval with a stochastic sampling strategy for 3D shapes and a binary coding strategy for learning discriminative binary codes. However, most of these retrieval methods have two operative networks which cause a high computational cost. Besides, since they directly mapped features into a joint embedding space, it is difficult to effectively reduce the domain discrepancy.

2.2 Teacher-Student Strategy in Metric Learning

Since Hinton et al. [14] showed that a complex and powerful teacher model can guide the training of a small student network which can decrease the inference time and improve its generalization ability, this teacher-student strategy has received attention in the field of metric learning. Chen et al. [18] proposed cross sample similarities for knowledge transfer in deep metric learning, and modified the classical list-wise rank loss to bridge teacher networks and student networks. Yu et al. [19] presented a network distillation to compute image embeddings with small networks and developed two loss functions to communicate teacher and student networks. For the sketch-based 3D shape retrieval, Dai and Liang [20] proposed a cross-modal guidance network by using teacher-student strategy and used pre-learned features of 3D shapes to guide the feature learning of 2D sketches. However, their method cannot effectively minimize between-class similarity as well as maximize within-class similarity.

3 Method

3.1 Network Architecture

The network architecture of our proposed sketch-based 3D shape retrieval method is described in Fig. 2, which consists of a teacher network and a student network. Since sketches are abstract simple lines with limited information and 3D shapes are realistic geometric objects with more details, we select 3D shapes as the input of the teacher network and extract the semantic features from them to guide the training of the student network that takes sketches as input. By using the similarity loss to measure the cosine distance between sketches and 3D shapes, the features of sketches are optimized and gradually close to the pre-learned semantic features of 3D shapes during the training process of the student network.

In the teacher network, we apply the MVCNN [15] architecture, including CNN_1 and CNN_2, which are connected by a view-pooling layer, to represent multi-views of 3D shapes and extract the semantic features. First, we render a 3D shape from 12 different views by placing 12 virtual cameras around it every 30 degrees. Since there is still a big discrepancy between rendered images and sketches, we adopt the classic canny edge detector [21] to extract the edges of rendered images which are similar to the sketch lines. After that, the edge images are passed through CNN_1 separately to obtain view based features. Note that all branches of CNN_1 share the same parameters. In order to synthesize the information from all views into a single, we use element-wise maximum operation across the views in the view-pooling layer. Finally, these pooled feature maps are passed through CNN_2 to obtain the shape descriptor. After finishing training the teacher network, we make all data of 3D shapes pass through the teacher network and obtain the pre-learned semantic features of 3D shapes.

In the student network, we adopt a transfer network CNN_3 to learn the semantic features of sketches. The input sketches are directly passed through

CNN_3 to obtain the features. The student network is trained according to the optimization objective function, i. e., the similarity loss, which is guided by the pre-learned semantic features of 3D shapes.

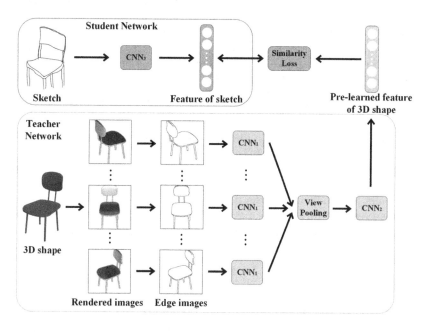

Fig. 2. The network architecture of our method.

3.2   Similarity Loss

In order to find the desired 3D shape, we always want that the extracted feature of the sketch is more similar to the same-class 3D shape and more dissimilar to the different-class 3D shape, i. e., maximizing the within-class similarity and minimizing the between-class similarity. However, a query sketch usually has tens or hundreds related 3D shapes with the same class label, and it is difficult to tell which 3D shape is more similar or dissimilar to the query sketch. Note that our aim is to find 3D shapes belonging to the class labels of query sketches rather than find the most similar 3D shapes. Therefore, we focus on extracting the class features rather than the individual features of 3D shapes. The class feature is the mean value of the pre-learned features of the 3D shapes in the same class. We use cosine similarity to measure the distance between a sketch and a 3D shape, which is defined as:

$$s = \frac{f_s \cdot f_c}{\|f_s\|_2 \|f_c\|_2} \tag{1}$$

where f_s is the sketch feature and f_c is the class feature of the 3D shape.

In a mini-batch with size N, we have N sketches and N_c pre-learned class features of 3D shapes. For each sketch i in the mini-batch, we calculate its cosine similarity with the class features of all 3D shapes. We denote the cosine similarity between the sketch and the class feature of the same-class 3D shape by s_p^i, i. e., the positive pair, and the cosine similarity between the sketch and the rest class features of 3D shapes by $s_n^i = \{s_1, s_2, \ldots, s_{Nc-1}\}$, i. e., the negative pairs. In order to maximize the similarity score of the positive pair and minimize the similarity score of the negative pair, the similarity loss function is defined as:

$$L = \frac{1}{N} \sum_{i=1}^{N} c[\max(s_n^i) - s_p^i + m]_+ \tag{2}$$

where $[]_+$ is a ramp function and m is a margin for a better similarity separation between positive and negative pairs.

The reason why we choose the maximum similarity score from the group of s_n^i to represent the negative pair in Eq. 2 is that it can ensure the scores of all negative pairs are smaller than the positive pair and also increase the difficulty of learning as the same effect of m. Since it is difficult to optimize the Eq. 2, we adopt a smooth approximation by using a LogSumExp function to replace $\max(s_n^i)$ and a softplus function to replace $[]_+$, and then obtain the smooth similarity loss function:

$$L_{smooth} = \frac{1}{N} \sum_{i=1}^{N} \log \left\{ 1 + \exp \left[\log \left(\sum_{n=1,n \neq p}^{N_c} \exp(rs_n^i) \right) - s_p^i + m \right] \right\} \tag{3}$$

where r is a scale factor. By training the student network with L_{smooth}, the sketch feature f_s is gradually close to the pre-learned class feature f_c of the same-class 3D shapes and keeps away from the different-class 3D shapes simultaneously.

4 Experiments

4.1 Datasets

We evaluate our proposed method on a frequently-used benchmark dataset, i. e., SHREC'13 [5], for sketch-based 3D shape retrieval. Some examples of sketches and corresponding 3D shapes in the dataset are shown in Fig. 1. The dataset is built by collecting large-scale hand-drawn sketches from TU-Berlin sketch dataset [22] and 3D shapes from Princeton Shape Benchmark [23], which consists of 90 classes including 7,200 sketches and 1,258 shapes. In each class, there are a total of 80 sketches, and 50 of which are for the training and the rest are for the test. The number of 3D shapes varies in different classes. For example, the largest class is 'airplane', which has 184 3D shapes, and there are 12 classes containing only 4 3D shapes.

4.2 Implementation Details

Our method is implemented on Pytorch with two NVIDIA GeForce GTX 2080 Ti GPUs.

Network Structure. The structure is illustrated in Fig. 2. The teacher network adopts the MVCNN [15] architecture and the CNN_1 and CNN_2 use the VGG-11 network [24]. In the student network, CNN_3 utilizes the ResNet-50 network [16].

Prepossessing. The prepossessing includes the network pre-training and data processing. The teacher network is pre-trained on ImageNet [25] with 1k categories, and then fine-tuned on all edge images of the 3D shapes. The student network is first pre-trained for the classification task based on a part of Quick-Draw dataset [26] with 3.45 million sketches in 345 categories, and then fine-tuned on the training dataset of sketches according to minimize Eq. 3. For the data processing, we uniformly resize the sketch images and the edge images of 3D shapes into a resolution of $224 \times 224 \times 1$.

Parameter Settings. In the teacher network, the learning rate and batch size are 5×10^{-5} and 8, respectively, and the number of training epochs is set to 20. In the student network, the learning rate and batch size are 1×10^{-4} and 48, respectively, and the number of training epochs is 10. Moreover, the margin m and the scale factor r are set to be 0.15 and 64, respectively. The Adam is employed as an optimizer for both networks and the weight decay is set to 0.

4.3 Experimental Results

We show some retrieval results on the SHREC'13 dataset in Fig. 3. The query sketches are listed on the left including the class of chair, bicycle, piano, table, palm tree and sea turtle, and their retrieved top 8 3D shapes are listed on the right according to the ranking of similarity scores. As shown in Fig. 3, our method is effective in retrieving the corresponding 3D shapes of the query sketches. The reasons for generating incorrect results are the limited number of 3D shapes (e.g., the classes of bicycle and sea turtle only contain 7 and 6 3D shapes in the dataset, respectively) and the high similarity score of similar shapes from different classes (e.g., the couch and bench shapes get high similarity scores according to the query sketch of piano).

In order to demonstrate the effectiveness of our proposed method, we compare our method with several state-of-the-art methods, including SBR-VC [5], Siamese [11], Shape2Vec [10], DCML [12], LWBR [9], DCA [8], SEM [7] and DSSH [13]. In addition, we adopt the widely-used evaluation metrics for the sketch-based 3D shape retrieval, including the nearest neighbor (NN), first tier (FT), second tier (ST), E-measure (E), discounted cumulated gain (DCG) and mean average precision (mAP) [4]. Table 1 shows the quantitative comparison of our method with the state-of-the-art methods on the SHREC'13 dataset. Except for the DSSH, it is clear to see that our method achieves the best performance than the state-of-the-art methods for all the evaluation metrics. Compared to the latest method DSSH, our method performs better or equally in the NN, E and DCG metrics.

We also visually compared our method with DSSH to show our advantages. As shown in Fig. 4, for the hand and horse sketch examples, our retrieved 3D

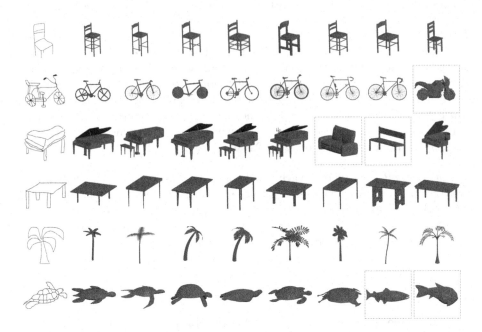

Fig. 3. Some examples of retrieval results. The left column is the query sketches and the right columns are the top 8 retrieved 3D shapes, and the wrong results are highlighted by red dashed squares. (Color figure online)

Table 1. The comparison of our method and the state-of-the-art methods on the SHREC'13 dataset.

Method	NN	FT	ST	E	DCG	mAP
Siamese [11]	0.405	0.403	0.548	0.287	0.607	0.469
Shape2Vec [10]	0.620	0.628	0.684	0.354	0.741	0.650
DCML [12]	0.650	0.634	0.719	0.348	0.766	0.674
LWBR [9]	0.712	0.725	0.785	0.369	0.814	0.752
DCA [8]	0.783	0.796	0.829	0.376	0.856	0.813
SEM [7]	0.823	0.828	0.860	0.403	0.884	0.843
DSSH [13]	0.831	**0.844**	**0.886**	0.411	0.893	**0.858**
Ours	**0.836**	0.833	0.883	**0.411**	**0.896**	0.853

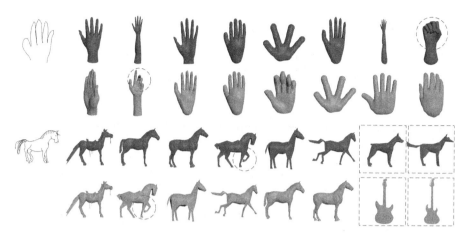

Fig. 4. The comparison of our method and DSSH [13] in two retrieval examples. The blue and gray colors denote the retrieval results of our method and DSSH, respectively, and the wrong results and mismatched details are highlighted by red dashed squares and circles, respectively. (Color figure online)

shapes are more accurate than DSSH. First, the retrieved 3D shapes with mismatched details have a low-ranking in our method. For example, an unextended hand is ranked last in our method but ranked second in DSSH, and a horse with a lifted leg is ranked fourth in our method but ranked second in DSSH. Second, our wrong results are similar to the right results. For example, the wrong shapes of DSSH are guitars which are extraordinarily different to the horse, whereas our retrieved dogs are similar to the horse. Therefore, compared with DSSH, our method is more suitable for measuring feature distance between sketches and 3D shapes.

5   Conclusion

In this paper, we propose a novel semantic similarity metric learning method for sketch-based 3D shape retrieval, and use a teacher-student strategy to obtain efficient networks for learning semantic similarity between sketches and 3D shapes. We first adopt the pre-trained classification network as the teacher network to extract the semantic features of 3D shapes, and then train the student network by using the pre-learned features of 3D shapes with a similarity loss function and finally learn the semantic features of sketches. As a result, our method effectively maximizes the within-class similarity and minimizes the between-class similarity. The experiments show that our method performs better than the state-of-the-art methods.

Acknowledgements. This research is supported by the PDE-GIR project which has received funding from the European Union's Horizon 2020 research and innovation programme under the Marie Skłodowska-Curie grant agreement No 778035.

References

1. Chen, D.Y., Tian, X.P., Shen, Y.T., Ouhyoung, M.: On visual similarity based 3D model retrieval. Comput. Graph. Forum **22**(3), 223–232 (2003)
2. Shih, J.L., Lee, C.H., Wang, J.T.: A new 3D model retrieval approach based on the elevation descriptor. Pattern Recog. **40**(1), 283–295 (2007)
3. Shao, T., Xu, W., Yin, K., Wang, J., Zhou, K., Guo, B.: Discriminative sketch-based 3d model retrieval via robust shape matching. Comput. Graph. Forum **30**(7), 2011–2020 (2011)
4. Li, B.: A comparison of methods for sketch-based 3D shape retrieval. Comput. Vis. Image Unders. **119**, 57–80 (2014)
5. Li, B., et al.: In: Biasotti, S., Pratikakis, I., Castellani, U., Schreck, T., Godil, A., Veltkamp R. (eds.) SHREC'13 Track: Large Scale Sketch-Based 3D Shape Retrieval, Eurographics Workshop on 3D Object Retrieval 2013, pp. 89–96 (2013)
6. Li, B., et al.: Shrec'14 track: extended large scale sketch-based 3d shape retrieval. In: 2014 Eurographics Workshop on 3D Object Retrieval, pp. 121–130 (2014)
7. Qi, A., Song, Y., Xiang, T.: Semantic embedding for sketch-based 3d shape retrieval. In: British Machine Vision Conference (2018)
8. Chen, J., Fang, Y.: Deep cross-modality adaptation via semantics preserving adversarial learning for sketch-based 3d shape retrieval. In: Proceedings of the European Conference on Computer Vision, pp. 605–620 (2018)
9. Xie, J., Dai, G., Zhu, F., Fang, Y.: Learning barycentric representations of 3D shapes for sketch-based 3d shape retrieval. In: Proceedings of the IEEE Conference on Computer Vision and Pattern Recognition, pp. 5068–5076 (2017)
10. Tasse, F.P., Dodgson, N.: Shape2vec: semantic-based descriptors for 3D shapes, sketches and images. ACM Trans. Graph. **35**(6), 1–12 (2016)
11. Wang, F., Kang, L., Li, Y.: Sketch-based 3D shape retrieval using convolutional neural networks. In: Proceedings of the IEEE Conference on Computer Vision and Pattern Recognition, pp. 1875–1883 (2015)
12. Dai, G., Xie, J., Zhu, F., Fang, Y.: Deep correlated metric learning for sketch-based 3D shape retrieval. In: Proceedings of the AAAI Conference on Artificial Intelligence, vol. 31 (2017)
13. Chen, J., et al.: Deep sketch-shape hashing with segmented 3d stochastic viewing. In: Proceedings of the IEEE Conference on Computer Vision and Pattern Recognition, pp. 791–800 (2019)
14. Hinton, G., Vinyals, O., Dean, J.: Distilling the knowledge in a neural network. arXiv preprint arXiv:1503.02531 (2015)
15. Su, H., Maji, S., Kalogerakis, E., Learned-Miller, E.: Multi-view convolutional neural networks for 3D shape recognition. In: Proceedings of the IEEE International Conference on Computer Vision, pp. 945–953 (2015)
16. He, K., Zhang, X., Ren, S., Sun, J.: Deep residual learning for image recognition. In: Proceedings of the IEEE Conference on Computer Vision and Pattern Recognition, pp. 770–778 (2016)
17. Zhu, F., Xie, J., Fang, Y.: Learning cross-domain neural networks for sketch-based 3D shape retrieval. In: Proceedings of the AAAI Conference on Artificial Intelligence, vol. 30 (2016)
18. Chen, Y., Wang, N., Zhang, Z.: Darkrank: Accelerating deep metric learning via cross sample similarities transfer. In: Proceedings of the AAAI Conference on Artificial Intelligence, vol. 32 (2018)

19. Yu, L., Yazici, V.O., Liu, X., Weijer, J.V.D., Cheng, Y., Ramisa, A.: Learning metrics from teachers: Compact networks for image embedding. In: Proceedings of the IEEE Conference on Computer Vision and Pattern Recognition, pp. 2907–2916 (2019)
20. Dai, W., Liang, S.: Cross-modal guidance network for sketch-based 3D shape retrieval. In: 2020 IEEE International Conference on Multimedia and Expo, pp. 1–6 (2020)
21. Canny, J.: A computational approach to edge detection. IEEE Trans. Pattern Anal. Mach. Intell. **8**(6), 679–698 (1986)
22. Eitz, M., Hays, J., Alexa, M.: How do humans sketch objects? ACM Trans. Graph. **31**(4), 1–10 (2012)
23. Shilane, P., Min, P., Kazhdan, M., Funkhouser, T.: The Princeton shape benchmark. In: Proceedings Shape Modeling Applications, pp. 167–178 (2004)
24. Simonyan, K., Zisserman, A.: Very deep convolutional networks for large-scale image recognition. arXiv preprint arXiv:1409.1556 (2014)
25. Krizhevsky, A., Sutskever, I., Hinton, G.E.: ImageNet classification with deep convolutional neural networks. Commun. ACM **60**(6), 84–90 (2017)
26. Ha, D., Eck, D.: A neural representation of sketch drawings. arXiv preprint arXiv:1704.03477 (2017)

ScatterPlotAnalyzer: Digitizing Images of Charts Using Tensor-Based Computational Model

Komal Dadhich, Siri Chandana Daggubati, and Jaya Sreevalsan-Nair$^{(\boxtimes)}$ ⓘ

Graphics-Visualization-Computing Lab,
International Institute of Information Technology, Bangalore (IIITB),
26/C, Electronics City, Bengaluru 560100, Karnataka, India
{komal.dadhich,daggubati.sirichandana}@iiitb.org,
jnair@iiitb.ac.in
http://www.iiitb.ac.in/gvcl

Abstract. Charts or scientific plots are widely used visualizations for efficient knowledge dissemination from datasets. Nowadays, these charts are predominantly available in image format in print media, the internet, and research publications. There are various scenarios where these images are to be interpreted in the absence of datasets that were originally used to generate the charts. This leads to a pertinent need for automating data extraction from an available chart image. We narrow down our scope to scatter plots and propose a semi-automated algorithm, ScatterPlotAnalyzer, for data extraction from chart images. Our algorithm is designed around the use of second-order tensor fields to model the chart image. ScatterPlotAnalyzer integrates the following tasks in sequence: chart type classification, image annotation, object detection, text detection and recognition, data transformation, text summarization, and optionally, chart redesign. The novelty of our algorithm is in analyzing both simple and multi-class scatter plots. Our results show that our algorithm can effectively extract data from images of different resolutions. We also discuss specific test cases where ScatterPlotAnalyzer fails.

Keywords: Chart images · Scatter plots · Multi-class bivariate data · Spatial locality · Local features · Chart data extraction · Positive semidefinite second-order tensor fields · Structure tensor · Tensor voting · Chart type classification · Convolutional Neural Network (CNN)

1 Introduction

Plots or charts are one of the simplest visualizations for data analysis, of which bar charts and scatter plots are commonly used. Today, most charts are widely available in image format in print media, the internet, research publications, etc.

Supported by the Machine Intelligence and Robotics (MINRO) Grant, Government of Karnataka.

M. Paszynski et al. (Eds.): ICCS 2021, LNCS 12746, pp. 70–83, 2021.
https://doi.org/10.1007/978-3-030-77977-1_6

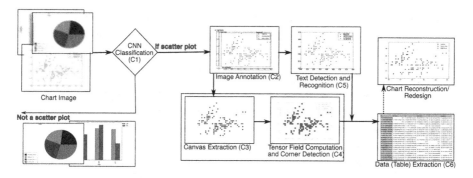

Fig. 1. Our proposed workflow for data extraction from a given image of a scatter plot with six components (C1–C6), where C1–C3 are preprocessing steps, C4 is the computational modeling step, C5 and C6 are postprocessing steps. The output is a data table, which can be optionally used for chart reconstruction or redesign.

Digitizing chart images is a pertinent requirement for automating chart interpretation, which has applications in assisted technologies for STEM (Science, Technology, Education, and Mathematics) education of the visually impaired. Recently, machine and deep learning solutions have been widely used for reasoning over [10, 13], and to a lesser extent, data extraction [16]. However, a generalized computational model for such images can also alternatively lead to data extraction and its applications, such as chart reconstruction for improving image resolution or quality or chart redesign for improving interpretation.

The simple design of bar charts and scatter plots provide high levels of interpretability for low-dimensional data, even with complex trends. However, uni- or bi-variate data analysis is less efficient in the light of multivariate data collections being the norm. Hence, more complex forms of these chart types are used to address the insufficiency of its simpler forms. While the simpler forms are used widely in the early stages of chart graphicacy in school education, widely available chart images tend to be complex variants. The studies on both bar charts and scatter plots publish results for accuracy in text detection and/or data extraction [16]. The complex forms of bar charts, such as grouped bar charts, have been used for chart image analysis [7, 18]. At the same time, limited studies demonstrate similar analysis of complex forms of scatter plots, where one such study looks at scatter points or marks with different types of formatting [6]. Overall, there is a gap in the analysis of complex variants of scatter plots. Here, we consider such scatter plots, particularly those with color, as a visual encoding.

The simple scatter plot is where scatter points represent (x, y) tuples, and its location encodes the data. Whereas the complex variants of scatter plots additionally use shape, size, and color of scatter points for encoding extra dimensions of the data. One such variant is where color is used to encode different classes or groups in the bi-variate data used, which we refer to as *multi-class scatter plots*. Multi-class scatter plots are predominantly used to organize the display of the clustering in datasets, correlation, etc. For example, such charts are used in machine learning to visualize multi-class classification. In visualization parlance, multi-class scatter plots can be considered to be two or more overlays of simple scatter plots using the same coordinate system and units and scales on the axes. Hence, we hypothesize that the data extraction algorithm for a simple scatter plot can be effectively extended to multi-class scatter plots using a data-parallel construct. As a result, we propose a generalized algorithm that performs data extraction from both simple and multi-class scatter plots.

Our algorithm uses perceptual information from the charts to extract data, where preprocessing is implemented using widely used image processing and artificial intelligence techniques. This perceptual information is represented using a second-order tensor field, which is a computational model, as has been done in our previous work [21]. These tensor fields exploit local geometry to extract objects such as scatter points from the chart in pixel space. The next step in data extraction is to devise an algorithm that uses these tensor fields to extract data in pixel space, which is further used with text extracted from the images to output a data table [7]. We propose such a data extraction algorithm from images for simple and multi-class scatter plots, ScatterPlotAnalyzer (Fig. 1).

2   Related Work

Chart analysis for data extraction depends on processes, such as chart type classification, chart image annotation, feature extraction, and text recognition. Several tools and techniques have already been devised to automatically interpret charts from their images, where the outcomes correspond to different aspects of chart analysis. ReVision is one of the earlier works that introduced the idea of using feature vectors and geometric structures to extract visual elements and data encoded in the chart [18]. In WebPlotDigitizer, the user is provided with an option to use the automated or manual procedure for data extraction from the given chart image [17]. The tool requires the user to select the chart type for the uploaded image from a given list, align the axis and mask the chart component by drawing multiple points on the graphical object.

Just as WebPlotDigitizer, Scatterscanner requires user interactivity for data extraction from scatter plots, specifically [4]. Scatteract is an automated system that extracts data from scatter plot images by mapping pixels to the coordinate system of the chart with the help of OCR [6].

Similar to many computer vision problems, machine and deep learning models have been introduced for chart classification and object detection. A web-based system Beagle takes charts in scalable vector graphics format and classifies charts found as visualizations [3]. ChartSense uses GoogleNet to perform chart classification for line, bar, pie, scatter charts, map, and table types [9]. Figure-Seer uses a similar fine-tuning approach [19] to localize and classify result-figures in research papers. Text interpretation is equally important for data extraction. A Darknet neural network as an object detection model in combination with OCR for text detection is used from bar charts, whereas the pixels of a specific color within the periphery of the circle are utilized for pie charts [5].

3 Tensor Field as a Computational Model

Tensor fields have been widely used to exploit geometric properties of objects for edge detection in natural images using structure tensor and tensor voting [12]. In our previous work, we have used tensor voting for extracting bars and scatter points in images of simple charts, thus generating a computational model using second-order tensor fields [21]. This model has shown promising results for data extraction using perceptual information in pixel space. Hence, we use this tensor-based computational model for extracting data from both simple and multi-class scatter plots. The computational model is one of the components of ScatterPlot-Analyzer, which completes the step of converting the extracted values in pixel space to data space.

Tensor Voting Using Gradient Tensor: We use a local geometric descriptor in the form of a second-order tensor, which is required for tensor voting computation [22]. Tensor voting itself returns a second-order tensor that is geometry-aware in a more global context [12]. Hence, the tensor voting field is used for scatter point extraction from a scatter plot image. Structure tensor T_s at a pixel provides the orientation of the gradient computed from the local neighborhood.

$$T_s = \mathcal{G}_\rho * (G^T G), \text{ where } G = \begin{bmatrix} \frac{\partial I}{\partial x} & \frac{\partial I}{\partial y} \end{bmatrix}$$

is the gradient tensor at the pixel with the intensity I; convolved (using $*$ operator) with Gaussian function $\mathcal{G}$ with zero mean and standard deviation ρ. The tensor vote cast at x_i by x_j using a second-order tensor K_j in d-dimensional space is, as per the closed-form equation [25]: $S_{ij} = c_{ij} R_{ij} K_j R'_{ij}$,

$$\text{where } R_{ij} = (I_d - 2r_{ij}r_{ij}^T); \; R'_{ij} = (I_d - \tfrac{1}{2}r_{ij}r_{ij}^T)R_{ij},$$

I_d is the d-dimensional identity matrix; unit vector of direction vector $r_{ij} = \hat{d}_{ij}$, with $d_{ij} = x_j - x_i$; σ_d is the scale parameter; and $c_{ij} = \exp\left(-\left(\sigma_d^{-1}.\|d_{ij}\|_2^2\right)\right)$. The gradient $>$ can be used as K_j [15].

Anisotropic Diffusion: As the tensor votes T_v in normal space has to encode object geometry in tangential space, we perform anisotropic diffusion to transform T_v to tangential space [21,22]. The eigenvalue decomposition of the two-dimensional T_v yields ordered eigenvalues, $\lambda_0 \geq \lambda_1$, and corresponding eigenvectors v_0 and v_1, respectively. Anisotropic diffusion of T_v using parameter δ is,

$$T_{v\text{-ad}} = \sum_{k=0}^{1} \lambda'_k . v_k v_k^T, \text{ where } \lambda'_k = \exp\left(-\tfrac{\lambda_k}{\delta}\right).$$

Diffusion parameter value ($\delta = 0.16$) is widely used [21,24].

Saliency Computation: The saliency of a pixel belonging to geometry features of line- or junction/point-type is determined by the eigenvalues of $T_{v\text{-ad}}$ [21]. We get the saliency maps at each pixel of an image of its likelihood for being a line- or junction-type feature, C_l and C_p, respectively, $C_l = \frac{\lambda_0 - \lambda_1}{\lambda_0 + \lambda_1}$ and $C_p = \frac{2\lambda_1}{\lambda_0 + \lambda_1}$, using eigenvalues of $T_{v\text{-ad}}$ of the pixel, such that, $\lambda_0 \geq \lambda_1$. The pixel with $C_p \approx$ 1.0 is referred to as a critical point or degenerate point, in the parlance of tensor fields. Our goal is to find all the critical points in the chart image in the C4 step (Fig. 1), as a few of them are significant for data extraction in pixel space.

4   Components of ScatterPlotAnalyzer

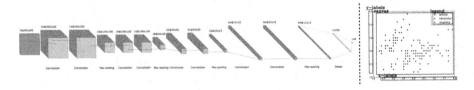

Fig. 2. The preprocessing steps of chart classification and annotation of ScatterPlot-Analyzer. (Left) The architecture diagram of our CNN-based classifier for identifying bar chart subtypes. (Right) Human-guided annotation of the chart image, using an example of scatter plot, to extract the chart canvas, legend, and text needed for object detection, multi-class inferences, and contextualization.

We propose an algorithm, ScatterPlotAnalyzer, that extracts data from a given chart image after determining it is a scatter plot. ScatterPlotAnalyzer has six main components to meet this requirement (Fig. 1), namely, chart type classification, chart image annotation, canvas extraction, tensor field computation, text recognition, and data table extraction.

Chart Type Classification (C1): Different types of charts, *e.g.*, bar chart, scatter plot, pie chart, etc., encode data differently. Hence, data extraction from their images needs to exploit these characteristics of visual encoding for reverse engineering. Visual encodings include the location of scatter point for data tuple (x, y), the height of the bar for y-value, and sector size in a pie chart for percentage value. Separating these visual encodings is important for abstractions leading to a generalized data extraction from different chart types, which unifies a data extraction system for all charts.

The chart objects such as bars, scatter points, and lines have specific geometric structure that can be exploited, unlike the objects found in natural images. The chart subtypes for bar charts are grouped, stacked, column, bar charts, and histograms as a special case. We now consider simple and multi-class scatter plots as chart subtypes in scatter plots. We observe that chart subtypes preserve the geometry in chart objects, as is in the parent chart type. While this similarity helps in generalizing data extraction workflow, the differences in geometry help in classification. Overall, the similarity in geometry across subtypes limit the applicability of contour-based techniques for chart type classification, which provides the granularity of subtypes. Hence, we use a convolutional neural network (CNN) for generic chart type classification, which is widely used for similar applications. Some of the pre-trained models have been used for chart type classification of images by imposing certain constraints, *e.g.*, training with a small image corpus. However, these classifiers have been found to perform with low accuracy. The classifier in ChartSense has used GoogleNet [9] and has been trained on different chart types, but performs well for six out of ten types (bar chart, line chart, map, pie chart, scatter plot, and table) with the small corpus.

We build our classifier using an architecture inspired by VGGNet (Fig. 2). VGGNet architecture is popularly used for object detection tasks, and it requires the convolutional layers to be stacked. We add convolutional layers, followed by pooling layers, and finally, fully connected layers, kernel size of (5, 5). We train the CNN model on our chart image dataset containing more than 1000 images of four chart types, namely, bar chart, scatter plot, pie chart and line chart. Our CNN based classifier is constrained by the requirement of input images of fixed size for training. Hence, we first resize the images in the dataset to 200×200 size. The image resizing and classifier implementation has been done using Python imaging (PIL) and Keras libraries, respectively.

Our classifier is sufficient for ScatterPlotAnalyzer to identify scatter plots, without the granularity of simple or multi-class typification. This is because this typification can be implemented more efficiently by interpreting the legend rather than using CNN.

Image Annotation (C2) and Canvas Extraction (C3): Image annotation is required to prepare a training dataset for object detection, segmentation, and similar computer vision applications. Image annotation provides labels to different regions of interest (ROI). The ROIs are detected and extracted using the predefined labels for such regions. Image annotation strictly requires contextual labels and appropriate associations between labels and ROIs. Automation of annotation is not a completely solved problem, owing to which human-guided annotation of images is widely used in practice.

Manual marking and annotation of bounding boxes for ROIs have been widely used for chart image analysis [5,13]. We specifically use the following labels for specific ROIs needed for tensor field computation. These labels include canvas, x-axis, y-axis, x-labels, y-labels, legend, title, x-title, and y-title. We use LabelImg [23], a Python tool, for marking and annotating bounding boxes for ROIs. LabelImg has a graphical user interface (GUI) to select an image, trace a bounding box for an ROI, and label the ROI appropriately. We label the ROI that contains the chart objects such as bars, lines, or scatter points as *Canvas*. The extracted ROI is referred to as *chart canvas*, which is one of the chart image components [21]. LabelImg generates the annotation in an XML file that can be used further to automate the extraction of the canvas region and text localization. The former is used for chart extraction (C3), and the latter for text detection (C5). Figure 2 (right) shows an annotated scatter plot.

The canvas extraction step (C3) is implemented using image preprocessing methods to remove the remaining elements other than chart objects such as gridlines, overlaid legends, etc. This is required as tensor field computation in C4 is sensitive to the presence of these extraneous elements. Marker-based watershed segmentation and contour detection algorithm have been used in combination to remove these extraneous components in the chart canvas effectively [21]. These steps also fill hollow bars and scatter points, as required, since the tensor field is computed effectively for filled chart objects. Highly pixellated edges in aliased images lead to uneven edges in each chart object. This issue is addressed by using the contour detection method to add a fixed-width border to chart objects [21]. Overall, the processes in chart canvas extraction are chosen specifically to extract chart objects using tensor fields.

Tensor Field Computation (C4): We use tensor voting on gradient tensor to compute a second-order tensor field T_v, which is further improved using anisotropic diffusion (Sect. 3). The resulting field, $T_{v\text{-ad}}$, is then analyzed by identifying its critical points using the saliency map values (C_l, C_p) at each pixel. We use a threshold on the trace of the tensor $T < 0.01$ for scatter plots to discard weak critical points. The thresholding is tensor-based, as the trace is a tensor invariant. CIE-Lab used for perceptual modeling of color is better suited for tensor voting than the RGB model [14]. Hence, the chart image is converted from the RGB model to the CIE-Lab model prior to the actual tensor field computation.

DBSCAN Clustering: We observe that the critical points of chart image computed from tensor field computation form sparse clusters along the boundary of a scatter point as well as it's interior [21]. Since the scatter point interior gives its location, that is the data, and we extract the cluster in the interior. We use density-based clustering, DBSCAN [8], to localize these clusters. We adjust the hyperparameters of DBSCAN clustering to give the best clusters corresponding to specific chart types, _e.g._, clusters in the corners of bars for a bar chart and clusters near the scatter point centroid for scatter plots. These hyperparameters influence the location of the cluster centroids. The cluster centroid gives the data tuple (x, y) corresponding to the scatter point but in pixel space.

Text Recognition (C5): The data extracted using tensor fields are in pixel space and have to be contextualized to the data space for extracting the data table. Hence, we now combine the data in pixel space with the text information in the image. We perform text detection to get x-axis and y-axis labels and compute the scale factor between the pixel and data spaces. The recognition of other textual elements, namely, plot title, legend, x-axis, and y-axis titles, also plays a crucial role in analyzing chart images.

We use deep-learning-based OCR, namely Character Region Awareness for Text Detection, CRAFT [2] for effective text area detection, including arbitrarily-oriented text. This approach is designed for relatively complex text in images, and it works by exploring each character region and considering the affinity between characters. A CNN designed in a weakly-supervised manner predicts the character region score map and the image's affinity score map. The character region score is used to localize individual characters and affinity scores to group each character to a single instance. So, the instance of text detected is not affected by its orientation and size. From the detected text boxes, its orientation is computed and then rotated.

A unified framework for scene text recognition that fits all variants of scenes called the scene text recognition framework, STR [1] is implemented subsequent to CRAFT. Being a four-stage framework consisting of transformation, feature extraction, sequence modeling, and prediction, STR resembles the combination of computer vision tasks such as object detection and sequence prediction task. Hence, it uses a convolutional recurrent neural network (CRNN) for text recognition. We find that the CRAFT model, along with the STR framework, works efficiently to retrieve labels and titles of the chart image better than other available text recognition tools, *e.g.,* Tesseract OCR [20].

Information Aggregation for Data Extraction (C6): For multi-class scatter plots, interpreting the legend is an essential step. In the absence of legend, we consider the chart to be a simple scatter plot. In the presence of a legend, the number of classes specified in it is used to determine if the chart is a simple or a multi-class scatter plot. We use CIE-Lab color space to identify the colors corresponding to the classes in the legend. The legend's colors are extracted using the morphological operations used in image preprocessing in the legend ROI, as done on the chart canvas in C3.

We use the pixel color at the cluster centroid extracted in C4 to indicate the class corresponding to the scatter point. The actual class label is extracted from the text recognition in the legend in C5. Thus the data table now has a class label along with the (x, y) tuple extracted in the pixel space.

As a final process in ScatterPlotAnalyzer, the data extracted in pixel space is transformed into data space by using the text recognized for axis and corresponding tick labels. The labels add appropriate textual information for the variable along each axis by providing its name and unit value. The unit value of the variables and scatter point location, along with its class information, are used to extract the data table in C6.

5   Experiments and Results

We have used ∼1000 images from the FigureQA dataset [10] containing bar charts and its subtypes, scatter plots, pie charts, and line charts for training our model. Our model works with ∼90% accuracy.

We perform experiments to analyze the performance of ScatterPlotAnalyzer that answer the following questions:

- Is the choice of the tensor field from tensor voting, $T_{\text{v-ad}}$, better than the conventionally used structure tensor T_s for data extraction?
- How accurate is the data reconstruction for both simple and multi-class scatter plots?

We consider the structure tensor T_s and tensor voting field $T_{\text{v-ad}}$ for different multi-class scatter plots by studying the shape of the tensor using ellipsoid glyphs [11]. Ellipsoid glyphs in three-dimensional space reduce to elliptical glyphs in two-dimensional, as in our case. The elliptical glyphs are drawn with the major and minor axes of the ellipses oriented along the major and minor eigenvectors of the tensor. The eigenvalues of the tensor give the major and minor lengths of axes of the ellipse. Thus, the degenerate points with almost equal eigenvalues are closer to the circular shape. Our results in Fig. 3 show that the degenerate points

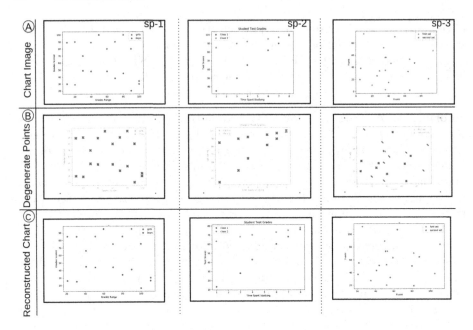

Fig. 3. The tensor field computation step in ScatterPlotAnalyzer (C4) uses the images of multi-class scatter plots (A) to compute the structure tensor T_s (B), and the tensor voting field we use, $T_{\text{v-ad}}$ (C). While (B) and (C) show tensor field visualization using ellipsoid glyphs, colored using saliency map, (D) shows the dot map of the saliency map. The color bar shows the coolwarm color mapping used for saliency map of the tensor field. The glyph visualization is demonstrated on a 1:3 subsampled image for clear visualization.

are stronger when using the tensor voting field as opposed to the structure tensor. We can conclude that the degenerate points are strengthened by performing tensor voting on gradient tensor and subsequent anisotropic diffusion.

In terms of the accuracy of the data table given by ScatterPlotAnalyzer, we analyze the qualitative and quantitative results. Figure 4 shows the different steps in ScatterPlotAnalyzer for both simple and multi-class scatter plots. We observe that the overlapping points in scatter plots do not get extracted accurately, as only perceptually visible scatter points are extracted. At the same time, we observe that the human eye can detect partial overlaps; however, our tensor field is not able to extract the overlapping points as multiple points. Hence, we observe *omission errors* or type-II errors, as has been reported in our previous work [21]. For the extracted points, the data extracted is mostly accurate. Our experiments include a simple scatter plot without legend, a 3-class scatter plot, and 2-class scatter plots, shown in Fig. 4(i)–(iv). We have highlighted the type-II errors in row D in Fig. 4 in translucent red highlights.

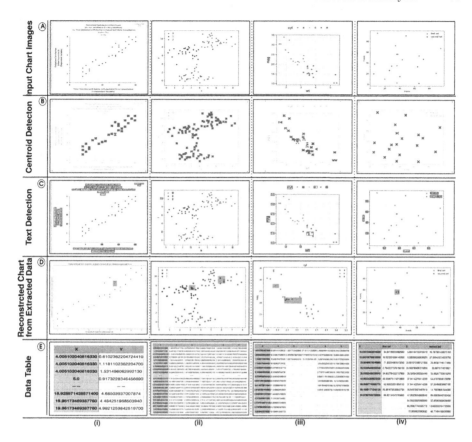

Fig. 4. The outcomes of the key steps in our ScatterPlotAnalyzer, namely, tensor field computation (C4), text detection (C5), and data extraction (C6), implemented on the source input chart images. The dataset includes available source images of (i) a simple scatter plot, and (ii–iii) multi-class scatter plots, and an image of multi-class scatter plot created by plotting available data in (iv). (Color figure online)

In terms of quantifying the error in our data extraction, we use synthetic datasets for both simple and multi-class scatter plots. We plot the data using `matplotlib`, a Python plotting library, extract the data table and reconstruct the image using ScatterPlotAnalyzer. We compute the Pearson's correlation of synthetic datasets and their extracted counterparts. We have reported the differences in correlation coefficient r for simple scatter plots in Fig. 5 and the same for multi-class scatter plots in Fig. 6. We observe that the errors in correlation coefficients are proportional to the density of scatter points in the plot. In the case of multi-class scatter plots, the errors in correlation coefficients are additionally proportional to the density of points in regions where both classes overlap in the image. We observe that comparing correlation coefficients in original and reconstructed charts helps in comparing the overall appearance of the charts, which is more significant for text summarization than exact data extraction.

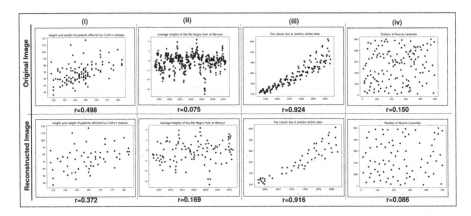

Fig. 5. Correlation coefficient (r) values in both original and ScatterPlotAnalyzer reconstructed images of simple scatter plots.

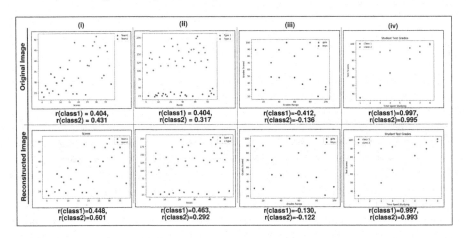

Fig. 6. Correlation coefficient (r) values in both original and ScatterPlotAnalyzer reconstructed images of multi-class scatter plots.

We observe that the quality of the images, irrespective of their source, perform the same using ScatterPlotAnalyzer. The images from the internet in Fig. 3 and Fig. 4(i)–(iii) show outputs comparable to the images generated by plotting from available datasets, in Figs. 4(iv), 5, and 6. Thus, our image preprocessing methodology helps in improving the quality of the image to be used with ScatterPlotAnalyzer, even though the images from the internet tend to have lower resolution, noise, and aliasing.

Scatteract reports a maximum of 89.2% success rate, where F_1 score > 0.8 for a plot implies success in data extraction. In our experiments, we do not get F_1 score distribution similar to that of Scatteract, i.e. very low and very high values. Instead, we get a continuous uniform distribution for $F_1 \geq 0.4$. Hence, we relax the constraint appropriately to $F_1 > 0.5$, then Scatteract with the use of RANSAC regression for mapping pixel-to-chart coordinates has 89.5% success rate for simple scatterplots for procedurally generated ones, 78% for simple scatterplots from the web, and with other regression methods has 70.3% at its best. In comparison, our method has 73.3% success rate for simple scatterplots, which improves to 93.3% for $F_1 > 0.4$. We can likewise improve the success rate of our method by refining the pixel-to-chart coordinate mapping using RANSAC regression. The chart classification accuracy for scatter plots is at 81% in our work, comparable to 86% in ChartSense [9], while the best is at 98% in the method by Choi et al. [5]. Our chart classifier, which has been trained and tested with images from the web, requires more pre-training to improve the accuracy rate. Despite the mismatches in quantitative assessment, our results are comparable to those from Scatteract and the method by Choi et al., qualitatively.

Pre-training the CNN is the most time-consuming process in our algorithm. The tensor field computation is also compute intensive. Its implement ion can be made more efficient by using parallel implementation, owing to its embarrassingly parallel characteristic. Sparsification of the degenerate points is required for improving the accuracy of our data extraction process. However, sparsification itself can be improved using an analytical choice of threshold, as opposed to the heuristic we use generically. Thus, the scope of our future work includes improving the overall performance and accuracy of ScatterPlotAnalyzer.

6 Conclusions

In this work, we have proposed ScatterPlotAnalyzer, an algorithm for extracting data from images of scatter plots. We have focused on both simple and multi-class scatter plots, where the class information of the scatter points is encoded using color. We have designed ScatterPlotAnalyzer with the central theme of second-tensor fields using tensor voting for geometry extraction. We use a Convolutional Neural Network inspired by VGGNet architecture to classify the scatter plots. We then use human-guided image annotation to extract the canvas containing chart objects, where the interactive annotation makes our algorithm semi-automated. We use image preprocessing to complete the extraction, and scatter points themselves are extracted from the topological analysis of the tensor field. The postprocessing involves the use of deep-learning OCR to localize and detect text. Text is an important ingredient for contextualizing the dataset and converting the extracted data from pixel to the original data space. For identifying the class information from the extracted scatter points, we use information extracted from the legend. Overall, ScatterPlotAnalyzer is an end-to-end algorithm for extracting data from images of scatter plots.

Performance characterization and improving accuracy are organically the next steps to improve ScatterPlotAnalyzer. We require metrics such as the density of scatter points and overlap between multiple classes to determine the accuracy of ScatterPlotAnalyzer implemented on such an image. ScatterPlot-Analyzer is a good proof-of-concept of exploiting image properties for extracting information from chart images. ScatterPlotAnalyzer will be beneficial for aiding other learning approaches as an integrated solution.

Acknowledgments. The authors are grateful to all members of the Graphics-Visualization-Computing Lab and peers at the IIITB for supporting this work. The authors would like to acknowledge that discussions with T. K. Srikanth, IIITB; Sindhu Mathai of Azim Premji University; Vidhya Y. and Supriya Dey of Vision Empower; Neha Trivedi, XRCVC; Vani, Pushpaja, Kalyani, and Anjana of Braille Resource Center, Matruchayya, have shaped this work. The authors are thankful for the helpful comments from anonymous reviewers.

References

1. Baek, J., et al.: What Is Wrong With Scene Text Recognition Model Comparisons? Dataset and Model Analysis. CoRR abs/1904.01906, 4714–4722 (2019). https://doi.org/10.1109/ICCV.2019.00481
2. Baek, Y., Lee, B., Han, D., Yun, S., Lee, H.: Character region awareness for text detection. CoRR abs/1904.01941 (2019). http://arxiv.org/abs/1904.01941
3. Battle, L., Duan, P., Miranda, Z., Mukusheva, D., Chang, R., Stonebraker, M.: Beagle: automated extraction and interpretation of visualizations from the web. In: Proceedings of the 2018 CHI Conference on Human Factors in Computing Systems, p. 594. ACM (2018)
4. Baucom, A., Echanique, C.: ScatterScanner: Data Extraction and Chart Restyling of Scatterplots (2013)
5. Choi, J., Jung, S., Park, D.G., Choo, J., Elmqvist, N.: Visualizing for the non-visual: Enabling the visually impaired to use visualization. In: Computer Graphics Forum, vol. 38, pp. 249–260. Wiley Online Library (2019)
6. Cliche, M., Rosenberg, D., Madeka, D., Yee, C.: Scatteract: automated extraction of data from scatter plots. In: Ceci, M., Hollmén, J., Todorovski, L., Vens, C., Džeroski, S. (eds.) ECML PKDD 2017. LNCS (LNAI), vol. 10534, pp. 135–150. Springer, Cham (2017). https://doi.org/10.1007/978-3-319-71249-9_9
7. Dadhich, K., Daggubati, S.C., Sreevalsan-Nair, J.: BarChartAnalyzer: digitizing images of bar charts. In: The Proceedings of the International Conference on Image Processing and Vision Engineering (IMPROVE 2021). INSTICC (2021, to appear). https://www.iiitb.ac.in/gvcl/pubs/2021_DadhichDaggubatiSreevalsanNair_IMPROVE_preprint.pdf
8. Ester, M., Kriegel, H.P., Sander, J., Xu, X., et al.: A density-based algorithm for discovering clusters in large spatial databases with noise. In: KDD 1996, pp. 226–231 (1996)
9. Jung, D., Kim, W., Song, H., Hwang, J.I., Lee, B., Kim, B., Seo, J.: ChartSense: interactive data extraction from chart images. In: Proceedings of the 2017 CHI Conference on Human Factors in Computing Systems, CHI 2017, pp. 6706–6717. Association for Computing Machinery, New York (2017). https://doi.org/10.1145/3025453.3025957
10. Kahou, S.E., Michalski, V., Atkinson, A., Kádár, Á., Trischler, A., Bengio, Y.: FigureQA: an annotated figure dataset for visual reasoning. arXiv preprint arXiv:1710.07300 (2017)

11. Kindlmann, G.: Superquadric Tensor Glyphs. In: Proceedings of the Sixth Joint Eurographics-IEEE TCVG Conference on Visualization, pp. 147–154. Eurographics Association (2004)

12. Medioni, G., Tang, C.K., Lee, M.S.: Tensor voting: theory and applications. In: Proceedings of RFIA, Paris, France 3 (2000)

13. Methani, N., Ganguly, P., Khapra, M.M., Kumar, P.: PlotQA: reasoning over scientific plots. In: The IEEE Winter Conference on Applications of Computer Vision, pp. 1516–1525, March 2020. https://doi.org/10.1109/WACV45572.2020.9093523

14. Moreno, R., Garcia, M.A., Puig, D., Julià, C.: Edge-preserving color image denoising through tensor voting. Comput. Vis. Image Underst. **115**(11), 1536–1551 (2011)

15. Moreno, R., Pizarro, L., Burgeth, B., Weickert, J., Garcia, M.A., Puig, D.: Adaptation of tensor voting to image structure estimation. In: Laidlaw, D., Vilanova, A. (eds.) New Developments in the Visualization and Processing of Tensor Fields, pp. 29–50. Springer, Heidelberg (2012). https://doi.org/10.1007/978-3-642-27343-8_2

16. Poco, J., Heer, J.: Reverse-engineering visualizations: recovering visual encodings from chart images. In: Computer Graphics Forum, vol. 36, pp. 353–363. Wiley Online Library (2017)

17. Rohatgi, A.: Webplotdigitizer (2011)

18. Savva, M., Kong, N., Chhajta, A., Fei-Fei, L., Agrawala, M., Heer, J.: Revision: automated classification, analysis and redesign of chart images. In: Proceedings of the 24th Annual ACM Symposium on User Interface Software and Technology, pp. 393–402. ACM (2011)

19. Siegel, N., Horvitz, Z., Levin, R., Divvala, S., Farhadi, A.: FigureSeer: parsing result-figures in research papers. In: Leibe, B., Matas, J., Sebe, N., Welling, M. (eds.) ECCV 2016. LNCS, vol. 9911, pp. 664–680. Springer, Cham (2016). https://doi.org/10.1007/978-3-319-46478-7_41

20. Smith, R.: An overview of the Tesseract OCR engine. In: Ninth International Conference on Document Analysis and Recognition (ICDAR 2007), vol. 2, pp. 629–633. IEEE (2007)

21. Sreevalsan-Nair, J., Dadhich, K., Daggubati, S.C.: Tensor fields for data extraction from chart images: bar charts and scatter plots. In: Hotz, I., Masood, T.B., Sadlo, F., Tierny, J. (eds.) Topological Methods in Visualization: Theory, Software and Applications (in press). Springer-Verlag, and arXiv preprint (2020). https://arxiv.org/abs/2010.02319

22. Sreevalsan-Nair, J., Kumari, B.: Local geometric descriptors for multi-scale probabilistic point classification of airborne LiDAR point clouds. In: Schultz, T., Özarslan, E., Hotz, I. (eds.) Modeling, Analysis, and Visualization of Anisotropy. MV, pp. 175–200. Springer, Cham (2017). https://doi.org/10.1007/978-3-319-61358-1_8

23. Tzutalin: Labelimg (2015). https://github.com/tzutalin/labelImg

24. Wang, S., Hou, T., Li, S., Su, Z., Qin, H.: Anisotropic elliptic PDEs for feature classification. IEEE Trans. Vis. Comput. Graph. **19**(10), 1606–1618 (2013)

25. Wu, T.P., Yeung, S.K., Jia, J., Tang, C.K., Medioni, G.: A Closed-Form Solution to Tensor Voting: Theory and Applications. arXiv preprint arXiv:1601.04888 (2016)

EEG-Based Emotion Recognition Using Convolutional Neural Networks

Maria Mamica, Paulina Kapłon, and Paweł Jemioło[(⊠)]

AGH University of Science and Technology, A.Mickiewicza 30, 30-059 Krakow, Poland
{mamica,pkaplon}@student.agh.edu.pl,
pawljmlo@agh.edu.pl

Abstract. In this day and age, Electroencephalography-based methods for Automated Affect Recognition are becoming more and more popular. Owing to the vast amount of information gathered in EEG signals, such methods provide satisfying results in terms of Affective Computing. In this paper, we replicated and improved the CNN-based method proposed by Li et al. [11]. We tested our model using a Dataset for Emotion Analysis using EEG, Physiological and Video Signals (DEAP) [9]. Performed changes in the data preprocessing and in the model architecture led to an increase in accuracy – 74.37% for valence, 73.74% for arousal.

Keywords: Deep learning · Convolutional Neural Networks, CNN · Electroencephalography · EEG · Emotion Recognition · Affective Computing

1 Introduction

As technology becomes more advanced and publicly available, the everyday recipients' expectations are growing exponentially fast. The services are supposed to conform to the needs of every customer. The market expands by incorporating Machine Learning to correctly recognize the emotions present while interacting with the services to meet the growing expectations.

If an efficient and highly accurate Emotion Recognition system were implemented, everyday life would be significantly improved. Not only would we be provided with genuinely entertaining products, but also approached more suitably. Thus, there is a big emphasis on studying human psychics and producing new, better methods for Emotion Recognition. The most efficient way is to discuss the findings and discoveries on the broad forum of scientists.

Nevertheless, we are currently witnessing a crisis in psychology. Studies that have been considered credible for decades have been criticized as they contained methodological flaws [3,14]. Unfortunately, a similar phenomenon also applies to Computer Science. Novel Artificial Intelligence articles are often lacking in description of parameters and architecture. Even the latest articles often overlook such important information [6]. It was an inspiration for the authors of this

© Springer Nature Switzerland AG 2021
M. Paszynski et al. (Eds.): ICCS 2021, LNCS 12746, pp. 84–90, 2021.
https://doi.org/10.1007/978-3-030-77977-1_7

paper. That is why we decided to replicate and, if possible, extend one of the articles on Emotion Recognition which has been published recently.

We strongly believe science should be replicable, and therefore, we include all the necessary information for anyone wanting to use the proposed methods. What is more, in addition to replicating the described methods, we also introduced some improvements that allowed us to increase accuracy in detecting affective states. We followed the approach introduced in the paper by Li et al. [11] and incorporated a few improvements. We selected the above study as our starting point since we were interested in the presented methodology, especially in processing signals onto the 4-channel images and Convolutional Neural Networks. What is more, the study was relatively recent at that moment and brought promising results.

The rest of the paper is organized as follows. Section 2 gives an overview of Affective Computing and presents the technologies used in this research. In Sect. 3, we described the model [11] and all introduced improvements. In Sect. 4, obtained results are provided and compared. Finally, in Sect. 5, we present the conclusions and plans for future developments related to this research.

2   Background

Affective Computing is a paradigm of Human-Computer Interaction that aims to recognize, interpret, process and simulate human emotions to adapt to a particular user in a specific emotional state. One of Affective Computing core concepts is Affective Loop [15]. In the Loop, emotions are seen as a process based on the interaction. The initial state of the Loop starts as a user begins the communication with a system. Then, the system responds in a way that affects and pulls the user into the following exchanges. The user is more engaged in the cooperation if the system can correctly read reactions in which Emotion Recognition is essential [15]. Nowadays, researchers focus on games as they represent the true potential of Affective Computing [8,13,19].

There are many different approaches to Emotion Recognition. It applies to both the computation and the extraction of the data. Among the many ways to gather needed data, the most popular are directly asking about them, using voice or facial expression, analyzing texts or speech [7]. Interpreting physiological signals like electrocardiography or electroencephalography is also growing wide as the technology required to collect the data is slowly becoming more available. This paper focuses on the latter.

EEG (Electroencephalography) is a primarily noninvasive method aiming to record the electrical activity of the brain. The examination is conducted using several electrodes placed directly on the scalp of the subject. These sensors register changes in the electric potential on the surface of the skin caused by the activity of the cerebral cortex. After triggering different emotions, for example, by showing emotion-inflicting images [12] or videos [9], we can gather EEG record corresponding to these emotions [17].

The reading of the electroencephalograph is mainly broken down into frequency, voltage, reactivity, synchrony and distribution [5]. The examination of

EEG signals usually involves the analysis of rhythmic activity. Then, such rhythmic activity can be divided into five bands based on the frequency. They consist of consequently: Delta, Theta, Alpha, Beta and Gamma bands, with the Delta band's frequency being the lowest [1].

In the paper [11], the most relevant channels are used: Gamma, Beta, Alpha, Theta. According to [1], the Delta waves are the slowest EEG waves and are only detected during deep sleep. Analyzing each bandwidth individually and four of them as a whole helps to map them to the emotions they represent.

As the human psyche is complex, and it is still unclear how emotions should be represented, one of the most popular approaches has been incorporated. It is called Arousal and Valence Space (AVS) [16]. The AVS allows categorizing the vast abstract, which is emotion, as one of the four classes (see Fig. 1(a)). The following parameters determine these classes:

– Arousal – indicates engagement and ranges from inactive to active.
– Valence – ranges from unpleasant to pleasant.

The arousal and valence are then used to define the four classes. For example, according to [10], such states can be represented as:

1. high arousal and high valence [HAxHV], e.g. excitement/delight/happiness
2. high arousal and low valence [HAxLV], e.g. anger/fear/startle
3. low arousal and low valence [LAxLV], e.g. sadness/boredom/misery
4. low arousal and high valence [LAxHV], e.g. pleasure/calmness/sleepiness

The AVS can be extended by adding another parameter – dominance, which could range between feeling powerless and being in control (see Fig. 1(b)).

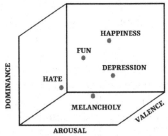
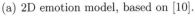

(a) 2D emotion model, based on [10].   (b) 3D emotion model, based on [18].

Fig. 1. Different emotion models visual representations.

3 Presented Approach

3.1 Dataset

As we aimed to replicate the paper by Li et al. [11] fully, we decided to develop and evaluate described models using the same reference dataset, DEAP. It consists of the EEG signals gathered from 32 participants while watching 40 one-minute music videos. The trials were chosen manually or with the use of affective tags. Additionally, it involves the ratings provided by volunteers indicating the level of valence, arousal and other dimensions [9].

3.2 Prepossessing

The model presented in this paper was inspired by another article [11]. Li et al. proposed not only the preprocessing approach but also the Neural Network architecture, which gave quite satisfying results. Their method is described next.

To retrieve more samples, all EEG signals were segmented using 8 s window with 50% overlapping. Furthermore, for each channel, 4 frequency bands were extracted (theta rhythm (4–8 Hz), alpha rhythm (8–13 Hz), beta rhythm (13–30 Hz) and gamma rhythm (30–40 Hz) and the averaged Power Spectral Density was calculated over the bands. In order to project 3D coordinates of EEG electrodes onto 2D space, the Azimuthal Equidistant Projection was performed. Authors [11] chose the Cz sensor as the centre point and computed the azimuth and the distance value – denoted as Θ and ρ respectively – for all the electrodes using the below equations:

$$\rho = arccos(sin\varphi_1 sin\varphi + cos\varphi_1 cos\varphi cos(\lambda - \lambda_0)) \qquad (1)$$

$$\Theta = arctan(\frac{cos\varphi sin(\lambda - \lambda_0)}{cos\varphi_1 sin\varphi - sin\varphi_1 cos\varphi cos(\lambda - \lambda_0)}) \qquad (2)$$

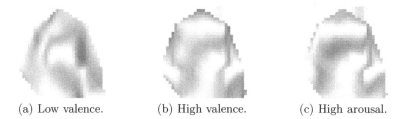

(a) Low valence. (b) High valence. (c) High arousal.

Fig. 2. Exemplary projections generated using the described method.

Herein, φ denotes latitude, and λ stands for longitude represented in the geographic coordinate system. (φ_1, λ_0) are geographic coordinates of the Cz point and can be found on the website of measuring equipment [4]. Finally, Cartesian coordinates (x,y) were computed using the below set of equations:

$$x = \rho sin\Theta \qquad (3)$$

$$y = -\rho cos\Theta \qquad (4)$$

To interpolate the calculated values over 32 × 32 mesh, the Clough-Tocher scheme was applied as the final step of the image generation operation [2]. Exemplary projections are presented in Fig. 2.

3.3 Replicated Model

Li et al. [11] proposed the Convolutional Neural Network (CNN) model. Its graphical representation can be found in the mentioned paper. It was built with three convolutional layers with 3×3 sized filters, each followed by a max-pooling layer with 2×2 sized blocks. Furthermore, the model was equipped with a fully connected layer with 256 units and a 0.5 dropout rate. Finally, 10-fold cross-validation was performed.

Unfortunately, in [11], the number of epochs and the batch size was not provided. Although it was impossible to replicate the model entirely, these parameters might be adjusted empirically without substantially impacting the results. Thus, we decided to set the batch size to 256 and the number of epochs to 200.

3.4 Improvements

In the DEAP dataset [9], participants' ratings are associated with the trials. While segmenting data, each new sample must be labelled with the emotion assigned to the example considered in a particular step. It means that too short segments may not involve specific brain responses associated with the examined emotion [18]. Having that in mind, we decided to increase the window size to 16 s. After this operation, the number of samples equals 7680 ($6 \times 32 \times 40$).

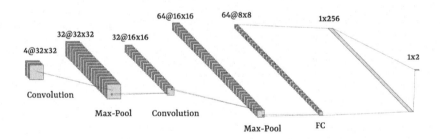

Fig. 3. The proposed Neural Network architecture.

To improve the model, various architectural approaches were tested, and finally, the Neural Network shown in Fig. 3 was chosen. It contains two convolutional layers with a 3×3 sized kernel and 1 pixel of padding, each followed by the max-pooling layer having 2×2 filters. The model has one fully connected layer equipped with 256 units and a 0.5 dropout rate. To improve the generalization of the model, an additional dropout with a 0.2 rate was applied directly before the output layer. For the activation, the *ReLU* function was chosen for all layers, except for the output one, for which the *softmax* was applied.

To measure the performance of the model during the training, the cross-entropy function was used. It is optimized by Adam optimizer with a learning rate set to 0.001. For the training, the number of epochs was adjusted to 200 and the batch size to 256. As the number of samples decreased comparing to the article [11], the number of folds used in cross-validation was set to 5 instead of 10. This operation increased the size of the validation set, and more authentic results might have been observed.

4 Preliminary Results

The model proposed in this article, and the model [11], which was the starting point of our research, were evaluated using the DEAP dataset [9]. Both methods consider CNN based binary classification for valence and arousal indicators with classes: low and high. Firstly, the method proposed by Li et al. [11] was successfully replicated. However, the researchers did not provide all necessary information. All required hyperparameters, which the authors did not provide, were adjusted empirically to give the best results.

In the article [11], the performance of the model was measured as the average of K-fold cross-validation. We followed this approach and achieved 70.76% for valence and 70.54% for arousal. The performance of the method proposed in this paper was also measured as the average of K-fold cross-validation. Accuracy gained for the valence classification equals 74.37% and for the arousal – 73.74%. The proposed method outperforms the study [11].

The improvement in the performance of the proposed method was, among others, caused by the increase of the segmenting window. Thus, all considered data samples carry more information, and they might be more accurately classified by the model. The application of dropout layers and the K-fold cross-validation technique enabled the generalization of the network. Hence, the model does not overfit and gives quite a high accuracy. Although the CNN architecture was simplified, it gives better results.

5 Conclusions and Future Work

In this paper, we replicated the method proposed in the article [11]. As not all parameters were provided, it was required to find them empirically. We strongly believe that authors should more carefully describe their work to keep the scientific development on a high level. Lack of some required parameters may be confusing for all researchers wanting to contribute or replicate the method. Nevertheless, we find the Li et al. approach promising. Even though some parameters were missing, we could adjust them empirically and fully replicate the model, achieving good results.

In the article, we additionally introduced improvements to the considered model. We observed an increase in the accuracy of the model – 74.37% for valence, 73.74% for arousal. In the future, we want to extend the architecture by applying more fully connected layers and potentially implementing multimodal fusion with other physiological (e.g. ECG, EDA) and behavioural (e.g. photo, video, game logs) data.

References

1. Abo-Zahhad, M., Ahmed, S., Seha, S.N.: A new EEG acquisition protocol for biometric identification using eye blinking signals. Int. J. Intell. Syst. Appl. **07**, 48–54 (05 2015)
2. Alfeld, P.: A trivariate clough–tocher scheme for tetrahedral data. Comput. Aided Geomet. Des. **1**(2), 169–181 (1984)
3. Bakker, M., Wicherts, J.: The (mis)reporting of statistical results in psychology. Behav. Res. Mthods **43**, 666–678 (April 2011)
4. BioSemi B.V.: Biosemi EEG ECG EMG BSPM NEURO amplifier electrodes
5. Duffy, F.H., Iyer, V.G., Surwillo, W.W.: Clinical Electroencephalography and Topographic Brain Mapping: Technology and Practice. Springer (2012). 10.1007/978-1-4613-8826-5
6. George, F.P., et al.: Recognition of emotional states using EEG signals based on time-frequency analysis and SVM classifier. Int. J. Electr. Comput. Eng. **9**, 1012 (04, 2019)
7. Imani, M., Montazer, G.A.: A survey of emotion recognition methods with emphasis on e-learning. J. Netw. Comput. Appl. (08, 2019)
8. Jemioło, P., Giżycka, B., Nalepa, G.J.: Prototypes of arcade games enabling affective interaction. In: International Conference on Artificial Intelligence and Soft Computing, pp. 553–563. Springer (2019)
9. Koelstra, S., et al.: Deap: a database for emotion analysis using physiological signals. IEEE Trans. Affect. Comput. **3**, 18–31 (12, 2011)
10. Kollias, D., et al.: Deep affect prediction in-the-wild: Aff-wild database and challenge. Int. J. Comput. Vis. **127**, (2019). 10.1007/s11263-019-01158-4
11. Li, C., Sun, X., Dong, Y., Ren, F.: Convolutional neural networks on EGG-based emotion recognition. In: Jin, H., Lin, X., Cheng, X., Shi, X., Xiao, N., Huang, Y. (eds.) Big Data, pp. 148–158. Springer Singapore, Singapore (2019)
12. Mikels, J., Fredrickson, B., Samanez-Larkin, G., Lindberg, C., Maglio, S., Reuter-Lorenz, P.: Emotional category data on images from the international affective picture system. Behav. Res. Methods **37**, 626–630 (12, 2005)
13. Nalepa, G.J., Kutt, K., Giżycka, B., Jemioło, P., Bobek, S.: Analysis and use of the emotional context with wearable devices for games and intelligent assistants. Sensors **19**(11), 2509 (2019)
14. Nuijten, M., et al.: The prevalence of statistical reporting errors in psychology (1985–2013). Behav. Res. Methods **48**,1205-1226 (10, 2015)
15. Picard, R.W.: Affective Computing. MIT Press, Cambridge (2000)
16. Russell, J.: A circumplex model of affect. J. Personal. Soc. Psychol. **39**, 1161–1178 (12, 1980)
17. SatheeshKumar, J., Bhuvaneswari, P.: Analysis of electroencephalography (EEG) signals and its categorization-a study. Proc. Eng. **38**, 525–2536 (09, 2012)
18. Yang, L., Liu, J.: EEG-based emotion recognition using temporal cnn. In: Data Driven Control and Learning Systems Conference. pp. 437–442 (2019)
19. Yannakakis, G.N., Martínez, H.P., Jhala, A.: Towards affective camera control in games. User Model Uiser Adapt. Int. **20**(4), 313–340 (2010)

Improving Deep Object Detection Backbone with Feature Layers

Weiheng Hong[1,2(✉)] and Andy Song[1]

[1] RMIT University, Melbourne, VIC 3000, Australia
andy.song@rmit.edu.au
[2] Xiamen Research Center of Urban Planning Digital Technology,
Xiamen 361015, Fujian, China

Abstract. Deep neural networks are the frontier in object detection, a key modern computing task. The dominant methods involve two-stage deep networks that heavily rely on features extracted by the backbone in the first stage. In this study, we propose an improved model, ResNeXt101S, to improve feature quality for layers that might be too deep. It introduces splits in middle layers for feature extraction and a deep feature pyramid network (DFPN) for feature aggregation. This backbone is neither much larger than the leading model ResNeXt nor increasing computational complexity distinctly. It is applicable to a range of different image resolutions. The evaluation of customized benchmark datasets using various image resolutions shows that the improvement is effective and consistent. In addition, the study shows input resolution does impact detection performance. In short, our proposed backbone can achieve better accuracy under different resolutions comparing to state-of-the-art models.

Keywords: Object detection · Deep learning · Deep neural networks · Input resolutions · Feature extraction · Feature learning

1 Introduction

As a longstanding and fundamental field in computer vision, object detection remains an active yet challenging area in modern AI [16,25]. The goal of object detection is to determine whether there are any objects of given categories (such as person, dog, car) in the given images, if present, to return the location and area of each object instance marked by a bounding box [20]. The recent success of deep learning has made significant advancements in object detection [24]. In general, there are two dominating deep network backbones, Faster RCNN (Region Based Convolutional Neural Networks) and Mask RCNN, proposed by Girshick *et al.* [5,19]. They achieved state-of-the-art performance on various datasets such as the MS COCO (Microsoft Common Objects in Context) dataset.

In this study, we aim to improve object detection by proposing alternative feature extraction layers for the existing backbones. The hypothesis is that features from the multi-resolution feature layers may have different importance towards

© Springer Nature Switzerland AG 2021
M. Paszynski et al. (Eds.): ICCS 2021, LNCS 12746, pp. 91–105, 2021.
https://doi.org/10.1007/978-3-030-77977-1_8

the final object objection performance. Features at certain layers may not be well captured if the layer is too deep. Hence redirecting the flow of feature learning may be more beneficial as feature quality may be improved. Consequently, the detection performance can be improved. The existing object detection framework could be enhanced by leveraging these features. In addition, the effect of input resolution is investigated in this study. Input size does not only affect network structure but also connects to detection accuracy. Existing work shows that low resolution may not negatively impact some vision tasks yet can significantly save computational cost [27]. A study by [17] shows that face recognition requires a minimum resolution of 32×32 pixels. However, the input size for object detection is relatively unexplored [21]. Hence the proposed improvement on object detection backbone accommodates that need so the input of different sizes can be used. To evaluating the effectiveness of the proposed improvement, customized benchmark COCO data are used in the following study. The proposed object detection backbone is beneficial as evidenced by the comparison with state-of-the-art. The details are presented in the following sections.

2 Background and Related Work

Object detection is defined as follows, to determine if or not there are instances of objects from given categories on a given image. If objects are present, the locations and areas of the detected instances should be marked. Although there are numerous kinds of objects that exist in our visible world, object detection research mainly studies methods for detecting highly structured objects and articulated objects rather than unstructured scenes. Structured objects such as faces, cars, ships, and airplanes, normally have a consistent shape. Articulated objects are usually living beings such as a person, a dog, and a bird. Different from these two types of objects, unstructured scenes are unpredictable in terms of shape, for example, sky, fire, and water. Four kinds of recognition can be derived from object detection. That includes image-level object classification, bounding box level object detection, pixel-wise semantic segmentation and instance-level semantic segmentation, as illustrated by [13] (Fig. 1). Surveys indicate the bounding box

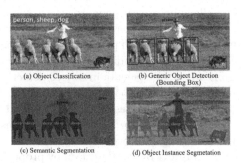

Fig. 1. Examples of four object detection tasks illustrated by Liu *et al.* [13].

object detection is the most widely used and is the basis for evaluating the performance of an object detection algorithm. Some object detection frameworks use a bonding box combined with others. For example, Mask RCNN uses pixel-wise segmentation, which is to assign each pixel in an image to a semantic class label for high-level detection. Object detection tasks can be handled by several types of methods including deep networks models [28], statistical models, and Genetic Programming. The currently best performing methods are deep networks based.

2.1 Object Detection Frameworks

Many object detection models have been proposed. They can be categorized into two groups: one-stage framework and two-stage framework. The latter is represented by region-based frameworks which in general achieved superior performance than other methods. This approach uses a CNN (Convolutional Neural Networks) backbone to generate category-independent regions from an image. Consequently feature extractors are embedded to find useful features from these regions. Based on these features a classifier then is applied to determine whether instances of a given class are present in region proposals. If present, category labels will be returned. This type of two-stage approach can be found in many object detection frameworks such as RCNN [4], Fast RCNN [3], Faster RCNN and Mask RCNN. On the MS COCO object detection competition, state-of-the-art Mask RCNN framework, Cascade Mask R-CNN (Triple-ResNeXt152, multi-scale), achieved top performance of 71.9% mAP with $IoU = 0.50$ [15].

Two-stage frameworks can achieve superior detection performance. In comparison, one-stage frameworks can often increase detection speed as they use a unified pipeline structure to directly predict class labels with a single neural network. Region proposal network and feature extractor are absent in this category. One-stage frameworks such as YOLO [18], SSD [14], and YOLO9000 have gained popularity in recent years, due to their simplicity and low cost. This approach makes real-time object detection possible especially under circumstances that computational resource is limited, such as droids and other embedded systems.

In summary, two-stage frameworks achieve state-of-the-art detection accuracy with complex neural network architectures, while one-stage framework using a simple elegant structure to achieve high detection speed. This study aims to improve object detection accuracy then focuses on the two-stage approach. In the family of two-stage object detection frameworks, RCNN is the leading model. Originating from the earliest RCNN model, the RCNN family has advanced to Faster RCNN and Mask RCNN, which are mentioned before as state-of-the-art in object detection. Hence the focus of this study is on Faster RCNN and Mask RCNN, improving their feature extraction backbone.

2.2 Feature Extractors

One of the major components in the object detection models is the backbone, which is responsible for extracting features for the subsequent classification stage. Feature extractor plays a crucial role in feature representation as well as in the whole detection task [23]. Deep neural networks (DCNNs) have been found capable of generating distinct features from raw images at a different level in multi-resolution pyramid representation. DCNN is present in many well-known deep models such as AlexNet [9], VGGNet [22], ResNet [6], ResNeXt [26], DenseNet [8], and MobileNet [7] *et al.* They are listed in Table 1 for comparison in terms of parameter size, the number of layers, and test errors on the benchmark.

Table 1. Comparison of representative DCNN for image classification

DCNN architecture	#Paras ($\times 10^6$)	#Layers (CONV + FC)	Test error (Top 5)
AlexNet [9]	57	5 + 2	20.91%
VGGNet19 [22]	134	13 + 2	9.62%
ResNet50 [6]	23	49	7.13%
ResNet101 [6]	42	100	6.44%
ResNeXt50 [26]	23	49	6.30%
ResNeXt-101 [26]	42	100	5.47%
DenseNet201 [8]	18	200	6.43%
MobileNet [7]	3.2	27 + 1	9.71%

From Table 1 a trend can be seen that in general deeper networks, meaning with more layers, can lead to better feature representation in CNN, hence to lower error rates. Another observation is the relation between model size or the number of parameters and the use of FC (Fully Connected) layers. A model with a '+' sign in Table 1 means FC is present. For example, AlexNet is 5+2, meaning 2 layers of FC are presented with 5 convolutional layers. From the table, we can see that AlexNet and VGGNet utilize FC layers which result in significantly more parameters onto the model. DenseNet and ResNet have fewer parameters by leaving out the FC layers. Therefore, avoiding the use of FC layers can lead to smaller models without damaging, if not improving, the detection accuracy.

After the introduction of Mask RCNN, a more effective feature extraction method is proposed by Lin *et al.*, call Feature Pyramid Network (FPN). FPN uses a top-down architecture with lateral connections in each layer of the backbone to build a feature pyramid and made predictions independently at all levels. These RoI (Region of Interest) features extracted from different layers contribute to the feature maps in various aspects. Mask RCNN produces excellent accuracy and efficiency largely due to the use of ResNet-FPN backbone [11].

2.3 Performance Evaluation

Most of the current work uses mAP for evaluation, especially after the introduction of the MS COCO dataset. Instead of using a fixed IoU (Intersection over Union) threshold, the MS COCO introduces various metrics to better measure the performance of a given object detection model. That includes AP ($IoU = 0.50 : 0.95$), AP ($IoU = 0.75$), and AP ($IoU = 0.5$). AP ($IoU = 0.50 : 0.95$) metric is primarily used in the MS COCO Dataset challenge. AP ($IoU = 0.75$) represents a more strict metric for evaluation. In this study, we use the most commonly used AP ($IoU = 0.5$) metric to evaluate our object detection model.

2.4 Datasets

For generic object detection, four famous datasets are utilized around the community, include PASCAL VOC [2], ImageNet [1], MS COCO [12] and Open Images [10]. In this research, in order to study the relationship between various image resolution and object detection models, MS COCO dataset with the highest image resolution and well organized could be the preference around four image datasets. In addition, object segmentation data make it possible to use in the Mask R-CNN model, which is designed for object detection and segmentation using segmentation-level detection technology.

2.5 Resolutions

The impact of resolution is relatively under-explored in machine vision, in particular object detection. Shivanthan et al. [27] report that using low-resolution grayscale (LG) images for saliency detection can lead to speedups in model training and detection time. Region Proposal Network (RPN), based on this novel saliency-guided selective attention theory, separates the objects' regions and background regions. Therefore, using LG images for object detection can greatly improve the efficiency of object detection and keep the object detection model in a small size. But experiments indicate this model usually fails to detect the main object when the size of the image is smaller than 64×64 pixels. A study on face recognition requires a minimum input of 32×32 pixels [17].

3 Methodology

The main methodology of this study is presented in this section. That includes dataset preparation for evaluation, the preliminary experiment on the chosen data, and the design of deep models.

3.1 Data Preparation

COCO 2017 data set is a leading benchmark for object detection[1]. Three types of datasets are included Training dataset, Validation dataset, and Testing dataset. The testing dataset is used for COCO competition that does not provide annotations for evaluation locally, so we use the Training dataset for model learning, and the Validation dataset to evaluate the model by computing the bounding box AP ($IoU = 0.50$) value. COCO data sets contain 200,000 images of 80 object categories. In this study, we group selected categories into three categories: rectangle object class (such as buses, vehicles), convex-polygon object class (such as

---

[1] Available on the MS COCO dataset website http://cocodataset.org.

dogs), and round objects (such as apples). These include the most representative classes of the COCO dataset. Such alternation is to facilitate the study especially the analysis as a benefit on backbone improvement should be independent of how to categorize target objects.

3.2 Preliminaries

As the RPN object detection model proposed by Shivanthan *et al.* failed to detect the object when the image is resized to 64×64 pixels, we set 64×64 as the definition of low resolution in our preliminary work to test the performance of different object detection model include Faster RCNN and Mask RCNN with various popular backbones. Model is trained with a training dataset of 64×64 resolution, and then tested on 64×64 validation dataset. The metric is the bounding box mAP ($IoU = 0.50$) resulting from our three-class COCO datasets. Table 2 shows the results from the preliminary study which involves a range of widely used deep models including AlexNet, MobileNet, DenseNet, VGG, ResNet models, and ResNeXt models. The suffix number after a model name is the depth of the model. For example, ResNet34 means that is a ResNet model with 34 layers. As can be seen from the table, with FPN, ResNet models and ResNeXt models achieved mAP values higher than other models in both Faster RCNN and Mask RCNN. ResNeXts performs slightly better than ResNets. The subsequent study is therefore based on ResNeXt models. In addition, Mask RCNN models, in general, perform better than their Faster RCNN counterparts. While ResNeXt50 with FPN gets the best result of mAP on both Faster RCNN and Mask RCNN frameworks, a natural question is that why ResNeXt101 with a deeper convolutional neural network was inferior to ResNeXt50. A similar

Table 2. Bounding box mAP (IoU = 0.50) value of object detection results on validation dataset under 64×64 resolution. The number in the backbone represents the number of the layers.

Backbone	Faster RCNN	Mask RCNN
AlexNet	9.7	9.9
MobileNet	14.6	14.2
Densenet201	15.7	16
VGGNet16	21.2	21.5
VGGNet19	19.8	20.7
ResNet34 + FPN	30	30.9
ResNet50 + FPN	30.8	31.7
ResNet101 + FPN	30.6	32.1
ResNet152 + FPN	31.7	31.6
ResNeXt50 + FPN	**32.5**	**33.4**
ResNeXt101 + FPN	31.4	32.2

phenomenon happens in VGGNet16 and VGGNet19. While VGGNet19 contains more layers and parameters than VGGNet16, it does not achieve a better result than VGGNet16. The analysis is that the deeper networks may result in better features but also may treat features indifferently through the deep layers. So deeper net may not be as helpful if features are not utilized well. These features may not represent some small and unnoticeable objects, especially from low-resolution images. If these objects are indeed targets, the generated features may not capture them leading to a slight decline in terms of mAP measure.

Table 3. ResNeXt50 structure, ResNeXt101 structure and ResNeXt101S structure. "C = 32" suggests grouped convolutions with 32 groups.

Layer name	ResNeXt50		ResNeXt101		ResNeXt101S (ours)	
conv1	7×7, 64, stride 2					
conv2_x	3×3 max pool, stride 2					
	$\begin{bmatrix} 1 \times 1, 128, \\ 3 \times 3, 128 \\ 1 \times 1, 256, \end{bmatrix}$	$\times 3$	$\begin{bmatrix} 1 \times 1, 128, \\ 3 \times 3, 128 \\ 1 \times 1, 256, \end{bmatrix}$	$\times 3$	$\begin{bmatrix} 1 \times 1, 128, \\ 3 \times 3, 128 \\ 1 \times 1, 256, \end{bmatrix}$	$\times 3$
conv3_x	$\begin{bmatrix} 1 \times 1, 256, \\ 3 \times 3, 256 \\ 1 \times 1, 512, \end{bmatrix}$	$\times 4$	$\begin{bmatrix} 1 \times 1, 256, \\ 3 \times 3, 256 \\ 1 \times 1, 512, \end{bmatrix}$	$\times 4$	$\begin{bmatrix} 1 \times 1, 256, \\ 3 \times 3, 256 \\ 1 \times 1, 512, \end{bmatrix}$	$\times 4$
conv4_x	$\begin{bmatrix} 1 \times 1, 512, \\ 3 \times 3, 512 \\ 1 \times 1, 1024, \end{bmatrix}$	$\times 6$	$\begin{bmatrix} 1 \times 1, 512, \\ 3 \times 3, 512 \\ 1 \times 1, 1024, \end{bmatrix}$	$\times 23$	$\begin{bmatrix} 1 \times 1, 512, \\ 3 \times 3, 512 \\ 1 \times 1, 1024, \end{bmatrix}$	$\times 6$
					$\begin{bmatrix} 1 \times 1, 512, \\ 3 \times 3, 512 \\ 1 \times 1, 1024, \end{bmatrix}$	$\times 17$
conv5_x	$\begin{bmatrix} 1 \times 1, 1024, \\ 3 \times 3, 1024 \\ 1 \times 1, 2048, \end{bmatrix}$	$\times 3$	$\begin{bmatrix} 1 \times 1, 1024, \\ 3 \times 3, 1024 \\ 1 \times 1, 2048, \end{bmatrix}$	$\times 3$	$\begin{bmatrix} 1 \times 1, 1024, \\ 3 \times 3, 1024 \\ 1 \times 1, 2048, \end{bmatrix}$	$\times 3$

Preliminary work shows ResNeXt50 with FPN performs better than ResNeXt101 with FPN in object detection of low-resolution images. The detail of our model design will be introduced in the first part. The experiment is set to evaluate our model in the second part.

3.3   Backbone Model Design

In this study, we propose an improved object detection backbone by using the deep feature pyramid network (DFPN) method which can enhance the expression of feature maps. There are a number of powerful backbones based on Faster RCNN and Mask RCNN with high performance, such as ResNet, RTesNeXt, and VGGNet. In this section, we first describe our backbone design and the two frameworks that are used, Faster RCNN and Mask RCNN respectively. Our backbone construction is based on ResNeXt that uses a parallel structure with 32 groups of the identical blocks of ResNet to construct its block.

The structure of ResNeXt50 and ResNeXt101 are presented in Table 3. The main difference between these two CNNs is the *conv4_x* layer. While ResNeXt50 uses 6 sequential blocks to extract features, ResNeXt101 uses a deeper block of 23 layers in the *conv4_x* layer. As we discussed in the preliminary work, that leads to the relative lower *mAP* as some small and unnoticed objects may not be captured by the extracted features.

The structure of our proposed backbone is also presented in Table 3 alongside with ResNeXt50 and ResNeXt101. We name it as ResNeXt101S. The major difference is the splitting of the *conv4_x* layer into two sub-layers. They are *conv4_x_6* layer and *conv4_x_17* layer respectively. The first sub-layer consists of 6 blocks of ResNeXt, while the second sub-layer is composed of 17 blocks of ResNeXt. As for the structure of each block, the same set of output channels is still maintained. They all behave as blocks of *conv4_x* of ResNeXt101 and are grouped by 32 parallel paths. A diagram of the proposed ResNeXt101 backbone can be seen in Fig. 2. Other than the splitting middle layer, a new Deep Feature Pyramid Network (DFPN) is also proposed to replace the region proposal network. It not only takes features from the basic layers of ResNeXt101S, but also takes features from the inner layer of *conv4_x* as shown in the figure. By using a 5-layer top-down architecture through the entire ResNeXt101S CNN, it extracts and generates a feature pyramid from basic layers and inner layers include *conv2_x*, *conv3_x*, *conv4_x_6* layer, *conv4_x_17* layer, and *conv5_x*. The features are extracted from each level of the feature pyramid to contribute towards the feature maps in which predictions are made at all levels. The aim is to compensate for lost features and to enable more prominent features to be

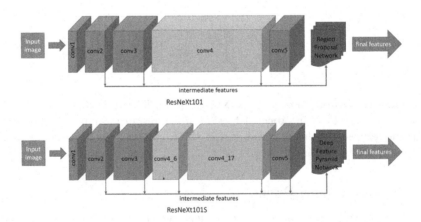

Fig. 2. Two Backbones: Top: ResNeXt101 with FPN. Bottom: ResNeXt101S with DFPN.

captured. Based on ResNeXt101S with DFPN, our backbone is adapted into both Faster RCNN and Mask RCNN frameworks.

With Faster RCNN: while the original Faster RCNN model using the RoIPool method and applying VGG16 as the backbone, we introduce our backbone based on the original Faster RCNN framework, but applying RoIAlign for pooling, which has been shown of being able to increase mAp in Mask RCNN.

The Anchor Generator as the top component of region proposal network (RPN) is used to generate a number of boxes (Anchors) to detect target objects in the image that the Anchor Box size in RPN can be associated with the input image size, thus we set the size with this equation:

$$area(x) = \left\{ (\tfrac{0.7x}{2^i})^2 \; i = 0, 1, 2, 3, 4 \right. \tag{1}$$

For the different input image sizes, we design different object detection with size-specific anchor boxes respectively, then apply for training and testing.

With Mask RCNN: while the original Mask RCNN model using ResNeXt-101 with FPN as a backbone to achieve state-of-the-art mAP performance, we introduce our proposed backbone on the original Mask RCNN Framework, but set the Anchor Box size also using the Eq. (1) shown above.

4 Experiments and Results

4.1 Experiment Settings

Experiments are set to verify and evaluate the performance of our proposed ResNeXt + DFPN backbone. During the training, each batch has 2 images in one GPU. The Adadelta algorithm for optimization is used with the learning rate of 0.3, which is decreased by 5 at the 30 iterations [29]. A coefficient of 0.9 is used. For each model, we train with one NVIDIA Tesla V100 GPU and select the results with the best bounding box AP value.

Firstly we compare the performances of object detection models on three-class datasets under 64×64 resolution, similar to the experiments in the preliminary work. Our ResNeXt101S + DFPN backbone with the backbones which perform quite well in the preliminary work include VGGNet16, ResNet + FPN and ResNeXt + FPN. All backbones are combined with Fater RCNN and Mask RCNN in the experiments by training with 64×64 training dataset and testing in 64×64 validation dataset. The detection accuracy is measured via the bonding box mAP (*IoU* = 0.50) value over three class datasets. As the Mask RCNN framework can give better detection ability than Faster RCNN due to its more informed learning style, we adopt the Mask RCNN framework to evaluate various backbones in terms of model size and detection accuracy.

In the second stage of experiments we investigated the effectiveness of our ResNeXt101S + DFPN backbone with Mask RCNN on three class datasets under various image resolutions. That includes 64×64, 128×128, 256×256,

512×512, and 1024×1024. Since the maximum resolution of original images is 640, we also test with the size of 640×640. We compare our backbone with ResNeXt50 + FPN and ResNeXt101 + FPN backbones. The performances are measured via bonding box mAP ($IoU = 0.50$) value over three class datasets.

The results are presented in the following two subsections. To compare the performance of our model with others, VGGNet16, ResNet50 models, ResNeXt models are also included.

4.2 Object Detection with Low-Resolution Images

While testing with "rectangle" class dataset, "convex polygon" class dataset, and "round" class dataset under 64×64 image resolution, Table 4 presents the boding box mAP ($IoU = 0.50$) value of the object detection results.

Table 4. Bounding box mAP ($IoU = 0.50$) value (%) of object detection results on dataset under 64×64 resolution. The experiment result of VGGNet16, ResNet + FPN, ResNeXt + FPN, and ResNeXt101S + DFPN backbones with Faster RCNN and Mask RCNN frameworks.

Backbone	Faster RCNN	Mask RCNN
VGGNet16	*21.2*	21.5
ResNet50 + FPN	30.8	31.7
ResNet101 + FPN	30.6	32.1
ResNet152 + FPN	31.7	31.6
ResNeXt50 + FPN	32.5	33.4
ResNeXt101 + FPN	31.4	*32.2*
ResNeXt101S + DFPN (ours)	**33**	**33.9**

As the Faster RCNN first proposed in 2015 using VGGNet16 achieves 21.2 mAP testing accuracy, Mask RCNN proposed in 2017 using ResNeXt101 with FPN with the detection accuracy increases to 32.2. Rather than testing on high-resolution images as in previous papers, we apply low-resolution images. It can be seen that ResNeXt50 with FPN performs better than ResNeXt101 with FPN in both RCNN frameworks. While ResNeXt50 with FPN achieves 32.5 mAP in Faster RCNN and 33.4 mAP in Mask RCNN. The proposed new backbone, ResNeXt101S with DFPN, actually increases 0.5 mAP value higher than them in both Faster RCNN and Mask RCNN frameworks.

The experiment results of bounding box mAP value of object detection on the dataset is presented in Table 5. That result shows DFPN used in ResNeXt101S is beneficial in terms of raising detection accuracy in low-resolution images. That observation verifies the analysis that ResNeXt101 does not give better performance than ResNeXt50 in low-resolution images because its deeper convolutional neural network structure is less effective in capturing features.

By splitting In this way, it increases 1.7 mAp point of detection accuracy to 33.9 compare with the original Mask RCNN model which uses ResNeXt101 with FPN as the backbone. While ResNeXt101S with DFPN in the Mask RCNN framework obtains the best mAP result in bounding box object detection in low-resolution images, the model size of it is very close to the ResNeXt101 with FPN. Compared with the model size of 430 MB of ResNeXt101 with FPN, our ResNeXt101S with DFPN is just 10 MB larger.

4.3 Object Detection with Various Resolutions

As ResNeXt101S with DFPN performed quite well in low-resolution images, we here evaluate whether it is still superior to others in different resolutions of images. The resolutions for test include 64×64, 128×128, 256×256, 512×512, 640×640, and 1024×1024.

Table 5. Bounding box mAP ($IoU = 0.50$) value (%) of object detection results on three class datasets under dataset-specific resolutions. The experiment result of ResNeXt50 + FPN, ResNeXt101 + FPN, and ResNeXt101S + DFPN backbones with Mask RCNN frameworks.

Resolution	ResNeXt50 with FPN	ResNeXt101 with FPN	ResNeXt101S with DFPN (ours)
64×64	33.4	32.2	**33.9**
128×128	42.3	41.5	**42.7**
256×256	54.1	55.4	**55.9**
512×512	58.9	60.2	**61**
640×640	61.1	63.6	**63.9**
1024×1024	63.8	65.1	**65.9**

Table 5 shows the results. Under the resolution of 64×64 and 128×128, ResNeXt101 has the worst result among three backbones, while ResNeXt101 performs slightly better than ResNeXt50, which means by using DFPN with ResNeXt101S in Mask RCNN, the model does improve feature quality in low-resolution images. For the other resolutions that are higher than 128×128, ResNeXt50 does not have better detection accuracy than ResNeXt101, which shows the advantage of a deep convolutional neural network. Since under the sufficient resolution condition, a deeper CNN extracts the shape of the feature with more distinct expression than others. In this case, using ResNeXt101 + FPN as the backbone is possible to achieve a better object detection performance than using ResNeXt50 + FPN. While ResNeXt101 + FPN performs quite well in high resolution, ResNeXt101S + DFPN can still achieve higher mAP result with slightly improvement in 256×256, 512×512, 640×640, and 1024×1024. Such an outcome indicates using ResNeXt101S with DFPN can still improve feature quality at high-resolution input.

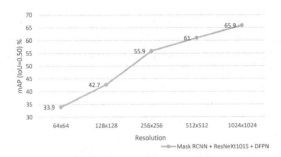

Fig. 3. Bounding box mAP ($IoU = 0.50$) value (%) vs. Dataset-specific resolutions. The experiment result of ResNeXt101S + DFPN backbone with Mask RCNN frameworks.

Overall by using ResNeXt101S with DFPN in Mask RCNN, our model achieved the best object detection performance for various resolutions of input images. An interesting observation that worth mentioning is comparing the performance under 640×640 with the performance under 1024×1024. While we upscale the original image with a maximum image size of 640 to a higher 1024, the object detection accuracy still increases with 2.0 mAP points rather than remain the same as 640×640 images despite the fact that no extra information was added in the up-scaling process.

Table 6. Bounding box AP ($IoU = 0.50$) value (%) of object detection results under 1024×1024 resolution on Rectangle class, Convex-polygon class and Round object class. The experiment result of ResNeXt50 + FPN, ResNeXt101 + FPN, and ResNeXt101S + DFPN backbones with Mask RCNN frameworks.

Backbone	Rectangle class	Convex-polygon class	Round class	Average mAP
ResNeXt50 + FPN	80.9	80	30.5	63.8
ResNeXt101 + FPN	81.7	81.2	32.4	65.1
ResNeXt101S + DFPN (ours)	**82.6**	**82.1**	**33.1**	**65.9**

Detected "Rectangle" Object   Detected "Convex-polygon" Object   Detected "Round" Object

Fig. 4. Example of prediction result.

Figure 3 shows the tendency graph of detection performance in relation with various resolutions. We collect the data with the sequence of the multiple of image resolution which contains the resolution include 64×64, 128×128, 256×256, 512×512, and 1024×1024. With increasing resolution, the accuracy of object detection increases. We can observe that the accuracy increasing rate is fastest from 128×128 to 256×256, and then become slow down in the 512×512 and 1024×1024. As a larger image needs more time-consuming and computational-consuming, an image resolution as small as possible is desirable. Therefore, 256×256 could be selected as the optimal resolution for object detection with good object detection accuracy and high efficiency.

For further investigation, we present the analysis of detection performance per class in Table 6. The table shows the mAPs of all three models tested under images of 1024×1024 in this study. As can be seen, our proposed ResNeXt101S + DFPN backbone performed best in all three classes. Its good performance is independent of class type. In terms of the classes, "Rectangle" objects and "Convex-polygon" objects can be much more accurately detected than "Round" objects. That is the case for all three models. The possible explanation is that round objects may be confused with non-target round objects. Our further study will address that to improve performance. Nevertheless, such results confirm that the good performance of our proposed backbone is not random. Examples of detected objects are illustrated in Fig. 4. They represent three categories of objects from the COCO dataset. The bounding boxes (in red) are the output from the Mask RCNN model using our proposed backbone. These boxes fit with the target object tightly showing the good performance of the detection model.

5   Conclusions

In this paper, we proposed an improved backbone for object detection with an innovative method that combines advantages from both backbones. The aim is to improve feature quality for deep layers. From experiments, it can be seen that deep models may not extract high-quality features if the layers are deep. The improved backbone split the feature layers and re-direct intermediate features to a proposed deep feature pyramid network (DFPN) for feature aggregation. This backbone can be integrated into leading frameworks including Faster RCNN and Mask RCNN and is applicable for handling a range of different image resolutions. With the improved backbone, better detection performance can be achieved on different resolutions comparing to state-of-the-art models. In conclusion, our method improves the object detection performance without increasing the number of parameters and computational complexity distinctly. The proposed backbone is beneficial in improving feature quality for object detection.

References

1. Deng, J., Dong, W., Socher, R., Li, L.J., Li, K., Fei-Fei, L.: ImageNet: a large-scale hierarchical image database. In: 2009 IEEE Conference on Computer Vision and Pattern Recognition, pp. 248–255. IEEE (2009)

2. Everingham, M., Eslami, S.A., Van Gool, L., Williams, C.K., Winn, J., Zisserman, A.: The pascal visual object classes challenge: a retrospective. Int. J. Comput. Vision **111**(1), 98–136 (2015)
3. Girshick, R.: Fast R-CNN. In: Proceedings of the IEEE International Conference on Computer Vision, pp. 1440–1448 (2015)
4. Girshick, R., Donahue, J., Darrell, T., Malik, J.: Rich feature hierarchies for accurate object detection and semantic segmentation. In: Proceedings of the IEEE Conference on Computer Vision and Pattern Recognition, pp. 580–587 (2014)
5. He, K., Gkioxari, G., Dollár, P., Girshick, R.: Mask R-CNN. In: Proceedings of the IEEE International Conference on Computer Vision, pp. 2961–2969 (2017)
6. He, K., Zhang, X., Ren, S., Sun, J.: Deep residual learning for image recognition. In: Proceedings of the IEEE Conference on Computer Vision and Pattern Recognition, pp. 770–778 (2016)
7. Howard, A.G., et al.: MobileNets: efficient convolutional neural networks for mobile vision applications. arXiv preprint arXiv:1704.04861 (2017)
8. Huang, G., Liu, Z., Van Der Maaten, L., Weinberger, K.Q.: Densely connected convolutional networks. In: Proceedings of the IEEE Conference on Computer Vision and Pattern Recognition, pp. 4700–4708 (2017)
9. Krizhevsky, A.: One weird trick for parallelizing convolutional neural networks. arXiv preprint arXiv:1404.5997 (2014)
10. Kuznetsova, A., et al.: The open images dataset v4: unified image classification, object detection, and visual relationship detection at scale. arXiv preprint arXiv:1811.00982 (2018)
11. Lin, T.Y., Dollár, P., Girshick, R., He, K., Hariharan, B., Belongie, S.: Feature pyramid networks for object detection. In: Proceedings of the IEEE Conference on Computer Vision and Pattern Recognition, pp. 2117–2125 (2017)
12. Lin, T.-Y., et al.: Microsoft COCO: common objects in context. In: Fleet, D., Pajdla, T., Schiele, B., Tuytelaars, T. (eds.) ECCV 2014. LNCS, vol. 8693, pp. 740–755. Springer, Cham (2014). https://doi.org/10.1007/978-3-319-10602-1_48
13. Liu, L., et al.: Deep learning for generic object detection: a survey. arXiv preprint arXiv:1809.02165 (2018)
14. Liu, W., et al.: SSD: single shot MultiBox detector. In: Leibe, B., Matas, J., Sebe, N., Welling, M. (eds.) ECCV 2016. LNCS, vol. 9905, pp. 21–37. Springer, Cham (2016). https://doi.org/10.1007/978-3-319-46448-0_2
15. Liu, Y., et al.: CBNet: a novel composite backbone network architecture for object detection. arXiv preprint arXiv:1909.03625 (2019)
16. Liu, Z., et al.: Swin transformer: hierarchical vision transformer using shifted windows. arXiv preprint arXiv:2103.14030 (2021)
17. Lui, Y.M., Bolme, D., Draper, B.A., Beveridge, J.R., Givens, G., Phillips, P.J.: A meta-analysis of face recognition covariates. In: 2009 IEEE 3rd International Conference on Biometrics: Theory, Applications, and Systems, pp. 1–8. IEEE (2009)
18. Redmon, J., Divvala, S., Girshick, R., Farhadi, A.: You only look once: unified, real-time object detection. In: Proceedings of the IEEE Conference on Computer Vision and Pattern Recognition, pp. 779–788 (2016)
19. Ren, S., He, K., Girshick, R., Sun, J.: Faster R-CNN: towards real-time object detection with region proposal networks. In: Advances in Neural Information Processing Systems, pp. 91–99 (2015)
20. Russakovsky, O., et al.: ImageNet large scale visual recognition challenge. Int. J. Comput. Vision **115**(3), 211–252 (2015)
21. Shekhar, S., Patel, V.M., Chellappa, R.: Synthesis-based robust low resolution face recognition. arXiv preprint arXiv:1707.02733 (2017)

22. Simonyan, K., Zisserman, A.: Very deep convolutional networks for large-scale image recognition. arXiv preprint arXiv:1409.1556 (2014)
23. Tan, M., Pang, R., Le, Q.V.: EfficientDet: scalable and efficient object detection. In: Proceedings of the IEEE/CVF Conference on Computer Vision and Pattern Recognition, pp. 10781–10790 (2020)
24. Tang, P., et al.: Weakly supervised region proposal network and object detection. In: Ferrari, V., Hebert, M., Sminchisescu, C., Weiss, Y. (eds.) ECCV 2018. LNCS, vol. 11215, pp. 370–386. Springer, Cham (2018). https://doi.org/10.1007/978-3-030-01252-6_22
25. Wang, C.Y., Bochkovskiy, A., Liao, H.Y.M.: Scaled-yolov4: scaling cross stage partial network. arXiv preprint arXiv:2011.08036 (2020)
26. Xie, S., Girshick, R., Dollár, P., Tu, Z., He, K.: Aggregated residual transformations for deep neural networks. In: Proceedings of the IEEE Conference on Computer Vision and Pattern Recognition, pp. 1492–1500 (2017)
27. Yohanandan, S., Song, A., Dyer, A.G., Tao, D.: Saliency preservation in low-resolution grayscale images. In: Ferrari, V., Hebert, M., Sminchisescu, C., Weiss, Y. (eds.) ECCV 2018. LNCS, vol. 11210, pp. 237–254. Springer, Cham (2018). https://doi.org/10.1007/978-3-030-01231-1_15
28. Zangeneh, E., Rahmati, M., Mohsenzadeh, Y.: Low resolution face recognition using a two-branch deep convolutional neural network architecture. arXiv preprint arXiv:1706.06247 (2017)
29. Zeiler, M.D.: Adadelta: an adaptive learning rate method. arXiv preprint arXiv:1212.5701 (2012)

Procedural Level Generation with Difficulty Level Estimation for Puzzle Games

Łukasz Spierewka, Rafał Szrajber[ID], and Dominik Szajerman[✉][ID]

Institute of Information Technology, Lodz University of Technology, Łódź, Poland
dominik.szajerman@p.lodz.pl

Abstract. This paper presents a complete solution for procedural creation of new levels, implemented in an existing puzzle video game. It explains the development, going through an adaptation to the genre of game of the approach to puzzle generation and talking in detail about various difficulty metrics used to calculate the resulting grade. Final part of the research presents the results of grading a set of hand-crafted levels to demonstrate the viability of this method, and later presents the range of scores for grading generated puzzles using different settings. In conclusion, the paper manages to deliver an effective system for assisting a designer with prototyping new puzzles for the game, while leaving room for future performance improvements.

Keywords: Procedural content generation · Puzzle game · Difficulty level estimation

1 Introduction

This paper presents a complete solution for procedural generation of levels in an existing puzzle video game called *inbento*. It is a mobile puzzle game released in September 2019 for the iOS and Android platforms. In this game the player is tasked with finishing over a hundred levels. They are introducing new concepts and gradually rising in difficulty.

The theme of the game is based on the idea of preparing bento: a type of Japanese cuisine where the meal is packed tightly in a container. Bento boxes can usually contain various ingredients: boiled rice, raw or cooked fish, prepared egg, vegetables, sandwiches or more [13].

1.1 Game Rules

inbento is centered around the idea of preparing a complete bento box in accordance with a recipe given to the player in each level. The game view consists of three main elements: the recipe book, the bento box and the cutting board, each serving a distinct purpose. Figure 1 presents three example game views.

© Springer Nature Switzerland AG 2021
M. Paszynski et al. (Eds.): ICCS 2021, LNCS 12746, pp. 106–119, 2021.
https://doi.org/10.1007/978-3-030-77977-1_9

The right side of the recipe book contains the desired end state of the puzzle – a solution that has to be replicated by the end user. In the middle of the game view sits the bento box. Figure 1 on the left shows the initial state of the level. The goal of the game is making this box look exactly like the reference image shown in the recipe book.

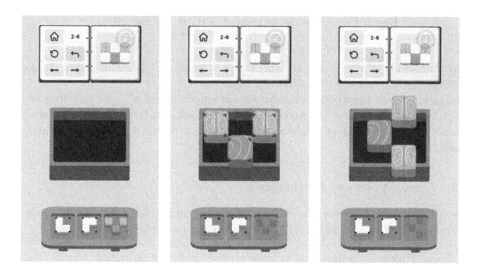

Fig. 1. Various stages of trying to place a piece into the box; Left: initial state (not picked up); Middle: valid placement (indicated by black markers displayed in the box); Right: invalid placement.

On the very bottom sits the cutting board, serving as a collection of interactive pieces: composites of one or more blocks that can be placed into the bento box. Using all available pieces to complete a puzzle is one of the requirements of the game. Upon dragging the piece into the box, the game validates whether all of the blocks fit within the boundaries of the container. If the result is negative (e.g. Fig. 1, right), the piece is sent back into the cutting board. If it is positive, piece blocks are placed into the box replacing previous content.

Boxes in the game can take on various sizes, from a single-cell grid (1×1) all the way up to a four-by-three grid (4×3). Similarly, the maximum size of the cutting board was limited to up to 8 pieces. The maximum piece size is 3×3 blocks. As seen on Fig. 2, each piece can be rotated to appear in one of four different states which can later be placed into the box. Because of this, each piece can allow the creation of up to four times as many different states upon placement, which is utilized to increase puzzle complexity.

Despite the apparent simplicity of the game, it turns out that even small differences in the size of the box, the number and size of pieces, the size of cutting board translate into an exponentially growing number of combinations of solutions, which makes the design process difficult. Procedural level generation

Fig. 2. Each piece can be rotated to 0, 90, 180 and 270°.

and the testing of its results are also made difficult by this. This paper shows these issues and our way of addressing them.

Since finished bento boxes rarely contain empty space, all final solutions for every level in the game do not contain any empty blocks. It was closely followed during the design stage of *inbento* and preserving it for automatically generated levels is also desirable.

1.2 Piece Types and Mechanics

In order to make the game more satisfying in later chapters, *inbento* continuously introduces new piece types (Fig. 3) that affect the gameplay in various ways:

1. **Food Piece** is the most basic piece type. Consists of ingredient blocks that replace any blocks they're placed on top of.
2. **Swap Piece** – by using it the user is able to switch positions of two blocks inside the box. The piece needs to consist of exactly two non-empty blocks, and at least one of them needs to be placed on a non-empty block.
3. **Move Piece** – upon use of it, all affected blocks will be moved by one grid cell in accordance with directions of the arrows visible on each piece block. After a block has been moved it leaves behind an empty space.
4. **Grab Piece** – after placement all affected box blocks are taken out of the container and sent back into the cutting board, creating a new food piece mirroring the shape of the affected box blocks. At least one targeted block must be non-empty for this piece to work.
5. **Copy Piece** consists of two block types – the "source" block and "target" blocks. Upon placement, all target blocks (regardless of whether they are empty or not) are replaced with the source block's type. For this piece to work, the source type must not be empty and the piece itself needs to contain at least one source block and one target block.
6. **Rotation Lock** is a special mechanic that can be used together with all other piece types. When enabled, the affected piece cannot be rotated and has to be placed in the box as-is. Since using this mechanic reduces the number of available states generated from each placement, it was used mainly to lower the complexity of earlier levels in order to maintain a smoother difficulty curve across chapters.

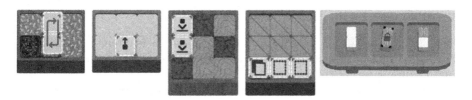

Fig. 3. Various piece types. From the left: swap, move, grab, copy, rotation lock.

2 Related Work

2.1 Procedural Content Generation

With the complexity of games being on the rise since the start of the industry there has also been an increasing demand for titles that provide more content, and thus more playtime for the end user. While companies at the highest tiers – the so-called AAA developers, which is a term referring to games with the largest budgets for creation and marketing [8] – can fill their titles with more things to do by scaling up the workforce, smaller developers often do not have the money or the time that would allow them to catch up to their larger competitors. Thus, the need for being able to generate new content without hand-authoring it was born.

Through the employment of Procedurally Generated Content (PCG) [18], developers can stretch out their game length almost indefinitely. Instead of creating all high-quality content such as levels, art assets or even game mechanics or NPCs behavior by hand, programmers can instead define the boundaries of a system governed by algorithms, which takes in certain pre-programmed inputs in order to create entirely new outputs [15,16].

There are various existing cases of using PCG to create large amounts of in-game content in titles both big and small.

"Rogue" is one of the best-known early examples of using generation to create unique maps on the spot, which resulted in a different playthrough for the player each time they launch the game. This mixture of theoretically infinite levels combined with challenging "permadeath" gameplay was so impactful that it launched an entirely new subgenre of games called roguelikes [14] featuring titles like "NetHack", "Ancient Domains of Mystery" and "Angband".

Another example of PCG can be found in No Man's Sky. This title is a space-exploration adventure game and one of the most recognizable releases of last few years, mainly due to the fact that the title's simulated universe contains 18 quintillion planets. Hand-crafting this amount of content would usually take an enormous amount of time, which is why the developers of the game have created a system that generates these planets based on an existing 64-bit seed value [11].

2.2 Difficulty Estimation

When planning out player progression in a game, designers often strive for opti-mal results in terms of introducing a new user into the experience and ensuring that they are enjoying themselves during the entire duration of the title. This desire is tightly coupled with the game's difficulty presented through challenges that are meant to counter the player's abilities and knowledge that are growing with each minute spent with the simulation.

The concept of flow state named by Csíkszentmihályi [5] outlines that the optimal state of happiness for a person appears when they are presented with challenges matching their skill level. If the challenge is too low, the user experi-ences apathy or boredom; if it is too high they might feel anxious and worried.

What sits in the middle of these two states is the flow channel. If the player skills are met with the right amount of challenge, they should feel stimulated and engaged by the game (Fig. 4, left) [2].

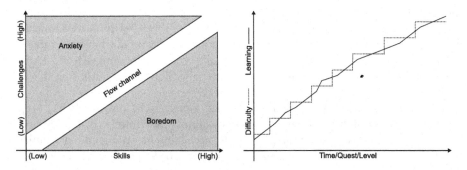

Fig. 4. Left: A visualization of the flow channel [5]; Right: A graph depiction of the difficulty and learning (player skill) curves; the sections when one of the curves rises above the other exemplifies maneuvering within the flow channel [2].

One tool that can help the designers with ensuring that the experience stays within the desired area for as long as possible is the concept of difficulty curves (Fig. 4, right), which serve as a graphical representation of how the game's diffi-culty changes over course of the playthrough [1,6]. Usually these are categorized into two main types: time-based (based on how long the user was playing) and distance-based (how much of the game has been finished by the user). Through mapping out the title's progression in this manner, the creators can aim to deliver an optimal first time user experience (FTUX) [7] – a concept that is especially important for games targeted for casual audience playing on platforms that do not garner long-term attention from users, such as smartphones.

The biggest issue in applying these methods comes down to the fact that often their data is supplied through user focus testing which can be time-consuming and costly. Difficulty curves are usually formulated by the designer and then refined through testing the game's content on potential players [17]. However, there have been multiple papers with different approaches to difficulty estimation that is less reliant on user data; through genetic algorithms [3], constraint satisfaction problem (CSP) solving [10] or calculating common features in games into a single difficulty function [12] in order to lessen the burden on the creators. The last-mentioned method has been adapted in our work to evaluate dfficulty during the procedural level generation.

3   Method

3.1   Procedural Level Generation

PCG is usually based on random numbers. Our initial approach was to randomly select from available piece types and try to generate a solution. The method attempted to insert the piece into the box a specified number of times at various positions and rotations, in order to try and naïvely match it to the grid. If the function did not succeed within this possibility space, it was assumed that the piece could not be properly placed under the existing conditions and discarded. The main advantages of this method are its reliability and speed. Since the result is generated through a simulated act of regular play (placing the pieces in the box one-by-one), the end state is guaranteed to be achievable by the rules of the game which avoids the need to verify whether a solution exists.

Unfortunately, because of the simplicity of this approach the finished levels are usually of a poor quality. Final solutions often contain empty spaces and are not visually pleasing. Furthermore, the possibility space of the level is not fully explored during generation, the algorithm has no way of ascertaining whether the end result can be achieved using a smaller amount of pieces than the entire inventory available to the player. This directly contradicts the rules and warranted the search for an alternative approach.

While the previous algorithm only tested a singular path from the initial state to the solution, the second iteration fully examines the total possibility space from the beginning to all potential end states of the graph tree in order to later be able to remove unwanted solutions.

Given an initial state the algorithm tries to explore it, creating child nodes in the tree and later examining them one-by-one, until all non-discarded states have zero remaining pieces, after which undesired states are removed from the final list.

The function is mapping out the entire tree of states using an approach similar to how a BFS (Breadth First Search) algorithm works for undiscovered graphs [9]. The algorithm keeps track of remaining nodes, removes elements from it one by one and tries to map all of its children that can be explored further, until the results (states without remaining pieces) are achieved. By using a queue, the tree is traversed in linear fashion, level sequence after level sequence until the bottom of the graph is reached.

During exploration the algorithm attempts to place each piece remaining in the state into the box using the placement sets. These sets represent pre-computed arrays of grid cells affected during piece placement (Fig. 5). By preparing this data beforehand for each piece the algorithm is able to perform faster compared to a version where during each iteration for a given piece it would have to search for all valid placements.

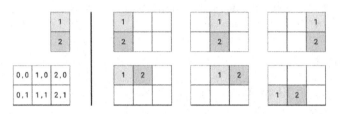

Fig. 5. Left: an example piece, and a 3 × 2 grid with coordinates in each cell; Right: example placement sets for different positions and rotations of the example piece.

If the placement was unsuccessful–for example, if the piece could not be placed in this location due to piece rules detailed in Subsect. 1.2 or because the move did not generate any changes in the box (i.e. replacing blocks with exact same ones) – the state is discarded.

However, if placement was successful the piece is removed from this newly created state and based on the number of remaining pieces it is either added back into the queue (if there are any pieces left) or added to the results (if there are none remaining). If the newly generated result has already been achieved, the duplicate counter of its predecessor is incremented in order to use this information later.

Once the queue is emptied, unwanted duplicates–states that have been found at earlier graph levels–are discarded in order to ensure that the final solution can only be achieved through using up all available pieces.

In order to get the final results a filtering and sorting algorithm has been implemented. This method takes in all gathered results, removes all solutions containing any empty fields, and finally orders the list starting with solution states containing the smallest number of duplicates.

Through testing it has been found that levels with the least amount of alternative paths towards the solution are usually the most interesting. The function also outputs all potential alternatives for the designer to select the desired end state.

The exhaustive nature of this solver-assisted generation method ensures that any final state can only be found at the requested depth and is able to filter the selection of results through the lens of solution uniqueness.

However, all of that comes at the expense of speed. The computational requirements of this BFS-inspired algorithm increase exponentially based on initial settings. Simple 2 × 2 levels with up to 4 pieces are usually generated in less than a second, but raising the amount of pieces above that number brings in a roughly tenfold increase in waiting times depending on pieces that are initially generated.

As the tool is not meant for consumer use and was designed mostly to assist game developers, it was assumed that this was an acceptable compromise in exchange for the improved end results.

3.2   Difficulty Level Estimation

In this work, a combination of all categories of difficulty measures [12] has been adapted through metrics which have been outlined:

1. **Search depth** v_{sd} is a metric of deductive iterations required to solve a puzzle [4], and was calulated from the maximum number of states s that can be generated for given puzzle:

$$v_{sd} = \min(1, \frac{\log_{100}(s)}{5}) \qquad (1)$$

2. **Palette** v_{pa} is determined by the number of different food types t available in the level:

$$v_{pa} = \min(1, \frac{\exp(t)}{50}) \qquad (2)$$

3. **Piece types** v_{pt} is based on the variety of different u mechanics present in the puzzle:

$$v_{pt} = \min(1, \frac{u}{4}) \qquad (3)$$

4. **Extra surface** v_{es} compares the total surface area a_p of available pieces (the number of non-empty blocks) with the level area a_l in order to determine how many supplementary tiles are contained in the user's inventory:

$$v_{es} = \min(1, \log_{64}(\max(a_p - a_l))) \qquad (4)$$

5. **Duplicate solutions** calculated using the BFS-inspired solver; first, the tree of possible results is fully explored in order to determine the number of paths (duplicates) d leading to the desired solution; afterwards, this value is compared with the total number of states s in order to calculate a perceived difficulty:

$$v_{ds} = (1 - \log_s(d)) \cdot \min(1, \frac{s}{1000}) \qquad (5)$$

The calculated values are later combined using a weighted linear function (Eq. 6) where each value is treated independently in order to achieve a result that is simple to understand.

$$\text{difficulty}(L) = \sum_{i \in \{sd,pa,pt,es,ds\}} (w_i \cdot v_i(L)) + w_0 \qquad (6)$$

In order to ascertain the difficulty grade for a level L in *inbento*, a number of variables V_i with values ranging from 0 to 1 have been multiplied by their respective weight values w_i, and summed together with a base weight w_0.

There are games and factors where a different relationship can be considered – quadratic or even exponential. However, in the case of *inbento*, after a few attempts, it turned out that the linear relationships are both simple and sufficient (Sect. 4). Moreover, the methods of calculating the coefficients v_i themselves contain non-linearities resulting from their characteristics.

The weights of this function have been selected in the process of optimizing the function through a set of levels hand-crafted by the game's design team. As these puzzles have been hand-crafted and tested in the months leading up to the game's release and iterated post-launch in order to bring the players up to speed with *inbento*'s mechanics, it is believed that this set serves as a decent example for the algorithm.

4   Results

In this section, an overview of the results of the method will be presented, starting with grading existing, hand-crafted levels to measure how close the results were to the intent of the original designer. This will be followed by grading of puzzles created by the generator in order to analyze the quality of the final solution.

First, the difficulty grades for man-made existing levels from Chaps. 1 through 4 of the game (36 levels in total) were measured.

The guiding principle behind the design of these levels was to present the player with an optimal FTUX [7] that would ease them into the game and explain mechanics without overly large increases in difficulty.

Each chapter in *inbento* contains 9 levels, only 7 of which need to be completed to be able to proceed into the next set. Furthermore, a chapter can also contain multiple "tutorial" levels that purposefully lower their difficulty in order to focus on explaining a new game system.

Table 1 shows that the difficulty grade does indeed gradually rise between each chapter, showing large dips for tutorial levels that are deliberately simpler in order to teach users new concepts. One outlier in that trend is the existence of two levels: 2-1 and 2-2, which have been specifically designed as more complex in order to highlight the difference between regular blocks, and ones with the rotation lock mechanic.

The method showed promising results when applying the algorithm to puzzles created by a human designer. This part focuses on applying it to a collection of levels generated using the second (BFS-based) algorithm.

To validate the solution a sample of initial generator settings has been selected. For each collection of settings, 20 complete levels (meaning that the level can be solved in at least one way) have been generated and the resulting grades have been presented in Table 2. The final grade is a median of all resulting scores and has been presented along with a median generation time for the chosen settings.

Table 1. Difficulty grading for Chapters 1–4. The charts are showing final weighted difficulty scores for each level. Green color indicates tutorial levels which in most cases are accompanied by a drastic drop in difficulty.

Level index	1-1	1-2	1-3	1-4	1-5	1-6	1-7	1-8	1-9
Tutorial stage	yes	-	-	yes	-	yes	-	-	-
Search depth score	0	0.06	0.12	0	0.12	0.09	0.09	0.18	0.27
Palette score	0.02	0.05	0.05	0.02	0.05	0.05	0.05	0.05	0.02
Piece types score	0	0	0	0	0	0	0	0	0
Extra surface score	0	0	0	0	0	0	0	0	0.26
Solution score	0	0.01	0.01	0	0.01	0.01	0.01	0.08	0.47
Final score	0.02	0.12	0.18	0.02	0.18	0.15	0.15	0.31	1.04
Final weighted score	0.03	0.35	0.64	0.03	0.64	0.49	0.49	1.00	2.22

Level index	2-1	2-2	2-3	2-4	2-5	2-6	2-7	2-8	2-9
Tutorial stage	yes	yes	-	yes	yes	-	-	-	-
Search depth score	0.21	0.21	0.18	0.06	0.12	0.09	0.27	0.3	0.46
Palette score	0.05	0.05	0.15	0.05	0.15	0.15	0.15	0.15	0.15
Piece types score	0	0	0	0	0	0	0	0	0
Extra surface score	0.26	0.26	0.17	0	0.17	0.33	0.17	0.26	0.17
Solution score	0.37	0.27	0.25	0.01	0.03	0.04	0.77	0.69	0.62
Final score	0.9	0.8	0.75	0.12	0.46	0.61	1.35	1.04	1.39
Final weighted score	1.85	1.73	1.56	0.35	0.99	1.09	2.62	2.08	3.33

Level index	3-1	3-2	3-3	3-4	3-5	3-6	3-7	3-8	3-9
Tutorial stage	yes	-	-	-	-	-	-	-	-
Search depth score	0.03	0.11	0.23	0.29	0.29	0.28	0.3	0.34	0.49
Palette score	0.05	0.05	0.05	0.15	0.15	0.15	0.15	0.4	0.4
Piece types score	0	0	0	0.25	0.25	0.25	0.25	0	0
Extra surface score	0	0	0	0.17	0	0.26	0	0	0
Solution score	0	0.01	0.05	0.46	0.7	0.69	0.82	0.41	0.56
Final score	0.08	0.17	0.33	1.31	1.39	1.63	1.52	1.15	1.45
Final weighted score	0.2	0.6	1.18	2.71	2.81	3.09	3.01	2.59	3.05

Level index	4-1	4-2	4-3	4-4	4-5	4-6	4-7	4-8	4-9
Tutorial stage	yes	-	-	-	-	yes	-	-	-
Search depth score	0.22	0.42	0.47	0.59	0.62	0.38	0.51	0.51	0.59
Palette score	0.05	0.15	0.15	0.4	0.15	0.15	0.15	0.4	0.15
Piece types score	0.25	0.25	0.25	0.25	0.25	0.25	0.25	0.25	0.25
Extra surface score	0	0	0	0	0	0	0	0	0
Solution score	0.04	0.78	0.95	0.82	0.76	0.8	0.87	0.92	0.91
Final score	0.56	1.06	1.81	2.06	1.78	1.58	1.78	2.08	1.09
Final weighted score	1.05	3.51	3.94	4.67	4.43	3.37	4.04	4.43	4.45

Table 2. Median grades and generation times for each collection of generator settings, averaged from 20 solvable, generated levels for each set.

Level size	2×1	2×1	2×2	2×2	2×2	3×2	3×2	3×2
Case	1	2	1	2	3	1	2	3
Food type count	2	2	2	2	4	2	2	2
Piece count	2	2	2	2	3	3	3	4
Piece type count	1	3	1	3	1	1	3	1
Median grade	0.42	0.82	0.38	0.96	1.83	0.97	2.51	2.05
Med. generation time [in ms]	0.02	0.03	0.14	0.22	2.47	4.43	7.37	168.07

Level size	3×2	3×2	3×2	3×3	3×3	3×3	3×3
Case	4	5	6	1	2	3	4
Food type count	4	4	3	2	2	4	4
Piece count	3	4	5	3	3	3	4
Piece type count	1	1	3	1	3	1	3
Median grade	2.63	3.64	4.69	2.59	3.42	3.34	4.67
Med. generation time [in ms]	6.99	240.98	16674.81	30.19	100.69	61.3	3169.38

5 Discussion

The results show that as has been anticipated, increasing level complexity through modifying the settings results in a rise in perceived puzzle difficulty. Different settings contribute to the final results in various ways. For example, while 2 × 1 and 2 × 2 levels have very similar perceived difficulty levels for basic settings (two types of food, two food pieces each), using a different set of values for the same sizes results in a much more noticeable difference that only grows as the scale of the puzzle grows (e.g. cases 1 and 2 for 3 × 2 level).

The largest jumps in difficulty grade usually occur with increasing the number of available pieces – as each additional option available to the player contributes greatly to the search depth measure, levels with more pieces usually score much higher compared to others. This can be seen in the comparison of cases 1 and 3 for level 3 × 2.

An interesting observation in the analysis of the 3 × 3 level is a similar jump in difficulty grade when comparing case 1 to case 2, to case 3 and to case 4, respectively. In each of these cases, only one of the three initial parameters was increased.

In turn, comparing the differences between cases 1 and 2 and the differences between cases 1 and 3 for 3 × 3 levels shows that it is possible to compensate for a change in one of the initial parameters by changing the other so that the final grade remains approximately the same. This allows the designer to experiment with the level parameters despite the need to set it at specific (increasing) levels.

Table 2 also includes median generation times for the levels with given parameters. It is obvious that the piece count parameter has a very large influence on the generation time. Unfortunately, in the case of using an algorithm that searches the entire solution space, which gives accurate results, it is inevitable. There are two possible solutions to this problem if it were to become significant.

Fig. 6. Example levels generated for each collection of settings from Table 2.

Heuristic approach, available as an option or technical based on parallel or GPU assisted processing.

Examples of the levels from each generator settings have been shown in Fig. 6.

6   Conclusions and Future Work

This work has proposed a complete solution for generating new puzzles for *inbento* and grading their perceived difficulty levels. The final result allows the designer to generate levels for the game using an easily modifiable set of parameters and receive information about the difficulty grade for that level.

Because the rules of the game and selected difficulty variables require full exploration of the possibility space of a given level, the algorithm's computation time increases exponentially as puzzle complexity goes up. Further work could focus on increasing the performance of this solution through data layout optimization and parallelizing the solver in order to cut down on the computation time. A best-case result here would allow for this solution to be used in the actual game as a separate mode that would allow players themselves the access to new, well-designed stages within a fraction of a second.

The future work could also include a comparison of the applied method with alternative methods for procedural generation. Modern methods using machine learning, for example, in LSTM or GAN networks, however, require much more expenditure on manual making of levels to prepare training and test datasets.

While the selection of variables constituting the final equation for weighted difficulty scoring looks satisfactory, there is space for deeper exploration of less obvious metrics such as the different effects of combining multiple mechanics in a single level and the relationship between the initial and final level states.

Another main conclusion of this study that could be learned from the process shown is to highlight places where the processing time is not crucial. While the runtime parts of game engines are thoroughly optimized, the tool side, which supports the designer's work, should rather focus on user convenience and maximum adjustment of the output data. PCG tools can take on as many operations as possible so that the runtime part should perform as few tasks as possible in order to run with the highest possible efficiency.

Acknowledgment. This work was supported by The National Centre for Research and Development within the project "From Robots to Humans: Innovative affective AI system for FPS and TPS games with dynamically regulated psychological aspects of human behaviour" (POIR.01.02.00-00-0133/16). We thank Mateusz Makowiec, Marcin Daszuta, and Filip Wróbel for assistance with methodology and comments that greatly improved the manuscript.

References

1. Andrzejczak, J., Osowicz, M., Szrajber, R.: Impression curve as a new tool in the study of visual diversity of computer game levels for individual phases of the design process. In: Krzhizhanovskaya, V.V., et al. (eds.) ICCS 2020. LNCS, vol. 12141, pp. 524–537. Springer, Cham (2020). https://doi.org/10.1007/978-3-030-50426-7_39
2. Aponte, M.V., Levieux, G., Natkin, S.: Measuring the level of difficulty in single player video games. Entertainment Computing 2(4), 205–213 (2011). https://doi.org/10.1016/j.entcom.2011.04.001
3. Ashlock, D., Schonfeld, J.: Evolution for automatic assessment of the difficulty of sokoban boards. In: IEEE Congress on Evolutionary Computation. IEEE (2010). https://doi.org/10.1109/cec.2010.5586239
4. Browne, C.: Metrics for better puzzles. In: Seif El-Nasr M., Drachen A., Canossa A. (eds.) Game Analytics, pp. 769–800. Springer, London (2013). https://doi.org/10.1007/978-1-4471-4769-5_34
5. Csikszentmihalyi, M.: Flow: the psychology of optimal experience. Harper Row 45(1), 142–143 (1990). https://doi.org/10.1176/appi.psychotherapy.1991.45.1.142

6. Diaz-Furlong, H.A., Solis-Gonzalez, C.A.L.: An approach to level design using procedural content generation and difficulty curves. In: 2013 IEEE Conference on Computational Intelligence in Games (CIG). IEEE (2013)

7. Feng, L., Wei, W.: An empirical study on user experience evaluation and identification of critical UX issues. Sustainability 11(8), 2432 (2019). https://doi.org/10.3390/su11082432

8. Hillman, S., Stach, T., Procyk, J., Zammitto, V.: Diary methods in AAA games user research. In: Proceedings of the 2016 CHI Conference Extended Abstracts on Human Factors in Computing Systems. ACM (2016). https://doi.org/10.1145/2851581.2892316

9. Holdsworth, J.J.: The nature of breadth-first search. School of Computer Science, Mathematics and Physics, James Cook University, Tech. rep. (1999)

10. Jefferson, C., Moncur, W., Petrie, K.E.: Combination. In: Proceedings of the 2011 ACM Symposium on Applied Computing - SAC 2011. ACM Press (2011). https://doi.org/10.1145/1982185.1982383

11. Kaplan, H.L.: Effective random seeding of random number generators. Behav. Res. Methods Instrument. 13(2), 283–289 (1981). https://doi.org/10.3758/bf03207952

12. van Kreveld, M., Loffler, M., Mutser, P.: Automated puzzle difficulty estimation. In: 2015 IEEE Conference on Computational Intelligence and Games (CIG). IEEE (2015). https://doi.org/10.1109/cig.2015.7317913

13. Nishimoto, H., Hamada, A., Takai, Y., Goto, A.: Investigation of decision process for purchasing foodstuff in the "bento" lunch box. Procedia Manuf. 3, 472–479 (2015). https://doi.org/10.1016/j.promfg.2015.07.210

14. Parker, R.: The culture of permadeath: roguelikes and terror management theory. J. Gaming Virtual World 9(2), 123–141 (2017). https://doi.org/10.1386/jgvw.9.2.123_1

15. Rogalski, J., Szajerman, D.: A memory model for emotional decision-making agent in a game. J. Appl. Comput. Sci. 26(2), 161–186 (2018)

16. Sampaio, P., Baffa, A., Feijo, B., Lana, M.: A fast approach for automatic generation of populated maps with seed and difficulty control. In: 2017 16th Brazilian Symposium on Computer Games and Digital Entertainment (SBGames). IEEE (2017). https://doi.org/10.1109/sbgames.2017.00010

17. Sarkar, A., Cooper, S.: Transforming game difficulty curves using function composition. In: Proceedings of the 2019 CHI Conference on Human Factors in Computing Systems. ACM (2019). https://doi.org/10.1145/3290605.3300781

18. Togelius, J., Kastbjerg, E., Schedl, D., Yannakakis, G.N.: What is procedural content generation? In: Proceedings of the 2nd International Workshop on Procedural Content Generation in Games - PCGames 2011. ACM Press (2011). https://doi.org/10.1145/2000919.2000922

ELSA: Euler-Lagrange Skeletal Animations - Novel and Fast Motion Model Applicable to VR/AR Devices

Kamil Wereszczyński[1]([✉])[iD], Agnieszka Michalczuk[1][iD], Paweł Foszner[1][iD], Dominik Golba[2], Michał Cogiel[2], and Michał Staniszewski[1][iD]

[1] Department of Computer Graphics, Vision and Digital Systems,
Faculty of Automatic Control, Electronics and Computer Science,
Silesian University of Technology, Akademicka 2A, 44-100 Gliwice, Poland
{kamil.wereszczynski,michal.staniszewski}@polsl.pl
[2] KP Labs, Gliwice, Poland

Abstract. Euler Lagrange Skeletal Animation (ELSA) is the novel and fast model for skeletal animation, based on the Euler Lagrange equations of motion and configuration and phase space notion. Single joint's animation is an integral curve in the vector field generated by those PDEs. Considering the point in the phase space belonging to the animation at current time, by adding the vector pinned to this point and multiplied by the elapsed time, one can designate the new point in the phase space. It defines the state, especially the position (or rotation) of the joint after this time elapses. Starting at time 0 and repeating this procedure N times, there is obtained the approximation, and if the $N \to \infty$ the integral curve itself. Applying above, to all joint in the skeletal model constitutes ELSA.

Keywords: Skeletal animation · Euler Lagrange equation · Partial differential equation · Computer animation · Key-frame animation

1 Introduction

Automatic animation generation in 3D computer graphics is used in many applications, e.g. Min et al. in [32] for generation of human animation, Spanlang et al. in [38] for mapping the motion obtained from acquisition system for avatar and robots or Li et al. in [28] for skinning the body of the skeletal models. Currently, the process of creating an animation is based primarily on remembering key poses for the skeletal model. The key-poses are made by animator, could be obtained from the motion acquisition systems or (in fact in most general cases) are the mixture of this to technique. This method of storing the skeletal animation applies more general technique of *keyframe animation*, described e.g. by Burtnyk and Wein in [8], Catmul in [10] or Parent in [35] (sec 3.5) It is driven from hand-made animation (e.g. Sito and Whitaker in book "Timing

© Springer Nature Switzerland AG 2021
M. Paszynski et al. (Eds.): ICCS 2021, LNCS 12746, pp. 120–133, 2021.
https://doi.org/10.1007/978-3-030-77977-1_10

for animation" [37]). The remaining poses are joined with different methods of interpolation e.g. Ali Khan and Sarfraz in [4] or Mukundan in [33]. Animations can be generated manually by appropriate software or by external tools such as Motion Capture [5], IMU [25] or markerless systems [43]. In any case, the volume of data that must be generated and saved in order to play the animation is a big limitation. There is also the problem of repeating animations and combining them. It is not always possible to combine animations at any point in time and it is necessary to wait until a previous animation ends. Therefore, the presented work proposes a new approach for animation generation, which is based on the connection of the articulated joints with the partial differential equations (PDE) Euler-Lagrange in the form of the ELSA algorithm: Euler-Lagrange Skeletal Animations. Potential applications in computer graphics, virtual (VR) and augmented reality (AR) systems were also indicated.

1.1   State of the Art

One of the possible applications of animation bases on human motion. Lobão et al. in [29] define two types of animations in that context: keyframed animation and skeletal animation (ibidem, pg. 299). The first name can be confusing, because, as they themselves wrote (ibidem, pg. 301), skeletal animation is also based on the key frames. The main difference lays not in the technique of the animation but the object that is animated: in the first case whole mesh is stored in the key frames, while in the second one - only articulated model, called *skeleton*, in which the whole process is based on a skeleton consisting of a combination of rigid bones and joints that connects those ones. Therefore we will name those techniques relatively: *mesh animation* and *skeletal animation*.

Currently the skeletal animation is applied in 3D computer graphics to animate articulated characters and enables to transfer the pose of a virtual skeleton into the surface mesh of a model. The skeletal animation technique can be divided into four parts: rigging, weighting, pose selection and skinning [3,6]. An extension of the classical skeletal animation pipeline was introduced by Sujar et al. in [41], which relies on dealing also with the internal tissue of a model and allows to adapt a virtual anatomy of a patient to any desired pose by application of the patient's bone and their skins. The animation of complex data can be based on low-dimensional position systems. A popular way of reaching that goal relies on recent marker-based animation methods described by Krayevoy and Sheffer in [23] or Stoll et al. in [40] and on concept of shape deformation methods, in which markers are used as control points in the character's mesh animation e.g. in works of: Zollhöfer et al. [47], Zhao and Liu [45] or Levi and Gotsman [27]. The whole animation pipeline basing on low number of positional constraints was presented by Le Naour et al. in [26] in the course of potential loss of precision and position information. The problem was solved by application of efficient deformation algorithm and an iterative optimization method.

Another approach dealing with skeletal animation implies 4-point subdivision to fulfill the whole animation frames [46], where instead of classical interpolation methods, an adaptive and high-performance algorithm of additional subdivision was used to create in-between frames in the skeletal animation. In the animation of motion, the cubic Cardinal spline can be used [21] in order to interpolate for each segment piece-wise cubicals with given endpoint tangents. Automatic skinning and animation of skeletal models can be achieved by not only dividing a character models into segments but also by subdividing each segment into several chunks [28] and introducing energy terms and deformation techniques. Correctly generated animation can be also applied for avatars and robots. In some examples, the robot's animation can be generated and modelled in a 3D program and dedicated software is prepared for translating 3D model to the physical robot [17]. For the purpose of the real-time whole-body motion of avatars and robots the issue of remapping the tracked motion was presented [38]. In the following approach, the tracked motion of the avatar or robot was combined with motions visually close to the tracked object.

Alternatively, animations can be generated on the base of mechanical rules. One of possible solution was presented in [22] relying on [19] by application of the diagonalized Lagrangian equations of motion for multi-body open kinematic chains with assumed N degrees of freedom [18]. The given methods focused on the introduced equations of motion, required diagnosing transformations and on the physical interpretation. In this case, diagonalization means that at each fixed point in time, the equation at each joint is out from all other joint equations. The *Partial Differential Equation* PDE (e.g. Evans in [13]) can be used for modelling the physics's phenomena within the motion models as well (e.g. Courant and Hilbert in [11]). This notion was utilized by Castro et al. in [9] for the cyclic animations by joining of mechanical description by PDEs and skeletal system. In order to define the boundary conditions, it is required to create the surface of a model from a set of predefined curves enabling solution of a number of PDEs. The first of presented approaches attached the set of curves into a skeletal model for the purpose of the animation holding along cyclic motion by use of a set of mathematical expressions. The second proposition implies the mathematically manipulated spine connected with the analytic solution of the PDEs treated as a driving mechanism in order to obtain a cyclic animation. The generated animation can be used in two different ways, as a surface structure of a model based on the PDEs or as an original model by means of a point to point map.

One of the widespread standard for exchanging the skeletal animations is *Biovision Hierarchy* - BVH (e.g. Dai et al. in [12]), which contains the bones' offset and the hierarchy of the model in the *header* and the Cardan angles (e.g. Tupling and Pierrynowski in [42]) for each joint in the *data description* section. The alternative are binary files: C3D (documentation in [1]) used e.g. by Madery et al. in [30] for creation of theirs kit database of human motion recorded on Kinect; or the *Acclaim Motion Capture (AMC)* described by Geroch in [14]. The number of bones in the skeletal model could be different depending on the demands, e.g. Mueller et al. in [34] uses 24 joints while in documentation of AMC viewer [2] there is 29 joints used. Nevertheless it can be seen that there is a huge number of values needed to be stored for even 30 s length animation

100 Hz key-frames frequency. For the model with 24 it is required $216k$ and for the one with 29 - $261k$ of values to be stored.

Against the above, in production solutions there are the same animation used for a number of characters. Therefore there appears the repeatability of the motion of different characters on the scene, which does not correspond to the reality. Exemplary, Mills in [31] shows experimentally that the R^2 measure of repeatability between two passes of walking is equal barely $R^2 = 0.531$ for knee abduction or $R^2 = 0.985$ for right knee flexion/extension and the mean for all human joint is $R^2 = 0.7797 \pm 0.153$. The confirmation of this fact can be found e.g. by White et al. in [44] or by Steinwender et al. in [39] in case of children. For the joints in the machines (e.g. industrial robot arm) the repeatability is on the higher level, but it is quite noticeable - exemplary Sepahi-Boroujeni in [36] shows that R^2 lays in the range of $[0.82, 0.84]$. Therefore introducing the repeatability to the animated characters and machines could improve the immersion of the user (player), especially in the context of Augmented and Virtual Reality industrial systems.

Except of that, the computational complexity of the algorithms is important matter, especially in the real-time application. It is unlikely to expect that a frame in a real-time solution with varying FPS, will appear in exactly the same moments as the key-frames, therefore in the vast majority of frames, the actual pose has to be computed with methods that can be theoretically sophisticated and computationally complex (e.g. Haarbach et al. [15] or mentioned Ali-Khan et al. [4]).

1.2   Motivation

The proposed notion ELSA (Euler-Lagrange Skeletal Animation) is a novel and original approach to the process of generate the poses of skeleton in consecutive frames and it is an opportunity to improve topics with data volume, computational complexity and animation repeatability. The algorithm uses a concept of Euler-Lagrange (or Lagrange second kind) PDE and its numerical solving using the notion of *phase space* (see e.g. Hand and Finch [16]) for modelling the motion of the joints in skeletal model. For the machines (like industrial robots) there are 8 types of joints (see Blake in [7], Hoffman et al. in [17]). For the skeletal human's models the most general observation is all joints behave like *ball joints* with 6 DoF *(Degrees of Freedom)* within some limitation - e.g. knee joint's mobility is on the Sagittal plane in general, but still there exists some small deviations in the remaining planes, even in case of healthy person. If one dispose the phase spaces for each of the joint type, the animation could be stored as the start and end point in the phase space and the remaining poses lays on integral curve connecting this two points, so called *phase space trajectory* and can be easily computed using the phase space mechanism, which will be described in the further part of this article.

ELSA also can introduce the limited distortions to the modelled motion. Namely, in a selected few steps it is enough to slightly disturb the acceleration component of vector pinned to the given position - velocity point in a phase space. It changes the rotation of the vector minimally, however the trend of the motion could be maintained. By the virtue of above each transition will be different - even though it is made of the same start and end poses. The distortion operator will behave differently in different kind of phase spaces recognizable in the light of the *Phase portrait* analysis (see Jordan and Smith [20]).

2   Materials and Methods

The skeletal model used in computer graphics and computer vision is a hierarchic tree structure, where the nodes are called *joints* and represents the connection between **rigid** bones, which could be identified as the edges of this tree. The main structure of the skeleton is defined by the starting pose (or T-pose, A-pose). In that pose the position of each joint can be defined as the *offset* from the parent joint, which is the vector in the coordinate system of parent joint. The *pose* is defined as the rotations' set of bones in the coordinate system of the joint that is the bone's begin. Therefore this rotation designates the position of the joint being the end of that bone, hence its position is generated by the rotation in the parent joint's coordinate system. One obtain the position of all joints in the skeleton by carrying such a procedure from the root to the leafs of the tree. The animation arises by computing the poses for each frame of the animation. The motion of the rigid body (in the case: bone), represented abstractly by a single point (in the case: joint), in the perspective of Lagrangian mechanics (see Hand and Finch, ibidem) can be described by the function:

$$q : \mathbb{R} \longrightarrow \overline{Q} = Q \times \dot{Q}, \wedge q(t \in \mathbb{R}) = \left[q_1(t), \ldots, q_n(t), \dot{q}_1(t), \ldots, \dot{q}_n(t)\right]^T, \quad (1)$$

where Q is called the *configuration space* of the joint. The point in the configuration space denotes the state of the joint defined by the quantities which are appropriate for the given type of joint (e.g. position in space or rotation). The point in the subspace $\dot{Q}$ denotes the change of the state in time (e.g. velocity, angular velocity). Thus, the space $\overline{Q} = Q \times \dot{Q}$ represents all possible joint's states and all possible changes for each position, which can be written as a set:

$$\overline{Q} = Q \times \dot{Q} = (X, \dot{X}) = (x_1, \ldots, x_n, \dot{x}_1, \ldots \dot{x}_n) \quad (2)$$

The q function, called *trajectory*, therefore represents a single joint motion, since it can be used as the representation of that joint's animation.

The Lagrange equation of that system has the general form as follows:

$$\mathcal{L} : Q \times \dot{Q} \times \mathbb{R} \longrightarrow \mathbb{R}, \wedge \mathcal{L}\left(q_1(t), \ldots, q_n(t), \dot{q}_1(t), \ldots, \dot{q}_n(t), t\right) = E_k - E_p, \quad (3)$$

where E_k is kinetic energy and E_p potential energy. In this work, it was assumed that the joints has unit mass, nevertheless it is possible to introduce the other masses of the arms tied to joints. Those can be computed basing on the material and volume fetched from the parameters of the mesh saved in the CAD file. The kinetic energy of the joint obviously depends on the position and velocity of the bone, while the potential energy depends on the applied force. In this paper, no gravity was assumed. Two types of force are taken into consideration: internal, resulting from the force moving the bone, and external, applied by the next element in the kinematic chain, or by the user interacting with the device.

The Euler - Lagrange equation takes the form of a system of partial differential equations:

$$\frac{d}{dt}\left(\frac{\partial \mathcal{L}}{\partial \dot{q}_1}\right) - \frac{\partial \mathcal{L}}{\partial q_1} = 0 \wedge \cdots \wedge \frac{d}{dt}\left(\frac{\partial \mathcal{L}}{\partial \dot{q}_n}\right) - \frac{\partial \mathcal{L}}{\partial q_n} = 0 \tag{4}$$

Consider the first term of each of these equations: $\frac{d}{dt}\left(\frac{\partial \mathcal{L}}{\partial \dot{q}_i}\right)$. $\dot{q}_i := \dot{q}_i(t) = \dot{x}_i$ is $(n+i)th$ argument of the Lagrangian. It computes the time derivative of this argument. This produces the second derivative $\ddot{x}_i$. In specific cases, this derivative can be equal to 0, but in general it can be assumed that, as a rule, each of the above equations can be transformed in such a way that there will be a second derivative on the left side of equation and the function φ_i on the right side, depending only on the arguments of the state space:

$$\ddot{x}_1 = \varphi_1(x_1,\ldots,x_n,\dot{x}_1,\ldots,\dot{x}_n) \wedge \cdots \wedge \ddot{x}_n = \varphi_n(x_1,\ldots,x_n,\dot{x}_1,\ldots,\dot{x}_n) \tag{5}$$

It can be written in a vector form:

$$\ddot{X} = \Phi(X,\dot{X}), \ X \in Q, \ \wedge \ \Phi(X,\dot{X}) = \left[\varphi_1(X,\dot{X}),\ldots\varphi_n(X,\dot{X})\right]^T \tag{6}$$

Thus a vector field can be created:

$$\mathbb{V}: Q \times \dot{Q} \longrightarrow \dot{Q} \times \ddot{Q}: \ \mathbb{V}(X,\dot{X}) = [\dot{X},\ddot{X}] = [\dot{X},\Phi(X,\dot{X})]. \tag{7}$$

Such a vector field is called *phase space* of the partial differential equation (PDE) from which it was defined, so it will be called *EL animation phase space*. Such fields have the property that the *Integral curve* of this field, starting at point $\overline{X} = [X,\dot{X}]$, determines the trajectory of the motion of an object whose initial state is equal to $\overline{X}$. While it denotes both the position in space and the current velocity, having such a vector field, it is easy to determine the further movement, i.e. the position in the next frame, according to the formula:

$$\overline{X} = x_1,\ldots,x_n,\dot{x}_1,\ldots\dot{x}_n, \ \overline{Y} = y_1,\ldots,y_n,\dot{y}_1,\ldots\dot{y}_n, \ \overline{Y} = \overline{X} + t\mathbb{V}(\overline{X}). \tag{8}$$

Thus, having the vector field $\mathbb{V}$ determined, computing the animation is very efficient, because it bases on multiplying the vector by the scalar and adding the vectors. Moreover, to remember the entire animation, regardless of its complexity, it is enough to remember the starting point and possibly the end point in the vector field. Summarizing this section, one can define:

Definition 1. *Euler-Lagrange animation, or in short* **EL animation** *of a rigid body, it is an integral curve of a vector field* $\mathbb{V}(X,\dot{X}) = [\dot{X},\Phi(X,\dot{X})]$ *starting at the* **animation start** *point.* **The end of the animation** *will be named as the point on the integral curve after which the animation will no longer take place.*

Starting from any point in the phase space, it is possible to determine the point at which the animated object will be located after the specified time, by formula 8. Applying it iterative, one can obtain a set of points representing the state of the animated object in successive moments of time. In the case of interval between frames is small enough ($\Delta t \rightarrow 0$), an integral curve in the phase space is created, which is called the *phase curve*. The projection of this curve onto the configuration space is the object's motion trajectory.

3   Results and Discussion

In this section we present the theoretical results in the form of the formal definition the Euler Lagrange Skeletal Animation (ELSA). Furthermore we present the implementation of the ELSA made in the Unity 3D environment. Finally we present some visual results on an example of two 3D models: parking gate and the industrial drill.

3.1   Euler-Lagrange Skeletal Animation Concept

The notion of skeletal model could be defined using the formal notion of the tree, proposed by Kuboyama in its doctoral dissertation [24]. On the other hand, the skeletal model can be mapped from a set of its joints to the tangent bundle of the phase spaces. This additional structure provides the skeletal model with a tool for animation construction.

Definition 2. *Euler-Lagrange Skeleton Model (ELSM) R is the system $R = (P, S, \varphi)$, where:*

1. *P is a set of joints: $P = \{p_1, ..., p_m, \forall i \ p_i \in \mathbb{R}^3\}$. The number m is a **size** of tree and is denoted by a symbol $|R|$.*
2. *S is a tree:*

$$S = (P, r), where: \quad r : P \rightarrow P, \quad r^0(p_k) = p_k, \quad \forall k \geq 0 \ \exists \delta : \ r^\delta(p_k) = p_1 \quad (9)$$

   *The p_k joint is a **child joint** of the joint p_i if and only if $r(p_k) = p_i$, which can be written equivalently: $p_i \rightarrow p_k$. The p_k joint is a **parent** of the joint p_i. The p_1 joint is the root. The natural number δ is called the joint depth.*
3. *φ is a mapping from a set of joints into a tangent bundle of the phase space:*

$$\varphi : P \rightarrow T[Q \times \dot{Q}] : \quad \varphi(p \in P) = \mathbb{V}_p(X_p, \dot{X}_p) = [\dot{X}_p, \ddot{X}_p]. \quad (10)$$

Given the point before a certain time interval Δt elapses from now (t), according to the Eq. 8 one can compute the point in the space $\varphi(p_k)$ in time t

$$\overline{X_p}(t) = \overline{X_p}(t - \Delta t) + \Delta t \varphi(p) \quad (11)$$

In some cases, e.g. in case of prismatic joint (see [7]), the X_p represents the position of joint directly. In other cases (e.g. ball joint like in case of human joints), the X_p represents the (Cardan) angles. In those cases the position of the joint $p \in P$ in time t can be computed as follows:

$$p(t) = p_{r(p)}(t) + \Pi(X_{r(p)}(t), p), \qquad (12)$$

where $\Pi(X, p)$ determines the vector which direction is given by rotation around the point $\mathbf{0}$ by an (Cardan) angle X and has the same magnitude as p.

Indeed, the formula above constitutes the *kinematic chain* for the joint p which, in each step of the recurrence, computes the rotation of the given point by the given (Cardan) angles around the parent's joint's coordinate system. It can be expressed without recurrence as follows:

$$p(t) = r^\delta(p) + \sum_{i=1}^{\delta} \Pi\left(X_{r^i(p)}, r^{i-1}(p)\right). \qquad (13)$$

Moreover, the set $P(t) = \{p_i(t) : \forall i \ p_i \in P\}$ determines the *pose* of the model in time t and $P(0) = P$ is the initial pose (A-Pose, T-Pose) that can be obtained from the BVH file. Hereby the *pose-driven initial configuration* can be constructed:

$$I_{PD} = \{\overline{X}_{p_1}(0), \ldots \overline{X}_{p_m}(0) : \forall k \ \overline{X}_{p_k} = [\mathbf{0}, \dot{X}_{p_k}(0)]\},$$

where one consider that all joints are not rotated in the initial pose and the initial distribution of joints in the 3D space is given by the initial Pose. Alternatively, one can define the initial position of the joint directly by the *explicit initial configuration*:

$$I_{EX} = \{\overline{X}_{p_1}(0), \ldots, \overline{X}_{p_m}(0)\},$$

from which the initial pose is obtained.

Definition 3. *Euler-Lagrange Skeletal Animation (ELSA)*, *embedded on given ELSM model* (P, S, φ), *started from the initial configuration* $IC = \{\overline{X}_{p_i} : \forall i \ p_i \in P\}$ *is the set of integral curves* $\Gamma = \{\gamma_i(t) : \forall i \ \gamma_i(0) = \overline{X}_{p_i}(0) \land t \in [0, L]\}$, *where* L *is the **length** of the animation. The length could be designated by the time the animation lasts or by the **ending pose** $P(L)$, which if approached, ends the animation.*

Having the pose in given time $P(t)$ one can construct the pose after a sufficient short time period elapses: $P(t + \Delta t)$ applying the Eq. 11 to each joint separately. In case joint's phase space describes rotation and angular velocities the Eqs. 12 or 13 should be used, while in opposite case the position of joint $p_k = X_k$ directly. To make each animation's performance unique the *distortion operator* acting on the function φ is introduced, defined as follows:

$$\mathbf{D}_\mu \varphi = [\dot{X}, \mu(\ddot{X})], s.t. : \forall \overline{X} \in U \subseteq Q : \lim_{t \to \infty} \gamma_{\mathbf{D}_\mu \varphi}(\overline{X}, t) = \gamma_\varphi(\overline{X}, t), \lor \qquad (14)$$

$$\forall \overline{X} \in U \subseteq Q : \exists t \leq L \ \forall T > t \ \gamma_{\mathbf{D}_\mu \varphi}(\overline{X}, T) = \gamma_\varphi(\overline{X}, T)$$

where $\gamma_\varphi(\overline{X}, t)$ is the integral curve on the field defined by φ starting on the point $\overline{X}$. The above definition characterize the distortion model applied to the skeletal model. Firstly operator acts on the acceleration (second derivative) only to ensure the consistency with the phase space definition, where the first part of the vector in the field has to be equal to the first derivative coordinate of the point it is pinned to. Secondly, the conditions ensure that the animation cannot "jump" to another, not convergent, phase curve for ever. In fact the first condition covers the area that are under the influence of approximately stable point or curve in the phase portrait, while the second one - the cases of stable (but not approximately) curves, unstable points and curves. The distortion operator does not have to act in every frame of the animation. Its acting throw the animation to other trajectory (e.g. randomly). In case it comes back to the first (2^0) or will be, after some time, sufficiently close to the original one (1^0), the operator could be applied again.

3.2 ELSA Implementation

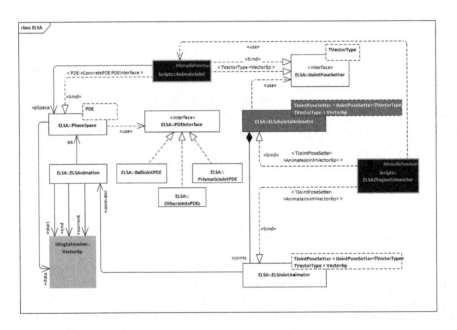

Fig. 1. UML diagram of the ELSA implementation.

ELSA was implemented in C# language as the set of scripts for the 3D graphics engine Unity3D using object oriented and generic programming method (Fig. 1). The instance of template class *ELSA:PhaseSpace* within the parameter *PDE* constrained to a class implementing the interface *PDEInterface* returns the vector $[\dot{X}, \ddot{X}]^T$ for the given argument $[X, \dot{X}]^T$ according to the concrete realization of that interface. Hence one can say, that the template declaration of this class creates the abstract structure like tangent bundle from the Definition 2 (pt. 3). Moreover, the declaration of the variable with the concrete PDE as the template parameter plays the role of the function φ from this definition. Such an implementation is not only consistent with the presented definition, but it realizes as well, the generic version of the strategy GoF design pattern "Strategy" which allows for flexible generation of the desired PDE for not implemented cases of joints yet. Within the presented implementation there are PDEs for each of 8 commonly known joint types delivered. On the diagram on mentioned figure, there are shown only two exemplary concrete PDEs: *BallJointPDE* and *PrismaticJointPDE*. The PhaseSpace has constructor hidden, hence it is created by static *Generator*. It is the first of two connectors between the ELSA module and the visualization part of an application.

Subsequently, a class *ELSAnimation* represents a single ELS animation defined by: (1) *PhaseSpace*, (2) Initial pose of the joint given by *Vector6p* and (3) the end given by the final pose. This class is instantiated by the static generator as well. It avails to fetch the next frame of an animation (meaning "after time $\Delta t > 0$" from now), the previous one ($\Delta t < 0$), rewind to initial and final pose and check if the animation is completed (meaning the final pose was obtained).

ELSJointAnimator is a template class which uses the interface *IJointPoseSetter* (which will be described further) for setting the pose of the joint supposed to be animated and existing on the scene being rendered. It uses the object of class *ELSAnimation* for computing the pose in current time and delivers the methods with real rotation and position change of object existing on the scene and for pose update exclusively. In also introduces the option of determination of the animation end by its length or defining the animation as *endless*, which is intended for periodic phase curves. *ELSJointAnimator*'s instances are aggregated through composition by the *ELScheletalAnimator*, which provides the same methods but applied to the whole skeletal model, defined outside the ELSA module.

The interface *IJointPoseSetter* is the second connection between ELSA functionalities and the visualization part of the application, made with Unity Engine. It defines one method *SetPose()*, which allows to set the pose of the object, that can be seen on the 3D scene (*Unity::GameObject*) considered as joint. Marking the *GameObject* as joints is made by connection the script realizing *IJointPoseSetter* as the *Unity component*. On the Fig. 2 there are selected frames fetched from an animation of bottom part of humanoid skeletal model. The animation is applied to two joints. One can make the observation that the mapping φ seems to be connected with some kind of fibration, which demand further theoretical research. If these assumptions prove to be confirmed, it would mean that each pose determines one fiber connected with φ. Hence the whole ELSA, as in the limit is the continuous, would designate the smooth curve on the fibration and finally on the manifold that would be the base of the fibration. The similar animation of the same class, will designates similar curves on that manifold

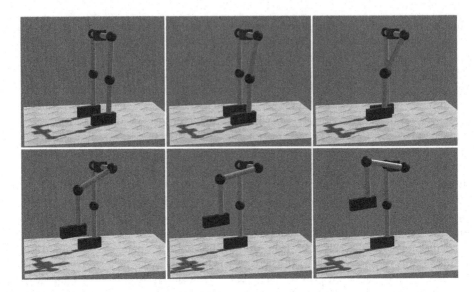

Fig. 2. Exemplary frames from animation of one leg in the bottom part of humanoid skeleton.

probably. The set of those curves could be utilised in many ways, e.g. as the feature vector for the classification.

4   Conclusions and Acknowledgments

In the presented paper, a novel and original approach to generate skeletal animations was proposed. The notion of ELSA, prepared in game engine environment, was applied in the process of generation of skeletal poses in consecutive frames. In contrast to previous works, the presented method bases on the Euler Lagrange equations of motion, configuration and phase space notion, which for our best knowledge was applied for the first time. It enables to define the whole skeletal animation by introducing initial and final poses without any additional operations. Those features highly improve the computational complexity and reduce data volume required to store in a memory. The additional benefit of ELSA approach lies in repeatability of any animation and its unique recreation. The presented work describes future research opportunities, practical applications and is available along with source code. The presented concept of animation generation was originally used in the Platform 4.0. That project was mainly designed to create an interactive training process by application of augmented, mixed and virtual reality. In that problem the input of a system usually consists of just raw 3D model saved in any CAD format. The task of a designer is to correctly prepare model and attach correct animations which is more complicated process. Having properly joint identified, user is able to create any kind of animation.

Presented work is a part of the project "Devising a modern platform for creating interactive training and servicing instructions for machinery and technical devices using AR, VR and MR technologies". Project is co-financed by the European Union through the European Regional Development Fund under the Regional Operational Programme for Slaskie Voivodeship 2014–2020.

References

1. C3D.ORG - The biomechanics standard file format. https://www.c3d.org/
2. CS 15-464/16-464 Technical Animation. http://graphics.cs.cmu.edu/nsp/course/15-464/Fall05/assignments/StartupCodeDescription.html
3. Abu Rumman, N., Fratarcangeli, M.: Position-based skinning for soft articulated characters. Comput. Graph. Forum **34**(6), 240–250 (2015). https://doi.org/10.1111/cgf.12533
4. Ali Khan, M., Sarfraz, M.: Motion tweening for skeletal animation by cardinal spline. In: Abd Manaf, A., Sahibuddin, S., Ahmad, R., Mohd Daud, S., El-Qawasmeh, E. (eds.) Informatics Engineering and Information Science, CCIS, pp. 179–188. Springer, Heidelberg (2011). https://doi.org/10.1007/978-3-642-25483-3-14
5. Asraf, S.M.H., Abdullasim, N., Romli, R.: Hybrid animation: implementation of motion capture. IOP Conf. Ser. Mater. Sci. Eng. **767**, 012065 (2020). https://doi.org/10.1088/1757-899x/767/1/012065
6. Baran, I., Popović, J.: Automatic rigging and animation of 3d characters. ACM Trans. Graph. **26**(3), 72-es (2007). https://doi.org/10.1145/1276377.1276467
7. Blake, A.: Design of mechanical joints. No. 42 in Mechanical engineering, M. Dekker, New York (1985)
8. Burtnyk, N., Wein, M.: Computer-generated key-frame animation. J. SMPTE **80**(3), 149–153 (1971). https://doi.org/10.5594/J07698
9. Castro, G.G., Athanasopoulos, M., Ugail, H.: Cyclic animation using partial differential equations. Vis. Comput. **26**(5), 325–338 (2010). https://doi.org/10.1007/s00371-010-0422-5. Company: Springer Distributor: Springer Institution: Springer Label: Springer Number: 5 Publisher: Springer-Verlag
10. Catmull, E.: The problems of computer-assisted animation **12**(3), 348–353 (1978). http://doi.acm.org/10.1145/800248.807414
11. Courant, R., Hilbert, D.: Methods of Mathematical Physics, vol. 2. Wiley-VCH, Weinheim (1989). oCLC: 609855380
12. Dai, H., Cai, B., Song, J., Zhang, D.: Skeletal animation based on BVH motion data. In: 2010 2nd International Conference on Information Engineering and Computer Science, Wuhan, China, pp. 1–4. IEEE, December 2010. https://doi.org/10.1109/ICIECS.2010.5678292
13. Evans, L.C.: Partial differential equations. No. v. 19 in Graduate studies in mathematics, American Mathematical Society, Providence, R.I (1998)
14. Geroch, M.S.: Motion capture for the rest of us. J. Comput. Sci. Coll. **19**(3), 157–164 (2004)
15. Haarbach, A., Birdal, T., Ilic, S.: Survey of higher order rigid body motion interpolation methods for keyframe animation and continuous-time trajectory estimation. In: 2018 International Conference on 3D Vision (3DV), Verona, pp. 381–389. IEEE, September 2018. https://doi.org/10.1109/3DV.2018.00051

16. Hand, L.N., Finch, J.D.: Analytical Mechanics. Cambridge University Press, Cambridge (1998)
17. Hoffman, G., Ju, W.: Designing robots with movement in mind. J. Hum. Robot Interact. **3**(1), 89 (2014). https://doi.org/10.5898/JHRI.3.1.Hoffman
18. Jain, A., Rodriguez, G.: Diagonalized dynamics of robot manipulators. In: Proceedings of the 1994 IEEE International Conference on Robotics and Automation, San Diego, CA, USA, pp. 334–339. IEEE Computer Society Press (1994). https://doi.org/10.1109/ROBOT.1994.351273
19. Jain, A., Rodriguez, G.: Diagonalized Lagrangian robot dynamics. IEEE Trans. Robot. Autom. **11**(4), 571–584 (1995). https://doi.org/10.1109/70.406941
20. Jordan, D.W., Smith, P.: Nonlinear Ordinary Differential Equations: An Introduction for Scientists and Engineers, 4th edn. Oxford University Press, Oxford (2007)
21. Ali Khan, M., Sarfraz, M.: Motion tweening for skeletal animation by cardinal spline. In: Abd Manaf, A., Sahibuddin, S., Ahmad, R., Mohd Daud, S., El-Qawasmeh, E. (eds.) ICIEIS 2011. CCIS, vol. 254, pp. 179–188. Springer, Heidelberg (2011). https://doi.org/10.1007/978-3-642-25483-3_14
22. Kozlowski, K.: Standard and diagonalized Lagrangian dynamics: a comparison. In: Proceedings of 1995 IEEE International Conference on Robotics and Automation, Nagoya, Japan, vol. 3, pp. 2823–2828. IEEE (1995). https://doi.org/10.1109/ROBOT.1995.525683
23. Krayevoy, V., Sheffer, A.: Boneless motion reconstruction. In: ACM SIGGRAPH 2005 Sketches, SIGGRAPH 2005, pp. 122-es. Association for Computing Machinery, New York (2005). https://doi.org/10.1145/1187112.1187259
24. Kuboyama, T.: Matching and learning in trees. Doctoral dissertation (2007)
25. Kuo, C., Liang, Z., Fan, Y., Blouin, J.S., Pai, D.K.: Creating impactful characters: correcting human impact accelerations using high rate IMUs in dynamic activities. ACM Trans. Graph. **38**(4) (2019). https://doi.org/10.1145/3306346.3322978
26. Le Naour, T., Courty, N., Gibet, S.: Skeletal mesh animation driven by few positional constraints. Comput. Anim. Virtual Worlds **30**(3–4) (2019). https://doi.org/10.1002/cav.1900
27. Levi, Z., Gotsman, C.: Smooth rotation enhanced As-Rigid-As-Possible mesh animation. IEEE Trans. Visual Comput. Graphics **21**(2), 264–277 (2015). https://doi.org/10.1109/TVCG.2014.2359463
28. Li, J., Lu, G., Ye, J.: Automatic skinning and animation of skeletal models. Vis. Comput. **27**(6), 585–594 (2011). https://doi.org/10.1007/s00371-011-0585-8. Company: Springer Distributor: Springer Institution: Springer Label: Springer Number: 6 Publisher: Springer-Verlag
29. Lobão, A., Evangelista, B., Leal de Farias, J., Grootjans, R.: Skeletal animation. In: Beginning XNA 3.0 Game Programming, pp. 299–336. Apress (2009). https://doi.org/10.1007/978-1-4302-1818-0-12
30. Mandery, C., Terlemez, O., Do, M., Vahrenkamp, N., Asfour, T.: The KIT whole-body human motion database. In: 2015 International Conference on Advanced Robotics (ICAR), Istanbul, Turkey, pp. 329–336. IEEE, July 2015. https://doi.org/10.1109/ICAR.2015.7251476
31. Mills, P.M., Morrison, S., Lloyd, D.G., Barrett, R.S.: Repeatability of 3D gait kinematics obtained from an electromagnetic tracking system during treadmill locomotion. J. Biomech. **40**(7), 1504–1511 (2007). https://doi.org/10.1016/j.jbiomech.2006.06.017
32. Min, J., Chen, Y.L., Chai, J.: Interactive generation of human animation with deformable motion models. ACM Trans. Graph. **29**(1), 1–12 (2009). https://doi.org/10.1145/1640443.1640452

33. Mukundan, R.: Skeletal animation. In: Advanced Methods in Computer Graphics: With examples in OpenGL, pp. 53–76. Springer, London (2012). https://doi.org/10.1007/978-1-4471-2340-8-4

34. Muller, M., Roder, T., Clausen, M., Eberhardt, B., Kruger, B., Weber, A.: Documentation Mocap Database HDM05, p. 34

35. Parent, R.: Computer Animation: Algorithms and Techniques. The Morgan Kaufmann Series in Computer Graphics and Geometric Modeling. Morgan Kaufmann Publishers, San Francisco (2002)

36. Sepahi-Boroujeni, S., Mayer, J., Khameneifar, F.: Repeatability of on-machine probing by a five-axis machine tool. Int. J. Mach. Tools Manuf. **152** (2020). https://doi.org/10.1016/j.ijmachtools.2020.103544

37. Sito, T., Whitaker, H.: Timing for Animation, 2nd edn. Focal, Oxford (2009). oCLC: 705760367

38. Spanlang, B., Navarro, X., Normand, J.M., Kishore, S., Pizarro, R., Slater, M.: Real time whole body motion mapping for avatars and robots. In: Proceedings of the 19th ACM Symposium on Virtual Reality Software and Technology - VRST 2013, p. 175. ACM Press, Singapore (2013). https://doi.org/10.1145/2503713.2503747

39. Steinwender, G., Saraph, V., Scheiber, S., Zwick, E.B., Uitz, C., Hackl, K.: Intrasubject repeatability of gait analysis data in normal and spastic children. Clin. Biomech. **15**(2), 134–139 (2000). https://doi.org/10.1016/S0268-0033(99)00057-1

40. Stoll, C., Aguiar, E.D., Theobalt, C., Seidel, H.: A volumetric approach to interactive shape editing. In: Saarbrücken: Max-Planck-Institut für Informatik (2007)

41. Sujar, A., Casafranca, J.J., Serrurier, A., Garcia, M.: Real-time animation of human characters' anatomy. Comput. Graph. **74**, 268–277 (2018). https://doi.org/10.1016/j.cag.2018.05.025

42. Tupling, S.J., Pierrynowski, M.R.: Use of cardan angles to locate rigid bodies in three-dimensional space. Med. Biol. Eng. Comput. **25**(5), 527–532 (1987). https://doi.org/10.1007/BF02441745. Company: Springer Distributor: Springer Institution: Springer Label: Springer Number: 5 Publisher: Kluwer Academic Publishers

43. Wang, J., Lyu, K., Xue, J., Gao, P., Yan, Y.: A markerless body motion capture system for character animation based on multi-view cameras. In: ICASSP 2019–2019 IEEE International Conference on Acoustics, Speech and Signal Processing (ICASSP), pp. 8558–8562 (2019). https://doi.org/10.1109/ICASSP.2019.8683506

44. White, R., Agouris, I., Selbie, R., Kirkpatrick, M.: The variability of force platform data in normal and cerebral palsy gait. Clin. Biomech. **14**(3), 185–192 (1999). https://doi.org/10.1016/S0268-0033(99)80003-5

45. Zhao, Y., Liu, J.: Volumetric subspace mesh deformation with structure preservation. Comput. Animat. Virtual Worlds **23**(5), 519–532 (2012). https://doi.org/10.1002/cav.1488

46. Zhou, F., Luo, X., Huang, H.: Application of 4-point subdivision to generate inbetween frames in skeletal animation. In: Technologies for E-Learning and Digital Entertainment, LNCS, vol. 3942, pp. 1080–1084. Springer, Heidelberg (2006). https://doi.org/10.1007/11736639-134

47. Zollhöfer, M., Sert, E., Greiner, G., Süßmuth, J.: GPU based ARAP deformation using volumetric lattices. In: Eurographics (2012)

Composite Generalized Elliptic Curve-Based Surface Reconstruction

Ouwen Li[1], Ehtzaz Chaudhry[1(✉)], Xiaosong Yang[1], Haibin Fu[1], Junheng Fang[1], Zaiping Zhu[1], Andres Iglesias[2], Algirdas Noreika[3], Alfonso Carriazo[4], Lihua You[1], and Jian Jun Zhang[1]

[1] The National Center for Computer Animation, Bournemouth University, Poole, UK
echaudhry@bournemouth.ac.uk
[2] Department of Applied Mathematics and Computational Sciences, University of Cantabria, 39005 Cantabria, Spain
[3] Indeform Ltd., Kaunas, Lithuania
[4] Department of Geometry and Topology, University of Seville, Seville, Spain

Abstract. Cross-section curves play an important role in many fields. Analytically representing cross-section curves can greatly reduce design variables and related storage costs and facilitate other applications. In this paper, we propose composite generalized elliptic curves to approximate open and closed cross-section curves, present their mathematical expressions, and derive the mathematical equations of surface reconstruction from composite generalized elliptic curves. The examples given in this paper demonstrate the effectiveness and high accuracy of the proposed method. Due to the analytical nature of composite generalized elliptic curves and the surfaces reconstructed from them, the proposed method can reduce design variables and storage requirements and facilitate other applications such as level of detail.

Keywords: Curve modelling · Composite generalized elliptic curves · Surface modelling · Composite generalized elliptic curve-based surface reconstruction · Analytical mathematical expressions

1 Introduction

Cross-section curves define two-dimensional contours of 3D objects. They have a lot of applications in many fields. Especially, they are used in medical imaging to reconstruct 3D models from cross-sections for visualization.

Cross-sections obtained from imaging techniques or 3D polygon models are defined by discrete points, which involve many design variables, require high storage costs, cause slow network transmission, and do not meet the requirements of many applications such as level of detail where different resolutions are used in different situations.

In order to tackle the above problems, this paper will propose composite generalized elliptic curves, present their analytical mathematical expressions, and investigate surface reconstruction from composite generalized elliptic curves. Examples will be given to demonstrate the effectiveness and accuracy of the proposed method.

© Springer Nature Switzerland AG 2021
M. Paszynski et al. (Eds.): ICCS 2021, LNCS 12746, pp. 134–148, 2021.
https://doi.org/10.1007/978-3-030-77977-1_11

2   Related Work

Various methods have been developed to reconstruct 3D models from cross-sections called cross-section curves in this paper. Here, we briefly review some of them.

Early work on surface reconstruction from cross-sections given in [1] found minimum cost cycles in a directed toroidal graph and presented a fast triangular tiling algorithm to reconstruct 3D surfaces consisting of triangular tiles from a set of cross-section curves. Using the contour information obtained from computed tomography, ultrasound examinations, and nuclear medicine examinations, the comparison among several triangular tiling algorithms are made in [2]. Through pruning the Delaunay triangulations obtained between two adjacent cross-sections, a 3D polyhedron model was constructed whose triangular faces intersects with the planes along the given cross-section curves [3]. Assuming that cross-sections are perpendicular to the z-axis and reconstructed surfaces can be represented in cylindrical coordinates, fitting tensor product splines to the given cross-sections with the least squares method was used to reconstruct open and closed 3D surfaces in [4]. A method was proposed in [5] to obtain a triangulated mesh from cross-sections and a piecewise parametric surface was fitted to the triangulated mesh to produce a reconstructed surface. 3D reconstruction was achieved by decomposing one branching problem into several single-branching problems and using the closest local polar angle method to connect single branch contours [6]. A reconstructed surface was obtained in [7] by imposing three constraints on it, deriving precise correspondence and tiling rules from the constraints, and making the reconstructed surface produce the original contours. In order to avoid the problems caused by correspondence, tiling, and branching, surface reconstruction was formulated in terms of a partial differential equation and its solution was derived from the distance transform for dense contours and by adding a regulation term to ensure smoothness in surface recovery for sparse contours [8, 9]. The Bernstein basis function (BBF) network, which is a two-layer basis function network, was used to fit closed parametric curves or contours to the edge points of planar images through the network training and reconstruct 3D surfaces from fitted parametric curves [10]. Using cross-section ellipses plus a displacement map for modelling and deformations of human arms and legs was given in [11]. The work in [11] was further extended to human deformations in [12]. Based on a stratification of polygons and anisotropic distance functions, an implicit surface was used to reconstruct a 3D model from cross-sections, which avoid correspondence and branching problems [13]. A G^1 continuous composite surface consisting of skinned, branched, and capped surfaces was used in [14] to reconstruct 3D models where each of skinned surfaces was represented by a B-spline surface approximating cross-sections and transformed into a mesh of rectangular Bezier patches and triangular G^1 surfaces were constructed in branched and capped regions. A new algorithm was proposed in [15] to deal with curve networks of arbitrary shape and topology on non-parallel cross-section planes with arbitrary orientations. How to provide topological control during surface reconstruction from an input set of planar cross-sections was investigated in [16]. A template model was used to better capture geometric features and deformed to fit a set of cross section curves in [17]. A globally consistent signed distance function was used in [18] to construct closed and smooth 3D surfaces from unorganized planar cross-sections. Based on the idea of incremental sampling and potential field iterative correction, radial basis function and signed distance field based surface reconstruction is used in orebody modelling [19].

Different from the above work, in this paper, we will first propose analytical mathematical representations of cross-section curves called composite generalized elliptic curves including generalized ellipses, generalized elliptic curves, and composite generalized elliptic segments. Then we give analytical mathematical formulae for surface reconstruction from composite generalized elliptic curves.

3   Generalized Ellipses (GEs) and Generalized Elliptic Curves (GECs)

Elliptic cross-sections have been used in sweeping surfaces to describe human arms, legs, torso and neck and their shape changes in [12] by Hyun et al. The elliptic cross section-based sweeping surfaces are mathematically formulated as [12]

$$S(u, v) = R(u)E_u(v) + C(u)$$

$$= \begin{bmatrix} r_{11}(u) & r_{12}(u) & r_{13}(u) \\ r_{21}(u) & r_{22}(u) & r_{23}(u) \\ r_{31}(u) & r_{32}(u) & r_{33}(u) \end{bmatrix} \begin{bmatrix} a(t)\cos(v) \\ b(t)\sin(v) \\ 0 \end{bmatrix} + \begin{bmatrix} x(u) \\ y(u) \\ z(u) \end{bmatrix} \tag{1}$$

where $S(u, v)$ represents a sweep surface, $R(u)$ and $C(u)$ are rotation and translation, respectively, and $E_u(v)$ defines a standard ellipse of variable size.

Human bodies have irregular cross-section shapes. They cannot be well approximated with standard ellipses, which lead to unrealistic human body modelling as shown in Fig. 1. In the figure, we give groundtruth cross-section curves of human left leg, right arm and torso in (a), (b) and (c), respectively, which are taken from a human body model created with polygon modelling and highlighted in red. Using standard ellipses to approximate these cross-section curves, we obtain those highlighted in blue in the figure. Comparing the red and blue curves in Fig. 1, we can conclude that very large errors are introduced by standard ellipses.

To tackle the above problem, we propose composite generalized elliptic curves to approximate cross-section curves accurately. Composite generalized elliptic curves contain generalized ellipses, generalized elliptic curves, and composite generalized elliptic segments. Generalized ellipses and generalized elliptic curves will be investigated in this section. Composite generalized elliptic segments will be developed in the following section.

Generalized ellipses are applicable to closed curves. Since open curves are also widely applied in geometric modeling and computer-aided design, generalized elliptic curves are proposed to approximate open curves.

Through rotation operations, cross-section curves can be changed to lie in one of x-y, y-z, and x-z planes. Taking closed cross-section curves in the x-y plane as an example, their generalized ellipses can be mathematically formulated as

$$x(v) = a_0 + \sum_{j=1}^{J} (a_{2j-1} cos2j\pi v + a_{2j} sin2j\pi v)$$

$$y(v) = b_0 + \sum_{j=1}^{J}(b_{2j-1}sin2j\pi v + b_{2j}cos2j\pi v)$$

$$z(v) - z_c = 0 \qquad\qquad (2)$$

where $0 \leq v \leq 1$, a_j and $b_j(j = 0, 1, 2, 3, \cdots, 2J)$ are unknown constants, which can be determined by fitting the generalized ellipse to the discrete points defining a closed cross-section curve.

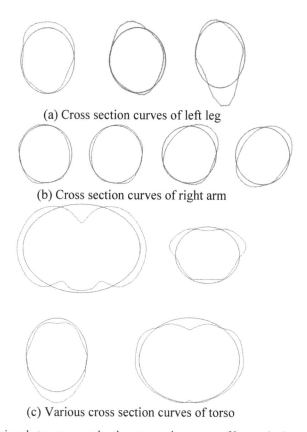

(a) Cross section curves of left leg

(b) Cross section curves of right arm

(c) Various cross section curves of torso

Fig. 1. Comparison between groundtruth cross-section curves of human body and approximated ellipses. (Color figure online)

For a given closed curve defined by $n + 1$ discrete points $X_i = \begin{bmatrix} x_i & y_i & z_i \end{bmatrix}^T (i = 0, 1, 2, \cdots, n)$, we first determine the total length L of the closed curve and the length L_i from the starting point X_0 to the point X_i. Then, we use $v_i = L_i/L$ to obtain the values of the parametric variable v at these points. Finally, we obtain all the unknown constants by using Eq. (2) to fit these $n + 1$ points.

For the groundtruth cross-section curves highlighted in red in Fig. 2, which are taken from a human body model, we use standard ellipses and generalized ellipses to approximate them. The obtained results are shown in the same figure where the blue colour shows the curves obtained from standard ellipses, the green colour shows the curves obtained from generalized ellipses, and J indicates the terms used in Eq. (2).

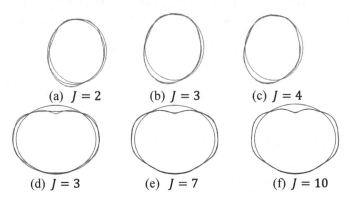

(a) $J = 2$ (b) $J = 3$ (c) $J = 4$

(d) $J = 3$ (e) $J = 7$ (f) $J = 10$

Fig. 2. Comparison of curves created by standard and generalized ellipses. (Color figure online)

Comparing the curves shown in Fig. 2, we can conclude: (1) there are very big differences between the standard ellipses and the groundtruth curves, (2) generalized ellipses approximate the groundtruth curves very well, (3) the groundtruth curves can be approximated accurately as the terms in Eq. (3) increase.

For closed curves approximated by Eq. (2), the starting point is the same as the ending point. For open curves, the starting point is different from the ending point and Eq. (2) is not applicable. Therefore, we define the following generalized elliptic curves to approximate open curves in the x-y plane.

$$x(v) = a_0 + \sum_{j=1}^{J}(a_{2j-1} \cos j\pi v + a_{2j} \sin j\pi v)$$

$$y(v) = b_0 + \sum_{j=1}^{J}(b_{2j-1} \sin j\pi v + b_{2j} \cos j\pi v)$$

$$z(v) - z_c = 0 \tag{3}$$

where $0 \leq v \leq 1$, a_j and $b_j (j = 0, 1, 2, 3, \cdots, 2J)$ are unknown constants.

The unknown constants in Eq. (3) can be determined as follows. For discrete points $X_i = \begin{bmatrix} x_i \ y_i \ z_i \end{bmatrix}^T (i = 0, 1, 2, \cdots, n)$ of a curve, we calculate the total length L of the curve and the length L_i from the starting point X_0 to the point X_i, and obtain $v_i = L_i/L$. Then, we use $x_0 = x(v_0)$ and $x_n = x(v_n)$ to determine two of the unknown constants $a_j (j = 0, 1, 2, \cdots, 2J)$ and $y_0 = y(v_0)$ and $y_n = y(v_n)$ to determine two of the unknown constants $b_j (j = 0, 1, 2, \cdots, 2J)$ in Eq. (3). All the remaining unknown constants in Eq. (3) are determined by using it to fit the remaining $n - 1$ points

$X_i(i = 1, 2, \cdots, n - 1)$. If tangential continuity is required, two more conditions of the first derivatives at the starting and ending points are used to determine two more unknown constants for each of the x and y components. Then, the remaining unknown constants are determined by curve fitting.

In Fig. 3, we use the generalized elliptic curve highlighted in red to approximate the original open curve highlighted in blue. The figure shows that when the total terms are not many, large errors are introduced as shown in Fig. 3(a). When more terms are used, the errors are reduced as shown by Fig. 3(d).

In order to give a quantity comparison, we introduce the maximum error and average error. If the points on the original curves are $O_n(n = 1, 2, \cdots, N)$ and the points on the generalized elliptic curves are $G_n(n = 1, 2, \cdots, N)$, the maximum error is defined as $E_r M = \max\{d_1, d_2, \cdots, d_N\}$ and the average error is defined as $E_r A = \sum_{n=1}^{N} d_n/N$ where d_n is the Euclidean distance between the point O_n and the point G_n.

Table 1. Errors between the original open curve and open GEC.

J	3	5	7	11	21	31
ErM	2.2492	1.8117	1.2384	0.6711	0.3047	0.2197
ErA	0.7459	0.4587	0.2519	0.1163	0.0460	0.0233
J	41	51	61	71	81	91
ErM	0.1417	0.0900	0.0608	0.0369	0.0101	1.34×10^{-4}
ErA	0.0138	0.0091	0.0072	0.0058	7.45×10^{-4}	4.35×10^{-6}

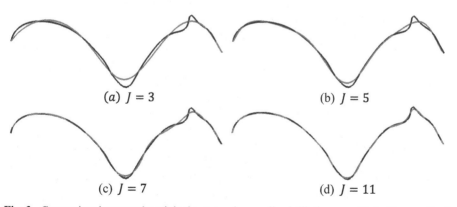

(a) $J = 3$ (b) $J = 5$

(c) $J = 7$ (d) $J = 11$

Fig. 3. Comparison between the original curve and generalized elliptic curve. (Color figure online)

In Table 1, we give a quantity comparison of the errors between the original open curve and the generalized elliptic curve. In the table, ErM and ErA indicate the maximum error and average error, respectively. For the original open curve shown in Fig. 3, the maximum error reduces more slowly than the average error. Only when many terms, i.e., $J = 91$, are used, small errors are obtained. It indicates the necessity of using composite generalized elliptic segments to reduce design variables and improve the accuracy of curve fitting, which will be investigated in the following section.

4   Composite Generalized Elliptic Segments (CGESs)

For an open or closed curve with a complicated shape, using a single generalized elliptic curve or a single generalized ellipse to approximate it may require many terms in Eq. (2) or (3). In order to reduce design variables, raise computational efficiency, and improve interactive modelling performance, it is worthy of developing a more efficient algorithm to cope with open and closed curves with complicated shapes and the corresponding surface reconstruction.

A complicated open and closed curve can be decomposed into several simpler ones. In order to approximate these decomposed curve segments, we use Eq. (3) to approximate each of them and name each segment as a composite generalized elliptic segment.

As an example, we consider how to use two composite generalized elliptic segments to approximate an open curve with a complicated shape highlighted in blue shown in Fig. 4(a). For the curves consisting of more than two composite generalized elliptic segments, the treatment is the same.

Firstly, we segment the blue open curve in Fig. 4(a) into two separate curves ABC and CDE shown in Figs. 4(b) and 4(c), respectively. With Eq. (3) and the method of determining its unknown constants given in Sect. 3, we can determine the two composite generalized elliptic segments highlighted in red in Figs. 4(d), 4(e), and 4(f). Then, we use one single generalized elliptic curve to approximate the open curve and depict the obtained curve highlighted in pink in Figs. 4(g), 4(h) and 4(i). In the figure, J indicates the terms defined in Eq. (3).

First, we investigate the accuracy of using two composite generalized elliptic segments and one single generalized elliptic curve when using almost the same number of unknown constants. The unknown constants used by two composite generalized elliptic segments and one single generalized elliptic curve shown in second and third rows of Fig. 4 are almost the same. For example, the total unknown constants for two composite generalized elliptic segments shown in Fig. 3(d) are $(2 \times 2 + 1) \times 4 = 20$, and the total unknown constants for the generalized elliptic curve shown in Fig. 4(g) are $(5 \times 2 + 1) \times 2 = 22$. In spite of almost the same number of unknown constants, the two composite generalized elliptic segments shown in Fig. 4(d) approximate the open curve accurately. In contrast, very large errors are caused by the generalized elliptic curve shown in Fig. 4(g), indicating that composite generalized elliptic segments are much more accurate than generalized elliptic curves.

Next, we set the errors of the two composite generalized elliptic segments to be roughly same as the errors of the generalized elliptic curve, and compare the number of the unknown constants in Table 2 where ErM and ErA indicate the maximum and

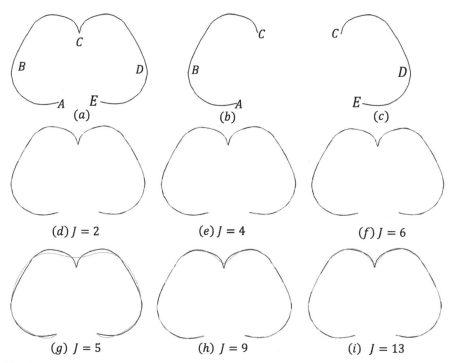

Fig. 4. Comparison of curve approximation between two CGESs and one GEC. (Color figure online)

average errors between the groundtruth curve and the approximated ones. The total unknown constants for $J = 5$ of the two composite generalized elliptic segments are $(5 \times 2 + 1) \times 4 = 44$. In contrast, the total unknown constants for $J = 25$ of the generalized elliptic curve are $(25 \times 2 + 1) \times 2 = 102$. In order to achieve roughly same maximum and average errors, the total unknown constants of the generalized elliptic curve are 2.32 times of the two composite generalized elliptic segments. It indicates that using multiple composite generalized elliptic segments to approximate a complicate curve can noticeably reduce the design variables in comparison with using a single generalized elliptic curve.

Table 2. Comparison of errors between two CGESs and one GEC.

	Composite generalized elliptic segments				Generalized elliptic curve			
	$J = 5$	$J = 7$	$J = 15$	$J = 39$	$J = 25$	$J = 40$	$J = 59$	$J = 79$
ErM	0.13	0.09	0.04	1.00×10^{-5}	0.49	0.23	0.08	5.30×10^{-5}
ErA	0.04	0.03	0.01	3.08×10^{-6}	0.06	0.03	0.01	3.57×10^{-6}

5   Composite Generalized Elliptic Curve-Based Surface Reconstruction

Since open and closed curves are described with generalized ellipses, generalized elliptic curves, and composite generalized elliptic segments, surface reconstruction is divided into generalized ellipse-based, generalized elliptic curve-based, and composite generalized elliptic segment-based, accordingly.

5.1   Generalized Ellipse or Generalized Elliptic Curve-Based Surface Reconstruction

If a surface is reconstructed from M generalized ellipses or generalized elliptic curves $X_m(v) = \left[x_m(v)\ y_m(v)\ z_m(v) \right]^T$ $(m = 0, 1, 2, \cdots, M)$, the surface can be mathematically formulated as

$$S(u, v) = \sum_{m=0}^{M} u^m G_m(v) \tag{4}$$

where $G_m(v)$ $(m = 0, 1, 2, \cdots, M)$ are unknown vector-valued functions, which can be determined by interpolating the $M + 1$ generalized ellipses or generalized elliptic curves.

When two surfaces are to be connected together with up to the tangential continuity, we can obtain the position function and first partial derivative function on the shared boundary curve from the surface, which has already been reconstructed. Then we can use the position function and first partial derivative function together with generalized ellipses or generalized elliptic curves to be interpolated to reconstruct a new surface.

With the above method, we reconstruct some surfaces from generalized ellipses and depict the reconstructed surfaces in Fig. 5. In the figure, (a) shows the surfaces totally determined by generalized ellipses, (b) shows the surface interpolating generalized ellipses and smoothly connecting to an adjacent surface, and (c) shows the surface interpolating generalized ellipses and smoothly connecting to two adjacent surfaces.

5.2   Composite Generalized Elliptic Segment-Based Surface Reconstruction

Whether the curves are converted into generalized ellipses, generalized elliptic curves or composite generalized elliptic segments, they must be divided into the same number of curve segments if a surface is to be constructed from them. Therefore, when a generalized ellipse or a generalized elliptic curve is to be combined with other composite generalized elliptic segments to reconstruct a surface, the generalized ellipse or generalized elliptic curve should be firstly divided into the same segments as those of composite generalized elliptic segments. For example, if we are required to reconstruct a surface from three composite generalized elliptic segments and a single closed generalized ellipse, we divide the generalized ellipse into three curve segments.

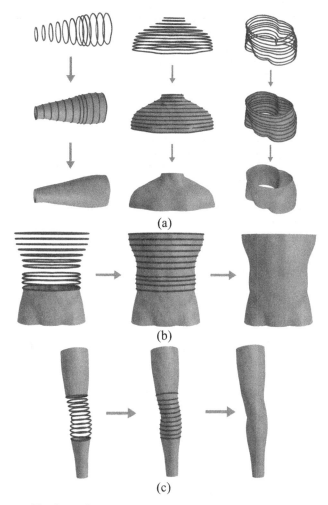

Fig. 5. Surfaces reconstructed from generalized ellipses.

The continuity between different surface patches along the v parametric direction can be achieved by taking the mathematical expressions for u parametric variable of the adjacent two surface patches to be the same form. In the following, we indicate that the surface patches reconstructed in this way can maintain the required continuity.

For $I + 1$ curves each of which consists of $J + 1$ segments, we use vector-valued position functions $X_{ij} (i = 0, 1, 2, \cdots, I; j = 0, 1, 2, \cdots, J)$ to represent the jth segment of the ith curve. Here X_{ij} has the three components x_{ij}, y_{ij} and z_{ij}. Since z_{ij} takes an identical value for all segments of a same curve, the continuity for this component is always ensured. Therefore, the following treatment is for x and y components. Here we only consider positional and tangential continuities. Of course, the treatment discussed here is also suitable for higher order continuities.

If up to tangential continuity is considered, both positional and tangential continuities at the joint between the jth segment X_{ij} and $(j+1)^{th}$ segment X_{ij+1} should be achieved when constructing these curve segments. That is

$$X_{ij}(v = v_{ij}) = X_{ij+1}(v = v_{ij})$$

$$\left\{\frac{\partial X_{ij}}{\partial v}\right\}_{v=v_{ij}} = \left\{\frac{\partial X_{ij+1}}{\partial v}\right\}_{v=v_{ij}}$$

$$(i = 0, 1, 2, 3, \cdots, I; j = 1, 2, 3, \cdots, J-1) \tag{5}$$

If the curves are closed, the positional and tangential continuities at the closure should also be introduced which lead to the additional equations below

$$X_{iJ}(v = 2\pi) = X_{i0}(v = 0)$$

$$\left\{\frac{\partial X_{iJ}}{\partial v}\right\}_{v=2\pi} = \left\{\frac{\partial X_{i0}}{\partial v}\right\}_{v=0}$$

$$(i = 0, 1, 2, 3 \ldots, I) \tag{6}$$

If $J+1$ surface patches are constructed from the above $I+1$ curves, we take their surface functions to be

$$S_j(u, v) = \sum_{m=0}^{I} u^m \bar{X}_{mj}(v)$$

$$(j = 0, 1, 2, 3, \cdots, J) \tag{7}$$

where $\bar{X}_{mj}(v)$ $(m = 0, 1, 2, 3, \ldots, I; j = 0, 1, 2, 3, \ldots, J)$ are unknown functions.

At the position u_i of the ith curve segment, the jth surface patch $S_j(u, v) = \sum_{m=0}^{I} u^m \bar{X}_{mj}(v)$ and the $(j+1)^{th}$ surface patch $S_{j+1}(u, v) = \sum_{m=0}^{I} u^m \bar{X}_{mj+1}(v)$ should pass through the jth and $(j+1)^{th}$ curve segments, respectively, i.e.,

$$\sum_{m=0}^{I} u_i^m \bar{X}_{mj}(v) = X_{ij}(v)$$

$$\sum_{m=0}^{I} u_i^m \bar{X}_{mj+1}(v) = X_{ij+1}(v)$$

$$(i = 1, 2, 3, \ldots, I) \tag{8}$$

Expanding Eqs. (6), (7) and (8) and rewriting them into the form of matrix, we obtain the following mathematical expressions from Eqs. (6) and (8) for open curves and from Eqs. (6), (7) and (8) for closed curves

$$[R_j(u_{im})]\{\bar{X}_j(v)\} = \{X_j(v)\}$$

$$[R_{j+1}(u_{im})]\{\bar{X}_{j+1}(v)\} = \{X_{j+1}(v)\} \tag{9}$$

where $[R_j(u_{km})]$ and $[R_{j+1}(u_{km})]$ are $K \times K$ square matrices with the elements $u_{km} = u_k^m$, $\{\bar{X}_j(v)\} = [\bar{X}_{0j}(v) \, \bar{X}_{1j}(v) \, \bar{X}_{2j}(v) \cdots \bar{X}_{Ij}(v)]^T$, $\{\bar{X}_{j+1}(v)\} = [\bar{X}_{0j+1}(v) \, \bar{X}_{1j+1}(v) \, \bar{X}_{2j+1}(v) \cdots \bar{X}_{Ij+1}(v)]^T$, $\{X_j(v)\} = [X_{0j}(v) \, X_{1j}(v) \, X_{2j}(v) \cdots X_{0j}(v)]$ and $\{X_{j+1}(v)\} = [X_{0j+1}(v) \, X_{1j+1}(v) \, X_{2j+1}(v) \cdots X_{Ij+1}(v)]^T$ are the vectors with K elements.

Using $[R_j(u_{im})]^{-1}$ and $[R_{j+1}(u_{im})]^{-1}$ to indicate the inverse matrices of $[R_j(u_{im})]$ and $[R_{j+1}(u_{im})]$, respectively, left multiplying both sides of the first equation of Eq. (9) by $[R_j(u_{im})]^{-1}$ and left multiplying both sides of the second equation of Eq. (9) by $[R_{j+1}(u_{im})]^{-1}$, we obtain the unknown functions with the following equations

$$\{\bar{X}_j(v)\} = [R_j(u_{im})]^{-1}\{X_j(v)\}$$

$$\{\bar{X}_{j+1}(v)\} = [R_{j+1}(u_{im})]^{-1}\{X_{j+1}(v)\}$$

$$(j = 0, 1, 2, 3, \ldots, J) \tag{10}$$

From Eq. (8), we know that $[R_j(u_{im})]^{-1}$ and $[R_{j+1}(u_{im})]^{-1}$ are identical which can be written as

$$R = [R_j(u_{im})]^{-1} = [R_{j+1}(u_{im})]^{-1} = [R_{ij}] \tag{11}$$

where R_{ij} $(i = 0, 1, 2, 3, \cdots, I; j = 0, 1, 2, 3, \cdots, I)$ are the elements of the square matrix R.

With Eq. (11), we can obtain the mathematical expressions of the elements in vectors $\bar{X}_j(v)$ and $\bar{X}_{j+1}(v)$

$$\bar{X}_{lj}(v) = \sum_{i=0}^{I} R_{li}X_{ij}(v)$$

$$\bar{X}_{lj+1}(v) = \sum_{i=0}^{I} R_{li}X_{ij+1}(v)$$

$$(l = 0, 1, 2, 3, \ldots, I) \tag{12}$$

Substituting the first of Eq. (12) into $S_j(u, v) = \sum_{m=0}^{I} u^m \bar{X}_{mj}(v)$ and the second of Eq. (12) into $S_{j+1}(u, v) = \sum_{m=0}^{I} u^m \bar{X}_{mj+1}(v)$, the position functions of the jth and $(j+1)^{th}$ surface patches are found to be

$$S_j(u, v) = \sum_{m=0}^{I} u^m \sum_{i=0}^{I} R_{mi}X_{ij}(v)$$

$$S_{j+1}(u, v) = \sum_{m=0}^{I} u^m \sum_{i=0}^{I} R_{mi} X_{ij+1}(v) \tag{13}$$

From Eq. (13), we can calculate the first partial derivatives of the jth and $(j+1)^{th}$ surface patches with respect to the parametric variable v. Considering the positional and tangential continuity conditions (5) and (6) of $X_{ij}(v)$ and $X_{ij+1}(v)$ at their joints v_{ij}, we find

$$S_j(u, v_{ij}) = S_{j+1}(u, v_{ij})$$

$$\left\{ \frac{\partial S_j(u, v)}{\partial v} \right\}_{v=v_{ij}} = \left\{ \frac{\partial S_{j+1}(u, v)}{\partial v} \right\}_{v=v_{ij}} \tag{14}$$

Equation (14) indicates that along the shared boundary curve between the jth and $(j+1)^{th}$ surface patches, both positional and tangential continuities are guaranteed.

In Fig. 6, we use the above method to reconstruct a surface from three composite generalized elliptic segments. It demonstrates the effectiveness of the above method.

Fig. 6. A surface reconstructed from three composite generalized elliptic segments.

5.3 Human Body Model Reconstruction with Composite Generalized Elliptic Curves

With the method developed above, the reconstruction of a human body model starts from obtaining cross-section curves shown in Fig. 7(a). These cross-section curves are converted into composite generalized elliptic curves. Then, the corresponding part models of the human body are reconstructed from the composite generalized elliptic curves and shown in Fig. 7(b).

For this example of human body model reconstruction, all the part models share the same boundary curves. When these part models are reconstructed, they are naturally connected together as shown in Fig. 8 where Fig. 8(a) shows the reconstructed 3D model with cross section curves, and Fig. 8(b) shows the rendered 3D model.

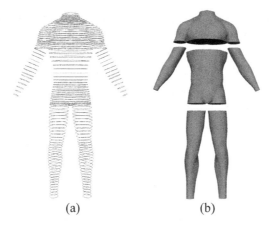

Fig. 7. Cross-section curves and part models of a human body.

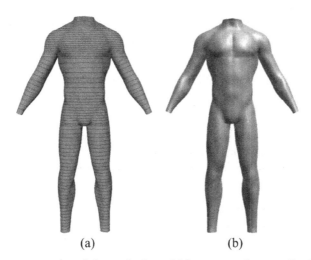

Fig. 8. Surface reconstruction of a human body model from composite generalized elliptic curves.

6 Conclusions

In this paper, we have proposed generalized ellipses, generalized elliptic curves, composite generalized elliptic segments to approximate open and closed curves, demonstrate their effectiveness and accuracy, and present mathematical expressions of surface reconstruction from composite generalized elliptic curves.

Compared to the sweep-based human deformation approach proposed in [12], the method proposed in this paper approximates cross-section curves accurately and avoids the use of a displacement map, leading to accurate and detailed reconstruction of 3D models.

Unlike polygon models, which involve a lot of design variables (vertices), the proposed method representing 3D models with analytical mathematical expressions involves much fewer design variables. It brings in the advantages of low storage costs, fast network transmissions, and easy generation of 3D models with arbitrary resolutions, which is suitable for many applications such as level of detail in online games.

Acknowledgements. This research is supported by the PDE-GIR project which has received funding from the European Union Horizon 2020 research and innovation programme under the Marie Skodowska-Curie grant agreement No. 778035.

References

1. Fuchs, H., Kedem, Z.M., Uselton, S.P.: Optimal surface reconstruction from planar contours. Commun. ACM **20**(10), 693–702 (1977)
2. Cook, P.N., Batnitzky, S., Lee, K.R., Levine, E., Price, H.I.: Three-dimensional reconstruction from serial sections for medical applications. Proc. SPIE **283**, 98–105 (1981)
3. Boissonnat, J.D.: Shape reconstruction from planar cross sections. Comput. Vis. Graph. Image Proc. **44**(1), 1–29 (1988)
4. Dierckx, P., Suetens, P., Vandermeulen, D.: An algorithm for surface reconstruction from planar contours using smoothing splines. J. Comput. Appl. Math. **23**, 367–388 (1988)
5. Meyers, D., Skinner, S., Sloan, K.: Surfaces from Contours. ACM Trans. Graph. **11**(3), 228–258 (1992)
6. Xu, M., Tang, Z.: Surface reconstruction for cross sectional data. J. Comput. Sci. Technol. **11**(5), 471–479 (1996)
7. Bajaj, C.L., Coyle, E.J., Lin, K.N.: Arbitrary topology shape reconstruction from planar cross sections. Graph. Models Image Proc. **58**(6), 524–543 (1996)
8. Cong, G., Parvin, B.: An algebraic solution to surface recovery from cross-sectional contours. Graph. Models Image Proc. **61**, 222–243 (1999)
9. Cong, G., Parvin, B.: Robust and efficient surface reconstruction from contours. Vis. Comput. **17**, 199–208 (2001)
10. Knopf, G.K., Al-Naji, R.: Adaptive reconstruction of bone geometry from serial cross-sections. Artif. Intell. Eng. **15**, 227–239 (2001)
11. Hyun, D.-E., Yoon, S.-H., Kim, M.-S., Jüttler, B.: Modeling and deformation of arms and legs based on ellipsoidal sweeping. In: Proceedings of 11th Pacific Conference on Computer Graphics and Applications, pp. 204–212. IEEE Computer Society (2003)
12. Hyun, D.-E., Yoon, S.-H., Chang, J.-W., Kim, M.-S., Jüttler, B.: Sweep-based human deformation. Vis. Comput. **21**, 542–550 (2005)
13. Akkouche, S., Galin, E.: Implicit surface reconstruction from contours. Vis. Comput. **20**, 392–401 (2004)
14. Park, H.: A hybrid approach to smooth surface reconstruction from 2-D cross sections. Int. J. Adv. Manuf. Technol. **25**, 1130–1136 (2005)
15. Liu, L., Bajaj, C., Deasy, J.O., Low, D.A., Ju, T.: Surface reconstruction from non-parallel curve networks. Comput. Graph. Forum **27**(2), 155–163 (2008)
16. Zou, M., Holloway, M., Carr, N., Ju, T.: Topology-constrained surface reconstruction from cross-sections. ACM Trans. Graph. **34**(4), 1–10 (2015). Article no. 128
17. Holloway, M., Grimm, C., Ju, T.: Template-based surface reconstruction from cross-sections. Comput. Graph. **58**, 84–91 (2016)
18. Sharma, O., Agarwal, N.: Signed distance based 3D surface reconstruction from unorganized planar cross-sections. Comput. Graph. **62**, 67–76 (2017)
19. Zhong, D.-Y., Wang, L.-G., Jia, M.-T., Bi, L., Zhang, J.: Orebody modeling from non-parallel cross sections with geometry constraints. Minerals **9**(4), 229:1–17 (2019)

Supporting Driver Physical State Estimation by Means of Thermal Image Processing

Paweł Forczmański$^{(\boxtimes)}$ and Anton Smoliński

Faculty of Computer Science and Information Technology, West Pomeranian University of Technology, Szczecin, Żołnierska Street 49, 71-210 Szczecin, Poland
{pforczmanski,ansmolinski}@wi.zut.edu.pl

Abstract. In the paper we address a problem of estimating a physical state of an observed person by means of analysing facial portrait captured in thermal spectrum. The algorithm consists of facial regions detection combined with tracking and individual features classification. We focus on eyes and mouth state estimation. The detectors are based on Haar-like features and AdaBoost, previously applied to visible-band images. Returned face region is subject to eyes and mouth detection. Further, extracted regions are filtered using Gabor filter bank and the resultant features are classified. Finally, classifiers' responses are integrated and the decision about driver's physical state is taken. By using thermal image we are able to capture eyes and mouth states in very adverse lighting conditions, in contrast to the visible-light approaches. Experiments performed on manually annotated video sequences have shown that the proposed approach is accurate and can be a part of current Advanced Driver Assistant Systems.

Keywords: Thermal imaging · Face detection · Eyes detection · Mouth detection · Haar-like features · Gabor filtering · Drowsiness estimation

1 Introduction

Most traffic accidents, according to the road safety research [33], come as a result of drivers' behaviour, often associated with fatigue and drowsiness [29]. According to [32], approximately 20% of accidents are caused by loss of concentration hindering drivers in terms of immediate and right decisions [19]. A majority of techniques aimed at driver fatigue assessment is based on an analysis of signals captured by external sensors that are carried by observed persons. On the other hand, there is also an alternative possibility to perform a continuous observation of the driver without his/her cooperation - it is often realized by machine vision techniques applied to video streams captured in the cabin. The research [19] show that some characteristic behaviours are associated with different activity of head, eyelids and mouth [9].

© Springer Nature Switzerland AG 2021
M. Paszynski et al. (Eds.): ICCS 2021, LNCS 12746, pp. 149–163, 2021.
https://doi.org/10.1007/978-3-030-77977-1_12

In this paper we present a research devoted to the methodology for acquiring, integrating and analysing selected visual data in the context of assessing the psychophysical state and the degree of fatigue of drivers or operators of motor vehicles. The proposed algorithm is based on the analysis of visual premises captured in uncontrolled environmental conditions (e.g. in severe lighting conditions or in complete darkness) [20]. While there are many algorithms that are able to detect driver fatigue and drowsiness based on eyes state and blink analysis (e.g. [8,19,24]), they work in visible light, or in infrared band, only. It should be noted, that during travelling, drivers encounter dynamic lighting conditions (blinding sun, low-level ambient illumination) and various environmental conditions impairing the observation (e.g. passing through a shaded forest or dark tunnel). Hence, the obvious weaknesses of traditional imaging means.

Typical machine vision-based techniques allowing for a constant observation of a driver work often in the following manner. The first stage is face detection followed by facial features extraction [22]. Such regions-of-interest are then analysed in order to collect relevant spatial and temporal characteristics which are subjected to inference mechanism. Face and face features detection problems are considered solved under controlled conditions [10,31], yet when we deal with severe lighting, there are still many difficulties to overcome.

It is evident, that the analysis of the visible band image of a face recorded in the unfavourable illumination leads to the high error rate or it can be completely impossible. On the other hand, under the same environmental conditions, using thermal imaging, we are able to detect eyes blinking and yawning. Hence, we propose a method to detect eyes and mouth state in video sequences captured in thermal spectrum. Although the detection and tracking of facial landmarks can be performed with almost 100% accuracy, it should be remembered, that they are oriented at visible band images. Their accuracy for other modes, e.g. thermal images or depth maps, is not often explored. It is caused by unpopular and expensive thermal acquisition devices. However, it may change in future.

The authors of [17] presented a method of detecting yawning in thermal images for estimating driver's fatigue level. They focus on detecting thermal anomalies in the image area where a mouth is located. This approach may be sensitive to an unexpected phenomenon, unforeseen while determining their parameters empirically. In our approach we detect mouth state based on visual cues only, assuming that the appearance of the lips changes during yawning. Another exemplary system presented in [24] uses OpenCV face and eye detectors supported with the simple feature extractor based on the two dimensional Discrete Fourier Transform (DFT) to represent eyes regions. Similarly, the fatigue of the driver determined through the duration of the eyes' blinks is presented in [8]. It operates in the visible and near infra-red (NIR) spectra allowing to analyse drivers state in the night conditions and poor visibility. A more complex, multi-modal platform to identify driver fatigue and interference detection is presented in [7]. It captures audio and video data, depth maps, heart rate, steering wheel and pedals positions. The experimental results show that the authors are able to detect fatigue with 98.4% accuracy. There are solutions based on mobile devices, especially smartphones and tablets, or based on dedicated hardware [14,18,34]. In [1] the authors recognize the act of yawning using a simple webcam. In [12] the authors proposed a dynamic fatigue detection model based on Hidden Markov Model (HMM). This model can estimate driver fatigue in a probabilistic way using various physiological and contextual information. In a subsequent work [3] authors monitor information about the eyes and mouth of the driver. Then, this information is transmitted to the Fuzzy Expert System, which classifies the true state of the driver. The system has been tested using real data from various sequences recorded during the day and at night for users belonging to different races and genders. The authors claim that their system gives an average recognition accuracy of fatigue close to 100% for the tested video sequences.

The analysis shows that many of current works are focused on the problem of recognizing driver's fatigue, yet there is no common methodology of acquisition of signals (from various sources) used to evaluate vehicle operator physical condition and fatigue level. Moreover, most of works focus on single feature, not taking into consideration joint observations. Even though, if joint features are used (e.g. [27]), the analysis is performed on visible-light-only images.

On the other hand, to achieve the most accurate results, researchers use additional data sources such as EEG, ECG, heart rate, vehicle behavior on the road, etc. Unfortunately, in real conditions, obtaining this type of data is difficult or even impossible. Therefore, in their research, the authors focus on non-invasive methods of image analysis of various spectra [15,28,30].

2   Method Description

2.1   Assumptions

Capturing image in visible-band light is the most straightforward and widespread method of visual data acquisition. The sensors of photo cameras are rather cheap and their performance can be high, in terms of sensitivity, dynamic range and spatial resolution. Unfortunately, their parameters are preserved only in good and stable lighting conditions. It should be remembered that car's cabin is often illuminated by directional light, coming from the windows (sun, street and lamps of other cars). What is more, additional (stable) lighting of driver's face during driving is impossible, since it could disturb his/her functioning. The natural way of solving this dilemma is using other capturing device, working in different lighting spectra, e.g. near infrared (NIR with wavelength of 0.75–1.4 μm) or thermal (also known as long-wavelength infrared - LWIR with wavelength of 8–15 μm). Since NIR is associated with so called reflective infrared, it requires an illumination with infrared illuminator. Such a prolonged illumination may be hazardous to the driver's health [4]. Therefore, the LWIR spectrum has been selected, since it is free of the above mentioned weak points [17]. Thermal sensor captures drivers's facial image in a completely passive manner and it does not depend on the lighting conditions, which is important when dealing with an observation of a driver in uncontrolled environment.

Using the same methodology, ass in case of visible-band video, certain facial areas in the thermal portraits can be detected and extracted using specially trained detectors based on hand-crafted features, e.g. Haar-like features [13], Histogram of Oriented Gradients [13] and Local Binary Patterns [34]. The detectors are often trained by means of some variant of AdaBoost approach [6,14]. It should be also noted that recently proposed deep-learning applications [26] require significant computing power to operate in near-real-time.

2.2   General Overview

The algorithm consists of six main modules: (i) face detection and tracking, (ii) eyes detection and tracking, (iii) eyes state analysis, (iv) mouth detection and tracking, (v) mouth state analysis, and (vi) data integration, leading to the estimation of driver state detection. It works in a loop iterated over the frames from the video stream (see Fig. 1).

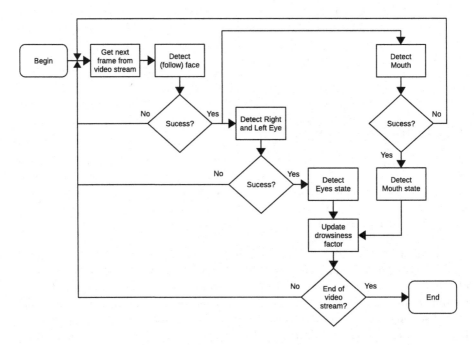

Fig. 1. Algorithm of drowsiness estimation by means of eyes and mouth state analysis.

2.3   Region Detectors

The input video stream is processed in frame-by-frame manner. Each frame is a subject of independent face, eyes and mouth detection. Firstly, a pretrained Viola-Jones detector [10] is used to detect faces. A successful detection triggers simple face tracker, that assumes only small movement of the head in consecutive frames and approximates position between successful detections. The previous research [10,11,25] showed, that it is possible to perform an efficient and precise face detection in images taken in non-visible bands. Previous experiments involving thermal images [10] showed also, that although traditional Viola-Jones detector using hand-crafted features is not so strong as many recently proposed deep-learning approaches, it requires less computations and is capable of working in real-time.

Detected face's boundaries are the subject of parallel eyes and mouth detectors. Actually, there are three separate detectors, one for each eye and for mouth, all based on Viola-Jones approach, trained on a set of over 1700 eyes and 1450 mouth samples extracted from benchmark video streams captured in our simulated cabin [20]. Since the detection may not always precisely estimate the position of eyes and mouth, we introduced a tracker that is based on position approximation [11], assuming that under regular driving conditions eyes and mouth positions should not change significantly across a small number of neighbouring frames.

Taking into consideration the above mentioned assumptions, the resulting coordinates of the face's, eyes' and mouth's bounding boxes are calculated over the averaged 10 past detections. Final confidence level is normalized to the range $\langle 0, 1 \rangle$. Some exemplary detection results are presented in Fig. 2. As it can be seen, face orientation does not influence the detection performance.

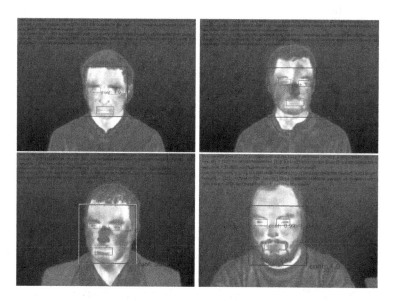

Fig. 2. Exemplary results of face tracking joint with eyes and mouth detection.

2.4 Eyes and Mouth State Estimation

Extracted eyes and mouth regions are then predicted as open or closed. Both stages work in the similar manner, hence they are described as a single universal procedure. The classification is performed in the following steps (for each eye and mouth separately):

1. Extract eye (mouth) region from detected face region;
2. Resample the image to a fixed rectangle of 100 × 50 pixels;
3. Filter the image with Gabor filter bank (30 sets of parameters resulting in 30 output images);
4. Calculate energy and standard deviation for each resulting image (resulting in 60 values);
5. Concatenate above coefficients into a singe feature vector;
6. Perform the classification.

Gabor filtering [2] is performed with kernel size equal to 20 × 10, standard deviation of the Gaussian envelope equal to 2.5, spatial aspect ratio equal to 5.0 and phase offset equal to $3\pi/2$. By altering the orientation angle of the normal to the parallel stripes of a Gabor function in the range $(0, 5\pi/6\rangle$ with a $\pi/6$ step and wavelength of the sinusoidal factor equal to $\{1, 2, 4, 8, 16\}$ we get 30 output images. These parameters are in line with the literature [5] and give the possibility to obtain a compact textural representation. For every filtering result we calculate energy (normalized by the image's dimensions) and standard deviation. Resulting feature vectors are then taken as input for the binary classifier.

Exemplary results of filtering for eye region are presented in Fig. 3. The left part of the image shows closed eye, while the right one - open eye, respectively. Similarly, the results of filtering for mouth region are presented in Fig. 4, where the left part of the image is devoted to the closed mouth, while the right one to the open mouth, respectively.

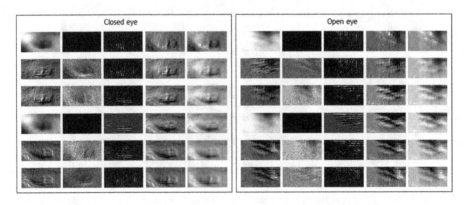

Fig. 3. Selected images of eyes regions after Gabor filtering.

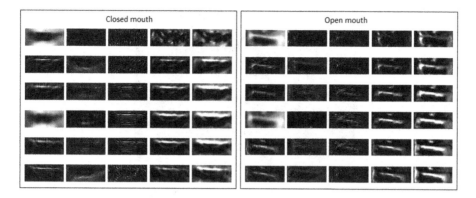

Fig. 4. Selected images of mouth regions after Gabor filtering.

2.5   Drowsiness Factor Estimation

We estimate fatigue state on a basis of eyelids closing time and mouth opening time. We adopted PERCLOS indicator (PERcentage of eye CLOSure) [16] in case of eye's analysis. Despite the more frequent opinion about PRECLOS cons, the method is still popular and the most accurate way to measure alertness bases only on visual data. More complex approaches require to use of inconvenient devices like EEG, EOG, heart rate monitors, etc. [23,30]. This indicator is represented as a proportion of time for which eyes are more than 80% closed (percentage of eye closure). It is usually calculated accumulatively over a predefined interval. As drowsiness often occurs after fatigue, yawning detection is an important factor to take into account because it is a strong signal that the driver can be affected by drowsiness in a short period of time. We estimate it by means of mouth opening time (we call it PEROPEN - PERcentage of mouth OPENing), as the number of frames where mouth state is predicted as closed, to the length of the examined time vector. The longer the time window is, the smoother the PERCLOS and PEROPEN functions are.

3   Experiments

3.1   Dataset

The data were gathered in Computer Vision Laboratory, Faculty of Computer Science and Information Technology, West Pomeranian University of Technology, using a proprietary simulation stand (see Fig. 5). The acquisition protocol of the benchmark data has been presented in [21]. All video materials have been recorded using FLIR SC325 camera working in LWIR band, equipped with 16-bit sensor of 320×240 and a lens of $25 \times 18.8°$ FOV.

Fig. 5. Experimental stand used for capturing video sequences

3.2   Evaluation Protocol

All algorithms have been implemented in the Python environment together with OpenCV, and SciKit-learn libraries. Original database includes records of 50 persons with different characteristics (women, men, people with and without beards, with and without glasses, young ones and older). We selected records of four persons of different physiognomy and manually annotated 10800 ground truth frames [21]. The set contained the following number of samples: left eye – 5558 (opened), 2414 (closed) and right eye – 8192 (opened), 4013 (closed). The respective numbers for mouth regions – 281 frames with open mouth, and 1170 with a closed mouth.

Initially, we experimented with typical binary classifiers on features extracted from eyes and mouth regions, independently. This part of the experiment was performed using Weka software. The hyperparameters for all classifiers was chosen according to typical and default settings. The mean values for 10-fold cross-validation for True Positive and False Positive rates are presented in Table 1. As a compromise between accuracy and computational overhead, we selected kNN ($k = 1$) to be implemented in further software.

The aim of both experiments was to verify the ability to identify images containing open and closed eyes/mouth regions in connection with face and eyes/mouth detection stages. The influence of the confidence level (responsible for the detection accuracy) on the eyes/mouth state detection was also investigated. It led to the observation that if we increase the confidence level (resulting in more eyes' candidates and more mouth's candidates being rejected), the classifier performs quite well. At the same time, if the confidence level drops, the overall accuracy also goes down.

3.3   Results

In comparison to the manually annotated video, our algorithm has an average accuracy near 70% for eyes and 87% for mouth state detection, respectively.

Table 1. The results of the experiments on the classification of eyes and mouth state (the best results are highlighted).

Classifier	True positive rate		False positive rate	
	Mouth	Eyes	Mouth	Eyes
1NN	0.989	0.962	0.035	0.053
Kstar	0.987	0.962	0.044	0.053
Random forest	0.992	0.947	0.029	0.089
Bagging	0.985	0.909	0.044	0.155
MultiLayerPerceptron	0.989	0.889	0.027	0.172
j48 (C4.5 decision tree)	0.980	0.887	0.040	0.153
Random tree	0.986	0.886	0.033	0.159
Classification via regression	0.988	0.876	0.025	0.193
REPTree	0.979	0.855	0.059	0.215
SimpleLogistic	0.989	0.748	0.032	0.435
SVM with SMO	0.989	0.728	0.032	0.586
NaiveBayes	0.975	0.637	0.052	0.283

It is important that most of the blinking eyes are detected, sometimes with a small delay, which is caused by a proposed buffer analysis. Likewise, for yawning detection algorithm we observe similar small delay, but comparing to the average blinking time, this delay is not crucial. In Fig. 6 the impact of buffer size on the detection accuracy is shown. For every single frame classification, the output value of analysis was the majority of states in the buffer. Using the buffer solves the problem of single frame misclassification, but in the long term it has a small negative impact on overall accuracy level (0.2%–0.6%). The accuracy results were calculated as a weighted average of accuracies for both states according to the number of frames.

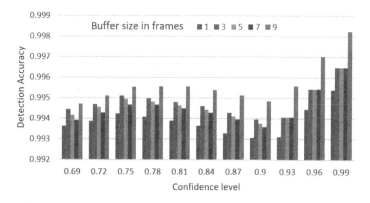

Fig. 6. The influence of mouth state buffer size, confidence level on detection accuracy

The accuracy changes following the detectors' confidence. As it was antici-pated, the confidence influences the detector performance. If the confidence is lower, then the total number of detected eye candidates is higher. The lower the detector confidence, the eyes state classifier's accuracy is also lower (see Fig. 7) since detected eyes are of poorer quality. Hence, if the eyes are detected with higher confidence, the eyes state classifier performs also with higher accuracy. The characteristic shape of accuracy curve for mouth state detection is caused by a significant decrease in the number of frames presenting open state at the classification stage (in the worst case 0.1% of all collected samples).

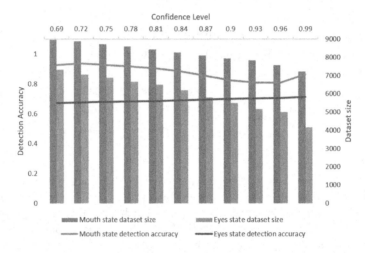

Fig. 7. The influence of dataset size (based on confidence level) on detection accuracy.

In Fig. 8 a visualization of mouth state detection accuracy is shown for a test sequence (frame numbers are on horizontal axis). The upper part of the figure presents the close state, and the bottom part shows the open mouth state. Blue blocks present ground true (annotated), while orange blocks present detected state. Visible three gaps in the further part of the video are caused by mouth occlusion (covering mouth with hand while yawning). The mouth was not detected there (only approximated by the tracker) and low confidence factor did not allow to make the prediction. An analogous visualization for eyes state detection is in Fig. 9.

In Fig. 10 the PERCLOS and PEROPEN functions calculated for the test sequence are presented. Given the short length of the test recording, a vector length of 800 frames was assumed. In case of a prolonged time of observation, longer buffers may be used. The analysis of above coefficients is based on the threshold which is responsible for estimating the drowsiness factor. By threshold-ing their values we can find the moments in time when the fatigue level increases and the drowsiness occurs. For both PERCLOS and PEROPEN, it is advisable

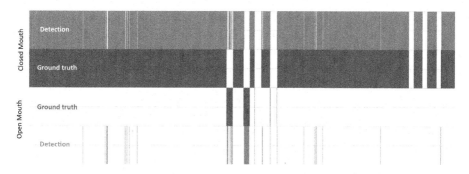

Fig. 8. Detection accuracy in time visualisation for mouth state detection. (Color figure online)

Fig. 9. Detection accuracy in time visualisation for eyes state detection. (Color figure online)

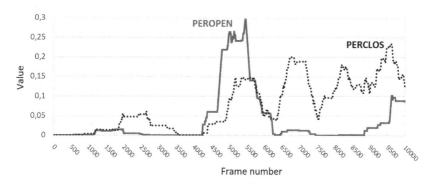

Fig. 10. PERCLOS and PEROPEN indicators for eyes and mouth states, respectively.

to set it to 0.15. As it can be seen, from frame nr 4500, the sequence of two yawns and more frequent eyes blinking begins, which may indicate actual drowsiness.

In order to evaluate the possibility to implement the algorithm in the mobile environment, we performed the tests involving three hardware platforms:

PC-based desktop (as a baseline) and two single-board computers (Raspberry PI 4 and NVidia Jetson AGX). It should be remembered that the algorithm was implemented in Python, not being the most optimal solution for end-user applications. Table 2 contains processing times of detection algorithm for the test sequence. As it can be seen, prototype implementation is not dedicated to single-board computers. On the other hand, for yawning detection only, the processing time is acceptable results. Unfortunately, for blinking detection, some optimizations should be performed to allow for real-time operation.

Table 2. Processing times on different hardware platforms

Platform/device	Time [s]	FPS
PC (Intel Xeon E5645 @2.4 GHz)	2665.59	4.06
nVidia Jetson AGX Xavier	3591.26	3.01
Raspberry pi4	6265.82	1.73

4   Conclusions

In the paper, we showed an algorithm of eyes and mouth state estimation based on thermal image analysis that may by used to detect driver's drowsiness. It consists of face, eyes and mouth detection and eyes and mouth state classification. It uses Haar-like features and Viola-Jones detector, while eyes and mouth states are classified using Gabor-filtered images with k-Nearest Neighbour classifier.

Application of thermal images here is a new approach, making the results resistant to many problems caused by visible spectrum imaging. A new contribution is also the introduction of the confidence coefficient that can be used to tune final system reliability. It should be remembered, the analysis of thermal images also has some limitations, e.g. the problem of eyes occlusion by thick transparent materials (e.g. glasses). In such case, we are forced to use mouth-only classifier.

The experiments showed that it is possible to detect eyes and mouth state in uncontrolled lighting conditions with significant accuracy. The results are competitive with the other proposals [7,8]. From the practical point of view, when fatigue is detected, the Advanced Driver Assistance System can stimulate the driver e.g. with acoustic or visual signals. What is more, in critical situations, after detecting severe fatigue or sleep, such system may even stop the vehicle.

In order to create a final solution, resistant to the problems presented above, future works in this area will include the integration of data from other imaging modalities, e.g. depth maps. It is also planned to analyse more types of driver's behaviour, e.g. head tilting, changing the focal point of sight or covering the mouth when yawning. These actions could be successfully recognized with help of current deep-learning based detectors, like CNNs or LSTMs. Since those approaches require significantly higher computing power, they were not considered at the present stage of development.

References

1. Alioua, N., Amine, A., Rziza, M.: Driver's fatigue detection based on yawning extraction. Int. J. Veh. Technol. **2014**, 7 (2014). Article ID: 678786
2. Andrysiak, T., Choras, M.: Image retrieval based on hierarchical Gabor filters. Int. J. Appl. Math. Comput. Sci. **15**(4), 471–480 (2005)
3. Azim, T., Jaffar, M.A., Mirza, A.M.: Fully automated real time fatigue detection of drivers through fuzzy expert systems. Appl. Soft Comput. **18**, 25–38 (2014)
4. Berglund, B., Rossi, G.B., Townsend, J.T., Pendrill, L.R.: Measurement with Persons: Theory, Methods, and Implementation Areas, p. 422. Psychology Press, New York (2013)
5. Bianconi, F., Fernandez, A.: Evaluation of the effects of Gabor filter parameters on texture classification. Pattern Recogn. **40**(12), 3325–3335 (2007)
6. Burduk, R.: The AdaBoost algorithm with the imprecision determine the weights of the observations. In: Nguyen, N.T., Attachoo, B., Trawiński, B., Somboonviwat, K. (eds.) ACIIDS 2014. LNCS (LNAI), vol. 8398, pp. 110–116. Springer, Cham (2014). https://doi.org/10.1007/978-3-319-05458-2_12
7. Craye, C., Rashwan, A., Kamel, M.S., Karray, F.: A multi-modal driver fatigue and distraction assessment system. Int. J. Intell. Transp. Syst. Res. **14**(3), 173–194 (2016). https://doi.org/10.1007/s13177-015-0112-9
8. Cyganek, B., Gruszczynski, S.: Hybrid computer vision system for drivers' eye recognition and fatigue monitoring. Neurocomputing **126**, 78–94 (2014)
9. Daza, I.G., Bergasa, L.M., Bronte, S., Yebes, J.J., Almazán, J., Arroyo, R.: Fusion of optimized indicators from Advanced Driver Assistance Systems (ADAS) for driver drowsiness detection. Sensors **14**, 1106–1131 (2014)
10. Forczmański, P.: Performance evaluation of selected thermal imaging-based human face detectors. In: Kurzynski, M., Wozniak, M., Burduk, R. (eds.) CORES 2017. AISC, vol. 578, pp. 170–181. Springer, Cham (2018). https://doi.org/10.1007/978-3-319-59162-9_18
11. Forczmański, P., Kutelski, K.: Driver drowsiness estimation by means of face depth map analysis. In: Pejaś, J., El Fray, I., Hyla, T., Kacprzyk, J. (eds.) ACS 2018. AISC, vol. 889, pp. 396–407. Springer, Cham (2019). https://doi.org/10.1007/978-3-030-03314-9_34
12. Fu, R., Wang, H., Zhao, W.: Dynamic driver fatigue detection using hidden Markov model in real driving condition. Exp. Syst. Appl. **63**, 397–411 (2016)
13. Hussien, M.N., Lye, M., Fauzi, M.F.A., Seong, T.C., Mansor, S.: Comparative analysis of eyes detection on face thermal images. In: 2017 IEEE International Conference on Signal and Image Processing Applications (ICSIPA), Kuching, pp. 385–389 (2017)
14. Jo, J., Lee, S.J., Park, K.R., Kim, I.J., Kim, J.: Detecting driver drowsiness using feature-level fusion and user-specific classification. Exp. Syst. Appl. **41**(4), 1139–1152 (2014)
15. Josephin, J., Lakshmi, C., James, S.: A review on the measures and techniques adapted for the detection of driver drowsiness. In: IOP Conference Series: Materials Science and Engineering, vol. 993, p. 012101 (2020). https://doi.org/10.1088/1757-899X/993/1/012101
16. Knipling, R.R., Wierwille, W.W.: Vehicle-based drowsy driver detection: current status and future prospects. In: Moving Toward Deployment. Proceedings of the IVHS America Annual Meeting. 2 Volumes, vol. 1, pp. 245–256 (1994)

17. Knapik, M., Cyganek, B.: Driver's fatigue recognition based on yawn detection in thermal images. Neurocomputing **338**, 274–292 (2019)
18. Kong, W., Zhou, L., Wang, Y., Zhang, J., Liu, J., Gao, S.: A system of driving fatigue detection based on machine vision and its application on smart device. J. Sens. **2015**, 11 (2015). https://doi.org/10.1155/2015/548602. Article ID: 548602
19. Krishnasree, V., Balaji, N., Rao, P.S.: A real time improved driver fatigue monitoring system. WSEAS Trans. Sig. Process. **10**, 146–155 (2014)
20. Małecki, K., Forczmański, P., Nowosielski, A., Smoliński, A., Ozga, D.: A new benchmark collection for driver fatigue research based on thermal, depth map and visible light imagery. In: Burduk, R., Kurzynski, M., Wozniak, M. (eds.) CORES 2019. AISC, vol. 977, pp. 295–304. Springer, Cham (2020). https://doi.org/10.1007/978-3-030-19738-4_30
21. Małecki, K., Nowosielski, A., Forczmański, P.: Multispectral data acquisition in the assessment of driver's fatigue. In: Mikulski, J. (ed.) TST 2017. CCIS, vol. 715, pp. 320–332. Springer, Cham (2017). https://doi.org/10.1007/978-3-319-66251-0_26
22. Mitas, A., Czapla, Z., Bugdol, M., Rygula, A.: Registration and evaluation of biometric parameters of the driver to improve road safety, pp. 71–79. Scientific Papers of Transport, Silesian University of Technology (2010)
23. Nissimagoudar, P.C., Nandi, A.V., Gireesha, H.M.: A feature extraction and selection method for EEG based driver alert/drowsy state detection. In: Abraham, A., Jabbar, M.A., Tiwari, S., Jesus, I.M.S. (eds.) SoCPaR 2019. AISC, vol. 1182, pp. 297–306. Springer, Cham (2021). https://doi.org/10.1007/978-3-030-49345-5_31
24. Nowosielski, A.: Vision-based solutions for driver assistance. J. Theor. Appl. Comput. Sci. **8**(4), 35–44 (2014)
25. Nowosielski, A., Forczmański, P.: Touchless typing with head movements captured in thermal spectrum. Pattern Anal. Appl. **22**(3), 841–855 (2018). https://doi.org/10.1007/s10044-018-0741-0
26. Poster, D., Hu, S., Nasrabadi, N.M., Riggan, B.S.: An examination of deep-learning based landmark detection methods on thermal face imagery. In: 2019 IEEE/CVF Conference on Computer Vision and Pattern Recognition Workshops (CVPRW), pp. 980–987 (2019)
27. Sacco, M., Farrugia, R.A.: Driver fatigue monitoring system using support vector machines. In: 2012 5th International Symposium on Communications, Control and Signal Processing, Rome, pp. 1–5 (2012)
28. Satish, K., Lalitesh, A., Bhargavi, K., Prem, M.S., Anjali, T.: Driver drowsiness detection. In: 2020 International Conference on Communication and Signal Processing (ICCSP), pp. 0380–0384 (2020)
29. Smolensky, M.H., Di Milia, L., Ohayon, M.M., Philip, P.: Sleep disorders, medical conditions, and road accident risk. Accid. Anal. Prev. **43**(2), 533–548 (2011)
30. Trutschel, U., Sirois, B., Sommer, D., Golz, M., Edwards, D.: PERCLOS: an alertness measure of the past. In: Proceedings of the Sixth International Driving Symposium on Human Factors in Driver Assessment, Training and Vehicle Design, pp. 172–179 (2017)
31. Viola, P., Jones, M.J.: Robust real-time face detection. Int. J. Comput. Vis. **57**(2), 137–154 (2004). https://doi.org/10.1023/B:VISI.0000013087.49260.fb

32. Virginia Tech Transportation Institute: Day or Night, Driving while Tired a Leading Cause of Accidents (2013). http://www.vtnews.vt.edu/articles/2013/04/041513-vtti-fatigue.html
33. Weller, G., Schlag, B.: Road user behavior model. Deliverable D8 project RIPCORD-ISERET, 6 Framework Programme of the European Union (2007). http://ripcord.bast.de
34. Zhang, Y., Hua, C.: Driver fatigue recognition based on facial expression analysis using local binary patterns. Optik Int. J. Light Electron Opt. **126**(23), 4501–4505 (2015)

Smart Events in Behavior of Non-player Characters in Computer Games

Marcin Zieliński, Piotr Napieralski⬤, Marcin Daszuta,
and Dominik Szajerman(✉)⬤

Institute of Information Technology, Lodz University of Technology, Łódź, Poland
`dominik.szajerman@p.lodz.pl`

Abstract. This work contains a solution improvement for Smart Events, which are one of the ways to guide the behavior of NPCs in computer games. The improvement consists of three aspects: introducing the possibility of group actions by agents, i.e. cooperation between them, extending the SE with the possibility of handling ordinary events not only emergency, and introducing the possibility of taking random (but predetermined) actions as part of participation in the event.

In addition, two event scenarios were presented that allowed the Smart Events operation to be examined. The study consists of comparing the performance of the SE with another well-known algorithm (FSM) and of comparing different runs of the same event determined by the improved algorithm.

Comparing the performance required proposing measures that would allow for the presentation of quantitative differences between the runs of different algorithms or the same algorithm in different runs. Three were proposed: time needed by the AI subsystem in one simulation frame, the number of decisions in the frame, and the number of frames per second of simulation.

Keywords: Smart Events · NPC behavior · Behavior trees

1 Introduction

Computer games are a special area of use of artificial intelligence. Interesting cases are open-world games, where the player can move freely in an environment inhabited by numerous NPCs. Often one of the main elements there are interactions with these NPCs. Typically in the open-world games such as "Grand Theft Auto V" or "Elex", most of the NPC logic is related to the use of artificial intelligence in combat. Less attention is paid to the behaviors related to their daily life, in situations when the player does not interact with them in any way. However, it is thanks to AI outside of combat that the game world seems more alive in the eyes of the player. One of the reasons for this is the high complexity when an increased number of NPCs/agents have many different behaviors depending on the context. In this situation, the game software is complicated.

© Springer Nature Switzerland AG 2021
M. Paszynski et al. (Eds.): ICCS 2021, LNCS 12746, pp. 164–177, 2021.
https://doi.org/10.1007/978-3-030-77977-1_13

This work focuses on one of the most frequently chosen directions in the design of artificial intelligence systems in games, and more specifically the "Smart" approach. A proposal was presented to expand the "Smart Events" solution, allowing for the elimination of its disadvantages.

2 Related Work

Embedding logic within the environment has become a well-accepted solution for managing the complexity of behaviors in both science and industry. The main example of this is Smart Objects (SO) introduced by Kallmann and Thalmann [7]. They are widely used in the computer game industry in a simplified form. The SO stores the information and is responsible for positioning and playing the animation of the character that uses it. An example of SO could be a lever object in a game scene. The character does not need to know what this object is. To change its state, it simply performs the action of using SO, and the lever object itself takes care of its positioning and animation. As a result, there can be many different types of SO in a scene and the character only needs one action to use them all properly. The disadvantage of this solution is the lack of support for interrupting the behavior of characters and no possibility of nesting them.

A solution close to SO, eliminating this problem, was implemented by the creators of "The Sims 4" game [6]. The gameplay, in simple terms, consists of selecting the appropriate SO in order to interact with them for the character under consideration, and thus to satisfy his needs. The character performs actions of his own choosing (low priority), as well as those ordered by the player (high priority). Interactions with SO can consist of smaller blocks that are non-breakable. However, it is possible to break all interactions that consist of multiple blocks.

In the game "Bioshock: Infinite" the developers used SO while modeling the behavior of the player's assistant – Elizabeth [5]. SO are responsible for her behavior. The designers filled the game world with invisible tags that allow Elizabeth to draw attention to individual places. Thanks to this, her character can move around the area, finding elements of interest to her and interacting with them.

Another way of using SO was proposed in the game "S.T.A.L.K.E.R." [4]. The most important feature of an artificial life is that each NPC has a life of their own. In order to achieve this, the designers provided the characters with the opportunity to move between the different levels in the game, as well as remember the information obtained during their existence. In order to diversify the behavior of NPCs and enliven the game world, the designers implemented a SO development solution, called by them "Smart Terrains". ST are responsible for assigning tasks and behaviors to individual characters located in the area defined by them. This allowed for the addition of fractions bases, characters sitting together by bonfires, etc. to the game.

A similar approach was proposed in [2]. The authors noticed the need to create a solution supporting the creation of complex NPC behaviors. Inspired by research in the field of crowd simulation, they adapted it for use in games and implemented a solution called Smart Areas. Objects of this type contain the logic responsible for deciding what behavior a NPC should adopt while in their area. In addition, it is possible to determine: how many agents can take a specific behavior at a given moment, what are the conditions imposed on an agent before taking an action, what actions are to be performed during various events (e.g. leaving the area/entering the Smart Area).

The authors of [11] proposed their solution in the form of the so-called Smart Events. It allows to simulate the reaction of a character to emergency events. The aim of the creators was to maintain a low demand for computing power and good scalability of the solution. To achieve this, they moved the logic responsible for choosing the behavior of a given character from the object representing it to an instance of an event. In order to differentiate behaviors for many agents, the authors referred to the fact that a person has different social roles defined by, for example, age, profession, gender or family relationships. In a given situation, a person can only assume one of their roles. The event object stores information about the behavior that an individual should adopt depending on the state of the event (e.g. the extent of fire spread) and its current major social role. This allows an individual to make a simple decision about their choice of behavior.

As can be seen, the AI behavior of agents in computer games can be solved in various ways. Another group are methods based on the agents' memories and even their emotions [9].

2.1 Primed Agent

Primed agent is a concept proposed in [11]. Based on research in the field of psychology, they introduced the so-called priming, i.e. putting one of the cultural concepts of himself in front of another. Such a change may occur in response to the language or context of the conversation. A man may have many concepts, including conflicting ones, but at a given moment only one of them becomes active. Expanding the proposals contained in the research, they distinguished eight features that are considered in the priming process (Table 1).

Table 1. Features considered in the priming process together with their representative values [11].

Trait	Possible values
Age	Child, adult, elder
Gender	Male, female
Ethnicity	American, European, Asian, African, Australian, Hispanic
Religion	Christian, Jewish, Muslim, Hindu, Buddhist
Vocation	Firefighter, policeman, teacher, student
Relational	Mother/father, daughter/son, husband/wife, friend/stranger
Political affiliation	Democrat, republican

Each agent has its own trait values and can have more than one trait value. Besides, some of them may have fields with additional data, e.g. for a parent they are the agent's children. Any of the traits can be brought to the fore. Both in [11] and in our solution, the priming process has been limited to appearing only when entering a location and when interacting with another agent.

2.2   Smart Events

The main idea behind the concept of Smart Events is to transfer the logic related to the reaction of agents to an event to the object representing this event [11]. This solution is an extension of the Smart Objects concept, in which the object informs the agent about how it should be used by this agent [7]. The developers defined the event as a planned or externally triggered addition or removal of a fact from the world model.

The Smart Event is represented by the following parameters: type, position, location, start time, end time, evolution (finite state machine), influence region, participants, event emergency level, corresponding (available) actions.

Communication between Smart Event and agents is carried out via the message board. The board is responsible for broadcasting and updating information about the evolution of the event. When an event starts, its information is sent to message boards according to the affected region. Then each board informs its subscribers about the event. Agents can be assigned to it statically – regardless of their location, or dynamically – by signing up when entering the board operation region and unsubscribing from it while leaving. An agent can choose whether he wants to react to the event he was informed about by comparing the level of emergency of the action he is currently performing to the emergency level of the event. If the current action turns out to be less important, it will ask the board to assign an appropriate, new action. Otherwise, the event will be ignored by him.

Smart Events may change over time. Associated agent's actions should also adapt to its development. FSM is used to model the evolution of an event. As it develops, the event will inform relevant message boards about changes in its status, including: emergency level, region of influence, and corresponding actions.

3   Method

Testing agent behavior algorithms in games is not easy. It is quite difficult to assess their plausibility, because a set of comparative data on human behavior may not be possible to prepare. Also, the sheer complexity of behaviors, changes, transitions between them, their pace does not facilitate such analyzes. At the same time, the examples should be relatively simple so that the mixing of various behaviors and various factors does not make observation impossible.

During this work, several tests were prepared consisting of scenes influencing the course of various elementary behaviors of agents. For the purposes of the presentation, two scenarios have been selected that cover the most representative and interesting cases possible. Elementary behaviors and their changes presented in them can be adapted to other situations. In this way, you can generalize the solution.

This section presents how Smart Events has been improved in this work. Then two scenarios to test it and measures that were used to compare the performance of the solutions were presented.

In their article, the creators of Behavior Objects described the lack of support for coordinating actions performed by agents as the main disadvantage of Smart Events. Another problem for them turned out to be that for each set of character traits, an event provides the same behavior [1].

3.1   Finite State Machine

FSM is a well-known algorithm used in computer games AI. It is used here for comparisons with the Smart Events method. An important element of the structure used is the message system. It allows agent to send immediate as well as delayed messages to a given machine. After receiving the message, it forwards it to the current state that can handle it based on the content it contains. Thanks to this solution, it is possible to send signals that may affect the operation of agents. Figure 1 presents an example of the firefighter's FSM.

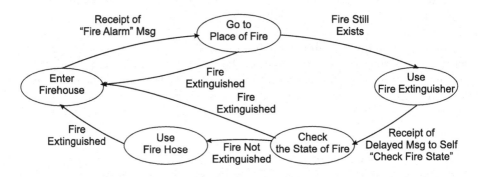

Fig. 1. An example of the firefighter's FSM.

3.2   The Proposed Solution

The extension of the Smart Events solution proposed in this chapter aims to eliminate the above-mentioned disadvantages. The task of the simulation using this extension is to show their improvement and to show the possibility of using the model in a non-emergency scenario.

The proposed development is largely based on parameterized behavior trees, which allows to eliminate the problem of the lack of coordination of agents' actions.

The improvement consists in replacing the action table, communication via the message board, and agent priming with the event behavior tree. When an event occurs, the agents taking part in it cease to perform their actions and are deprived of their autonomy. The event behavior tree takes control over them. The leaves of this tree represent actions performed by controlled agents.

The Smart Events extension proposed in this chapter uses the event behavior tree, but leaves its foundations intact only by extending it. The structure of the event object has been extended to include a set of related group actions. The group behavior table differs from the basic one by adding a column specifying the role in which the agent can act in a given situation.

The very process of assigning of action to an agent has changed. When an event occurs, the agent is assigned behavior according to the rules known from Smart Events. However, when all agents participating in the event are informed about it and receive selected actions, the event starts an attempt to perform a group action. For this purpose, it looks for the behavior in the table for which all roles can be filled by its participants. In addition to meeting the conditions in the table, in order for an agent to be taken into account in assigning roles, the importance of the previously assigned individual action must be lower than or equal to the emergency level of the event. This solution ensures that the agent chooses the action that is more important at a given moment, regardless of its type.

The second problem – a small variety of selected actions, was taken into account in the simulation when designing behavior trees. In addition to standard control nodes, a random selection node was used that allows to perform an action drawn from a given set. In the second simulation scenario presented in Sect. 3.4, it allowed to diversify the behavior of the shopkeeper while working at the cash register, and the customer while looking around the store. Another issue that helps to avoid low variety is the random success node shown in the behavior tree. It returns a positive execution status with a given probability, and thus decides about interrupting or further executing the actions that follow. It was used during the haggling of agents in order to simulate the decisions about lowering the price made by the shopkeeper.

3.3   Experiment: The "School" Scenario

The tests performed in this scenario are designed to compare two AI methods in the same task. They are aimed at using all elements of the Smart Events solution. The tests performed with the use of the FSM method have been developed in such a way as to obtain runs as similar as possible to Smart Events.

During the simulation, an emergency event scenario in the form of a fire at school was used, created by the authors of the Smart Events model [11]. It uses of all the model's most important elements, which translates into the possibility of a good comparison of both tested solutions.

The structure of the world model was prepared in a way that allows to reproduce the course of the presented scenario. It has the necessary components, such as the fire department, police building and school, as well as all the agents used. The designed section of the city, which is a model of the simulation world, consists of three buildings (Fig. 2):

- fire brigade – a place where a firefighter is, equipped with a desk, fire extinguisher and an alarm bell,
- police station – place where a policeman is, equipped with a desk and an alarm bell,
- school – the place where the teacher works and where the pupil is, mother and child go to this building. It consists of a hall, bathroom and one classroom. There are benches, a blackboard and a bucket in the classroom. This is where the fire breaks out. The bathroom is equipped with a towel and a tap with a sink. In the hall there is a fire extinguisher, a fire hose and an alarm bell.

Fig. 2. Simulation world model with buildings listed: fire station, police station, and school.

In the simulation scenario, there were six agents with different roles:

1. Firefighter – Occupation: Fireman, Workplace: Fire Department Building, Initial Behavior: Sit at the desk.
2. Policeman – Occupation: Policeman, Workplace: Police Station, Initial Behavior: Sit at Desk.
3. Teacher – Occupation: Teacher, Workplace: School Building, Initial Behavior: Teach at the blackboard.
4. Student – Occupation: Student, Workplace: School Building, Initial Behavior: Sit at the School Desk.
5. Son – Relationship: Agent "Mothers's" Son, Initial Behavior: Follow Mother.
6. Mother – account: agent "Son's" mother, initial behavior: take her son to school.

Agents perform their initial actions until a fire breaks out. When an event begins, they begin to implement behaviors related to it. All actions occurring during a fire along with the conditions under which they occur are presented in the table shown in Fig. 3.

Trait	Value	Priming	EmergencyLevel	Action
= Vocation	Teacher	Enter(School)	Low	PourWater ()
= Vocation	Teacher	Enter(School)	Medium	ExtinguishFire ()
= Vocation	Teacher	Enter(School)	High	LeadAwayFromSchool ()
= Vocation	Student	Enter(School)	Low	StareAtFire ()
= Vocation	Student	Enter(School)	Medium	FollowLeader ()
= Vocation	Student	Enter(School)	High	FollowLeader ()
= Vocation	Firefighter	Enter(emergency)	Low	Smoother ()
= Vocation	Firefighter	Enter(emergency)	Medium	ExtinguishFire ()
= Vocation	Firefighter	Enter(emergency)	High	FightFire ()
= Vocation	Policeman	Enter(emergency)	Low	Smoother ()
= Vocation	Policeman	Enter(emergency)	Medium	ExtinguishFire ()
= Vocation	Policeman	Enter(emergency)	High	ManageCrowd ()
= Relational	Son	Interact(myParent)	Low	StareAtFire ()
= Relational	Son	Interact(myParent)	Medium	FollowLeader ()
= Relational	Son	Interact(myParent)	High	FollowLeader ()
= Relational	Mother	Interact(myChild)	Low	Smoother ()
= Relational	Mother	Interact(myChild)	Medium	LeadAwayFromSchool ()
= Relational	Mother	Interact(myChild)	High	LeadAwayFromSchool ()

Fig. 3. The action table used during the fire in the simulations.

3.4 Experiment: The "Shop" Scenario

The purpose of this scenario was to explore the new possibilities offered by the improvement of the SE system.

The simulation scenario was inspired by [10] where the parameterization of behavior trees is considered. The situation of a customer coming to the store and trying to buy goods at a satisfactory price is presented there. A similar scenario takes place in the case of the simulation in this work.

Upon entering the building, the customer looks around him looking for interesting products. Then he walks over to the shopkeeper and he is trying to haggle a lower price for the pot, which he wants to buy. The shopkeeper may agree to a discount by reducing the price by a given value. If the customer is satisfied with the price, he agrees to it, then takes the pot and leaves the store. Otherwise, the customer tries to persuade the shopkeeper to reduce the price. If the shopkeeper does not agree to the discount, the customer looks around again.

The location used in the simulation shows the shop building. It is a shopkeeper's workplace, equipped with shelves with products, a counter and pedestals on which pots stand. The simulation world model is presented in Fig. 4.

During the test, two simulation runs with a different value of the random seed were performed. This allowed to show the differences in the actions performed by agents, and thus draw attention to their variety.

At the very beginning of the simulation, both agents enter the store to adopt their initial behavior. The customer goes to the store shelves downstairs and starts looking around. He had a choice of three places in the store. The one in which it performs this action has been selected using the previously mentioned

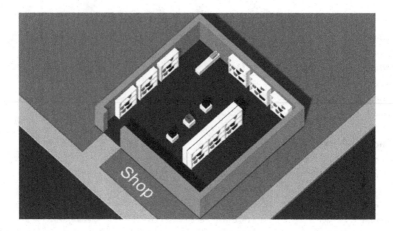

Fig. 4. A simulation world model that is a store where the scenario takes place.

random selection node. At that time, the shopkeeper is standing at the counter. From time to time, he whistles and sings (shown as text). His actions are similarly selected randomly from the pool. When the customer completes his look action, he triggers a haggle event. As both agents meet the conditions to be placed as shopkeeper and customer, they stop performing their actions and the group event behavior tree takes control of them.

Then, the customer comes to the counter and the shopkeeper greets him. Since the price of the pot he wants to buy is 6 coins and his budget is 4, he begins to haggle. The customer asks for a price reduction, then the shopkeeper decides to refuse, and the event ends. Such a verdict was established by the shopkeeper using a random success node with a 50% chance of a successful action. When the event is over, the agents resume performing their individual actions.

The customer selects the same location again where he performs a look action. The shopkeeper, on the other hand, makes a random decision to do something on his own. After the end of his behavior, the customer starts the haggling event again. This time, however, he manages to get a lower price. As the shopkeeper lowered the price of the pot to 5 coins, the customer still cannot afford it. So he continues to try to get the lower price. The shopkeeper again agrees to the reduction, thus meeting the conditions of the customer. The customer then agrees to buy a pot, takes it from the store and leaves, ending the event and simulation.

3.5   Performance Measures

Measurement data was taken every frame. In order to reduce the effect of the difference in the number of frames on the obtained results, they were given in the form of arithmetic means and maximum values.

All tests were conducted using a computer with an Intel Core i7-2670QM quad-core processor with a base clock of 2.20 GHz, Nvidia GeForce GT 540M graphics card and 8 GB of RAM.

The first of the performance measures was the time needed by the AI system in one simulation frame. A similar measurement method was used in the work of the authors of the artificial intelligence system, which was ultimately to be used in a high-budget computer game with an open world [8].

Another measured value is the number of decisions the AI system has to make in the frame, and also completely throughout the simulation. The authors of the Smart Events solution used a similar comparison in their article, but without giving specific numbers [11]. The total number of decisions taken for the same scenario is also a value independent of any offsets of agents' execution, which makes it good for performance comparisons.

The last value tested in the experiment is the number of frames per second achieved during the simulation. Measurements of this type were carried out for several solutions with different numbers of agents belonging to the crowd in the simulations conducted in [3].

4   Results and Discussion

4.1   Comparison to FSM

As previously mentioned, it is difficult to directly assess the believability (realism) of agent's behavior. However, the proposed solution you can be compared with the one widely used in games, e.g. FSM.

Figure 5 shows comparison of the run of the "School" scenario according to the FSM and SE methods:

1. top-left: SE-driven simulation after the first twenty seconds,
2. top-right: SE-driven simulation after the first sixty seconds,
3. bottom-left: FSM-driven simulation after the first twenty seconds,
4. bottom-right: FSM-driven simulation after the first sixty seconds.

The two methods are compared by putting together the paths that the agents travel through the event sequence. For each agent and each time instant t_i, the distance (in meters) between its locations (p_i) was calculated in both simulations: FSM and SE (Eq. 1). The average and maximum distances calculated for the agents participating in the scenario are presented in Table 2. As can be seen, the measured values also confirm a very high similarity of behaviors.

$$d(t_i) = ||\boldsymbol{p}_i^{FSM}(t_i) - \boldsymbol{p}_i^{SE}(t_i)|| \tag{1}$$

Performance tests were conducted using measures introduced in Subsect. 3.5.

Comparing the performance results obtained in the tests, it can be concluded that the solution using Smart Events is more efficient (Table 3). Despite the slight differences caused mainly by the small scale of the simulation, the percentage gain in efficiency is noticeable.

The number of decisions made by the AI system decreased significantly, which translates to the time it takes in a single simulation frame. This, in turn, directly affects the time that can be spent on other systems during game development.

Fig. 5. Comparison of the two steps of "School" scenario with both methods: SE and FSM. The agent positions for both algorithms in the same phases of the event are so similar that the differences are difficult to spot.

Table 2. Summary of calculated distances for each agent.

Agent	Average d [m]	Maximum d [m]
Firefighter	0.032772	0.37537
Policeman	0.009708	0.162012
Teacher	0.174445	0.527891
Student	0.104061	1.155141
Son	0.18145	1.142276
Mother	0.136615	0.958957

Table 3. Summary of performance test results for both compared methods.

Tested method	Time in one frame (μs)		Number of decisions in a frame		Total number of decisions	Average number of frames per second
	Average	Maximum	Average	Maximum		
FSM	0.517186	1602.909	0.128684521	184	764	79.5774344
SE	**0.338778**	**1255.364**	**0.121447455**	**118**	**735**	**80.57757943**
Difference	−34%	−22%	−6%	−36%	−4%	1%

The frame rate increase achieved in each simulation run was insignificant. It can be assumed that for more complex scenarios with more agents the performance differences would be noticeable to a greater extent.

The results achieved in the performance tests conducted allowed for the conclusion that the Smart Events solution is less demanding than the one using the FSM.

4.2 Improved Smart Events

The improved SE action is shown on the basis of the "Shop" scenario. The stages of haggling between agents are presented in Fig. 6. This is result for the second of the described in Subsect. 3.4 runs. There can be observed the key events from the point of view of the simulation of the group action driven by improved SE:

1. top-left: the customer triggers the haggle event, and shopkeeper speaks to him,
2. top-right: the customers begins to haggle,
3. bottom-left: the effect of random success node is positive and shopkeeper agrees to lower the price,
4. bottom-right: the price has reached a value that is satisfactory for the customer, so the whole event ends with a purchase.

Except that both runs end differently, the differences between them can also be seen when comparing their performance using a method analogous to the one used in the comparison of SE with FSM. Table 4 shows the results. Lengthening the duration of the simulation through a longer haggling process meant that in

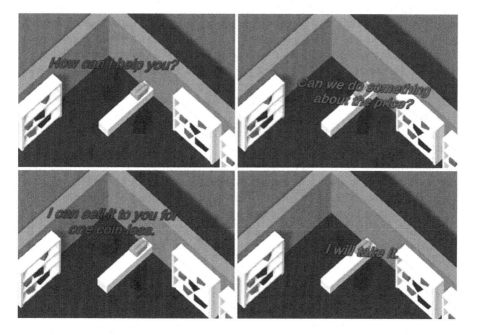

Fig. 6. Subsequent stages of agent interactions in the process of haggling.

the second run, the total number of decisions was doubled. The simulation in both cases showed the same scenario with two agents, therefore the maximum number of decisions in a single frame did not change. Similar to previous tests, due to the small scale, the percentage difference in average frames per second is close to zero.

Table 4. Summary of performance test results for both simulation runs using the improved method.

Test	Time in one frame (μs)		Number of decisions in a frame		Total number of decisions	Average number of frames per second
	Average	Maximum	Average	Maximum		
First	0.353066	1466.216	0.026282051	44	123	115.9219803
Second	0.421891	3085.444	0.044221479	44	357	116.1455381
Difference	19%	110%	68%	0%	190%	0%

5   Conclusions

The aim of this work was to show a new approach to smart events and to present the overall performance of this method. The tests conducted in the first scenario allowed to confirm that the tested solution give similar results and has lower performance requirements than a competitive solution based on FSMs. The second scenario – simulation of the customers's haggling with the shopkeeper made it possible to test the new approach in practice. It also showed the possibility of applying the tested solution in events not related to emergencies. The use of a group behavior tree in the event allowed for the creation of conditions for designing coordinated actions of agents. On the other hand, the use of nodes related to randomness in character behavior trees resulted in a significant diversification of the course of the scenario. The proposed extension of Smart Events gave satisfactory results in the simulation.

The measurement method prepared as part of the work allowed to examine the efficiency of the presented SE solution. The obtained results made it possible to perform two tests: the comparison of the SE and the popular FSM method as well as the comparison of the two simulation runs of improved SE, additionally confirming their differentiation.

SE allows developers to easily model the behavior of secondary NPC characters in the event of various types of emergency and non-emergency events that can be easily applied in a game.

Taking into account the development of the smart approach in modeling artificial intelligence in games so far, it can be concluded that this is a good direction for the development of this area of game development.

Acknowledgment. This work was supported by The National Centre for Research and Development within the project "From Robots to Humans: Innovative affective AI system for FPS and TPS games with dynamically regulated psychological aspects of human behaviour" (POIR.01.02.00-00-0133/16). We thank Mateusz Makowiec and Filip Wróbel for assistance with methodology and comments that greatly improved the manuscript.

References

1. Cerny, M., Plch, T., Marko, M., Gemrot, J., Ondracek, P., Brom, C.: Using behavior objects to manage complexity in virtual worlds. IEEE Trans. Comput. Intell. AI Games **9**(2), 166–180 (2017). https://doi.org/10.1109/tciaig.2016.2528499
2. Cerny, M., Plch, T., Marko, M., Ondracek, P., Brom, C.: Smart areas - a modular approach to simulation of daily life in an open world video game. In: Proceedings of the 6th International Conference on Agents and Artificial Intelligence. SCITEPRESS - Science and and Technology Publications (2014). https://doi.org/10.5220/0004921107030708
3. Gu, Q., Deng, Z.: Generating freestyle group formations in agent-based crowd simulations. IEEE Comput. Graphics Appl. **33**(1), 20–31 (2013). https://doi.org/10.1109/mcg.2011.87
4. Iassenev, D., Champandard, A.J.: A-life, emergent AI and S.T.A.L.K.E.R. (2008). https://aigamedev.com/open/interviews/stalker-alife/
5. IGN: Bioshock Infinite - The Revolutionary AI Behind Elizabeth (2013). https://www.youtube.com/watch?v=2viudg2jsE8
6. Ingebretson, P., Rebuschatis, M.: Concurrent interactions in the SIMS 4. In: Game Developers Conference (2014). http://www.gdcvault.com/play/1020190/Concurrent-Interactions-in-The-Sims
7. Kallmann, M., Thalmann, D.: Modeling behaviors of interactive objects for real-time virtual environments. J. Visual Lang. Comput. **13**(2), 177–195 (2002). https://doi.org/10.1006/jvlc.2001.0229
8. Plch, T., Marko, M., Ondracek, P., Cerny, M., Gemrot, J., Brom, C.: An AI system for large open virtual world. In: Proceedings of the 10th Annual AAAI Conference on Artificial Intelligence and Interactive Digital Entertainment (2014)
9. Rogalski, J., Szajerman, D.: A memory model for emotional decision-making agent in a game. J. Appl. Comput. Sci. **26**(2), 161–186 (2018)
10. Shoulson, A., Garcia, F.M., Jones, M., Mead, R., Badler, N.I.: Parameterizing behavior trees. In: Allbeck, J.M., Faloutsos, P. (eds.) MIG 2011. LNCS, vol. 7060, pp. 144–155. Springer, Heidelberg (2011). https://doi.org/10.1007/978-3-642-25090-3_13
11. Stocker, C., Sun, L., Huang, P., Qin, W., Allbeck, J.M., Badler, N.I.: Smart events and primed agents. In: Allbeck, J., Badler, N., Bickmore, T., Pelachaud, C., Safonova, A. (eds.) IVA 2010. LNCS (LNAI), vol. 6356, pp. 15–27. Springer, Heidelberg (2010). https://doi.org/10.1007/978-3-642-15892-6_2

Place Inference via Graph-Based Decisions on Deep Embeddings and Blur Detections

Piotr Wozniak[2] and Bogdan Kwolek[1(✉)]

[1] AGH University of Science and Technology, 30 Mickiewicza,
30-059 Kraków, Poland
[2] Rzeszów University of Technology, Al. Powstańców Warszawy 12,
35-959 Rzeszów, Poland
bkw@agh.edu.pl
http://home.agh.edu.pl/~bkw/contact.html

Abstract. Current approaches to visual place recognition for loop closure do not provide information about confidence of decisions. In this work we present an algorithm for place recognition on the basis of graph-based decisions on deep embeddings and blur detections. The graph constructed in advance permits together with information about the room category an inference on usefulness of place recognition, and in particular, it enables the evaluation the confidence of final decision. We demonstrate experimentally that thanks to proposed blur detection the accuracy of scene recognition is much higher. We evaluate performance of place recognition on the basis of manually selected places for recognition with corresponding sets of relevant and irrelevant images. The algorithm has been evaluated on large dataset for visual place recognition that contains both images with severe (unknown) blurs and sharp images. Images with 6-DOF viewpoint variations were recorded using a humanoid robot.

Keywords: Visual place recognition · CNNs · Images with unknown blur

1 Introduction

Simultaneous localization and mapping (SLAM) is the computational problem aiming at constructing and updating a map of an unknown environment while simultaneously keeping path of an agent's location within it [1]. Although SLAM is used in many practical applications, several challenges prevent its wider adoption. Since SLAM is based on sequential movement and measurements that are contaminated by some margin of error, the error accumulates over time, causing substantial deviation from actual agent's locations. This can in turn lead to map distortion or even collapse and thus making subsequent searches difficult. Loop closure is a task consisting in recognition of previously-visited location and updating the constructed map accordingly. Therefore, detecting loop closure (or

© Springer Nature Switzerland AG 2021
M. Paszynski et al. (Eds.): ICCS 2021, LNCS 12746, pp. 178–192, 2021.
https://doi.org/10.1007/978-3-030-77977-1_14

previously visited places) in order to correct the accumulated error during the exploration is very important task [2]. This permits the SLAM system to relocalize the sensor after a tracking failure, which might happen in unfavorable circumstances, like severe occlusion or abrupt movements.

The aim of visual place recognition (VPR) is to retrieve correct place matches under viewpoint and illumination variations, while requiring as less as possible computational power and memory storage [3]. Over the past years several methods for visual place recognition have been developed [2,3]. Although most of visual place recognition methods were developed for SLAM, VPR algorithms also found applications in monitoring of electricity pylons using aerial imagery [4], brain-inspired navigation [5], and image-search based on visual content [6]. VPR is very challenging problem because images of the same place but taken at different times may differ notably from each other. The differences can be caused by factors such as varying illumination, shadows as well as changes resulting from different passing the same route.

In robotics, most of evaluations of VPR systems were performed using data acquired by ground-based mobile platforms or robots. The degree of viewpoint variation that takes place during scene perception by a humanoid robot is far more complex than viewpoint variations experienced by mobile robots [7]. When a humanoid robot is walking, turning, or squatting, its head mounted camera moves in a jerky and sometimes unpredictable way [8]. Motion blur, one of the biggest problems for feature-based SLAM systems, causes inaccuracies and location losses during map construction. Most of datasets for visual place recognition provide lateral or 3D variations of viewpoint. The 24/7 Query dataset [9] contains outdoor images with 6-DOF viewpoint variation. Recently, the Shopping street dataset targeted for aerial place recognition with 6-DOF viewpoint change has been introduced in [10]. Most of VPR benchmark data are time-based, as frames are acquired and stored at a fixed FPS (frames per second) rate of a video camera. Typically, they are recorded under assumption of non-zero speed of the robot. In [11] a frame is picked every few meters to represent a new place. A disadvantage of both time- and distance-based approaches are huge requirements for data storage. Moreover, they lead to visually similar frames at different places and thus to inaccuracies and impracticality for long-term robot missions.

VPR is typically cast as image retrieval problem. Several handcrafted local and global feature descriptors were proposed for place recognition [3]. CNNs for visual place recognition were proposed in [12]. Since publication of this seminal work, more and more data-driven image description approaches have emerged. Performance of these algorithms has been studied in [13]. In [14], a VLAD [15] layer for CNN architecture that could be trained in end-to-end fashion, specifically for place recognition task has been proposed. The experimental results achieved by NetVLAD on very challenging datasets significantly outperformed results achieved by pre-trained CNNs. Very high potential of VLAD has recently been confirmed in [16], where a comprehensive comparison of 10 VPR systems identified the NetVLAD as the best overall performing method.

Motivated by lack of a dataset with variations arising during typical movement of humanoid and walking robots, particularly containing images with severe (unknown) blurs we recorded a dataset using camera mounted on head of humanoid robot. To cope with place recognition on the basis of images with unknown blur we propose an effective algorithm for blur detection. We demonstrate experimentally that the proposed algorithm considerably outperforms state-of-the-art algorithms on images with severe and unknown motion blur. We demonstrate also that owing to use of the proposed algorithm, considerable gains of performance in scene categorization can be achieved. We employ minimum spanning tree (MST) for place recognition purposes and show its usefulness. Thanks to information extracted on the basis of MST like proportion of images belonging to given class with respect to number of images from remaining classes in a given tree branch the system can infer about confidence of place recognition.

2   Relevant Work

Scene recognition is very challenging problem [17,18] and variety of approaches have been proposed during the last years. The most frequently used hand-crafted global descriptor is GIST [19]. With the rise of deep learning, learned features become increasingly widely used in localization algorithms. This resulted in a paradigm shift in VPR research consisting in focusing on neural network activations-based descriptors. Considerable potential of features extracted from CNN layers and used as global descriptors has been demonstrated in [20]. Scale Invariant Feature Transform (SIFT) and Speeded Up Robust Features (SURF) are two of the most commonly used local descriptors [21]. These local techniques extract invariant keypoints from an image and provide descriptions of these keypoints by an underlying low-level gradient-based descriptors. They have been applied in several algorithms for visual place recognition [3]. However, as observed in [9], SIFT can cope with large changes in appearance and illumination, but only when there is no large view point change [22]. On the other hand, geometric features like vertical lines can be very useful to represent buildings [23] or objects like doors in outdoor/indoor environments.

Pretrained CNN-based approaches to VPR can be roughly divided into two main categories in which: (i) responses from convolutional layers are extracted on the basis of the entire image [12], (ii) salient regions are identified through distinguishing patterns on the basis of convolutional layers responses to entire image [24]. High level features like object proposals have demonstrated remarkable potential in VPR [25]. Philbin et al. [26] learn a non-linear transformation model for descriptors that leads to greatly better matching performance. Tolias et al. [27] use max-pooling on cropped areas in CNN layers' features in order to extract ROIs. Mao et al. [28] propose multi-scale, non-rigid, pyramidal fusion of local features to improve VPR. In [29] a global matching-based, less-intensive place candidates selection is followed by local feature-based, more-intensive final candidate selection with focus on spatial constraints. Deep neural networks such as GoogLeNet, ResNet-152, VGG-16 and DenseNet-161 achieved classification accuracies of 53.6%, 54.7%, 55.2% and 56.1%, respectively on challenging Places-365 dataset [17]. The classification accuracies are lower in comparison to accuracies achieved by those networks on ImageNet dataset. The images acquired by mobile robots, and in particular humanoid robots or drones are even harder to classify. In [30], the transfer learning technique to retrain the VGG-F network in order to categorize places among 16 rooms on images acquired by a humanoid robot has been discussed.

3 Algorithm and Experimental Setup

At the beginning of this Section we propose an algorithm for blur detection. Afterwards, we present minimum spanning tree for place recognition. Then, in the next subsection we describe our dataset. In the last subsection we present the whole algorithm for place recognition.

3.1 Blur Detection

The basic idea of current approaches in robotics to visual place recognition is to search a database of indoor images and return the best match. Considerable attention is devoted to algorithms trained in end-to-end manner. Despite considerable research efforts, robust place recognition in indoor environments on the basis of on-board robot camera is an unsolved problem. The classification accuracies achieved by deep neural networks on challenging Places-365 dataset are lower in comparison to accuracies achieved by those networks on ImageNet dataset. The accuracies on real images acquired during robot motion are either too low for the purposes of loop-closure or are obtained with a high computational cost that prevents real-time applications. A dominating approach consists in learning or embedding features. One of the exceptions is a recent approach [29] in which a global matching-based, less-intensive place candidates selection is realized in advance, and then a local feature-based, more-intensive final candidate selection with focus on spatial constraints is executed. It is also worth noting that most of the approaches to visual place recognition do not consider scenarios with significant motion blur or, as a last resort, neglect motion blur, especially when the robot or camera rotates.

   At the beginning we generated a dataset with images contaminated by motion blur. We employed MIT Indoor scene database [31] that consists of 15620 images with 67 indoor categories. The number of examples varies across categories, but there are at least 100 images per category. A Matlab function fspecial has been used to approximate the linear motion of a camera with provided lengths $(5, \ldots, 10)$ and directions $(0, \ldots, \pi/2)$. Motivated by recent research findings showing that CNN-based description of places or images using only regions of interest (ROI) leads to enhanced performance compared to whole-image description [32] we based our algorithm on such an approach. In [32] the ROI-based vector representation is proposed to encode several image regions with simple aggregation. An approach proposed in [24] employs a late convolutional layer as a landmark detector and a prior one in order to calculate local descriptors for matching such detected landmarks. For such a regions-based feature encoding a $10k$ bag-of-words (BoW) [33] codebook has been utilized. The proposed approach to blur detection is based on salient CNN-based regional representations. The layers conv5_3 and conv5_2 of VGG-16, pre-trained on ImageNet dataset were used to extract the features representing regions. This means that in our approach we perform blur detection not on the whole image but instead we employ only salient CNN-based regional representations of the image. As in [24] we utilize a higher convolutional layer to guide extraction of local features and to create multiple region descriptors representing each image. At the training stage for each image with and without blur we extracted ten descriptors of size equal to 512, representing image regions with highest average activations. We trained a neural network with one hidden layer to classify the mentioned above image descriptors into two categories. The number of neurons in the hidden layer was equal to 20. The trained neural network has then been used to detect the noise. In testing stage for each image we extracted 200 descriptors as a representation of image regions with highest average activations. The responses of the neural network for such descriptors were averaged. The average values were then used to label the images as blurred or sharp. For visualization purposes the outputs of the classifiers were also projected onto the input images, see Fig. 1 that depicts sample images. For the discussed images the averaged outputs are equal to 0.1476, 0.5333 and 0.8532, respectively.

Fig. 1. Heat maps of images with increasing blur intensity.

We experimented with various numbers of descriptor vectors extracted on the test images. Figure 2 depicts sample images with some considered number of descriptors. As we can observe, the depicted heat maps change depending on number of descriptor vectors. Thus, we experimentally determined the number of descriptors leading to best blur detections and then determined the threshold to decide on the basis of averaged predictors if image is blurred or sharp one. This problem is an example of multi-objective optimization and in a future work the trade-off between number of descriptors and noise level will be determined automatically.

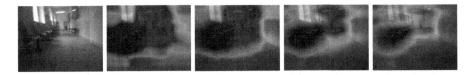

Fig. 2. Blurry input image (left) and heat maps for various number of descriptor vectors (50, 100, 200 and 300) extracted on the blurred image.

3.2   Minimum Spanning Tree-Based Place Recognition

By constructing a minimum-spanning tree the original dense graph is simplified into a minimum weight subgraph, which greatly reduces the number of edges and provides subgraphs of vertices of different degrees. Conventional minimum spanning tree-based clustering algorithms employ information about edges contained in the tree to partition a data set. A minimum spanning tree is a subset of edges of undirected graph that connects all vertices together, without any cycles and with the minimum total edge weight [34]. The property that there are no cycles means that there is only one path among any two nodes in the tree. In this work we compute a MST that connects all images of the training set. Nodes are connected by edges while weights express similarities between them. The edges were determined upon cosine similarity between global descriptors of images. In the proposed approach the MST has been utilized to support the place recognition. For each landmark place a number of relevant images has been determined. The MST has been built upon a selected global descriptor of the images. Given a MST created in advance on the training dataset, for each new image the algorithm seeks for the MST edge that is closest to this new image. The query images were classified as relevant or irrelevant on the basis of their similarities with the closest edges of the tree. Additional information about the room as well as blur of the images has been considered to enhance the place recognition. Moreover, a confidence of the place recognition has also been estimated.

3.3   The Dataset

The dataset has been recorded using a RGB camera mounted on the head of a humanoid robot. The dataset contains 9000 images, which were acquired in nine indoor rooms. Each image has been manually classified as sharp or blurred or considerably blurred. The training sequence contains 5287 blurred images and 1913 sharp images. A test sequence contains 1366 blurred images as well as 434 sharp images. For place recognition we also manually determined twenty two reference images with corresponding relevant and irrelevant images.

3.4   Algorithm

We trained the neural network to estimate the blur intensity and then used its outputs to detect if the input image is blurry or sharp one. Having on regard that the NetVLAD offers a powerful pooling mechanism with learnable parameters that can be easily plugged into any other CNN architecture or classifier we trained and then evaluated a set of classifiers for room recognition. A selected classifier is then used to recognize the room. We utilized VGG16 and added the NetVLAD layer after the conv_5 layer in order to extract the VLAD features. Given this and other selected features we precalculated the minimum spanning trees and evaluated them for place recognition.

Given all N training images and global descriptors, a pairwise similarity matrix of size $N \times N$ is determined for each descriptor. Afterwards, a MST is built on NetVLAD descriptor. The edges are determined upon cosine similarity between global descriptors of images. Then, blur information as well as room class are included in nodes of the MST built on the NetVLAD. Subsequently, the stored MST tree is processed using query images. Given a query image, only nodes of degree higher than two are assessed with respect to similarity with the query descriptor. Only 0.3 of the most similar nodes with the query descriptor are retained for further analysis. Afterwards, on the basis of the NetVLAD the most similar forty descriptors to the query descriptor together with corresponding node information are selected. Only nodes labeled as sharp as well as with the same class as the query image are included in the subset mentioned above. Such descriptors (images) are then sorted with respect to similarity with the query descriptor (image). Three sorted lists of images are determined for the NetVLAD descriptor and two additional global descriptors. Finally, the order of the images is updated upon the similarities of three global descriptors with the query image. As a result, the two descriptors (for instance Resnet-50 and GoogleNet), which individually get worse results than NetVLAD, in tandem may provide more relevant images to the query images and thus improve the average precision (AP) score of place recognition for a given query image.

Let us assume that we have a sorted list of similarities between the NetVLAD descriptors for the query image and the most relevant images. Let us also assume that we have also an ordered list of similarities between the ResNet50 descriptors and the most relevant images as well as ordered list of similarities between the GoogleNet descriptors and the most relevant images. For the image corresponding to the most similar NetVLAD descriptor with the query descriptor we determine the positions (indexes) in the ordered lists of ResNet50 and GoogleNet descriptors, which were determined for this considered image. We repeat this operation for the remaining descriptors and store indexes in subsequent rows of three column table. After computing the averages for all rows we obtain values which are used to reorder the relevant images with the query image.

We experimented with various configurations of the algorithm to evaluate the usefulness of blur detection as well as influence of classification scores on the performance of place recognition. We observed that knowledge about motion blur and room category has considerable influence on the final decision because in rooms like corridors the place recognition performance and ability do precisely determine the previously visited place for loop closure is lower. Finally, a classifier built on the MST has been utilized in image retrieval for the most similar image.

This means that final decision is taken using high-level information from noise detector, room recognition and information extracted on the basis of the MST.

By calculating the similarity measures between descriptor extracted from the current image and descriptors from the edges we can quickly determine the relevant sub-tree. Usually, descriptors in the same cluster have similar properties and tend to be in the same class. However, when in the same cluster there are exemplars belonging to different classes then the confidence of final decision is lowered. In our approach the confidence of place recognition is determined using the most relevant image found in the place recognition. Using the global descriptor of this image we searched for fifty most similar images. Such a pool of the most similar images has been determined on the basis of the MST edges holding cosine similarities between NetVLAD descriptors. When the decision confidence is below a threshold it is marked as not valuable for the loop-closure. In the basic approach the confidence has been determined as the ratio of sharp images to total number of images in the pool. We investigated also approaches combining blur information with class information. The MST have been calculated using dd tools [35]. Aside of the NetVLAD we employed the descriptors extracted from Resnet-50 and GoogleNet backbones.

4   Experimental Results

At the beginning we conducted experiments consisting in motion blur detection as well as deblurring real-world images. We ran our algorithm for blur detection on real images with severe (unknown) blurs and compared it with state-of-the-art algorithms, including [36, 37]. Table 1 presents experimental results that were achieved on test sequence Seq. #2 from our dataset. As we can observe, the best results were achieved by our algorithm. Taking into consideration that the decision whether the image is sharp is done on the basis of averaging the classifier output we evaluated also SVM with the calibrated output as well as the logistic regression (LR), which generates the calibrated output by default. It is also worth mentioning that the results achieved by CNNs specialized for non-uniform blur detection [38] are better in comparison to results achieved on the basis of method [37]. The discussed result has been achieved using neural network trained in 50 epochs. It has been trained on about 250 000 image descriptors randomly selected from the whole pool of training descriptors, whereas SVM and LR classifiers were trained on 50 and 150 thousand of descriptors, respectively. A recently proposed algorithm [39] achieved accuracy equal to 85.6%.

Afterwards, we determined descriptors representing images and calculated minimum spanning trees. The MSTs were visualized for images from each category as well as all images from the training set. Figure 3 depicts a sample MST that was obtained on the NetVLAD descriptor on all images from the training subset. We calculated, visualized and analyzed minimum-spanning trees on all images, images classified as sharp, and only blurry images. The discussed analysis of linkage maps was conducted with aim to collect the knowledge about dataset, and in particular to investigate influence of the blur on the performance of scene classification as well as place recognition on images with severe blurs.

Table 1. Blur detection on images from Seq. #2 with severe (unknown) blur.

Method	Accuracy	Precision	Recall	F1-score
var. Laplacian	0.8589	0.8114	0.7931	0.8015
SVM calibrated	0.9078	0.8650	0.8992	0.8798
Logistic regression	0.9194	0.8984	0.8770	0.8870
MB-det-CNN [37]	0.8720	0.8412	0.8231	0.8126
Our method	**0.9206**	**0.8869**	**0.9005**	**0.8934**

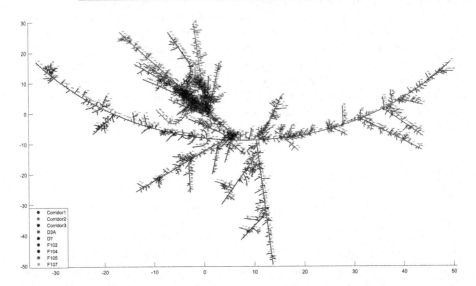

Fig. 3. Minimum spanning tree determined on NetLAD descriptor from training subset (plot best viewed in color).

Next, we evaluated state-of-the-art global descriptors in indoor scene recognition, where the set of scenes was a list of nine different room types. Table 2 presents experimental results which were achieved on sequence #2 from our dataset. We compared the performances achieved by the SVM with the linear kernel as well as k-NN. Table 2 presents only better result for each considered case. As we can observe, the categorization performance achieved on the basis of HOG and LBP descriptors is worse in comparison to remaining results. Classification performances achieved in transfer-learning based approach [30] are far better, see results C-E. Moreover, accuracies achieved upon the ReNet50 and SVM are noticeably better in comparison to results achieved on the basis of other deep neural architectures, including GoogleNet trained on Places-365 dataset. The classification results achieved by the k-NN on NetVLAD features are better in comparison to results mentioned above. The features were calculated using VGG-16, NetVLAD with whitening, trained on Tokyo Time Machine dataset [14] (downloaded from https://www.di.ens.fr/willow/research/netvlad/). The recognition of rooms only on images without blur, i.e. images automatically classified as non-blurry leads to considerable improvement of the results. This means that in such a scenario the robot first classifies the acquired image as blurry or non-blurry and then in case the image is blurry it acquires next one. As we can observe, costly and time consuming deblurring images with severe (unknown) blurs did not lead to better results. The discussed results were achieved using recently proposed deblurring algorithm [40]. Blur detection and then deblurring the images contaminated by blurs leads only to slightly better results, see results in the last row.

Table 2. Performance of room categorization on Seq. #2 from our dataset.

	Accuracy	Precision	Recall	F1-score
[A]HOG+SVM	0.6872	0.7063	0.6872	0.6921
[B]LBP+SVM	0.7639	0.7867	0.7639	0.7655
[C]VGG19+SVM	0.9056	0.9072	0.9056	0.9050
[D]GoogleNet Places-365+SVM	0.8939	0.8956	0.8939	0.8936
[E]ResNet50+SVM	0.9428	0.9474	0.9428	0.9434
[F]NetVLAD+KNN	0.9583	0.9600	0.9583	0.9583
[G]NetVLAD+MST	0.9544	0.9567	0.9544	0.9545
[H]NetVLAD+SVM+BlurDet.	**0.9652**	**0.9687**	**0.9652**	**0.9662**
[I]NetVLAD+SVM+Deblur	0.9528	0.9570	0.9528	0.9532
[J]NetVLAD+SVM+BlurDet.+Deblur	0.9550	0.9585	0.9550	0.9556

In last part of experiments we focused on place recognition. As mentioned above, basic idea of current image-based approaches to place recognition is to search a repository of indoor images and return the best match. In the first phase of this part of the research, we analyzed the performance of place recognition on images from Seq. #2 using the NetVLAD, GoogleNet and ResNet50 features. The NetVLAD features have been extracted using VGG-M network trained on TokyoTM dataset. The size of the feature vector extracted upon conv5_3 layer is 1×4096. We utilized GoogleNet trained on Places-365 dataset and ResNet50 trained on the ImageNet. The size of the GoogleNet-based feature vector is 1×1024 and it was extracted from pool5-7x7_s1 layer. The ResNet50-based feature is of size 1×2048 and it was extracted from GlobalAveragePooling2DLayer, avg_pool layers. Table 3 presents mean average precision (mAP) scores as well as their average values, which were achieved in recognition of 22 places in nine rooms. The last two columns of the table contain the results that were achieved using the MST and a combined descriptor. For each descriptor we determined the pairwise similarity matrix. The similarity matrixes have then been normalized to 0–1 range. Afterwards an average similarity matrix for all descriptors has been calculated. Finally, we determined the MST of a complete undirected graph with weights given by the averaged similarity matrix. As we can observe, such an algorithm achieved the best mAP scores. Thanks to considering information about blur far better mAP scores can be obtained in place recognition.

Table 3. Performance of place recognition (bd. - blur detection).

	k-NN VGG-M NetVLAD		k-NN GoogleNet Places-365		k-NN ResNet50		MST combined desc.	
	bd.	-	bd.	-	bd.	-	bd.	-
Cor_1	1.0000	0.8638	0.9205	0.6742	0.9135	0.8242	1.0000	0.8633
Cor_2	1.0000	0.5804	1.0000	0.8029	1.0000	0.7501	1.0000	0.5804
Cor_3	0.9750	0.8007	1.0000	0.7369	0.7667	0.7962	1.0000	0.7857
D3A	0.6549	0.7832	0.6147	0.6403	0.5939	0.6906	0.6612	0.7852
D7	0.8193	0.8016	0.8570	0.7277	0.9810	0.8071	0.8193	0.8041
F102	1.0000	0.7144	0.6293	0.4635	0.9167	0.5764	1.0000	0.7833
F104	0.8537	0.8504	0.4815	0.4471	0.7704	0.7114	0.8537	0.8632
F105	1.0000	0.8813	0.8296	0.6392	0.9722	0.6380	1.0000	0.8851
F107	0.8772	0.7239	0.2875	0.2526	0.5326	0.4596	0.8963	0.7224
av. mAP	0.9089	0.7778	0.7356	0.5983	0.8274	0.6948	**0.9145**	0.7859

Fig. 4 depicts precision-recall plots for selected rooms. The precision is the fraction/percentage of retrieved images that are relevant. The recall is the fraction/percentage of relevant images that were retrieved. For the analyzed rooms: D3A, D7, F104 and F107 the number of landmark points was equal to three. For the remaining rooms the precision-recall curves were perfect.

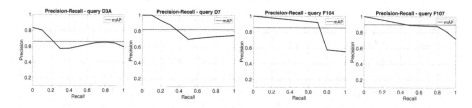

Fig. 4. Precision-recall plots for selected rooms (D3A, D7, F104, F107).

First row of Fig. 5 depicts query image and then relevant images, which are sorted from most similar to less similar. Second row contains example irrelevant images. The discussed images except query one were manually selected taking into account perceptual similarity/dissimilarity with the query image. Third row shows some correctly matched reference images with the query image, i.e. retrieved relevant images.

Fig. 5. Query image and relevant images (upper row), irrelevant images (second row), images retrieved using NetVLAD features (images acquired in room F102).

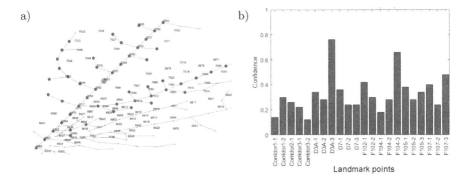

Fig. 6. Sub-tree with most similar image to the query image (green) and fifty most similar images with it (red), a), estimated confidence for all landmark points, b).

In the second phase of this part of the research, we performed experiments consisting in estimating the confidence of place recognition. The confidence of place recognition has been estimated as ratio of sharp images to total number of images in a pool of fifty most similar images with the image most similar to the query image, see Fig. 6a that contains sample sub-tree. Figure 6b illustrates the estimated confidence for all 22 landmark points. It turned out also that information about room category is important in context of confidence of robot decisions since spatial accuracy of place recognition in long and narrow corridors with similar scene content is much smaller. As we can observe, the lowest confidences are for Corridor_3 in 2nd landmark point and for Corridor_1, 1st landmark point. Experiments consisting in determining if acquired image is representative enough for scene recognition were conducted as well. For instance, if acquired image is blurry but the robot knows that it belongs to tree branch in which there are only images belonging to single room then it can decide to perform deblurring the image and then use it for place recognition.

5   Conclusions

In this work we introduce an algorithm for blur detection on images with severe (unknown) blur. We demonstrate experimentally that the proposed algorithm outperforms recent algorithms. We propose a new algorithm which on the basis of graph-based decisions on deep embeddings and blur detections permits robust place recognition as well as delivers decision confidences. The algorithm has been evaluated on challenging dataset for visual place recognition with images acquired by a humanoid robot.

Acknowledgment. This work was supported by Polish National Science Center (NCN) under a research grant 2017/27/B/ST6/01743.

References

1. Cadena, C., Carlone, L., Carrillo, H., Latif, Y., Scaramuzza, D., Neira, J., Reid, I., Leonard, J.: Past, present, and future of simultaneous localization and mapping: towards the robust-perception age. IEEE Trans. Robot. **32**(6), 1309–1332 (2016)
2. Cebollada, S., Paya, L., Flores, M., Peidro, A., Reinoso, O.:A state-of-the-art review on mobile robotics tasks using artificial intelligence and visual data. Expert Syst. Appl. **167**, 114–195 (2020)
3. Lowry, S., Sünderhauf, N., Newman, P., Leonard, J., Cox, D., Corke, P., Milford, M.J.: Visual place recognition: a survey. IEEE Trans. Robot **32**, 1–19 (2016)
4. Odo, A., McKenna, S., Flynn, D., Vorstius, J.: Towards the automatic visual monitoring of electricity pylons from aerial images. In: 16th International Joint Conference on Computer Vision, Imaging and Computer Graphics Theory and Applications (VISAPP) (2020)
5. Zhao, J., et al. J.:Place recognition with deep superpixel features for brain-inspire dnavigation. Rev. Sci. Instrum. **91**(12), 125110 (2020)
6. Tolias, G., Avrithis, Y., Jégou, H.: Image search with selective match kernels: aggregation across single and multiple images. Int. J. Comput. Vision **116**(3), 247–261 (2015)
7. Ovalle-Magallanes, E., Aldana-Murillo, N.G., Avina-Cervantes, J.G., Ruiz-Pinales, J., Cepeda-Negrete, J., Ledesma, S.: Transfer learning for humanoid robot appearance-based localization in a visual map. IEEE Access **9**, 6868–6877 (2021)
8. Pretto, A., Menegatti, E., Bennewitz, M., Burgard, W., Pagello, E.: A visual odometry framework robust to motion blur. In: IEEE International Conference on Robotics and Automation, pp. 2250–2257(2009)
9. Torii, A., Arandjelović, R., Sivic, J., Okutomi, M., Pajdla, T.: 24/7 place recognition by view synthesis. In: Conference on Computer Vision and Pattern Recognition (CVPR), pp. 1808–1817 (2015)
10. Maffra, F., Chen, Z., Chli, M.: Viewpoint-tolerant place recognition combining 2D and 3D information for UAV navigation. In: IEEE International Conference on Robotics and Automation (ICRA), pp. 2542–2549(2018)

11. Garg, S., Milford, M.: Straightening sequence-search for appearance-invariant place recognition using robust motion estimation. In: Proceedings of Australasian Conference on Robotics and Automation (ACRA), pp. 203–212 (2017)
12. Chen, Z., Lam, O., Adam, J., Milford, M.: Convolutional neural network-based place recognition. In: Proceedings of Australasian Conference on Robotics and Automation, pp. 1–8 (2014)
13. Suenderhauf, N., Shirazi, S., Dayoub, F., Upcroft, B., Milford, M.: On the performance of ConvNet features for place recognition. In: IEEE/RSJ International Conference on Intelligent Robots and Systems (IROS), pp. 4297–4304 (2015)
14. Arandjelovic, R., Gronat, P., Torii, A., Pajdla, T., Sivic, J.: NetVLAD: CNN architecture for weakly supervised place recognition. IEEE Trans. Pattern Anal. Mach. Intell. **40**(6), 1437–1451 (2018)
15. Arandjelovic, R., Zisserman, A.: All About VLAD. In: IEEE Conference on Computer Vision and Pattern Recognition, pp. 1578–1585. IEEE Computer Society (2013)
16. Zaffar, M., Khaliq, A., Ehsan, S., Milford, M., McDonald-Maier, K.: Levelling the playing field: a comprehensive comparison of visual place recognition approaches under changing conditions. CoRR abs/1207.0016 (2019)
17. Zhou, B., Lapedriza, A., Khosla, A., Oliva, A., Torralba, A.: Places: A 10 million image database for scene recognition. IEEE Trans. Pattern Anal. Mach. Intell. **40**(6), 1452–1464 (2018)
18. López-Cifuentes, A., Escudero-Viñolo, M., Bescós, J., Álvaro García-Martín: Semantic-aware scene recognition. Pattern Recogn. **102**, 107256 (2020)
19. Oliva, A., Torralba, A.: Modeling the shape of the scene: a holistic representation of the spatial envelope. Int J. Comput. Vision **42**(3), 145–175 (2001)
20. Yandex, A.B., Lempitsky, V.: Aggregating local deep features for image retrieval. In: IEEE International Conference on Computer Vision (ICCV), pp. 1269–1277 (2015)
21. Ma, J., Jiang, X., Fan, A., Jiang, J., Yan, J.: Image matching from handcrafted to deep features: a survey. Int. J. Comput. Vision **129**(1), 23–79 (2020)
22. Kwolek, B.: Visual odometry based on gabor filters and sparse bundle adjustment. In: Proceedings IEEE International Conference on Robotics and Automation, pp. 3573–3578 (2007)
23. Arth, C., Pirchheim, C., Ventura, J., Schmalstieg, D., Lepetit, V.: Instant outdoor localization and SLAM initialization from 2.5d maps. IEEE Trans. Visual. Comput. Graph. **21**(11), 1309–1318 (2015)
24. Chen, Z., Maffra, F., Sa, I., Chli, M.: Only look once, mining distinctive landmarks from ConvNet for visual place recognition. In: IEEE/RSJ International Conference on Intelligent Robots and Systems (IROS). pp. 9–16 (2017)
25. Hou, Y., Zhang, H., Zhou, S.: Evaluation of object proposals and ConvNet features for landmark-based visual place recognition. J. Intell. Robot. Syst. **92**(3–4), 505–520 (2017)
26. Philbin, J., Isard, M., Sivic, J., Zisserman, A.: Descriptor learning for efficient retrieval. In: Daniilidis, K., Maragos, P., Paragios, N. (eds.) ECCV 2010. LNCS, vol. 6313, pp. 677–691. Springer, Heidelberg (2010). https://doi.org/10.1007/978-3-642-15558-1_49
27. Tolias, G., Avrithis, Y., Jégou, H.: To aggregate or not to aggregate: selective match kernels for image search. In: IEEE International Conference on Computer Vision, pp. 1401–1408 (2013)
28. Mao, J., Hu, X., He, X., Zhang, L., Wu, L., Milford, M.J.: Learning to fuse multi-scale features for visual place recognition. IEEE Access **7**, 5723–5735 (2019)

29. Camara, L.G., Přeučil, L.: Visual place recognition by spatial matching of high-level CNN features. Robot. Auton. Syst. **133**, 103625 (2020)
30. Wozniak, P., Afrisal, H., Esparza, R.G., Kwolek, B.: Scene recognition for indoor localization of mobile robots using deep CNN. In: Chmielewski, L.J., Kozera, R., Orłowski, A., Wojciechowski, K., Bruckstein, A.M., Petkov, N. (eds.) ICCVG 2018. LNCS, vol. 11114, pp. 137–147. Springer, Cham (2018). https://doi.org/10.1007/978-3-030-00692-1_13
31. Quattoni, A., Torralba, A.: Recognizing indoor scenes. In: IEEE Conference on Computer Vision and Pattern Recognition, pp. 413–420 (2009)
32. Tolias, G., Sicre, R., Jégou, H.: Particular object retrieval with integral max-pooling of CNN activations. In: International Conference Learning Representations (ICLR 2016) (2016)
33. Sivic, J., Zisserman, A.: Video Google: a text retrieval approach to object matching in videos. In: IEEE International Conference on Computer Vision, pp. 1470–1477(2003)
34. Zhong, C., Malinen, M., Miao, D., Fränti, P.: A fast minimum spanning tree algorithm based on k-means. Inf. Sci. **295**(C), 1–17 (2015)
35. Tax, D.M.: Data description toolbox - dd tools, ver. 2.1.3. https://github.com/DMJTax/dd_tools (2021)
36. Narvekar, N., Karam, L.: A no-reference image blur metric based on the cumulative probability of blur detection (CPBD). IEEE Trans. Image Process. **20**(9), 2678–2683 (2011)
37. Pech-Pacheco, J.L., Cristobal, G., Chamorro-Martinez, J., Fernandez-Valdivia, J.: Diatom autofocusing in brightfield microscopy: a comparative study. In: Proceedings of the 15th International coneference on Pattern Recognition, vol. 3, pp. 314–317 (2000)
38. Sun, J., Wenfei Cao, Zongben Xu, Ponce, J.: Learning a convolutional neural network for non-uniform motion blur removal. In: IEEE Conference on Computer Vision and Pattern Recognition (CVPR), pp. 769–777 (2015)
39. Cun, X., Pun, C.M.: Defocus blur detection via depth distillation. In: European Conference on Computer Vision (ECCV), pp. 747–763. Springer (2020)
40. Tao, X., Gao, H., Shen, X., Wang, J., Jia, J.: Scale-recurrent network for deep image deblurring. In: IEEE Conference on Computer Vision and Pattern Recognition, pp. 8174–8182 (2018)

Football Players Movement Analysis in Panning Videos

Karol Działowski🆔 and Paweł Forczmański$^{(\boxtimes)}$🆔

Faculty of Computer Science and Information Technology, West Pomeranian University of Technology, Szczecin, Żołnierska Street 49, 71-210 Szczecin, Poland
pforczmanski@wi.zut.edu.pl

Abstract. In this paper, we present an end-to-end application to perform automatic multiple player detection, unsupervised labelling, and a semi-automatic approach to finding homographies. We incorporate dense optical flow for modelling camera movement and user-assisted calibration on automatically chosen key-frames. Players detection is performed with a pre-trained YOLOv3 detector and player labelling is done using features in HSV colorspace. The experimental results demonstrate that our method is reliable with generating heatmaps from players' positions in case of moderate camera movement. Major limitations of proposed method are the necessity of manual calibration of characteristic frames, inaccuracy with fast camera movements, and small tolerance of vertical camera movement.

Keywords: Sports video analysis · Soccer player tracking · Camera calibration · Pitch modelling

1 Introduction

Soccer is one of the most popular sport arts watched by millions of people around the world. This popularity led to a growing demand by sports professionals and fans for gathering various data related to the players and the game itself. Recently a lot of research was done in the field of *soccer video analysis*. Such systems can be used for getting insights about whole team or individual player performance, they can support referees in decision making, automatically extract highlights or intelligently control the broadcasting camera [9,16].

Systems for gathering data about team or player performance can reveal aspects that are hidden and not so obvious to the human eye. Such systems can measure the distance covered by players, their speed, average position on the pitch, etc. This data can be used by staff members or professional analysts to improve the team performance or by experts in television [1].

There are many approaches to get accurate time-related positions of the players that change at the ground level. Wearable tracking devices are recently the first choice in collecting such data and are used by the majority of elite teams [17]. Complex camera systems mounted around the stadium are another

M. Paszynski et al. (Eds.): ICCS 2021, LNCS 12746, pp. 193–206, 2021.
https://doi.org/10.1007/978-3-030-77977-1_15

option. Multiple fixed cameras can cover the entire pitch, detect the player's position and project them into a virtual top-view image via homography of the ground between the camera image and the top-view image [1,10]. However, those options are complex, expensive and often not affordable for the average lower league team.

Taking into consideration the above facts, there is a demand for an affordable data gathering system that does not require complex infrastructure and sophisticated sensors. We have observed that many teams, from amateur to professional levels, record their performance with a single video camera. Those materials are typically taken from a fixed position with a simple horizontal panning and can be used to measure team/player performance [19,31].

In this paper, we propose a semi-automatic system for tracking players on a soccer pitch using single panning camera. The proposed approach consists of two elements. Firstly, the camera motion is modelled with a dense optical flow which is used for selecting characteristic frames required for user-assisted camera calibration and finding homography. Camera calibration is performed on characteristic frames by manually selecting corresponding points between the camera image and pitch model It is done once for the whole video stream. Then transformation matrices are found for every camera angle with simple linear interpolation. Then we gather some number of frames from camera input. In each frame, we detect players bounding boxes with YOLOv3 and perform feature extraction by calculating hue and saturation histograms. Unsupervised learning is used for creating a model that discriminates players into five classes.

The second stage is responsible for player detection and player classification. Firstly, the pitch pixels segmentation is done by masking a particular color represented in Hue-Saturation-Value color space, then we use general YOLOv3 detector for getting bounding boxes that describe player positions. We do not train the detector but use network weights from *cvlib* library trained for general-purpose applications [23]. For each player, we perform feature extraction by calculating hue and saturation histograms and classification using the model learned at the previous stage. Each player position is projected into a 2D pitch model with a transformation matrix found at the first stage.

The paper is organized as follows. Related works are presented in Sect. 2. Section 3 describes our method with focus on pitch segmentation, camera calibration and modelling, players detection, players classification and players position tracking. Experimental results are presented in Sect. 4, while the conclusions and future research directions are presented in Sect. 6.

2   Related Works

2.1   Field Registration

Systems based on a single camera require pitch registration for representing players' positions in two dimensions. One way of acquiring this relation is by finding the homography placing a camera view into a two-dimensional view assuming the playing surface is flat.

In exemplary approaches [12,24], a user manually calibrates several reference images, then the system calibrates remaining images by finding correspondences from reference images. Recently, fully-automatic methods emerged, as they require no or fewer user interactions [4,28].

Sharma et al. [28] proposed a solution for the registration problem defined as the nearest neighbour search over edge maps and synthetically generated homography pairs. Extracting edge data from input images was done with 3 different approaches: histogram of oriented gradients (HOG) features, chamfer matching, and convolution neural networks (CNN).

Chen and Little [4] improved the idea of using synthetic data incorporating Sharma et al. work with generative adversarial network (GAN) model for detecting field marking and siamese networks for finding closest pair of synthetic edge map and homography.

2.2   Player Detection and Reidentification

Player detection can be done in multiple ways. Santhosh and Kaarthick [27] researched the use of HOG color-based detector. Johnson [15] based his work on an open-source multiperson pose estimator named Alpha Pose. Using CNN-base multibox detector for player detection was researched by Ramanathan et al. [25].

Another important aspect is player classification into five main classes corresponding to two teams, two goalkeepers, and referee, namely, player labelling which is usually done with color features [20].

Player tracking (reidentification) solves the problem of temporal correspondence. This step is required for tracking consistent trajectories for each player. A large number of tracking algorithms were used for solving this problem, such as Kalman filter [21], particle filter [1], mean-shift [6], etc.

3   Method Description

In our case, we use a panning camera which means swivelling a video camera horizontally from a fixed position. This motion is similar to the motion of a person when he/she turns his/her head on a neck from the left to the right.

The term panning is derived from panorama, suggesting an expansive view that exceeds the gaze, forcing the viewers to turn their heads in order to take everything in. Panning, in other words, is a device for gradually revealing and incorporating off-screen space into the image.

In this case, panning refers to the horizontal scrolling of an image wider than the display.

Proposed solution works only in an offline mode as camera motion modelling is required step in the process. This drawback could be solved using more advanced techniques as explained in Sect. 5.

The system is composed of two stages. In the first stage, we model camera angle changes using dense optical flow. Then semi-automatic camera calibration is performed and the team classification model is learned on few examples.

The second stage is fully automatic and consists of pitch detection, player detection, players team classification, and projection of positions from image coordinates to pitch coordinates.

3.1 Pitch Segmentation

Pitch segmentation is commonly used to eliminate spectator regions and reduce false alarms in player detection. A pitch mask is used in the camera angle modelling process using dense optical flow. Pitch segmentation is performed by selecting color ranges in the HSV color scheme.

Color range was chosen empirically and is defined as a pair of lower bound $L = (35, 50, 50)$ and upper bound $U = (75, 255, 255)$. This creates a problem when pitch color is not in defined color range or non-pitch areas are also green. This problem can be solved with other methods, e.g. segmentation GAN [4].

There are objects such as players, referees, ball and lines on the pitch. To eliminate those objects from the mask we use morphological operations, e.g. closing operation which consists of dilatation and erosion. Figure 1 shows the process of pitch segmentation.

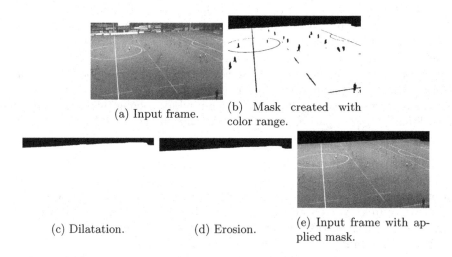

(a) Input frame.

(b) Mask created with color range.

(c) Dilatation.

(d) Erosion.

(e) Input frame with applied mask.

Fig. 1. Process of pitch segmentation.

3.2 Camera Angle Modelling

Analysis of camera angle is accomplished with the use of dense optical flow (calculated for every pixel in the image). It describes changes between the subsequent frames which are a result of moving objects or change of camera parameters [2]. Optical flow works well if pixel intensities of moving object do not change between consecutive frames and neighbouring pixels have similar motion.

These assumptions are met in case of football field observation. The classical equation of optical flow is defined as:

$$I(x, y, t) - I(x + \Delta x, y + \Delta y, t + \Delta t) = 0, \tag{1}$$

where pixel with intensity $I(x, y, t)$ has moved by Δx, Δy and Δt between two image frames [8].

After finding the pitch mask we inverse it and calculate optical flow for each frame. This allows us to reduce noise generated by objects moving on the pitch, i.e. players, referee and ball. For every frame, we calculate the mean optical flow vector which describes camera movement in a given frame. The accumulated sum of those means approximates camera angle in relation to the first frame. Example values for footage of panning to the left and then to the right are shown in Fig. 2. Based on this analysis we can choose characteristic key frames with the maximum and minimum value of the camera swing angle in comparison to the first frame of the input video.

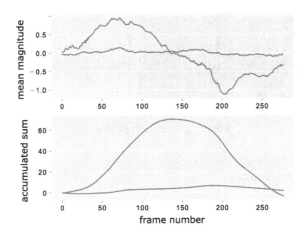

Fig. 2. Camera angle approximation with optical flow. Red line represents horizontal (panning) motion, blue line represents vertical motion. Accumulated sum represents camera angle in given frame compared to first frame of input video. (Color figure online)

3.3   Camera Calibration

Characteristic key frames are used for projective transformation. It allows for a mapping of one plane to another. In our algorithm, we use it to find ground position of players seen in different camera views [22]. Projecting transformation is expressed by the equation:

$$\begin{bmatrix} x_1 \\ y_1 \\ 1 \end{bmatrix} = \begin{bmatrix} h_{11} & h_{12} & h_{13} \\ h_{21} & h_{22} & h_{23} \\ h_{31} & h_{32} & h_{33} \end{bmatrix} \begin{bmatrix} x_2 \\ y_2 \\ 1 \end{bmatrix}, \tag{2}$$

where x_1 and y_1 are the coordinates of a single point on the input plane, x_2 and y_2 are the coordinates on the output plane, H is a transformation matrix.

In order to calculate H by means of least squares method we need to collect four pairs of so called calibration points. Such an approach is a compromise between computational complexity and the quality of the resulting transformation. Exemplary calibration points are presented in Fig. 3.

Knowing transformation matrices H_n and H_m corresponding to frames with maximum and minimum horizontal angle, respectively, we can interpolate transformation matrix for each frame using values from maximum – minimum range, as follows:

$$H_k = H_n + \frac{k(H_m - H_n)}{(m - n)}, \tag{3}$$

where k is camera angle, and $n < k < m$.

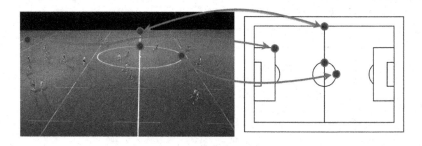

Fig. 3. Calibration points

3.4   Players Detection

Players detection is performed by the YOLOv3 detector. This was a state-of-the-art detection method at the time of developing our method. The YOLO (You Only Look Once) algorithm proposed by Joseph Redmon and Ross Girshick solves object detection as a regression problem and outputs the location and class of an input object on an end-to-end network in one step [26,30].

We perform detection using the YOLOv3 model trained on the COCO dataset capable of detecting 80 common objects [18] (e.g. cars, cats, dogs, pedestrians, etc.). We used detector *as is* with pre-trained weights from *cvlib* library [23].

Detection is done within a segmented pitch in every frame. For each bounding box recognized as a person, we extract color-based features and perform players' team assignments. We use the previously found transformation matrix H_k to project player position into pitch coordinates. The exemplary results are shown in Fig. 4.

Fig. 4. Classification and labelling step. Each player is labelled with an unique id and assigned to a team with decision tree based on extracted hue and saturation histograms.

3.5   Players Classification

Players classification can be done by means of various color and textural features [11]. In our approach we focused on typical color-based features. Firstly, at the initial stage, we collect 10 subsequent frames and perform player detection. Then for every player, we extract features that are concatenated hue and saturation histograms. We observed better results after splitting horizontally each player into halves and calculating features for each half independently. This approach takes into account the difference between shirts and shorts of soccer players.

With a sample set of features calculated over 10 evenly distributed frames, we perform agglomerative hierarchical clustering with specified 5 clusters, each for two teams, two goalkeepers, and a referee. Knowing labels for each feature we train decision tree on that data. This decision tree will be used in the principal stage of our method.

During classification, we calculate features for each player in the same way, i.e. by calculating hue and saturation histograms. We classify players using the previously trained decision tree model (Fig. 5).

After team assignment each player is given unique identifier based on identification step in the previous frame, i.e. we give the same identifier to the player with closest spatial distance from the preceding frame.

Having this information we apply post-processing for team assignment step. Dominant team label from previous 11 frames is accepted as a valid label in the current frame. This approach reduces false alarms in the team assignment step.

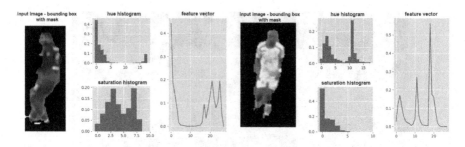

Fig. 5. Feature extraction for team assignment based on hue and saturation histograms.

4   Experimental Results

4.1   Experimental Environment

We prepared several separate applications for the experimental part of the work:

- *Pitchmap* - main program, which performs the entire process of calibration, detection and projection on the pitch map,
- *Annotator* - a program for manual calibration and marking of players in the input material, which are later used for comparison with our algorithm (Fig. 6),
- *Comparator* - a program for comparing positions from *Pitchmap* and positions from *Annotator* (Fig. 7),
- *Heatmap* - a program for generating heatmaps and path comparisons.

The output from *Heatmap* application is shown in Fig. 8a. and Fig. 8b. It contains heatmap generated from 18-seconds-long footage and single player trajectory, respectively.

The experimental protocol is as follows. Firstly, we annotate benchmark videos presenting short fragments of football matches (7–30 s long). Two videos have been recorded during amateur league match, while one video material comes from a television broadcast. The first video is rather steady and contains eight changes of a viewpoint. The camera in the second one has higher dynamics, it slowly moves towards right pitch side and then returns with a significantly higher speed. The last material contains continuous camera movement in one direction, after some time, the camera returns to its initial position. What is a little problematic, the camera's viewpoint moves slightly up and down.

During experiments we calculated paths of moving players and the heatmaps of their presence on the pitch. Finally, we compared the results of automatic procedure with manual annotations by means of objective Structural SIMilarity and mean distance between positions of individual players.

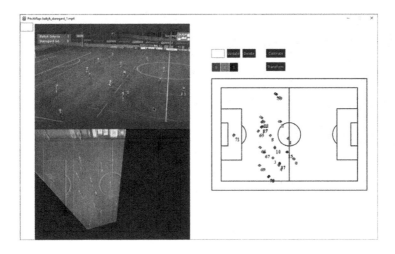

Fig. 6. Main window of *Annotator* application.

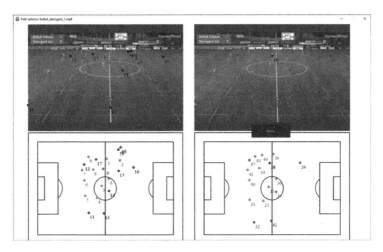

Fig. 7. Main window of *Comparator* application.

4.2   Results

The main focus has been put on the calibration procedure, since it has the greatest influence on the algorithm results. The following calibration methods has been be compared:

1. Calibration with three characteristic frames - calibration is performed on two frames with the greatest camera inclination and on a frame with central position. All intermediate tilt angles are interpolated.
2. Calibration with two characteristic frames - calibration is performed on the two frames with the greatest camera inclination. All intermediate tilt angles are interpolated.

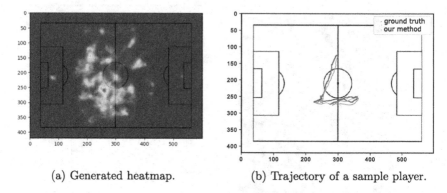

(a) Generated heatmap.   (b) Trajectory of a sample player.

Fig. 8. Exemplary output generated based on data aggregated by proposed method from 18 s footage with horizontal panning.

3. Calibration with manual feature frames - the user selects two key frames. All intermediate frames are interpolated.

The evaluation was performed using two methods: by means of image similarity comparison between heatmaps and by means of individual player path comparison. The image similarity was estimated using Structural Similarity Index Measure (SSIM) method [29], assuming the perfect match is represented by value close to one. The paths were compared using mean distance between vectors P and Q (of n frames) representing annotated (ground truth) position and estimated one, respectively:

$$d = \frac{1}{n} \sum_{i=0}^{n-1} ||(P_i - Q_i)||^2. \tag{4}$$

The results of heatmap comparison are presented in Table 1. It contains values of SSIM for three benchmark videos and two calibration methods. For the comparison purpose, the result of manual calibration was also given, yet it should be noted that is was estimated for a reduced material time-span (due to complex process of manual annotation of longer video sequences).

As it was observed, SSIM can not always be mapped onto a subjective assessment of similarity. The results show also that the calibration with three characteristic frames usually gives better results than calibration with two characteristic frames (Fig. 9).

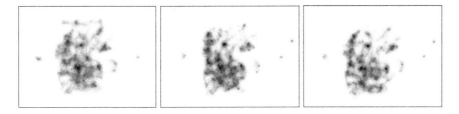

Fig. 9. Exemplary heatmaps for video nr.1: Manually annotated (left), semi-automatic with three key-frames (middle), semi-automatic with two key-frames (right).

Table 1. SSIM values for heatmaps generated using different calibration methods

Calibration method	Video material		
	1	2	3
Three key frames	0.875	0.871	0.831
Two key frames	0.857	0.858	0.834
Manual (for reduced time span)	0.866	0.852	0.904

The results of paths comparison are presented in Table 2. As in case of SSIM, it contains values of distance for three benchmark videos and two calibration methods. For the comparison purpose, the result of manual calibration was also given, yet it should be noted that is was estimated for a reduced material time-span (due to complex process of manual annotation of longer video sequences).

Figure 10 presents a projection onto X and Y axis for player path, for two calibration methods. Subjectively, the method with two key frames gives the better results, however the closer look at the objective measures (see Table 2) unveils that the method with three key frames is better.

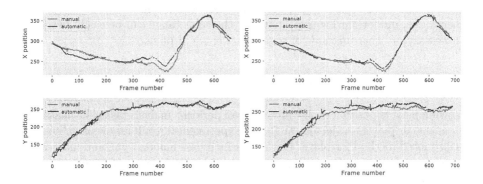

Fig. 10. Exemplary paths for individual player for video nr.1: semi-automatic calibration with three key-frames (left), semi-automatic with two key-frames (right).

Table 2. Mean distances (in pixels) between paths travelled by selected player and different calibration methods

Calibration method	Video material		
	1	2	3
Three key frames	9.694	9.816	27.302
Two key frames	9.908	13.399	28.444
Manual (for reduced time span)	18.063	10.562	9.438

5   Limitations and Future Work

Our proof-of-concept work can be improved in some areas. Replacing the manual calibration step with a fully automatic one [4,5,13] could significantly improve usability and allow on-site analysis. Manual calibration step can also reduce calibration accuracy in comparison to automatic solutions.

The second limitation is input constraints. Our method accepts only continuous footage with a horizontal pan so it cannot be applied on raw broadcast footage. However, this can be improved with automatic scene detection [14].

It should be noted that the proposed system is not robust in all situations. We have observed major errors when the camera angle changes frequently with significant speed. The proposed solution has a small tolerance for vertical panning movement which imposes a requirement for only horizontal panning footage. Those problems can be solved by slicing input video (e.g. by tensor-based methods [7]) or incorporating more sophisticated methods of camera calibration (e.g. [3]).

Another limitation is the requirement that the football pitch is green. The color range used for segmentation was chosen arbitrarily so proposed system is not universal for all kinds of surfaces (e.g. snow conditions, artificial surface in other colors, indoor hard courts), however the color range can be easily tweaked if necessary or other methods of segmentation could be used.

6   Conclusions

Knowing players position on the pitch we can perform analysis tasks like heatmap generation or players trajectory analysis. Experimental results show usefulness of the proposed method for player tracking, which uses homography and integrates the player information from single camera on the virtual ground image.

By having players positions in the soccer scene it could be used in some applications such as team strategy analysis, scene recovery and measuring individual players performance.

References

1. Baysal, S., Duygulu, P.: Sentioscope: a soccer player tracking system using model field particles. IEEE Trans. Circuits Syst. Video Technol. **26**(7), 1350–1362 (2015)
2. Beauchemin, S.S., Barron, J.L.: The computation of optical flow. ACM Comput. Surv. (CSUR) **27**(3), 433–466 (1995)
3. Chen, J., Little, J.J.: Sports camera calibration via synthetic data. CoRR abs/1810.10658 (2018). http://arxiv.org/abs/1810.10658
4. Chen, J., Little, J.J.: Sports camera calibration via synthetic data. In: Proceedings of the IEEE/CVF Conference on Computer Vision and Pattern Recognition Workshops (2019)
5. Chen, J., Zhu, F., Little, J.J.: A two-point method for PTZ camera calibration in sports. In: 2018 IEEE Winter Conference on Applications of Computer Vision (WACV), pp. 287–295. IEEE (2018)
6. Chiang, T.K., Leou, J.J., Lin, C.S.: An improved mean shift algorithm based tracking system for soccer game analysis. In: Proceedings: APSIPA ASC 2009: Asia-Pacific Signal and Information Processing Association, 2009 Annual Summit and Conference, pp. 380–385 (2009)
7. Cyganek, B., Woźniak, M.: Tensor-based shot boundary detection in video streams. New Gener. Comput. **35**(4), 311–340 (2017). https://doi.org/10.1007/s00354-017-0024-0
8. Dalka, P.: Methods of algorithmic analysis of the video image for applications in traffic monitoring [in Polish: Metody algorytmicznej analizy obrazu wizyjnego do zastosowań w monitorowaniu ruchu drogowego]. Ph.D. thesis, Gdansk University of Technology (2015)
9. D'Orazio, T., Leo, M.: A review of vision-based systems for soccer video analysis. Pattern Recogn. **43**(8), 2911–2926 (2010)
10. Enomoto, A., Saito, H.: AR display for observing sports events based on camera tracking using pattern of ground. In: Shumaker, R. (ed.) VMR 2009. LNCS, vol. 5622, pp. 421–430. Springer, Heidelberg (2009). https://doi.org/10.1007/978-3-642-02771-0_47
11. Frejlichowski, D.: A method for data extraction from video sequences for automatic identification of football players based on their numbers. In: Maino, G., Foresti, G.L. (eds.) ICIAP 2011. LNCS, vol. 6978, pp. 356–364. Springer, Heidelberg (2011). https://doi.org/10.1007/978-3-642-24085-0_37
12. Ghanem, B., Zhang, T., Ahuja, N.: Robust video registration applied to field-sports video analysis. In: IEEE International Conference on Acoustics, Speech, and Signal Processing (ICASSP), vol. 2. Citeseer (2012)
13. Homayounfar, N., Fidler, S., Urtasun, R.: Sports field localization via deep structured models. In: Procedings of the IEEE Conference on Computer Vision and Pattern Recognition, pp. 5212–5220 (2017)
14. Wang, J., Chng, E., Xu, C.: Soccer replay detection using scene transition structure analysis. In: Proceedings (ICASSP 2005). IEEE International Conference on Acoustics, Speech, and Signal Processing, 2005, vol. 2, pp. ii/433–ii/436 (2005). https://doi.org/10.1109/ICASSP.2005.1415434
15. Johnson, N.: Extracting player tracking data from video using non-stationary cameras and a combination of computer vision techniques. In: Proceedings of the 14th MIT Sloan Sports Analytics Conference, Boston, MA, USA (2020)
16. Larson, N.G., Stevens, K.A.: Automated camera-based tracking system for sports contests, US Patent 5,363,297, 8 November 1994

17. Leveaux, R., Messerschmitt, M.: The changing shape of sport through information technologies. In: Proceedings of the 26th International Business Information Management Association Conference-Innovation Management and Sustainable Economic Competitive Advantage: From Regional Development to Global Growth, IBIMA 2015 (2015)

18. Lin, T.-Y., et al.: Microsoft COCO: common objects in context. In: Fleet, D., Pajdla, T., Schiele, B., Tuytelaars, T. (eds.) ECCV 2014. LNCS, vol. 8693, pp. 740–755. Springer, Cham (2014). https://doi.org/10.1007/978-3-319-10602-1_48

19. Mackowiak, S., Konieczny, J.: Player extraction in sports video sequences. In: 2012 19th International Conference on Systems, Signals and Image Processing (IWSSIP), pp. 409–412. IEEE (2012)

20. Manafifard, M., Ebadi, H., Moghaddam, H.A.: A survey on player tracking in soccer videos. Comput. Vis. Image Underst. **159**, 19–46 (2017)

21. Najafzadeh, N., Fotouhi, M., Kasaei, S.: Multiple soccer players tracking. In: 2015 The International Symposium on Artificial Intelligence and Signal Processing (AISP), pp. 310–315. IEEE (2015)

22. Nowosielski, A., Frejlichowski, D., Forczmanski, P., Gosciewska, K., Hofman, R.: Automatic analysis of vehicle trajectory applied to visual surveillance. In: Choras, RS (ed.) Image Processing And Communications Challenges 7. Advances in Intelligent Systems and Computing, vol. 389, pp. 89–96 (2016). https://doi.org/10.1007/978-3-319-23814-2_11

23. Ponnusamy, A.: cvlib - high level computer vision library for Python (2018). https://github.com/arunponnusamy/cvlib

24. Puwein, J., Ziegler, R., Vogel, J., Pollefeys, M.: Robust multi-view camera calibration for wide-baseline camera networks. In: 2011 IEEE Workshop on Applications of Computer Vision (WACV), pp. 321–328. IEEE (2011)

25. Ramanathan, V., Huang, J., Abu-El-Haija, S., Gorban, A., Murphy, K., Fei-Fei, L.: Detecting events and key actors in multi-person videos. In: Proceedings of the IEEE Conference on Computer Vision and Pattern Recognition, pp. 3043–3053 (2016)

26. Redmon, J., Farhadi, A.: YOLOV3: an incremental improvement (2018). arXiv preprint arXiv:1804.02767

27. Santhosh, P., Kaarthick, B.: An automated player detection and tracking in basketball game. Comput. Mater. Continua **58**(3), 625–639 (2019)

28. Sharma, R.A., Bhat, B., Gandhi, V., Jawahar, C.: Automated top view registration of broadcast football videos. In: 2018 IEEE Winter Conference on Applications of Computer Vision (WACV), pp. 305–313. IEEE (2018)

29. Wang, Z., Simoncelli, E.P., Bovik, A.C.: Multiscale structural similarity for image quality assessment. In: The Thrity-Seventh Asilomar Conference on Signals, Systems Computers, 2003. vol. 2, pp. 1398–1402 (2003). https://doi.org/10.1109/ACSSC.2003.1292216

30. Yi, Z., Yongliang, S., Jun, Z.: An improved tiny-yolov3 pedestrian detection algorithm. Optik **183**, 17–23 (2019)

31. Zhu, G., et al.: Trajectory based event tactics analysis in broadcast sports video. In: Proceedings of the 15th ACM International Conference on Multimedia, pp. 58–67 (2007)

Shape Reconstruction from Point Clouds Using Closed Form Solution of a Fourth-Order Partial Differential Equation

Zaiping Zhu[1]($\boxtimes$), Ehtzaz Chaudhry[1], Shuangbu Wang[1], Yu Xia[1], Andres Iglesias[2], Lihua You[1], and Jian Jun Zhang[1]

[1] The National Center for Computer Animation, Bournemouth University, Poole, UK
`s5319266@bournemouth.ac.uk`
[2] Department of Applied Mathematics and Computational Sciences, University of Cantabria, 39005 Cantabria, Spain

Abstract. Partial differential equation (PDE) based geometric modelling has a number of advantages such as fewer design variables, avoidance of stitching adjacent patches together to achieve required continuities, and physics-based nature. Although a lot of papers have investigated PDE-based shape creation, shape manipulation, surface blending and volume blending as well as surface reconstruction using implicit PDE surfaces, there is little work of investigating PDE-based shape reconstruction using explicit PDE surfaces, specially satisfying the constraints on four boundaries of a PDE surface patch. In this paper, we propose a new method of using an accurate closed form solution to a fourth-order partial differential equation to reconstruct 3D surfaces from point clouds. It includes selecting a fourth-order partial differential equation, obtaining the closed form solutions of the equation, investigating the errors of using one of the obtained closed form solutions to reconstruct PDE surfaces from differential number of 3D points.

Keywords: Shape reconstruction · Fourth-order partial differential equation · Closed-form solutions · Error analysis

1 Introduction

Shape reconstruction has a lot of applications in many fields. Various surface reconstruction methods have been developed. These methods include polygon-based, implicit surface-based, and parametric surface-based. In addition, soft computing is also introduced into parametric surfaces to improve shape reconstruction from point clouds.

Shape reconstruction uses polygon meshes, implicit surfaces, and existing parametric surfaces such as Bézier, B-spline, and NURBS surfaces has some weaknesses. They include bid data, heavy geometry processing, high data storage cost, and slow data transmission over computer networks. How to address these weaknesses is an unsolved topic.

In contrast, PDE-based shape reconstruction has the following advantages. First, a single PDE surface patch can describe a complicated shape leading to smaller data

© Springer Nature Switzerland AG 2021
M. Paszynski et al. (Eds.): ICCS 2021, LNCS 12746, pp. 207–220, 2021.
https://doi.org/10.1007/978-3-030-77977-1_16

than NURBS, polygon and subdivision modelling techniques. Second, adjacent PDE surface patches naturally maintain position, tangent, or higher continuities and no manual operations are required to stitch different PDE patches together. Third, any irregular boundaries can be quickly specified by drawing a closed curve on 3D models, different sculpting forces can be applied to achieve the expected shape, and global shape manipulations can be easily obtained through shape control parameters etc., leading to more efficient shape manipulations.

However, a main difficulty for PDE-based shape manipulation is how to solve partial differential equations. Due to this difficulty, most studies investigated implicit PDE-based shape reconstruction which involves numerically solving partial differential equations. Although some research studies investigated explicit PDE-based shape reconstruction by interpolating four curves or satisfying the constraints on two opposite boundaries of a PDE patch, few studies presented closed form solutions of partial differential equations for 4-sided PDE patches. In this paper, we will propose a mathematical model, derive its closed form solutions, and use one of the closed form solutions to achieve shape reconstruction from point clouds.

2 Related Work

There are a lot of work of investigating shape reconstruction from point clouds. A comprehensive literature survey has been made in [1, 2]. Among these shape reconstruction methods, PDE-based shape reconstruction has also been investigated. The existing shape reconstruction methods can be divided into polygon-based, implicit surface-based, explicit surface-based, and soft computing and parametric surface-based. Shape reconstruction using explicit surfaces from solutions to partial differential equations was summarized in [3]. In this section, we briefly review some work on shape reconstruction from point clouds.

Polygon-based shape reconstruction is most popular. Many of them are based on the Delaunay triangulation. For each initial border edge in triangulated reconstruction, Boissonnant estimated a tangential plane and took a vertex of a surface triangle to be the sample point which maximizes the angle between the vertex and the k-nearest neighbors projected to the tangential plane [4], Hoppe et al. used k-nearest neighbors to find a tangential plan of every sample point and the marching cubes algorithm and the signed distance of the sample point closest to the tangential plan to reconstruct 3D surfaces [5]. Oblonšek and Guid presented a procedure to triangulate the input scattered point set, extract features, and fair the triangular mesh to achieve surface reconstruction [6]. Bernardini et al. proposed a ball-pivoting algorithm to interpolate a given point cloud to reconstruct a triangle mesh where a triangle is formed with three points when a ball touches them [7]. Gopi et al. projected the neighborhood of each of sample points on a tangential plan, derived the 2D Delaunay triangulation on the tangential plan, and mapped the result back to the 3D space, and reconstructed a 3D surface from the mapping [8]. Lee et al. reconstructed a 3D surface through repeated subdivision, displacement, and resampling of a control mesh model [9]. Jeong and Kim constructed a coarse base mesh from a bounding cube containing the input point cloud, and successively subdivide, smooth, and project it to the input point cloud to obtain shape reconstruction [10]. Not fitting dense smooth polygonal surfaces, Nan and Wonka used simple polygonal surfaces for reconstruction of piecewise planar objects [11].

Implicit surface-based shape reconstruction has also been extensively investigated. Duan et al. proposed a PDE-based deformable surface to reconstruct 3D models from volumetric data, unorganized point clouds and multi-view 2D images [12]. Linz et al. developed new formulations and fast algorithms to conduct implicit partial differential equation-based surface reconstruction [13]. Franchini et al. used the numerical technique based on an efficient semi-implicit scheme and finite volume space discretization to solve a time-dependent partial differential equation for implicit shape reconstruction [14]. Pana and Skala introduced an energy functional to combine flux-based data-fit measures and proposed a regulation term and a continuous global optimization method to carry out surface reconstruction from an oriented point cloud [15]. Using an implicit and continuous representation of reconstructed surfaces and optimizing a regularized fitting energy, Liu et al. developed a level-set based surface reconstruction method to process point clouds captured by a surface photogrammetry system [16]. He et al. presented two fast algorithms, one uses the semi-Implicit method and the other is based on the augmented Lagrangian method, to improve computational efficiency of surface reconstruction from point clouds [17].

Some researchers investigated explicit surface-based shape reconstruction. In order to solve the heavy computational cost of finite element methods or finite difference methods in solving the elliptic partial differential equation for 3D shape reconstruction, Li ang Hero developed a fast spectral method to improve the performance [18]. Ugail and Kirmani proposed an explicit partial differential equation-based reconstruction method which interpolates four curves parametrized in terms of the parametric variable v to obtain a reconstructed surface [19]. This method was also used to obtain reconstruction of 3D human facial images [20]. Rodrigues et al. obtained an analytical solution of a Laplace equation and used it in 3D data compression and reconstruction [21]. Sheng et al. integrated a point cloud update algorithm, a rapid iterative closest point algorithm, and an improved Poisson surface reconstruction algorithm together to improve the efficiency of surface reconstruction [22].

Soft computing has also been introduced into parametric surfaces to optimize shape reconstruction. Iglesias et al. introduced function networks into B-spline surfaces to solve the problem of shape reconstruction [23]. Gálvez and Iglesias integrated an iterative two-step genetic-algorithm and polynomial B-spline surfaces for efficient surface reconstruction from point clouds [24]. They proposed a particle swarm optimization approach to reconstruct non-uniform rational B-spline surfaces from 3D point clouds [25].

Since few research studies investigated shape reconstruction using explicit PDE surfaces, we will tackle this issue in this paper. The PDE mathematical model and its closed form solutions will be investigated in Sect. 3. The proposed PDE-based shape reconstruction and error analysis are examined in Sect. 4. Finally, the conclusion is drawn and some future research directions are discussed in Sect. 5.

3 Mathematical Model and Closed Form Solution

Partial differential equation-based shape reconstruction can be roughly divided into two categories: one uses implicit solutions of partial differential equations and the other uses explicit solutions of partial differential equations. Shape reconstruction using implicit solutions of partial differential equations involves a lot of numerical calculations, causing slow shape reconstruction which is not suitable for many applications requiring real-time performance. In contrast, shape reconstruction using explicit solutions of partial differential equation is based on accurate analytical or approximate analytical solutions of partial differential equations, which involves fewer calculations and is more efficient than shape construction using implicit solutions of partial differential equations.

However, a main problem for shape reconstruction using explicit solutions of partial differential equations is how to obtain accurate analytical or approximate analytical solutions of partial differential equations. Since solving partial differential equations analytically is not an easy task, the current explicit solutions of partial differential equations used for PDE-based geometric modelling and shape reconstruction mainly deal with two boundaries of a PDE surface patch, i.e., accurately satisfy partial differential equations and continuity constraints at two opposite boundaries of a PDE surface patch. How to obtain accurate analytical solutions of partial differentia equations which exactly satisfy partial differential equations and continuity constraints on four boundaries of a PDE surface patch is an important topic.

A vector-valued partial differential equation used to describe a 3D surface patch involves two parametric variables u and v. The four boundaries of the 3D surface patch are defined by $u = 0$, $u = 1$, $v = 0$, and $v = 1$. In order to satisfy positional continuities, four unknowns should be included in a closed form solution to a vector-valued partial differential equation to satisfy four positional functions, i.e. boundary curves at the four boundaries of a 3D surface patch. Similarly, in order to satisfy up to tangential continuities, eight unknowns should be involved in a closed form solution of a vector-valued partial differential equation to satisfy four positional functions and four tangential functions at the four boundaries of a 3D surface patch. From the theory of partial differential equation, the closed form solution to a second-order partial differential equation of parametric variables u and v has four unknowns, and the closed form solution to a fourth-order partial differential equation has eight unknowns.

Up to tangential continuities is most popularly used to create smooth 3D models. Taking all of these factors and a closed form solution into account, we propose to use the following vector-valued fourth-order partial differential equation for shape reconstruction

$$a_1 \frac{\partial^4 \mathbf{X}(u, v)}{\partial u^4} + a_2 \frac{\partial^4 \mathbf{X}(u, v)}{\partial u^4} = \mathbf{F}(u, v) \tag{1}$$

where a_1 and a_2 are called vector-value shape control parameters, which can be used to change the shape of a PDE surface without changing boundary continuities and each of which has three components a_{ix}, a_{iy}, and $a_{iz}(i = 1, 2)$ with $a_1 = \begin{bmatrix} a_{1x} \ a_{1y} \ a_{1z} \end{bmatrix}^T$ and $a_2 = \begin{bmatrix} a_{2x} \ a_{2y} \ a_{2z} \end{bmatrix}^T$, u and v are two parametric variables defined by $0 \le u \le 1$ and $0 \le v \le 1$, $\mathbf{X}(u, v)$ is a vector-valued position function used to define a

PDE surface patch, which has three components $x(u, v)$, $y(u, v)$, and $z(u, v)$ with $\mathbf{X}(u, v) = \left[x(u, v)\, y(u, v)\, z(u, v) \right]^T$, and $\mathbf{F}(u, v)$ is a vector-valued sculpting function, which also has three components $f_x(u, v)$ $f_y(u, v)$ and $f_z(u, v)$ with $\mathbf{F}(u, v) = \left[f_x(u, v)\, f_y(u, v)\, f_z(u, v) \right]^T$. The components of each of the two vector-valued shape control parameters cab be taken to be the same, i.e., $a_{ix} = a_{iy} = a_{iz}(i = 1, 2)$ or different, i.e., $a_{ix} \neq a_{iy} \neq a_{iz}(i = 1, 2)$.

In order to simplify mathematical notations, we define the following mathematical operations in this paper

$$a_1 a_2 = \left[a_{1x} a_{2x}\ a_{1y} a_{2y}\ a_{1z} a_{2z} \right]^T$$

$$\frac{a_1}{a_2} = \left[\frac{a_{1x}}{a_{2x}}\ \frac{a_{1y}}{a_{2y}}\ \frac{a_{1z}}{a_{2z}} \right]^T$$

$$e^{a_1} = \left[e^{a_{1x}}\ e^{a_{1y}}\ e^{a_{1z}} \right]^T$$

$$\sqrt[n]{\frac{a_1}{a_2}} = \left[\sqrt[n]{\frac{a_{1x}}{a_{2x}}}\ \sqrt[n]{\frac{a_{1y}}{a_{2y}}}\ \sqrt[n]{\frac{a_{1z}}{a_{2z}}} \right]^T$$

$$cosa_1 = \left[cosa_{1x}\ cosa_{1y}\ cosa_{1z} \right]^T$$

$$sina_1 = \left[sina_{1x}\ sina_{1y}\ sina_{1z} \right]^T \tag{2}$$

In this paper, we look for closed form solutions of the homogeneous form of the partial differential equation (1) and use one of them for shape reconstruction from point clouds. In the extended version of this paper, we will investigate the particular solution of the partial differential equation (1) and use it to develop a more powerful shape reconstruction tool.

We use the method of separation of variables to solve the homogeneous form of the vector-valued partial differential equation (1) and obtain its four closed form solutions. The details of solving the homogeneous form of the vector-valued partial differential equation (1) will be given in the extended version of this paper. In what follows, we use one closed form solution among the four obtained closed form solutions to demonstrate shape reconstruction using the closed form solutions of the vector-valued fourth-order partial differential equation (1) and investigate the errors of shape reconstruction. The closed form solution to be used can be written in the following form

$$\mathbf{X}(u, v) = \sum_{j=1}^{16} d_j \mathbf{f}_j(u, v) \tag{3}$$

where

$$f_1(u, v) = e^{q_2 u} e^{q_4 v} cos q_2 u cos q_4 v$$

$$f_2(u, v) = e^{q_2 u} e^{q_4 v} cos q_2 u sin q_4 v$$

$$f_3(u, v) = e^{q_2 u} e^{q_4 v} sin q_2 u cos q_4 v$$

$$f_4(u, v) = e^{q_2 u} e^{q_4 v} sin q_2 u sin q_4 v$$

$$f_5(u, v) = e^{q_2 u} e^{-q_4 v} cos q_2 u cos q_4 v$$

$$f_6(u, v) = e^{q_2 u} e^{-q_4 v} cos q_2 u sin q_4 v$$

$$f_7(u, v) = e^{q_2 u} e^{-q_4 v} sin q_2 u cos q_4 v$$

$$f_8(u, v) = e^{q_2 u} e^{-q_4 v} sin q_2 u sin q_4 v$$

$$f_9(u, v) = e^{-q_2 u} e^{q_4 v} cos q_2 u cos q_4 v$$

$$f_{10}(u, v) = e^{-q_2 u} e^{q_4 v} cos q_2 u sin q_4 v$$

$$f_{11}(u, v) = e^{-q_2 u} e^{q_4 v} sin q_2 u cos q_4 v$$

$$f_{12}(u, v) = e^{-q_2 u} e^{q_4 v} sin q_2 u sin q_4 v$$

$$f_{13}(u, v) = e^{-q_2 u} e^{-q_4 v} cos q_2 u cos q_4 v$$

$$f_{14}(u, v) = e^{-q_2 u} e^{-q_4 v} cos q_2 u sin q_4 v$$

$$f_{15}(u, v) = e^{-q_2 u} e^{-q_4 v} sin q_2 u cos q_4 v$$

$$f_{16}(u, v) = e^{-q_2 u} e^{-q_4 v} sin q_2 u sin q_4 v \tag{4}$$

and

$$q_2 = \frac{\sqrt{2}}{2} \sqrt[4]{\left| \frac{c_0}{a_1} \right|}$$

$$q_4 = \frac{\sqrt{2}}{2} \sqrt[4]{\left| \frac{c_0}{a_2} \right|} \tag{5}$$

where $d_j (j = 1, 2, 3, \cdots, 16)$ are the vector-valued unknowns, and c_0 is a vector-valued constant.

$X(u, v)$ in Eq. (3) defines a PDE surface patch. In the following section, we will discuss how to use it to achieve shape reconstruction for point clouds.

4 Shape Reconstruction and Error Analysis

Shape reconstruction from point clouds is to find the 16 vector-valued unknowns $d_j (j = 1, 2, 3, \cdots, 16)$ which make the PDE surface patch $X(u, v)$ best fit the points in the region to be reconstructed. For any unorganized point clouds, we can find the points close to each of a set of planes, which define a curve close to the plane. From these curves, we can obtain the values of the parametric variable u for these curves. From the points on each of the curves, we can obtain the values of the parametric variable v for all the points on the curve. For the points not on these curves, we can use the geometric relationships of their positions relative to the points on the curves to obtain the values of parametric variables u and v. By doing so, the values of the parametric variables u and v for all the points to be used for shape reconstruction are obtained. That is, for each point X_n, we obtain its parametric values u_n and v_n.

If N points $X_n(n = 1, 2, 3, \cdots, N)$ are to be used to reconstruct a PDE surface patch $X(u, v)$ we can calculate the squared sum of the errors between the known points $X_n(n = 1, 2, 3, \cdots, N)$ and the unknown points $X(u_n, v_n)$ with the following equation

$$E = \sum_{n=1}^{N} [X(u_n, v_n) - X_n]^2$$

$$= \sum_{n=1}^{N} \left[\sum_{j=1}^{16} d_j f_j(u_n, v_n) - X_n \right]^2 \tag{6}$$

The least squares are used to minimize the error E and find the 16 vector-valued unknowns with the equation below

$$\frac{\partial E}{\partial d_k} = 0$$

$$(k = 1, 2, 3, \cdots, 16) \tag{7}$$

Substituting Eq. (6) into Eq. (7), we obtain the following equations which can be used to determine the 16 vector-valued unknowns $d_j(j = 1, 2, 3, \cdots, 16)$

$$\sum_{j=1}^{16} d_j \sum_{n=1}^{N} f_j(u_n, v_n) f_k(u_n, v_n) = \sum_{n=1}^{N} X_n f_k(u_n, v_n)$$

$$(k = 1, 2, 3, \cdots, 16) \tag{8}$$

Equation (8) involves 16 equations which can be used to determine the 16 vector-valued unknowns $d_j(j = 1, 2, 3, \cdots, 16)$.

$f_j(u_n, v_n)$ and $f_k(u_n, v_n)$ in Eq. (8) involve the constants q_2 and q_4, which are determined by c_0, a_1 and a_2, respectively. a_1 and a_2 are vector-valued shape control parameters. They can be optimized to obtain the optimal PDE surface patch which best fits to the points $X_n(n = 1, 2, 3, \cdots, N)$.

However, if we take q_2 and q_4 as design variables, Eq. (8) becomes nonlinear. Solving Eq. (8) is to solve 16 nonlinear equations which makes the determination of the 16 vector-valued unknowns $d_j(j = 1, 2, 3, \cdots, 16)$ more difficult. Instead of optimizing q_2 and q_4 to find their optimal values, we set q_2 and q_4 to different values, solve the 16 linear algebra equations of Eq. (8), and find that $q_2 = 0.1$ and $q_4 = 0.1$ give good results.

In what follows, we use Eq. (3) to reconstruct a 3D PDE surface from different points, and compare the surface defined by the original points and the reconstructed PDE surface. In order to quantify the differences between the two surfaces, we calculate the maximum error and the average error between the two surfaces with the following equations

$$ErrM = max \left\{ |X_1 - X(u_1, v_1)| \, |X_2 - X(u_2, v_2)| \, \cdots \, |X_N - X(u_N, v_N)| \right\} \tag{9}$$

$$ErrA = \frac{1}{N} \sum_{n=1}^{N} |X_n - X(u_n, v_n)| \qquad (9)$$

where $ErrM$ indicates the maximum error between the two surfaces, $ErrA$ indicates the average error between the two surfaces, $|\cdot|$ indicates the distance between the correspondent points of the two surfaces.

Firstly, we consider reconstructing a surface from 16 points give in Table 1. Since a PDE surface patch (3) involves 16 unknowns, the PDE surface patch should pass through the 16 points if interpolation operation is used to determine the 16 unknowns. In this paper, the fitting operation defined by Eq. (8), not interpolation operation, is used to determine the 16 unknowns. Although the interpolation operation is not used, it is expected that the fitting operation should give high accuracy if not passing the 16 points.

Table 1. 16 points used to define the surface in Fig. 1.

(x, y, z)	(x, y, z)	(x, y, z)	(x, y, z)
$(0.05, -0.07, -1.12)$	$(0.07, -0.07, -1.11)$	$(0.10, -0.07, -1.11)$	$(0.13, -0.07, -1.12)$
$(0.04, -0.10, -1.10)$	$(0.07, -0.10, -1.09)$	$(0.10, -0.10, -1.10)$	$(0.13, -0.10, -1.10)$
$(0.04, -0.13, -1.08)$	$(0.07, -0.13, -1.06)$	$(0.10, -0.13, -1.07)$	$(0.13, -0.13, -1.09)$
$(0.04, -0.16, -1.05)$	$(0.07, -0.16, -1.04)$	$(0.10, -0.16, -1.04)$	$(0.13, -0.16, -1.06)$

Setting $N = 16$ in Eq. (8) and solving the 16 linear algebra equations, we obtain the 16 vector-valued unknowns. With the original 16 points, we create the surface depicted in Fig. 1(a). Substituting the obtained 16 vector-valued unknowns back into Eq. (3), we use Eq. (3) to create the PDE surface patch shown in Fig. 1(b). The maximum error and the average error between the two surfaces are given in Table 2.

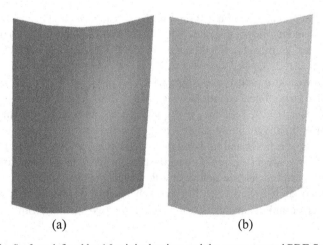

(a)   (b)

Fig. 1. Surface defined by 16 original points and the reconstructed PDE Surface.

Comparing the surface in Fig. 1(a) and the surface in Fig. 1(b), we could not find any differences, indicating the reconstructed PDE surface is the same as the original surface defined by the 16 points. This observation is also supported by the maximum error and average error given in Table 2. The maximum error between the two surfaces is 5.52×10^{-5} and the average error between the two surfaces is 2.31×10^{-5}. Both errors are very small, which indicating high accuracy of the fitting operation.

Table 2. Maximum errors and average errors between the two surfaces

N	16	25	36	49	64	81
ErrM	5.52×10^{-5}	1.56×10^{-3}	2.96×10^{-3}	9.68×10^{-3}	1.40×10^{-2}	2.02×10^{-2}
ErrA	2.31×10^{-5}	4.47×10^{-4}	1.33×10^{-3}	3.80×10^{-3}	5.76×10^{-3}	8.16×10^{-3}

Secondly, we consider reconstructing a surface from 25 points consisting of those in Table 1 and Table 3. Setting $N = 25$ in Eq. (8) and solving the 16 linear algebra equations, we obtain the 16 vector-valued unknowns. With the original 25 points, we create the surface depicted in Fig. 2(a). Substituting the obtained 16 vector-valued unknowns back into Eq. (3), we use Eq. (3) to create the PDE surface shown in Fig. 2(b). The maximum error and the average error between the two surfaces are given in Table 2.

Comparing the surface in Fig. 2(a) and the surface in Fig. 2(b), we still could not find any differences, indicating the reconstructed PDE surface is very similar to the original surface defined by the 25 points. As indicated in Table 2, the maximum error between the two surfaces is 1.56×10^{-3} and the average error between the two surfaces is 4.47×10^{-4}, indicating small errors between the original surface and the reconstructed PDE surface.

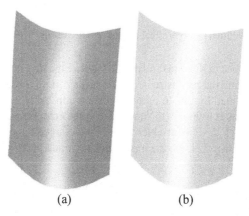

(a)   (b)

Fig. 2. Surface defined by 25 original points and the reconstructed PDE surface.

Table 3. Points used with those in Table 1 to define the surface in Fig. 2

(x, y, z)	(x, y, z)	(x, y, z)	(x, y, z)
(0.02, −0.07, −1.16)	(0.02, −0.10, −1.16)	(0.07, −0.19, −1.01)	(0.10, −0.19, −1.01)
(0.01, −0.19, −1.05)	(0.04, −0.19, −1.02)	(0.10, −0.10, −1.10)	(0.13, −0.10, −1.10)
(0.13, −0.19, −1.03)			

Thirdly, we consider reconstructing a surface from 36 points consisting of those in Table 1, Table 3, and Table 4. Setting $N = 36$ in Eq. (8) and solving the 16 linear algebra equations, we obtain the 16 vector-valued unknowns. With the original 36 points, we create the surface depicted in Fig. 3(a). Substituting the obtained 16 vector-valued unknowns back into Eq. (3), we use Eq. (3) to create the PDE surface shown in Fig. 3(b). The maximum error and the average error between the two surfaces are given in Table 2.

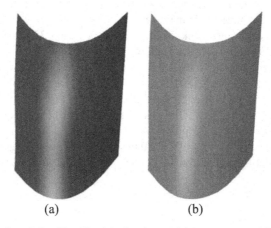

(a)   (b)

Fig. 3. Surface defined by 36 original points and the reconstructed PDE surface.

Comparing the surface in Fig. 3(a) and the surface in Fig. 3(b), no obvious differences could be found, indicating the reconstructed PDE surface gives a good approximation to the original surface defined by the 36 points. As shown in Table 2, the maximum error between the two surfaces is 2.96×10^{-3} and the average error between the two surfaces is 1.33×10^{-3}. Although design variables have been reduced by more than a half, i.e., from $36 \times 3 = 108$ to $16 \times 3 = 48$, the maximum error and average error are small.

Table 4. Points used with those in Tables 1 and 3 to define the surface in Fig. 3

(x, y, z)	(x, y, z)	(x, y, z)	(x, y, z)
(0.02, −0.04, −1.20)	(0.05, −0.04, −1.15)	(0.07, −0.04, −1.14)	(0.10, −0.04, −1.14)
(0.13, −0.04, −1.15)	(0.17, −0.03, −1.20)	(0.17, −0.07, −1.17)	(0.16, −0.10, −1.14)
(0.16, −0.13, −1.12)	(0.16, −0.16, −1.10)	(0.16, −0.19, −1.07)	

Fourthly, we consider reconstructing a surface from 49 points. Setting $N = 49$ in Eq. (8) and solving the 16 linear algebra equations, we obtain the 16 vector-valued unknowns. With the original 49 points, we create the surface depicted in Fig. 4(a). Substituting the obtained 16 vector-valued unknowns back into Eq. (3), we use Eq. (3) to create the PDE surface patch shown in Fig. 4(b). The maximum error and the average error between the two surfaces are given in Table 2.

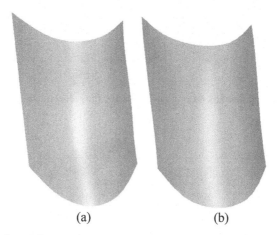

(a)   (b)

Fig. 4. Surface defined by 49 original points and the reconstructed PDE surface.

The surface in Fig. 4(a) looks like the surface in Fig. 4(b). As indicted in Table 2, the maximum error between the two surfaces is 9.68×10^{-3} and the average error between the two surfaces is 3.80×10^{-3}. For this case, the design variables are reduced by two thirds, i.e., from $49 \times 3 = 147$ to $16 \times 3 = 48$, the maximum error and average error are still small.

Fifthly, we consider reconstructing a surface from 64 points. Setting $N = 64$ in Eq. (8) and solving the 16 linear algebra equations, we obtain the 16 vector-valued unknowns. With the original 64 points, we create the surface depicted in Fig. 5(a). Substituting the obtained 16 vector-valued unknowns back into Eq. (3), we use Eq. (3) to create the PDE surface shown in Fig. 5(b). The maximum error and the average error between the two surfaces are given in Table 2.

The surface in Fig. 5(a) still looks like the surface in Fig. 5(b). The maximum error between the two surfaces is 1.40×10^{-2} and the average error between the two surfaces is 5.76×10^{-3}. For this case, the design variables are reduced by three fourths, i.e., from $64 \times 3 = 192$ to $16 \times 3 = 48$, the maximum error and average error are not big.

Finally, we consider reconstructing a surface from 81 points. Setting $N = 81$ in Eq. (8) and solving the 16 linear algebra equations, we obtain the 16 vector-valued unknowns. With the original 81 points, we create the surface depicted in Fig. 6(a). Substituting the obtained 16 vector-valued unknowns back into Eq. (3), we use Eq. (3) to create the PDE surface shown in Fig. 6(b). The maximum error and the average error between the two surfaces are given in Table 2.

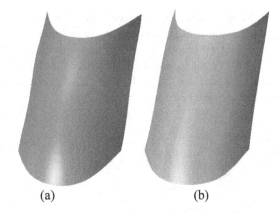

Fig. 5. Surface defined by 64 original points and the reconstructed PDE surface.

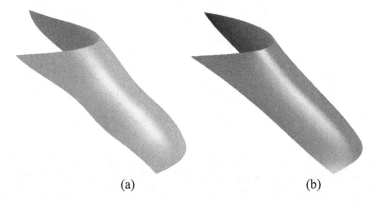

Fig. 6. Surface defined by 81 original points and the reconstructed PDE surface.

Since the design variables have been reduced from $81 \times 3 = 243$ to $16 \times 3 = 48$, i.e., reduced by four fifths, some differences can be observed from the two surfaces depicted in Fig. 6(a) and 6(b). As shown in Table 2, the maximum error between the two surfaces becomes 2.02×10^{-2} and the average error between the two surfaces becomes 8.16×10^{-3}, which are not big.

5 Conclusion

In this paper, we have proposed a new method for shape reconstruction from point clouds. It is based on accurate closed form solutions to a vector-valued fourth-order partial differential equation. Due to the nature of explicit analytical expressions of the reconstructed surface, the proposed method is very efficient in reconstructing 3D surfaces from point clouds. The error analysis given in this paper demonstrates good approximation of the reconstructed surface from the proposed PDE-based shape reconstruction method to point clouds.

There are a number of directions that can be investigated in the following work. First, the particular solution of the nonhomogeneous equation of Eq. (2) should be developed to make the proposed PDE-based shape reconstruction more powerful. Second, how to use the obtained solutions of the proposed vector-valued fourth-order partial differential equation to reconstruct unorganized point clouds with noise and missing data should be examined. In this paper, only shape reconstruction by using one single PDE surface patch was discussed. For complicated 3D models, a single PDE surface patch is unable to reconstruct complicated 3D models. How to use multiple PDE surface patches to reconstruct complicated 3D models from point clouds should be considered. The vector-valued shape control parameters involved in the PDE (2) affects the shape of reconstructed surfaces. How to identify an efficient and effective optimization method and combine it with a suitable method of solving the nonlinear Eqs. (8) to find optimal values of the two parameters q_2 and q_4, and obtain the 16 vector-valued unknowns to maximum the potential of the proposed PDE-based shape reconstruction should also be explored. Another interesting direction is using the obtained closed form solutions to accurately satisfy all the constraints on four boundaries of a PDE patch. To do this, Eq. (3) can be extended into a series $\mathbf{X}(u, v) = \sum_m^M \sum_{j=1}^{16} d_{mj} f_{mj}(u, v)$ with q_2 and q_4 being replaced by q_{2m} and q_{4m}. We will investigate this in our following work.

Acknowledgements. This research is supported by the PDE-GIR project which has received funding from the European Union Horizon 2020 research and innovation programme under the Marie Skodowska-Curie grant agreement No 778035. Zaiping Zhu is also sponsored by China Scholarship Council.

References

1. Berger, M., et al.: State of the art in surface reconstruction from point clouds. Eurographics 2014 - State of the Art Reports, pp. 161–185 (2014)
2. Berger, M., et al.: A survey of surface reconstruction from point clouds. Comput. Graph. Forum **36**(1), 301–329 (2017)
3. Othman, M.N.M., Yusoff, Y., Haron, H., You, L.H.: An overview of surface reconstruction using partial differential equation (PDE). IOP Conf. Ser. Mater. Sci. Eng. **551**, 1–5 (2019)
4. Boissonnant, J.D.: Geometric structures for three-dimensional shape reconstruction. ACM Trans. Graph. **3**(4), 266–289 (1984)
5. Hoppe, H., DeRose, T., Duchamp, T., McDonald, J., Stuetzle, W.: Surface reconstruction from unorganized points. In: Proceedings of SIGGRAPH 1992, pp. 71–78 (1992)
6. Oblonšek, Č., Guid, N.: A fast surface-based procedure for object reconstruction from 3D scattered points. Comput. Vis. Image Underst. **69**(2), 185–195 (1998)
7. Bernardini, F., Mittelman, J., Rushmeier, H., Silva, C., Taubin, G.: The ball-pivoting algorithm for surface reconstruction. IEEE Trans. Vis. Comput. Graph. **5**(4), 349–359 (1999)
8. Gopi, M., Krisnan, S., Silva, C.: Surface reconstruction based on lower dimensional localized Delaunay triangulation. Comput. Graph. Forum **19**(3), 467–478 (2000)
9. Lee, A., Moreton, H., Hoppe, H.: Displaced subdivision surfaces. In: Proceedings of SIGGRAPH 2000, pp. 85–94 (2000)
10. Jeong, W.K., Kim, C.H.: Direct reconstruction of displaced subdivision surface from unorganized points. Graph. Models **64**(2), 78–93 (2002)

11. Nan, L., Wonka, P.: PolyFit: polygonal surface reconstruction from point clouds. In: 2017 IEEE International Conference on Computer Vision (ICCV), Venice, pp. 2372–2380 (2017)
12. Duan, Y., Yang, L., Qin, H., Samaras, D.: Shape reconstruction from 3D and 2D data using PDE-based deformable surfaces. In: Pajdla, T., Matas, J. (eds.) ECCV 2004. LNCS, vol. 3023, pp. 238–251. Springer, Heidelberg (2004). https://doi.org/10.1007/978-3-540-24672-5_19
13. Linz, C., Goldlücke, B., Magnor, M.: A point-based approach to PDE-based surface reconstruction. In: Franke, K., Müller, K.-R., Nickolay, B., Schäfer, R. (eds.) DAGM 2006. LNCS, vol. 4174, pp. 729–738. Springer, Heidelberg (2006). https://doi.org/10.1007/11861898_73
14. Franchini, E., Morigi, S., Sgallari, F.: Implicit shape reconstruction of unorganized points using PDE-based deformable 3D manifolds. Numer. Math. Theory Methods Appl. 3(4), 405–430 (2010)
15. Pana, R., Skala, V.: Continuous global optimization in surface reconstruction from an oriented point cloud. Comput. Aided Des. 43, 896–901 (2011)
16. Liu, W., Cheung, Y., Sabouri, P., Arai, T.J., Sawant, A., Ruan, D.: A continuous surface reconstruction method on point cloud captured from a 3D surface photogrammetry system. Med. Phys. 42(11), 6564–6571 (2015)
17. He, Y., Huska, M., Kang, S.H., Liu, H.: Fast algorithms for surface reconstruction from point cloud, pp. 1–16 (2019). https://arxiv.org/abs/1907.01142
18. Li, J., Hero, A.O.: A fast spectral method for active 3D shape reconstruction. J. Math. Imaging Vis. 20, 73–87 (2004)
19. Ugail, H., Kirmani, S.: Method of surface reconstruction using partial differential equations. In: Proceedings of the 10th WSEAS International Conference on COMPUTERS, Vouliagmeni, Athens, Greece, 13–15 July, pp. 51–56 (2006)
20. Elyan, E., Ugail, H.: Reconstruction of 3D human facial images using partial differential equations. J. Comput. 2(8), 1–8 (2007)
21. Rodrigues, M., Osman, A., Robinson, A.: Partial differential equations for 3D data compression and reconstruction. Adv. Dyn. Syst. Appl. 8(2), 303–315 (2013)
22. Sheng, B., Zhao, F., Yin, X., Zhang, C., Wang, H., Huang, P.: A lightweight surface reconstruction method for online 3D scanning point cloud data oriented toward 3D printing. Math. Probl. Eng. 2018, 1–16 (2018). Article no. 4673849
23. Iglesias, A., Echevarría, G., Gálvez, A.: Functional networks for B-spline surface reconstruction. Future Gener. Comput. Syst. 20(8), 1337–1353 (2004)
24. Gálvez, A., Iglesias, A.: Iterative two-step genetic-algorithm-based method for efficient polynomial B-spline surface reconstruction. Inf. Sci. 182(1), 56–76 (2012)
25. Gálvez, A., Iglesias, A.: Particle swarm optimization for non-uniform rational B-spline surface reconstruction from clouds of 3D data points. Inf. Sci. 192(1), 174–192 (2012)

Data-Driven Computational Sciences

Addressing Missing Data in a Healthcare Dataset Using an Improved kNN Algorithm

Tressy Thomas$^{(\boxtimes)}$ and Enayat Rajabi

Shannon School of Business, Cape Breton University, Sydney, NS, Canada

Abstract. Missing values are ubiquitous in many real-world datasets. In scenarios where a dataset is not very large, addressing its missing values by utilizing appropriate data imputation methods benefits analysis significantly. In this paper, we leveraged and evaluated a new imputation approach called k-Nearest Neighbour with Most Significant Features and incomplete cases (KNNI$_{MSF}$) to impute missing values in a healthcare dataset. This algorithm leverages k-Nearest Neighbour (kNN) and ReliefF feature selection techniques to address incomplete cases in the dataset. The merit of imputation is measured by comparing the classification performance of data models trained with the dataset with imputation and without imputation. We used a real-world dataset, "very low birth weight infants", to predict the survival outcome of infants with low birth weights. Five different classifiers were used in the experiments. The comparison of multiple performance metrics shows that classifiers built on imputed dataset produce much better outcomes. KNNI$_{MSF}$ outperformed in general than the k-Nearest Neighbour Imputation using the Random Forest feature weights (KNNI$_{RF}$) algorithm with respect to the balanced accuracy and specificity.

Keywords: Missing value · Data imputation · kNN · Classification · Healthcare

1 Introduction

Data has quality if it satisfies the requirements of its intended use [1]. Real-world data are typically incomplete and incompleteness can impair the knowledge discovery process. In order to make data useful for the purpose of analyzing methods, such as data mining and machine learning, a significant amount of time is spent on pre-processing of data. For medical datasets, missing values are unfortunately unavoidable [2]. Incorrect imputation of missing values could lead to inaccurate research as well as wrong predictions [2]. Due to the nature of the domain and significance of applications, it is very important to have highly accurate results.

The missing mechanism is an important concept that defines the connection between the observed and missing variables in a dataset. The missing mechanism

© Springer Nature Switzerland AG 2021
M. Paszynski et al. (Eds.): ICCS 2021, LNCS 12746, pp. 223–230, 2021.
https://doi.org/10.1007/978-3-030-77977-1_17

gives an account on possible relationships between measured variables and the probability of missing data [3]. According to Little and Rubin's [4] taxonomy, missing mechanisms are missing completely at random (MCAR), missing at random (MAR) and not missing at random (NMAR). In MCAR, the reason of missingness is unrelated to any other observation. MAR arises if the reason for dropout depends on the observed outcomes. The MNAR mechanism depends, in whole or in part, on unobserved measurements itself. Details about the different missing mechanisms can be referenced in many published studies [3,4].

The complete case analysis is a way of treating missing values in a dataset by ignoring the incomplete cases in the dataset. In many cases, especially in medical domains, this approach can result in loss of information [5]. As the information in the incomplete cases in a dataset is not made useful, the statistical inferences or the model performance may not result in meaningful insights and predictions, particularly when the size of a dataset is not very large. There exists many missing value imputation techniques which can estimate the missing values so that the incomplete cases can be repaired and used for analysis or data modeling purposes without losing information or adding bias to dataset [5]. Replacing the missing data with an appropriate value derived from an observed data is called missing value imputation. Leveraging machine learning algorithms to impute missing values is getting popular due to its applicability. k-Nearest Neighbours (kNN), being one of the simplest and non-parametric instance-based approaches, is widely used in missing value imputation problems [6]. kNN based imputation methods are easy to implement and perform well in a variety of scenarios [6]. The basic kNN based imputation method uses the 'k' nearest neighbor's value to estimate the missing value. In this study, we evaluate the kNN imputation method ($KNNI_{MSF}$) on a healthcare dataset with a high level of missingness. This new imputation approach utilizes the most significant features with respect to the missing attributes and considers incomplete cases as well to estimate the missing values.

The merit of imputation is evaluated by comparing the performance of classifier algorithms with the dataset without any imputation treatment, with dataset after undergoing $KNNI_{MSF}$ imputation and with dataset imputed using another well-known Weighted kNN imputation based on Random Forest ($KNNI_{RF}$). The rest of the paper is arranged as follows: Sect. 2 presents the previous evaluation studies on missing value imputation using healthcare datasets. It also presents why this study is different from those previous studies and recommendations. Section 3 provides the details regarding our proposed imputation approach. The next section presents the experimentation settings, the dataset we used in this study along with the evaluation metrics. It is followed by Sect. 5 where the results and inferences regarding the experiments are discussed. Our paper ends with the conclusion section which presents the significant observations and future scope for this research.

2   Related Works

The missing data problem is crucial in healthcare domain. Hence, several published studies addressed the missing data problem. In one such study [7] the influence of missing value imputation on the classification accuracy was discussed. Globally average value, average value within a cluster and average value within the class were the missing value imputation techniques used. The missing values were artificially induced in four healthcare datasets and then imputed before evaluating the impact of missing value imputation experimentally. The comparison of classifier accuracy on different imputed datasets with the complete dataset in this study showed that there can be under – or overestimation of classification accuracy caused by choosing wrong method [7].

Machine learning techniques were found to perform better than the standard mean imputation technique in [13]. Cardiovascular data with missing value frequency up to 30% was used in the experiments [13]. Another study on missing healthcare data imputation is presented in [8]. This research implemented three algorithms in real healthcare dataset and concluded that MICE(Multiple Imputation by Chained Equations) algorithm performs better than Amelia and fuzzy unordered rule induction algorithm imputation(FURIA) [8].

In our study, we have utilized a real-world healthcare dataset with missing values at a higher percentage. The statistical imputation methods recommended in the mentioned studies assume that the missing data are Missing At Random (MAR). We are interested in an imputation technique that can be more generalized but also usable in critical domain applications. It is due to this fact that we are employing one of the easiest non-parametric algorithms (kNN) to implement missing value imputation. We compared our method with $KNNI_{RF}$ algorithm which in general outperforms the other kNN based imputation methods, based on our previous experimentation. $KNNI_{RF}$ is a weighted kNN imputation technique based on Random Forest where the weights for each variable are obtained using Random Forest approach [9] and these weights are used in the distance calculation.

We are presenting a new imputation approach based on kNN algorithm. Our approach considers the incomplete cases also for the estimation but only the relevant features are used for imputing the missing values. This approach is suitable for handling the missing values in small datasets with high missing percentages which we are evaluating in the experiments.

3   Methods

Our approach for the missing value imputation considers the significant features that are relevant for estimating that particular attribute. The steps used in the approach are as follows. The process starts with identifying the attributes with at least one missing value. For missing attribute, feature quality estimation algorithm ReliefF is executed to get the most significant features in the dataset that can predict the missing attribute [10]. ReliefF algorithm accounts the correlation and interaction between the attributes. This is important in estimating the missing values and helps estimate the correct value to replace the missingness. Only complete cases are used for the purpose of selecting the relevant features that can estimate the missing attribute. For each of the missing value of this attribute, Gower distance gd between the instances is calculated based on the Eq. 1 [11].

$$gd_{ij} = \frac{\sum_p(\delta_{ijp}d_{ijp}^f)}{\sum_p(\delta_{ijp})} \tag{1}$$

Where x_i is the missing vector and x_j is observed vector, k is the attribute. For numerical attributes in the instance (d_{ijp}^f) is calculated by

$$d_{ijp}^f = \frac{|x_i - x_j|}{|(max_N(x) - min_N(x))|} \tag{2}$$

where N is the total number of instances in the dataset. For other attributes (d_{ijp}^f) is 1 when the x_i and x_j attribute value differs. Otherwise it is set to 0.

For distance calculation, the features selected from the previous step are used. But for estimation of the missing value, all instances that have selected features and the missing feature present are considered. In the traditional kNN approach only complete cases are used. Utilizing the incomplete cases but with relevant features will provide better estimations, especially when multiple variables are missing in an instance. Similarity between each data point with the missing value instance is calculated as:

$$Sim = \frac{1}{gd + 1} \tag{3}$$

Then the weight for the 'k' neighbour instances are calculated based on the similarity:

$$Wt_k = \frac{Sim_k}{\sum_1^k Sim}. \tag{4}$$

For the estimation of numerical missing value, weighted sum of the nearest neighbour attribute values are used. For nominal values, mode is used to impute the missing value. The steps are iterated for all the missing attribute and its missing values.

4   Experimentation Set-up

For the experiments, we used the 'Very Low Birth Weight Infants' dataset. Data on 671 infants with very low (less than 1600 g) birth weight from 1981 to 87 were collected at Duke University Medical Center by Dr. Michael O'Shea [12]. There are 671 observations and 26 variables in the dataset. The details of the number of missing attributes by their missing value percentages are given in the Table 1. 78.54% of instances are labelled with the survival outcome as alive and 21.46% are with survival outcome as dead. There are only 174 cases which have all the 25 attributes present. Since most machine learning algorithms typically cannot utilize incomplete cases, in this dataset, about 75% of data will be lost if only complete cases are used for classification. This loss of data can result in a very significant loss of information. This is an example case of how we can utilize imputation to get more data and thus more information to achieve better classification or prediction.

Table 1. Missing value percentage in the dataset

Missing percentage	No of attributes
1–3%	4
3–9%	11
6–15%	3
15–30%	5
30–60%	2

To evaluate the merit of imputation, we have conducted classification using five different classifiers. First, data model was trained and evaluated using only the complete cases from the dataset. Then the dataset was imputed using two different missing value imputation methods $KNNI_{RF}$ and $KNNI_{MSF}$. The entire dataset was split into training and test in 70:30 ratio. Five fold cross validation was repeated five times for training the model. Model evaluation and comparison is done with the test data. The classifiers used here for the prediction of survival outcome of the infant are Logistic Regression (LR), Support Vector Machine using Radial basis function kernel (SVM), k-Nearest Neighbour Classifier (kNN), Gradient Boosting Machine(GBM) and Decision Tree Classifier (CT). The positive class in the classification model is the survival outcome 'live' and negative class is 'dead'.

Evaluation Metrics. Since the cost of miss-classification is very determinant factor in the evaluation of the model due to the nature of medical domain, we have used multiple metrics such as Accuracy, Balanced Accuracy, Sensitivity and Specificity for the model evaluation. To compare the performance of imputation method, Wilcoxon signed rank test is performed at $\alpha = 0.1$ with null hypothesis: $H_0 =$ The performance of classifiers using $KNNI_{RF}$ imputed dataset is equal to that of $KNNI_{MSF}$ imputed dataset. Alternative hypothesis: $H_1 =$ the performance of classifiers using $KNNI_{MSF}$ imputed data and that using $KNNI_{RF}$ imputed data are not equal.

5 Results

The performance of five classifiers were measured and the evaluation metrics are presented in Tables 2 for the three cases. First is 'Complete cases' where only complete cases from the original dataset was used in modelling and evaluation. The other two, $KNNI_{RF}$ and $KNNI_{MSF}$, represents the dataset imputed using

the KNNI$_{RF}$ and KNNI$_{MSF}$ imputation method respectively. The survival out-
come(alive or dead) prediction accuracy for each of the model is given in Table 2.
It can be seen from the results that the train models with imputed datasets perform better than that used without any imputation. Also, KNNI$_{MSF}$ resulted in
better accurate prediction than KNNI$_{RF}$ in most classifiers. It is evident from
the results that the balanced accuracy is very poor for the model trained with
the complete cases. The classifiers trained with imputed datasets performed relatively much better with respect to the balanced accuracy. Sensitivity measure,
which is the measure of how well the classifier predict the positive cases (alive),
also suggest a better performance of model trained with imputed data.

Table 2. Comparison of evaluation metrics of the classifiers

Missing value handling	LR	SVM	kNN	GBM	CT
Accuracy score of the classifiers					
Complete cases	82.35	82.35	92.16	90.20	86.27
KNNI$_{RF}$	88.56	88.56	87.06	91.54	89.05
KNNI$_{MSF}$	87.56	89.55	88.06	94.53	89.05
Balanced accuracy score of the classifiers					
Complete cases	43.75	43.75	48.96	47.92	45.83
KNNI$_{RF}$	80.87	73.15	81.72	85.31	77.80
KNNI$_{MSF}$	81.09	74.63	84.04	88.90	77.80
Sensitivity of the classifiers					
Complete cases	87.50	87.50	97.92	95.83	91.67
KNNI$_{RF}$	94.30	93.67	97.47	96.20	97.47
KNNI$_{MSF}$	92.41	93.67	98.10	98.73	97.47
Specificity of the classifiers					
Complete cases	0	0	0	0	0
KNNI$_{RF}$	67.44	69.77	48.84	74.42	58.14
KNNI$_{MSF}$	69.77	74.42	51.16	79.07	58.14

Specificity metric, in this case, shows that the model with complete cases is
not useful at all in predicting the minority class (survival outcome = dead). The
classifier models with KNNI$_{MSF}$ imputed data performed better in predicting the
minority class related to that of KNNI$_{RF}$ imputed data. The Wilcoxon signed
rank test shows fair evidence against null hypothesis which confirms the performance metrics(Balanced Accuracy and Specificity) of classifiers using KNNI$_{MSF}$
imputed data is greater that that using KNNI$_{RF}$ imputed data. Overall, the performance of KNNI$_{MSF}$ is either comparable or superior to KNNI$_{RF}$ with respect
to the evaluated metrics and is a good approach to be used for small datasets
with high missingness.

6   Conclusion and Future Works

Missing value datasets treated using imputation methods can result in better utilization of all available information for data modeling and statistical inferences. Especially in medical domain this can add much value and benefit. Our proposed missing value imputation method was tested on a healthcare dataset with high missingness percentage. The evaluation showed the merit of imputation with improved classifier performance. The comparison of classifiers trained with both complete cases and imputed datasets indicated that the proposed model performance is much better for the classifiers trained with imputed dataset. Also, the $KNNI_{MSF}$ imputation method performed better in general from the accuracy, balanced accuracy and specificity perspectives than the $KNNI_{RF}$ method. It can be concluded that $KNNI_{MSF}$ missing value imputation can treat missing values appropriately and the use of imputed datasets result in better data model training and model performance. In future, this new approach can be tested on more healthcare datasets with missing values to validate its performance.

References

1. Han, J., Kamber, M., Pei, J.: 3 - Data preprocessing. In: Han, J., Kamber, M., Pei, J. (eds.) Data Mining, 3rd edn., pp. 83–124. Morgan Kaufmann (2012). https://doi.org/10.1016/B978-0-12-381479-1.00003-4
2. Schmidt, D., Niemann, M., Lindemann-Von Trzebiatowski, G.: The handling of missing values in medical domains with respect to pattern mining algorithms. In: CEUR Workshop Proceedings, vol. 1492 (2015)
3. Enders, C.K., Craig, K.: Applied Missing Data Analysis. The Guilford Press. New York, London (2010)
4. Rubin, D.B.: Inference and missing data. Biometrika **63**(3), 581–592 (1976). https://doi.org/10.1093/biomet/63.3.581
5. Bartlett, J.W., Harel, O., Carpenter, J.R.: Asymptotically unbiased estimation of exposure odds ratios in complete records logistic regression. Am. J. Epidemiol. **182**(8), 730–736 (2014). https://doi.org/10.1093/aje/kwv114
6. Jadhav, A., Pramod, D., Ramanathan, K.: Comparison of performance of data imputation methods for numeric dataset. Appl. Artif. Intell. **33**(10), 913–933 (2019). https://doi.org/10.1080/08839514.2019.1637138
7. Orczyk, T., Porwik, P.: Influence of missing data imputation method on the classification accuracy of the medical data. J. Med. Inform.Technol. **22**, 111–116 (2013)
8. Chowdhury, M.H., Islam, M.K. Khan, Islam, S.: Imputation of missing healthcare data. In: IEEE 2017 20th International Conference of Computer and Information Technology (ICCIT) - Dhaka, Bangladesh, 22.12.2017–24-12-2017, pp. 1–6 (2017). https://doi.org/10.1109/ICCITECHN.2017.8281805
9. Kowarik, A., Templ, M.: Imputation with the R package VIM. J. Stat. Softw. **74** (2016). https://doi.org/10.18637/jss.v074.i07
10. Kononenko, I.: Estimating attributes: analysis and extensions of RELIEF. In: Bergadano, F., De Raedt, L. (eds.) ECML 1994. LNCS, vol. 784. Springer, Heidelberg (1994). https://doi.org/10.1007/3-540-57868-4
11. Gower, J.C.A.: General coefficient of similarity and some of its properties. Biometrics **27**(4) (1971). https://doi.org/10.2307/2528823

12. O'Shea, M., Savitz, D.A., Hage, M.L., Feinstein, K.A.: Prenatal events and the risk of subependymal/intraventricular haemorrhage in very low birthweight neonates. Paediatr Perinat Epidemiol. **6**(3), 352–62 (1992). https://doi.org/10.1111/j.1365-3016.1992.tb00775.x
13. Mostafizu , R., Davis, D.N.: Machine learning based missing value imputation method for clinical datasets. IAENG Trans. Eng. Technol. **229** (2012). https://doi.org/10.1007/978-94-007-6190-2_19

Improving Wildfire Simulations by Estimation of Wildfire Wind Conditions from Fire Perimeter Measurements

Li Tan[1], Raymond A. de Callafon[1(✉)], Jessica Block[2], Daniel Crawl[2], and Ilkay Altıntaş[2]

[1] Department of Mechanical and Aerospace Engineering,
University of California San Diego, La Jolla, CA, USA
{ltan,callafon}@eng.ucsd.edu
[2] San Diego Supercomputer Center, University of California San Diego,
La Jolla, CA, USA
{jblock,lcrawl,altintas}@ucsd.edu

Abstract. This paper shows how a gradient-free optimization method is used to improve the prediction capabilities of wildfire progression by estimating the wind conditions driving a FARSITE wildfire model. To characterize the performance of the prediction of the perimeter as a function of the wind conditions, an uncertainty weighting is applied to each vertex of the measured fire perimeter and a weighted least-squares error is computed between the predicted and measured fire perimeter. In addition, interpolation of the measured fire perimeter and its uncertainty is adopted to match the number of vertices on the predicted and measured fire perimeter. The gradient-free optimization based on iterative refined gridding provides robustness to intermittent erroneous results produced by FARSITE and quickly find optimal wind conditions by paralleling the wildfire model calculations. Results on wind condition estimation are illustrated on two historical wildfire events: the 2019 Maria fire that burned south of the community of Santa Paula in the area of Somis, CA, and the 2019 Cave fire that started in the Santa Ynez Mountains of Santa Barbara County.

Keywords: Wildfire · FARSITE · Uncertainty · Interpolation · Gradient-free optimization

1 Introduction

Fire is critical for healthy ecosystems around the world. With the increased and inevitable occurrence of wildfires, more accurate and responsive prediction of the wildfire propagation is important for resource allocation in fire fighting efforts. The fire growth modeling software FARSITE is widely used by the U.S Forest

Work is supported by WIFIRE Commons and funded by NSF 2040676 under the Convergence Accelerator program.

© Springer Nature Switzerland AG 2021
M. Paszynski et al. (Eds.): ICCS 2021, LNCS 12746, pp. 231–244, 2021.
https://doi.org/10.1007/978-3-030-77977-1_18

Service to simulate the propagation of wildfires [7], and is characterized by the ability to estimate the wildfire propagation under heterogeneous conditions of terrain, fuels and weather. Crucial sources of information in the modeling of fire progression are the prevailing wind conditions characterized by average wind speed and wind direction that determine the overall direction and rate of spread of the wildfire.

Considerable research has been done on studying the growth and behavior of wildfire. Rothermel introduced the mathematical model for predicting fire spread [12], and experiments have been conducted to analyse the influence of fuel and weather on the spread of fires [1]. Further steps in the study of the wildfire behavior were achieved by steering the model using real-time data [5,10,11]. Data assimilation by combining FARSITE and ensemble Kalman filter has been done in [6,13–15]. Furthermore, unmanned aerial vehicles have been applied to better monitor the large-scale wildfire [9,16]. As mentioned in [1], among the numerous factors that can affect the spread of the wildfire, wind speed and wind direction play the critical roles. Unfortunately, wind conditions are available only from sparsely placed weather stations.

Detailed studies are available on learning the (non-linear) relationship between the properties of the fuel and the wildfire progression [2–4], but often only limited information on wind speed and wind direction can be used. This means that the quality of the prediction is extremely dependent on the quality of an empirical estimate of the wind conditions obtained from geometrically spaced weather station. In reality, information of the actual wind conditions at the boundary of the wildfire is unavailable due to limited number of weather stations and the turbulent atmosphere caused by wildfire. As a result, significant and compounding errors can occur in the prediction of the wildfire propagation. A first step is to estimate the best initial wind conditions before any data assimilation procedure. In this situation, the error caused by an erroneous measurement of the wind conditions can be reduced, and the accuracy of the prediction by data assimilation techniques can be greatly improved.

In this paper a gradient-free optimization is used to provide an estimate of the initial wind conditions. The gradient-free optimization is based on iteratively refining a grid of possible wind speed and wind direction conditions and simulating wildfire progression through FARSITE. Since each grid point provides an independent wildfire simulation, the computations can be executed in parallel and also provides robustness to possible erroneous fire perimeter produced by FARSITE under certain wind conditions. For each grid, the optimal wind condition is estimated by a weighted least-squares error between a uncertainty weighted measured fire perimeter and the simulated fire perimeter. Additional refined gridding around the optimal wind conditions provides additional accuracy on the estimate. Due to the spread of the wildfire, it is highly possible that measured wildfire perimeters at different times are described by polygons with different numbers of vertexes compared to a simulated fire perimeter. Interpolation is then needed in order that different polygons have the same number of vertexes. Linear interpolation of the fire perimeters is used to guarantee the weighted least-squares error can always be computed. Furthermore, the weighting in the least-squares computation is adjusted to account for unevenly distributed polygons to allow an evenly distributed weighting of the complete fire perimeter.

The paper is organized as follows. Section 2 presents the polygon data model of the wildfire perimeters and the method to compare two different wildfire perimeters via a polygon interpolation and a weighted least-squares error computation. Section 3 outlines the gradient-free optimization based on iterative gridding to estimate the optimal initial wind speed and wind direction to match predicted and measured fire perimeters on the basis of the weighted least-squares error between polygons. Section 4 shows the numerical results for the estimation of wind conditions for two use cases of wildfires in California: the 2019 Maria fire that burned south of the community of Santa Paula and the 2019 Cave fire that started in the Santa Ynez Mountains of Santa Barbara County. Conclusions are summarized in Sect. 5.

2 Wildfire Perimeter and Error Quantification

2.1 Uncertainty Characterization

A wildfire may cover multiple disjoint burned areas. For simplicity of the analysis presented in this paper, the notion of wildfire progression is characterized by a wildfire perimeter that is considered to be a single closed polygon. The analysis presented here can be applied to each of the closed-polygons in case a wildfire does cover multiple disjoint burned areas. The single closed polygon describing the wildfire perimeter is an ordered sequence of N vertexes and N piece-wise linear line segments. The vertexes of the approximated polygon are located by the Eastern and Northern coordinate pairs $(e(k), n(k))$, $k = 1, 2, \ldots, N$.

Measurements of the wildfire perimeters can be a combined data collection effort from satellite imagery, aerial surveillance or manually mapped observations with different quality assessments [8]. Therefore, it is important to consider the two-dimensional (2D) uncertainty for each vertex of the closed polygon that describes the measured wildfire perimeter. The general description of the 2D uncertainty on a vertex $(e(k), n(k))$ is a rotated ellipse, where the semi-major axis $a(k)$, semi-minor axis $b(k)$, and the rotation angle $\alpha(k)$ collectively reflect the variance in the horizontal direction and vertical direction. Such detailed information may not be available and therefore the uncertainty on a vertex $(e(k), n(k))$ is expressed by a circle around each vertex with a radius $r(k)$, where the value of $r(k)$ is proportional to the uncertainty of the vertex on the polygon.

However, it is very likely that a measured fire perimeter comes with no uncertainty characterization. In that case, the assumption is made that the uncertainty on each vertex is proportional to the (smallest) distance to the neighboring vertex on the polygon. Formally this uncertainty is described by

$$
\begin{aligned}
r(k) &= \max(\min(l(k), l(k-1)), r_{min}) \\
l(k) &= \sqrt{(e(k+1) - e(k))^2 + (n(k+1) - n(k))^2}
\end{aligned}
\tag{1}
$$

for $k = 1, 2, \ldots, N$, where $r(k)$ is the assumed uncertainty, $l(k)$ is the distance between neighboring vertexes $(e(k+1), n(k+1))$, $(e(k), n(k))$, and r_{min} is a user-defined minimum value of uncertainty radius. The value of r_{min} is used to avoid the condition in which two adjacent vertexes are extremely close to each other, and can be determined by the accuracy of measuring method used to acquire the polygon of the fire perimeter. An illustration of the uncertainty assignment for a measured fire perimeter is given in Fig. 1.

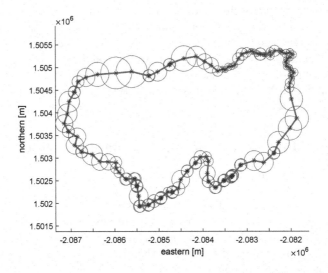

Fig. 1. Assignment of uncertainty radii $r(k)$ (circles) of a measured fire perimeter with vertexes $(e(k), n(k))$ (stars) and the resulting closed polygon (blue lines). (Color figure online)

2.2 Perimeter Interpolation

With the spread of a wildfire, the corresponding closed polygon describing the measured fire perimeter typically becomes larger and the number N_m of vertexes of the measured closed polygon $(e_m(k), n_m(k))$, $k = 1, 2, \ldots N_m$ increases accordingly. Similarly, the number of vertexes N_s on a simulated fire perimeter $(e_s(k), n_s(k))$, $k = 1, 2, \ldots N_s$ obtained with fire modeling software such as FARSITE will also increase, but in general $N_m \neq N_s$. Next to difference in number of vertexes, the ordering of the vertexes $(e_m(k), n_m(k))$, $k = 1, 2, \ldots N_m$ of the measured fire perimeter and $(e_s(k), n_s(k))$, $k = 1, 2, \ldots N_s$ are not the same and a direct comparison between a pair $(e_m(k), n_m(k))$ and $(e_s(k), n_s(k))$ is incorrect.

The solution to this problem is to first interpolate one of the fire perimeters to the same and higher number $N = \max(N_m, N_s)$ of vertexes of the other fire perimeter. Subsequently, when comparing pairs $(e_m(k), n_m(k))$ and

$(e_s(k), n_s(k))$, the starting vertex at $k = 1$ of one of the fire perimeters will be re-ordered to obtain the smallest weighted least-squares error between the polygons. In this paper, interpolation of the fire perimeter is done with standard 2D linear interpolation, where interpolated vertexes are introduced on the straight lines connecting the original vertexes of the closed polygon, and the procedure of linear interpolation is summarized in Algorithm 1.

Algorithm 1. Linear interpolation of wildfire polygon

Input: Vertexes of the original approximated polygon
Output: Newly constructed vertexes of the interpolated polygon
  1: Calculate the length of each side of the polygon.
  2: Calculate the cumulative side length from the starting point.
  3: Find locations with equally distributed length along the side of polygon from the starting point.
  4: Construct new polygon vertexes

Similarly, uncertainties of the original vertexes can also be interpolated with respect to the cumulative side length from the starting point. Due to the fact that the interpolation is related to the distance from the starting point, it is easy to verify that interpolation from different starting points will lead to different results. This will be considered in the subsequent section when the weighted least squares are calculated.

2.3 Weighted Least Squares Error

With an interpolated (and properly ordered) closed polygons of the simulated fire perimeter $(e_s(k), n_s(k))$, and the measured fire perimeter $(e_m(k), n_m(k))$ with an uncertainty $r(k)$ on each vertex $k = 1, 2, \ldots, N$, a weighted least-squares error

$$\frac{1}{N} \sum_{k=1}^{N} w(k)^2 \left[(e_s(k) - e_m(k))^2 + (n_s(k) - n_m(k))^2 \right], \quad w(k) = \frac{1}{r(k)} \quad (2)$$

can be used to define the distance between the fire perimeters. The weighting $w(k) = 1/r(k)$ ensures measurements with a large uncertainty $r(k)$ are weighted less in the error characterization. However, even with uncertainty radii defined by (1) with a minimum value r_{min}, the weighted least-squares error in (2) will be skewed and emphasizes parts of the closed-loop polygon where vertexes are closely clustered and have only small distances with respect to each other, as also illustrated in Fig. 1. The reasons are clear:

– Small uncertainty radii $r(k)$ due to (1) will result in a larger weighting $w(k) = 1/r(k)$ on the regions of the polygon where vertexes are closely clustered.
– More vertexes in areas of the polygon where vertexes are clustered further accentuates the weighting on these regions of the polygon.

To solve the problem of the skewed emphasis of the weighted least-squares error, the weighting $w(k)$ for each vertex is skew compensated by a weighting computed via

$$\tilde{w}(k) = w(k)c_w(k)u_w(k), \quad w(k) = \frac{1}{r(k)} \tag{3}$$

where $c_w(k)$ is a concentration weighting for each vertex used to account for clustering of vertexes on the closed polygon and the weighting $u_w(k)$ is the user-defined weighting for each vertex, used to actually emphasize certain vertexes on the closed polygon. The weighting $c_w(k)$ is defined as

$$c_w(k) = \frac{1}{m(k)} \tag{4}$$

where $m(k)$ is the number of successive vertexes around the k_{th} vertex with a small adjacent distance $l(k)$. A small adjacent distance $l(k)$ is defined by the relative distance condition

$$\frac{l(k)}{l_{mean}} < 0.2, \quad l_{mean} = \frac{1}{N}\sum_{k-1}^{N} l(k)$$

where $l(k)$ was defined in (1). The weighting $u_w(k)$ is defined to be 0 for the barrier points, defined as vertexes where the fire perimeter has not changed, and 1 for the other vertexes.

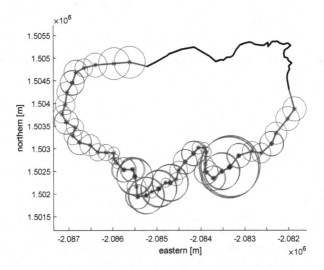

Fig. 2. Weighting radii $1/\tilde{w}(k)$ (circles) for skew-compensated least-squares compensation of a closed polygon of a measured fire perimeter consisting of vertexes $(e(k), n(k))$ (stars) and barrier vertexes (black line). (Color figure online)

An illustration of the skew compensation is show in Fig. 2. On account of the fact that barrier points will not move with the spread of the wildfire, a zero value weighting is assigned to each barrier point. Hence, the weighting radii of barrier points are infinitely large, and not included in Fig. 2.

Finally, to also address the re-ordering of the vertexes of the closed polygon, consider the short-hand notation based on complex numbers

$$
\begin{aligned}
x(k) &= e_s(k) + j \cdot n_s(k), \quad k = 1, 2, \ldots, N \\
y(k,q) &= e_m(k) + j \cdot n_m(k), \ k = q, q+1, \ldots, N, 1, \ldots, q-1
\end{aligned} \tag{5}
$$

where $x(k) \in \mathbb{C}$ for $k = 1, 2, \ldots, N$ represents the 2D coordinates of vertexes of a closed polygon of a simulated fire perimeter starting at index $k = 1$ and $y(k,q) \in \mathbb{C}$ represents the 2D coordinates of vertexes of a closed polygon of a measured (and possibly interpolated) fire perimeter, but reordered to start at index q. The ability to adjust the starting point $k = q$ of the closed polygon now allows for the definition of the skew compensated weighted least-squares error

$$
s = \min_q \frac{1}{N} \sum_{k=1}^{N} \tilde{w}(k)^2 |y(k,q) - x(k)| \tag{6}
$$

where $\tilde{w}(k)$ is defined in (3). The starting point $k = q$ is used to remove the dependency of cyclical ordering of complex points describing the closed polygon.

3   Wind Condition Estimation with FARSITE

3.1   Forward Simulations

In this study, FARSITE is used for the forward simulation of the simulated fire perimeter $x(k)$ as a function of the prevailing wind conditions u. FARSITE can be considered as a non-linear mapping $\rho(\cdot)$ for fire progression, simplified to

$$
x(k) = \rho(p(k), u, \theta, \Delta_T) \tag{7}
$$

where the input $p(k) \in \mathbb{C}^{N_p}$ is a closed polygon of N_p vertexes representing the initial fire perimeter. The simulated output $x(k) \in \mathbb{C}^{N_x}$, defined earlier in (5), is the closed polygon of $k = 1, 2, \ldots, N_x$ vertexes representing a simulated fire perimeter obtained after a time step of Δ_T. The additional inputs u represents the wind conditions, and θ denotes a parameter representing fuel content, fuel moisture and topography, all assumed to be constant over the time step of Δ_T.

Unknown wind conditions influence the interpolated and re-ordered vertexes of the measured fire perimeter represented by the closed polygon $y(k,q)$ defined in (5). Prevailing wind conditions in terms of wind speed and wind direction combined in a two dimensional input u will also influence the vertexes of the simulated fire perimeter represented by the closed polygon $x(k)$. Along with the definition of the weighting $\tilde{w}(k)$ in (3), it is expected that a minimization of s in (6) as a function of u will provide the best wind conditions to minimize the distance between $x(k)$ and $y(k)$.

3.2   Wind Speed and Wind Direction Optimization

The formal problem of finding an estimate of the prevailing wind conditions on the basis of a wildfire measurement $y(k)$ can be stated as the optimization

$$\min_u s(u), \quad s(u) = \min_q \frac{1}{N} \sum_{k=1}^{N} \tilde{w}(k)^2 |y(k,q) - x(k)| \tag{8}$$
$$x(k) = \rho(p(k), u, \theta, \Delta_T)$$

where $\tilde{w}(k)$ is defined in (3) and $y(k,q)$ is defined in (5). Although the $\min_q y(k,q)$ problem defined earlier in (6) is only a combinatorial problem and the optimization is a standard weighted least-squares problem, the non-linearity and non-convex mapping of $\rho(\cdot)$ requires a non-linear and iterative optimization, typically using the sensitivity or the gradient.

For FARSITE, that is responsible for the mapping in (7), the sensitivity or gradient $\frac{\partial}{\partial u}\rho(p(k), u, \theta, \Delta_T)$ is unknown. Numerical evaluation of the gradient is computationally expensive and moreover, FARSITE is known to produce occasional erroneous results at some initial wind conditions due to numerical problems in interpolation and reconstruction of the main fire perimeter (as will be shown later). These reasons motivate the use of a gradient-free optimization and the 2 dimensional size of u motivates a simple 2D gridding procedure over which $s(u)$ in (8) is evaluated. The 2D grid of u can be updated and refined iteratively to improve the accuracy of the final optimized solution for u. The pseudo-code for the iterative optimization can be summarized in Algorithm 2.

Algorithm 2. Optimizing algorithm

Input: θ, $p(k)$, $y(k)$, Δ_T, minimum wind condition perturbation λ and stopping criterion ε.

Output: Optimized $u \in \mathbb{R}^{2 \times 1}$

1: Create n^2 points of a symmetric 2D grid $u_{i,j}$ over a desired range $i = 1, 2, \ldots, n$ and $j = 1, 2, \ldots, n$ around an initial estimate u_0 of the wind conditions.
2: Parallel simulation in FARSITE with $p(k)$, $u_{i,j}$, θ and Δ_T to obtain $x_{i,j}(k)$ for each grid point.
3: Compute the n^2 weighted least squares errors $s(u_{i,j})$ defined in (8) over the grid $i = 1, 2, \ldots, n$ and $j = 1, 2, \ldots, n$
4: Find the smallest value $\hat{i}, \hat{j} = \min_{i,j} s(u_{i,j})$ to select the optimized wind condition $u_{\hat{i},\hat{j}}$
5: Set $u_0 = u_{\hat{i},\hat{j}}$ and stop when $|s(u_0 + \lambda) - s(u_0)| \leq \varepsilon$ or go back to step 1 to refine grid around u_0.

The weighted least-squares error is used to determine the difference between the simulated polygon and the measured polygon of wildfire. Simulations can be performed in parallel to speed up the process of finding the optimal initial wind conditions with the above mentioned algorithm.

4   Numerical Results

4.1   Maria Fire

The Maria Fire ignited in the evening hours of Thursday, October 31, 2019 and consumed well over 4,000 acres (16 km^2) within its first several hours of burning. The optimization of the wind conditions is performed for this fire at four different time stamps where measurements of the fire perimeter $y(k)$ were available. The objective of the optimization is to improve the fire simulations of the fire perimeters $x(k)$ with FARSITE in comparison with the observations $y(k)$ obtained at four time stamps.

error between simulation and measurement

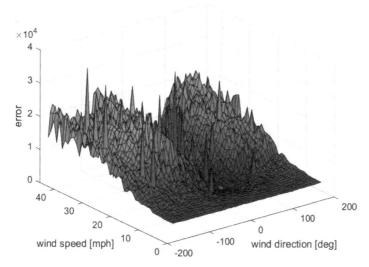

Fig. 3. Evaluation of weighted least-squares error $s(u_{i,j})$ between the simulated $x(k)$ and measured $y(k)$ fire perimeter at one particular time stamp of the Maria Fire using $n^2 = 100$ points of a symmetric 2D grid for the wind conditions $u_{i,j}$. The optimal wind condition with the lowest value of $s(u_{i,j})$ is indicated with a red dot. (Color figure online)

First we illustrate the results of the gradient-free optimization algorithm summarized in Algorithm 2 in Fig. 3. The numerical evaluation of the weighted least-squares error $s(u_{i,j})$ over $n^2 = 100$ points of a symmetric 2D grid $u_{i,j}$ in Fig. 3 clearly shows the non-differential behavior of $s(u)$, motivating the use of a gradient-free optimization. Sporadic large values for $s(u_{i,j})$ for certain wind conditions $u_{i,j}$ are explained by erroneous results due to numerical problems in interpolation and reconstruction of the main fire perimeter by FARSITE, as illustrated in Fig. 4. The simulation results show very similar fire perimeters for two wind conditions that are very close to an erroneous result.

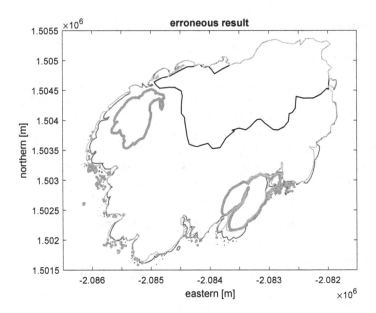

Fig. 4. Simulation of the predicted fire perimeter $x(k)$ with wind speed = 21 mph, wind direction = 34 deg (red), wind speed = 21 mph, wind direction = 35 deg (green), and wind speed = 21 mph, wind direction = 36 deg (cyan), with initial fire perimeter $p(k)$ (black). (Color figure online)

Based on gradient-free optimization algorithm summarized in Algorithm 2, the optimization can correct wildfire simulations when the initial guesses of the prevailing wind conditions are not correct. Correction of the wildfire simulations for the four different time stamps where measurements of the Maria fire perimeter $y(k)$ were available are summarized in Fig. 5. For each time stamp, the simulated fire perimeter $x(k)$ (green lines) based on an initial estimate u_0 of the wind conditions obtained from a weather station can be improved (yellow lines) by the optimization of the wind condition via Algorithm 2. It can be observed that the optimized wind conditions provide simulations of $x(k)$ that are closer to the measurements $y(k)$ (red lines).

4.2   Cave Fire

Although the accuracy of the simulation is improved by using the optimized wind conditions, there are still some parts of the optimized simulation $x(k)$ that are somewhat far from the measurement $y(k)$. One reason may be the measurement accuracy, as the combination of aerial surveillance and manually mapped observations is likely to introduce measurement errors. It can also be observed that as the fire perimeter becomes large enough, only using a prevailing wind direction is inadequate for the accurate prediction of the wildfire propagation as wind flow is shaped by topography and atmospheric interaction.

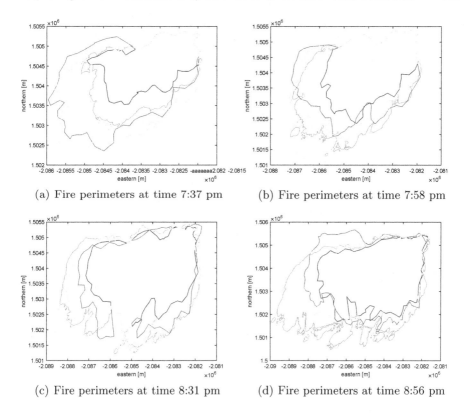

(a) Fire perimeters at time 7:37 pm

(b) Fire perimeters at time 7:58 pm

(c) Fire perimeters at time 8:31 pm

(d) Fire perimeters at time 8:56 pm

Fig. 5. Comparison of measured $y(k)$ and FARSITE simulated $x(k)$ fire perimeters for initial u_0 and optimized u wind conditions via Algorithm 2 for the Maria fire at four different time stamps. Initial ignition (blue); Simulation with initial guess u_0 (green); Simulations with optimized wind conditions u (yellow); Measurement at next time step (red). (Color figure online)

The measurement data available for the Cave Fire included here can better demonstrate the two issues of measurement errors and the assumption of a single prevailing wind condition. The 2019 Cave Fire started on November 25 and burned a total of 3,126 acres before being contained on December 19. As shown in Fig. 6(a), the top part of the first measurement (after the initial ignition) can be assumed to be wrongly characterized when compared to the second measurement. To be able to account for such errors on the measurement $y(k)$, the weighting radii $\tilde{w}(k)$ defined in (3) on the vertexes in the top part of the first measurement are adjusted to be zero. The effect of the weighting radii is illustrated in Fig. 6(b)) and it can be seen that the measurements in the top part of the first measurement are weighted with 0, while still allowing the remaining points $y(k)$ to be used for optimization of the prevailing wind conditions at this time stamp.

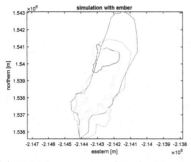

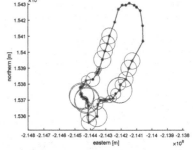

(a) Initial perimeter (blue); Measurement at the first step (red) and the second step (cyan) after the initial ignition. Simulations with optimized wind conditions (yellow);

(b) Weighting radii $\tilde{w}(k)$ (red circles) on the measurement of the vertexes $y(k)$ at the first time step at 03:48 a.m (blue stars).

Fig. 6. Simulation and measurements of the Cave Fire with measurement errors in the measured fire perimeter $y(k)$ at the first time stamp after the initial fire perimeter. (Color figure online)

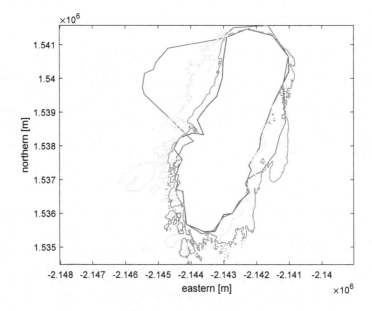

Fig. 7. Comparison of measurement $y(k)$ and FARSITE simulation when the Cave fire reaches a large dimension. Initial ignition at time 05:15 a.m (blue); Simulation with initial guess of wind conditions (green); Simulation $x(k)$ with optimized wind conditions (yellow); Measurement of fire perimeter $y(k)$ (red). (Color figure online)

When the Cave fire grows to a large dimension, as illustrated in Fig. 7, it becomes difficult to match the measured fire perimeter $y(k)$ with a simulated fire perimeter via single prevailing wind direction. The gradient-free optimization of Algorithm 2 does a better job covering the east side of the fire, but the west side of the fire cannot be covered with a single prevailing wind condition due to the topography and atmospheric wind shear effects acting on the fire. This illustrated the limitations of the proposed approach of optimizing only a prevailing wind condition.

5   Conclusions

This paper shows how fire perimeter measurements can be used to improve the accuracy of a wild fire perimeter simulation, by using the measurements to estimate and correct the prevailing wind speed and wind direction for the simulation. The estimation is based on a carefully defined weighted least-squares error that measures the discrepancy between closed polygons. The weighting in the least-squares error can account for measurement accuracy, and be adjusted for a skewed weighting caused by unequally distributed vertexes on the closed polygon of the fire perimeter. Using a gradient-free optimization that exploits a grid of the two dimensional wind conditions and parallel computations with FARSITE fire modeling, optimal wind conditions are obtained by an iterative grid assignment approach. Numerical results on actual fire perimeter data obtained from two recent destructive fires in California confirm the improvement of the accuracy of a wild fire perimeter simulation. Limitations of the proposed method are due to the optimization of a single prevailing wind condition – an assumption that may not hold when a wild fire covers a large area with varying topographical features.

References

1. Cheney, N., Gould, J., Catchpole, W.: The influence of fuel, weather and fire shape variables on fire-spread in grasslands. Int. J. Wildland Fire **3**(1), 31–44 (1993)
2. Cruz, M.G., Hurley, R.J., Bessell, R., Sullivan, A.L.: Fire behaviour in wheat crops-effect of fuel structure on rate of fire spread. Int. J. Wildland Fire **29**(3), 258–271 (2020)
3. Cruz, M.G., Sullivan, A.L., Gould, J.S.: The effect of fuel bed height in grass fire spread: addressing the findings and recommendations of Moinuddin et al. Int. J. Wildland Fire **30**, 215 (2018)
4. Cruz, M.G., Sullivan, A.L., Gould, J.S., Hurley, R.J., Plucinski, M.P.: Got to burn to learn: the effect of fuel load on grassland fire behaviour and its management implications. Int. J. Wildland Fire **27**(11), 727–741 (2018)
5. Douglas, C.C., et al.: Demonstrating the validity of a wildfire DDDAS. In: Alexandrov, V.N., van Albada, G.D., Sloot, P.M.A., Dongarra, J. (eds.) ICCS 2006. LNCS, vol. 3993, pp. 522–529. Springer, Heidelberg (2006). https://doi.org/10.1007/11758532_69
6. Fang, H., Srivas, T., de Callafon, R.A., Haile, M.A.: Ensemble-based simultaneous input and state estimation for nonlinear dynamic systems with application to wildfire data assimilation. Control Eng. Pract. **63**, 104–115 (2017)

7. Finney, M.A.: FARSITE, Fire Area Simulator - model development and evaluation, vol. 4. US Department of Agriculture, Forest Service, Rocky Mountain Research Station (1998)
8. Kolden, C.A., Weisberg, P.J.: Assessing accuracy of manually-mapped wildfire perimeters in topographically dissected areas. Fire Ecol. **3**, 22–31 (2007)
9. Lin, Z., Liu, H.H., Wotton, M.: Kalman filter-based large-scale wildfire monitoring with a system of UAVs. IEEE Trans. Ind. Electron. **66**(1), 606–615 (2018)
10. Mandel, J., et al.: Towards a dynamic data driven application system for wildfire simulation. In: Sunderam, V.S., van Albada, G.D., Sloot, P.M.A., Dongarra, J.J. (eds.) ICCS 2005. LNCS, vol. 3515, pp. 632–639. Springer, Heidelberg (2005). https://doi.org/10.1007/11428848_82
11. Mandel, J., et al.: A note on dynamic data driven wildfire modeling. In: Bubak, M., van Albada, G.D., Sloot, P.M.A., Dongarra, J. (eds.) ICCS 2004. LNCS, vol. 3038, pp. 725–731. Springer, Heidelberg (2004). https://doi.org/10.1007/978-3-540-24688-6_94
12. Rothermel, R.C.: A mathematical model for predicting fire spread in wildland fuels, vol. 115. Intermountain Forest and Range Experiment Station, Forest Service, United. . . (1972)
13. Srivas, T., Artés, T., De Callafon, R.A., Altintas, I.: Wildfire spread prediction and assimilation for farsite using ensemble Kalman filtering. Procedia Comput. Sci. **80**, 897–908 (2016)
14. Srivas, T., de Callafon, R.A., Crawl, D., Altintas, I.: Data assimilation of wildfires with fuel adjustment factors in farsite using ensemble Kalman filtering. Procedia Comput. Sci. **108**, 1572–1581 (2017)
15. Subramanian, A., Tan, L., de Callafon, R.A., Crawl, D., Altintas, I.: Recursive updates of wildfire perimeters using barrier points and ensemble Kalman filtering. In: Krzhizhanovskaya, V.V., et al. (eds.) ICCS 2020. LNCS, vol. 12142, pp. 225–236. Springer, Cham (2020). https://doi.org/10.1007/978-3-030-50433-5_18
16. Xing, Z., Zhang, Y., Su, C.Y., Qu, Y., Yu, Z.: Kalman filter-based wind estimation for forest fire monitoring with a quadrotor UAV. In: 2019 IEEE Conference on Control Technology and Applications (CCTA), pp. 783–788. IEEE (2019)

Scalable Statistical Inference
of Photometric Redshift
via Data Subsampling

Arindam Fadikar[1]([✉]) [ID], Stefan M. Wild[1] [ID], and Jonas Chaves-Montero[2] [ID]

[1] Mathematics and Computer Science Division, Argonne National Laboratory,
Lemont, IL 60439, USA
{afadikar,wild}@anl.gov
[2] Donostia International Physics Centre, Paseo Manuel de Lardizabal 4,
20018 Donostia-San Sebastian, Spain
jonas.chaves@dipc.org

Abstract. Handling big data has largely been a major bottleneck in traditional statistical models. Consequently, when accurate point prediction is the primary target, machine learning models are often preferred over their statistical counterparts for bigger problems. But full probabilistic statistical models often outperform other models in quantifying uncertainties associated with model predictions. We develop a data-driven statistical modeling framework that combines the uncertainties from an ensemble of statistical models learned on smaller subsets of data carefully chosen to account for imbalances in the input space. We demonstrate this method on a photometric redshift estimation problem in cosmology, which seeks to infer a distribution of the redshift—the stretching effect in observing the light of far-away galaxies—given multivariate color information observed for an object in the sky. Our proposed method performs balanced partitioning, graph-based data subsampling across the partitions, and training of an ensemble of Gaussian process models.

Keywords: Gaussian process · Data subsampling · Photometric redshift

1 Introduction

Data analysis techniques have become an essential part of a scientist's toolbox for making inferences about an underlying system or phenomenon. With the advancement of modern computing, processing power has gone up many fold, and data volumes have grown at a similar pace. Thus, there is a growing demand for scalable machine learning (ML) and statistical techniques that can handle large data in an efficient and reasonable way. Recent trends include the use of ML and statistical regression techniques such as deep neural networks [18,21],

M. Paszynski et al. (Eds.): ICCS 2021, LNCS 12746, pp. 245–258, 2021.
https://doi.org/10.1007/978-3-030-77977-1_19

tree-based models [16], and Gaussian process (GP) models [14,28]. Each technique has advantages and disadvantages that promote or limit its use for a specific application. For example, deep neural networks are known for achieving superior prediction accuracy but at the cost of significant data and compute cost for training. Statistical models such as GPs approximate the global input-output relationship and provide full uncertainty estimates at a fraction of the data required by deep neural networks. However, statistical models do not generally scale well with data size. Some GP models have been proposed to find workarounds such as introducing sparse approximation of large correlation matrices [20], using locally learned models [14] or considering only a subset of the data [1].

In this paper we propose a statistical modeling framework that can handle large training data by leveraging data subsampling combined with advanced statistical regression techniques to provide a full density estimate. The basic idea entails using smaller subsets of data to train regression models and building an ensemble of these models. Our modeling approach differs from other approximations in that we attempt to learn the global input-output relationship in each individual subsample and model. This is counter intuitive and opposite approaches that seek to build accurate local response surfaces [14,20]. However, uncertainty estimates can be undesirably constricted in locally learned models. Furthermore, data partitioning and subsampling in our proposed approach are driven entirely by data distribution and are free from model influences present in other locally learned models. We focus on the analysis paradigm in cosmology that deals with estimation of the redshifts of objects (e.g., galaxies) as a motivating application, which is discussed in the following section.

The rest of the paper is structured as follows. In Sect. 2 we describe the estimation problem and data used. We present our proposed methodology in Sect. 3 and numerical results in Sect. 4. In Sect. 5 we summarize our approach and its benefits.

2 Photometric Redshift Estimation

The cosmological analysis of galaxy surveys, from gathering information on dark energy to unveiling the nature of dark matter, relies on the precise projection of galaxies from two-dimensional sky maps into the three-dimensional space [20,32]. However, measuring the distance of distant objects in the Universe from the Earth is challenging. Furthermore, the accelerated expansion of the Universe causes the wavelength of light from a distant object to be stretched or *redshifted*. Interestingly, the redshift of an object is proportional to its receding speed and can be used to estimate the radial distance to this source. For an accurate redshift estimation, one would have to obtain the full spectrum of each galaxy at a very high resolution, which is a demanding and time-consuming task [3,19]. An alternative to this approach is to infer the redshift based on the intensity of galaxy light observed through a reduced number of wavebands [2]. Such an approximation is known as photometric redshift, whereas the redshift obtained from the full spectrum is called spectroscopic redshift.

Photometric redshift estimation methods can be divided into two categories: spectral energy distribution template fitting [10,25] and statistical regression or ML methods [7,11,20]. Supervised ML methods such as deep neural networks have recently seen success in approximating the mapping between broadband fluxes and redshift [3]. Many of the successes of ML methods, however, rely on the availability of large training datasets coupled with large computing power. Because of their increased computational complexity, statistical regression methods have not been a preferred choice, even when the data size is moderately large.

The case study we consider is the publicly available data release 7 of the Sloan Digital Sky Survey (SDSS) [33], which contains approximately 1 million galaxies with spectroscopic redshifts estimates. From this initial sample, we select $\simeq 100\,000$ galaxies with signal-to-noise ratio larger than 10 in all photometric bands, clean photometry, and high-confident spectroscopic redshift below $z = 0.3$. The photometry of each SDSS galaxy consists of flux measurement in five broadband filters u, g, r, i, and z, which serve as input to the predictive model for redshift.

In general we consider a dataset of $N(=\ 10^5)$ scattered data points $\mathbf{x}_1, \ldots, \mathbf{x}_N$ contained in a compact input space $\mathcal{X} \subset \mathbb{R}^d$, with corresponding logged redshifts $y_1, \ldots, y_N$. Without loss of generality, we assume that the input space has been normalized so that $\mathcal{X} = [0, 1]^d$ is the unit cube and the logged redshift values are transformed to mean zero and a standard deviation of 1. In our case study, d is 5, with the dimensions corresponding to four colors computed by taking the ratio of the flux consecutive filters and the magnitude in the i-band. In what follows, we refer to these variables as color space.

Approximating the relationship between broadband filter values and the photometric redshift is challenging because of several factors. For example, lack of coverage of the input space in the training dataset results in poor predictions at the extremes, as well as degeneracies arising in the prediction of the redshift in some cases. Hence, simple Gaussian predictive uncertainty may not accurately represent the distribution of the redshift given a set of colors. For example, a galaxy at a high redshift with high luminosity and a galaxy at a low redshift with low luminosity may register the same flux values to a telescope. Considering such uncertainty in the predictive distribution would require discovering and modeling these latent processes. Mixed-density networks [9] and full Bayesian photometric redshift estimation [4] are examples that model the predictive distribution by a finite number of Gaussians. While our objective is similar to [9], we give greater importance to obtaining a full predictive distribution that can accurately reflect multiple predictive modes.

3   Statistical Methodology

Our proposed approach consists of three intermediate stages: (1) partitioning the input color space $\mathcal{X}$ into a user-defined number of partitions in order to ensure balance among the input data, (2) subsampling datapoints from these partitions, and (3) training a regression model on the sampled data. The final predictive model is then based on an ensemble of models obtained by repeated execution of the latter two stages. Although we present the basic implementations here for simplicity, scalability is emphasized in each stage: the partitioning can exploit domain decomposition parallelism, the subsampling allows for model construction using datasets of user-specified greatly reduced size, and the ensemble is naturally parallelizable.

3.1 Partitioning the Input Space

We begin by partitioning the color space $\mathcal{X}$ into a set of mutually exclusive subsets $\mathcal{C} = \{\mathcal{C}_k : k \in \mathcal{I}\}$ such that $\cup_{k \in \mathcal{I}} \mathcal{C}_k = \mathcal{X}$ and $\mathcal{C}_{k_i} \cap \mathcal{C}_{k_j} = \emptyset$ for $k_i \neq k_j$.

We especially target datasets that are highly nonuniform, including those arising in redshift estimation, and thus we seek balanced partitions. By balanced, here we intend for the number of training points in all partitions to satisfy $|\mathcal{C}_k| \approx \frac{N}{|\mathcal{I}|}$, where $|\mathcal{C}_k|$ denotes the cardinality of $\mathcal{C}_k$.

The number of partitions (m) is not predetermined for our balanced hyperrectangle partitioning algorithm. However, the values of the inputs to the algorithm, $N_{\min}$ and $N_{\max}$—minimum and maximum number of datapoints in each partition, respectively—are influenced by a desired m. To achieve a (n approximately) balanced hyperrectangle partition, we propose a three-step procedure that consists of an initialization of coarse partitions, pruning to satisfy the minimality condition $\min_{k \in \mathcal{I}} |\mathcal{C}_k| \geq N_{\min}$, and then splitting to satisfy the maximality condition $\max_{k \in \mathcal{I}} |\mathcal{C}_k| \leq N_{\max}$. We also require $N_{\max} \geq 2N_{\min}$ to guarantee termination of our procedure.

Initialize. Interval boundaries $0 = x_0^p < x_1^p < \ldots < x_{m_p}^p = 1$ of size m_p are defined along each input dimension $p = 1, \ldots, d$. Then an initial set of hyperrectangle partitions $\mathcal{C}$ in the d-dimensional input space is constructed as the Cartesian product of all $1D$ intervals:

$$\mathcal{C} = \left\{ A_{j_1}^1 \times \cdots \times A_{j_d}^d : j_p = 1, \ldots, m_p, \, p = 1, \ldots, d \right\}. \tag{1}$$

This leaves open a choice of interval boundary values along each dimension. In this study, we have opted for quantile-based splits. The result of the initialization step is $\prod_{p=1}^d m_p$ hyperrectangles, some of which may be empty.

Merge. In the next step, partitions with cardinality less than $N_{\min}$ are merged successively with their neighbors until the minimality condition is satisfied. Note that at the end of the merge step, some partitions may have cardinality greater than $N_{\max}$. The merging algorithm is briefly described below.

We define $\mathcal{S}_{\min} = \{\mathcal{C}_k \in \mathcal{C} : |\mathcal{C}_k| < N_{\min}\}$ to be the set of partitions with cardinality less than $N_{\min}$. We begin by identifying a target partition $\mathcal{C}_{(0)} \in \arg \min_{\mathcal{C}_k \in \mathcal{S}_{\min}} |\mathcal{C}_k|$, namely, a partition with the smallest cardinality. We also define the directional neighborhood function $\mathcal{N}_{\omega}^p(\cdot)$, which represents neighbors of $\cdot$ along dimension $p \in \{1, \ldots, d\}$ and where $\omega \in \{\text{lower} \equiv l, \text{upper} \equiv u\}$ denotes the relative position of the neighbors with respect to dimension p.

Given $\mathcal{C}_{(0)}$, the selection of partition(s) to merge with is equivalent to finding an appropriate dimension and direction to merge along. At most $2d$ such merging choices exist. We require that the newly formed partition be a hyperrectangle. For a given partition $\mathcal{C}_{(0)}$, the merging dimension and direction are selected according to following condition:

$$(p^*, \omega^*) \in \arg \min_{\substack{p=1,\dots,d \\ \omega \in \{l,u\}}} \sum_{\mathcal{C}_k \in \mathcal{N}_\omega^p(\mathcal{C}_{(0)})} |\mathcal{C}_k|, \tag{2}$$

in other words, the dimension and direction for which the updated partition contains the least number of datapoints among all possible combinations. In the case of multiple (p^*, ω^*) possibilities in (2), the one that results in the most uniform sides after merging is selected. In particular, we choose the merger that results in the smallest ratio between the longest and shortest sides. Once the merged hyperrectangle achieves at least cardinality $N_{\min}$, the set $\mathcal{S}_{\min}$ is updated. The merging step continues until $\mathcal{S}_{\min}$ is empty.

Split. In the split step, partitions with cardinality greater than $N_{\max}$ are successively broken into smaller partitions until the maximality condition is satisfied. We define $\mathcal{S}_{\max} = \{\mathcal{C}_k : |\mathcal{C}_k| > N_{\max}\}$ to be the set of hyperrectangles with cardinality greater than $N_{\max}$. A partition with highest cardinality in $\mathcal{S}_{\max}$ is selected for a split, $\mathcal{C}_{(m)} \in \arg \max_{\mathcal{C}_k \in \mathcal{S}_{\max}} |\mathcal{C}_k|$. A split of a hyperrectangle is defined as breaking it into two hyperrectangles along one dimension. Hence, for any split operation the appropriate dimension needs to be identified. The break point in that dimension can be any location for which the two new partitions satisfy the minimality condition (such a location exists because we require $N_{\max} \geq 2N_{\min}$). Our case study uses the median as the break point. To promote uniformity in the shape of hyperrectangle partitions, we always select the dimension corresponding to the longest side to perform the split, unless any of the resulting hyperrectangles breaks the minimality conditions, in which case we move to the next best dimension. Successive splits are carried out until $\mathcal{S}_{\max}$ is empty.

3.2   Conditional Sampling from Partitions

After assigning N datapoints to m partitions $\mathcal{C}_{k_1}, \dots, \mathcal{C}_{k_m}$, we move to the next stage where one or more samples are drawn from each partition according to our proposed sampling rule. The primary goal of such a sampling scheme is to explore and discover latent processes by sequentially sampling from the partitions obtained in the previous step.

Induced Graph on the Partitions. An essential ingredient of our proposed sampling technique is a graph structure based on the partitions on $\mathcal{C}$. We define an undirected graph $\mathcal{G} = (V, E)$ induced by the partitions, where nodes are defined to be m partitions $V = \{\mathcal{C}_{k_1}, \dots, \mathcal{C}_{k_m}\}$ and there is an undirected edge between $\mathcal{C}_{k_i}$ and $\mathcal{C}_{k_j}$, $k_i \neq k_j$ if their closures share more than one point. In other words

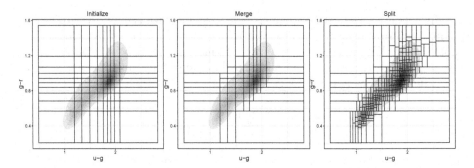

Fig. 1. Illustration of hyperrectangle partitioning of a $2D$ space containing approximately 80,000 datapoints, with input $(N_{\min}, N_{\max}) = (100, 300)$. At the beginning, 100 partitions are initialized based on marginal percentiles $10\%, \cdots, 90\%$ (left); merging then occurs for partitions with cardinality less than $N_{\min}$ (middle); and lastly partitions with cardinality more than $N_{\max}$ are successively split into smaller partitions, resulting in 393 partitions (right).

$|cl(\mathcal{C}_{k_i}) \cap cl(\mathcal{C}_{k_j})| > 1$. This criterion means that partitions that share only a corner are not considered neighbors. Note that the edges can entirely be determined during the partition phase for hyperrectangle partitions.

Other forms of graphs are also possible and sometimes necessary to reflect the known underlying manifold structure of the input space. Our proposed sampling procedure is independent of the graph-generating process and hence allows for more flexibility and adaptability to diverse arrays of applications.

The Algorithm. We define importance sampling [24] alike strategy that leverages the spatial dependence among the input values via the partition-induced graph $\mathcal{G}$. The basic idea is a sampler that traverses through the partitions via edges E in $\mathcal{G}$ and successively draws a datapoint from a partition by conditioning on the previously sampled datapoints from neighbor partitions. Such a strategy is motivated primarily by the intent of untangling convoluted latent processes that the data is arising from, without adding an extra computational burden. We note that since a single datapoint is drawn from each partition, the overall complexity of the sampling depends on N/m and m rather than N.

Without loss of generality, we assume that $\mathcal{G}$ is connected. If it is not, then the sampling stage can be carried out independently for each connected component. The sampling stage is an iterative process. We denote the datapoints in partition $\mathcal{C}_{k_i}$ as $\{(\mathbf{x}_1^{(i)}, y_1^{(i)}), \ldots, (\mathbf{x}_{n_i}^{(i)}, y_{n_i}^{(i)})\}$. The sampler is initialized by randomly sampling a datapoint $(\mathbf{x}_*^{(1)}, y_*^{(1)})$ from a partition $\mathcal{C}_{k_1}$. (The choice of k_1 is discussed later.) Next, the sampler moves to a partition that is an unsampled neighbor of $\mathcal{C}_{k_1}$. If there is more than one neighbor to choose from, the sampler moves to an available neighbor according to a criterion similar to the initialization step. The chosen partition is denoted by $\mathcal{C}_{k_2}$, and the datapoints $\{(\mathbf{x}_1^{(2)}, y_1^{(2)}), \ldots, (\mathbf{x}_{n_2}^{(2)}, y_{n_2}^{(2)})\}$ in $\mathcal{C}_{k_2}$ are weighted by a symmetric Gaussian kernel denoted by $K(\cdot|y_*^{(1)})$ centered at $y_*^{(1)}$. Let $f_y^{(i)}$ denote the distribution of y in $\mathcal{C}_{k_i}$. Then the weighted sampling distribution $g_y^{(i)}$ is defined as

$$g_y^{(i)}(y|\mathcal{C}_{k_i}) \propto f_y^{(i)}(y|\mathcal{C}_{k_i}) \times \prod_{j \in \mathcal{A}_i} K(y|y_*^{(j)}), \quad i = 2, \ldots, m, \tag{3}$$

where $\mathcal{A}_i = I\left(\mathcal{N}(\mathcal{C}_{k_i}) \cap \mathcal{V}_{(i-1)}\right)$, $\mathcal{N}(\cdot)$ are the neighbors of $\cdot$ in $\mathcal{G}$, $\mathcal{V}_{(i)}$ is the set of the first i partitions visited by the sampler, $I(\cdot)$ is the index set of $\cdot$, and $K(y|y_*^{(j)}) \propto \exp\{-\frac{1}{2\eta^2}(y - y_*^{(j)})^2\}$. η controls the width of the kernel K. Then a sample is drawn from $\mathcal{C}_{k_2}$ according to $g_y^{(2)}$, and the sampler walks through the partitions until all partitions are visited. We note that in practice we set η to a value less than the standard deviation of $y_*^{(j)}$ in (3).

As previously noted, one of the objectives of such a sampling scheme is to discover and model latent processes that the data might be arising from. Our initialization criterion is geared toward facilitating this goal. The initial partition is chosen to be one for which the variance of y within the partition is maximal. In other words,

$$\mathcal{C}_{k_1} \in \arg \max_{\mathcal{C}_k \in \mathcal{C}_{k_1}, \ldots, \mathcal{C}_{k_m}} \mathrm{Var}(y|\mathcal{C}_k).$$

Subsequent selection of partitions is carried out in a similar way. At any iteration, the next sampling partition is chosen from the unvisited neighbors of sampled partitions that have maximum variance,

$$\mathcal{C}_{k_i} \in \arg \max_{\mathcal{C}_k \in \mathcal{N}(\mathcal{V}_{(i-1)}) \cap \mathcal{V}_{(i-1)}^C} \mathrm{Var}(y|\mathcal{C}_k), \quad i = 2, \ldots, m.$$

This process continues until samples are drawn from all partitions. The weighting strategy of datapoints based on its neighbors encourages the sampling scheme to discover latent global structures in the input-output relationship.

In a complete setup, this sampling is performed multiple times (independently and in parallel), generating multiple datasets to train our statistical model.

3.3   Modeling via Ensembles

After a set of training data is generated, the last stage is to train a regression model. Our model of choice is Gaussian process [28], which is a semi-parametric regression model, fully characterized by a mean function and a covariance function. Historically GP models have been popular in both ML and statistics because of their ability to fit a large class of response surfaces [6, 13, 23, 29, 31]. The covariance function in a GP model often does the heavy lifting of describing the response variability by means of distance-based correlations among datapoints. Certain classes of GP covariance structures ensure smoothness and continuity in the response surface.

We define y to be a noisy realization from GP z. Then the data model can be written as

$$y(\mathbf{x}) = z(\mathbf{x}) + \epsilon, \quad \epsilon \overset{iid}{\sim} \mathrm{N}(0, \sigma^2) \qquad z(\mathbf{x}) \sim GP(0, \mathcal{C}_\Phi(\mathbf{x}, \mathbf{x}')),$$

where $\mathcal{C}_\Phi(\cdot, \cdot)$ is a covariance function with length-scale parameters Φ. The likelihood of the data is then given by the probability density function of a multivariate normal distribution:

$$\mathbf{y}^T = (y_1, \ldots, y_n)^T \sim \mathrm{MVN}(\mathbf{0}, \mathcal{C}_n), \tag{4}$$

where $\mathcal{C}_n$ is an $n \times n$ matrix, obtained by $\mathcal{C}_n = \left[\mathcal{C}_\Phi(\mathbf{x}_i, \mathbf{x}_j) + \sigma^2 \delta_{i=j} \right]_{1 \leq i,j \leq n}$. Distribution of z at an untried input setting $\mathbf{x}^*$ conditioned on n observations is also Gaussian, with mean and covariance given by

$$\begin{aligned}
\mathrm{E}(y(\mathbf{x}^*)|\mathbf{x}, \mathbf{z}, \cdot) &= c_n(\mathbf{x}^*)^T C_n^{-1} \mathbf{z}, \\
\mathrm{Var}(y(\mathbf{x}^*)|\mathbf{x}, \mathbf{z}, \cdot) &= \mathcal{C}_n - c_n(\mathbf{x}^*)^T C_n^{-1} c_n(\mathbf{x}^*),
\end{aligned} \tag{5}$$

where $c_n(\mathbf{x}^*) = (\mathcal{C}_\Phi(\mathbf{x}^*, \mathbf{x}_1), \ldots, \mathcal{C}_\Phi(\mathbf{x}^*, \mathbf{x}_n))$. Our implementation uses a scaled separable Gaussian covariance kernel $\mathcal{C}_\Phi(\mathbf{x}, \mathbf{x}') = \exp\left(-\sum_{p=1}^d \frac{(x_p - x_p')^2}{\phi_p} \right)$. The length-scale parameter $\Phi = (\phi_1, \cdots, \phi_d)$ controls the correlation strength along each dimension. Despite a GP's attractive properties, a major drawback of the standard GP model is the associated computational cost for estimating its parameters (i.e., Φ and σ). Each evaluation of the likelihood (4) involves inverting $\mathcal{C}_n$—an operation of $O(n^3)$ complexity. Hence, model training (i.e., estimation of Φ) becomes computationally infeasible as n grows. Alternatives have been proposed to deal with large n, including local GP approximations [14], knot-based low-rank representation [1,30], process convolution [17], and compactly supported sparse covariance matrices [20]. While all these methods achieve a certain computational efficiency at large n, none is intended to discover and model all the latent stochastic processes that the data is possibly arising from. Dirichlet process-based mixture models attempt to solve this problem [27], but scalability remains a challenge. In contrast, our proposed method uses $n = m \ll N$ training samples for each GP model, and the sampling scheme actively searches for all latent smooth processes that can be inferred from the data.

Full Predictive Model. Our final predictive model is constructed by taking an ensemble of N_m trained GPs models (each of these models trained with m samples as above). Denoting the predictive distribution at a new input $\mathbf{x}^*$ as $f_i(y^*|\mathbf{x}^*)$, we define the ensemble predictor as

$$\hat{f}(y^*|\mathbf{x}^*) = \frac{1}{N_m} \sum_{i=1}^{N_m} f_i(y^*|\mathbf{x}^*). \tag{6}$$

Each f_i has a Gaussian distribution with mean and variance given by (5). Prediction intervals are computed based on this mixed Gaussian distribution. The

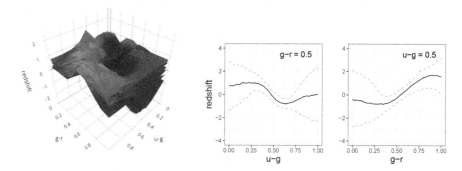

Fig. 2. Example ensemble of size 10 of mean GP surfaces; inputs are scaled to $[0,1]^2$, and the response is transformed to mean 0 and standard deviation 1. A 1D view of ensemble predictions of redshift at different values of u-g and g-r while keeping the other fixed at 0.5 is shown at the bottom. Black solid lines and red dashed lines represent median prediction and 90% confidence interval, respectively.

individual GPs can be trained independently and hence can be done in parallel. The choice of N_m can be guided by the computing budget and the choices of $N_{\min}, N_{\max}$ (which induce a value m). Figure 2 shows an illustration of 10 GP surfaces on a 2D color space, each trained on 50 datapoints sampled from data of size 2,000. Predicted distribution of redshift at any tuple $(u$-g, g-$r)$ is then a combination of Gaussian distributions from the 10 GP models. We note that the estimated prediction uncertainty from ensemble models would be much larger compared with that of a single model on the full data. Moreover, in a single GP case, the family of predictive distributions is restricted to the family of uni-modal symmetric Gaussian distributions, which is often inappropriate for noisy data. In contrast, the ensemble estimate can be interpreted as an approximation of the unknown distribution of the redshift conditional on the colors in a semi-parametric way.

4   Results

In this section we discuss numerical results from our proposed approach to model the redshift as a function of colors. After removing outliers, the size of the final dataset was reduced to 99,826, of which 20% were held out for out-of-sample prediction. Predictive distributions for the redshift conditional on colors were constructed by using our implementation of balanced hyperrectangle partitioning and sampling in R [26]; GP models were fitted by using the *mleHomGP* function in the `hetGP` package [5].

Partition and Sample. Inputs to the hyperrectangle partitioning scheme $N_{\min}$ and $N_{\max}$ were set at 50 and 150, respectively. These yielded a total of 749 nonempty rectangular partitions with an average cardinality of 107 in the 5D color space. Considering the tradeoff of execution time in the partition step and

type · balanced · imbalanced

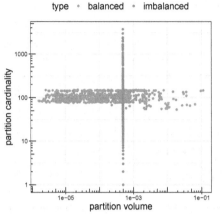

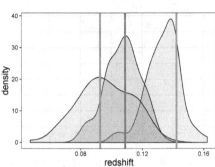

Fig. 3. Properties obtained from balanced hyperrectangle (in red) and equal-volume partitioning (in cyan), with 749 and 711 nonempty partitions, respectively. (Color figure online)

Fig. 4. Predicted density of redshift at three different color inputs (each indicated by a different shading in the figure). The vertical lines represent the truth for each color input. (Color figure online)

accuracy of the final prediction, we arrived at this particular choice of $N_{\min}$ and $N_{\max}$ after a few trials. At each iteration of sampling and modeling, we drew one sample per partition. The samples were used to train a zero-mean GP model with a nugget term [5]. Effectively, each individual GP model was trained on 749 examples. In contrast to partitions being balanced in terms of cardinality, partitions can be constructed to be balanced with respect to volumes, which for hyperrectangles equates to dividing the input space into equal-volume hyperrectangles. The scatter plot in Fig. 3 shows the cardinality and volume of partitions as points on $\mathbb{R}^2$ from balanced hyperrectangle and equal-volume partitions. By fixing the number of partitions and their volume, equal-volume partitioning returned 5,769 empty and 711 nonempty partitions with cardinality ranging from 1 to 10,000. Datapoints were not uniformly distributed along any dimension, thus making the equal-volume partitions have highly imbalanced data compositions. A sample from equal-volume partitions would always result in a biased training set for a model. In contrast, balanced hyperrectangle partitioning optimizes over cardinality by adaptively forming the hyperrectangles that enforce roughly homogeneous data size across partitions.

Model and Prediction. As described in Sect. 3.3, zero-mean GP models were fitted to each data subsample for $N_m = 50$ times. Maximum likelihood estimates of length-scale parameters Φ and the noise variance σ^2 were obtained and used to construct the ensemble estimate in (6). Figure 4 shows examples of predicted distributions at three different 5D color inputs and the true redshifts. To obtain smooth empirical density estimates, we sampled from the distribution given by

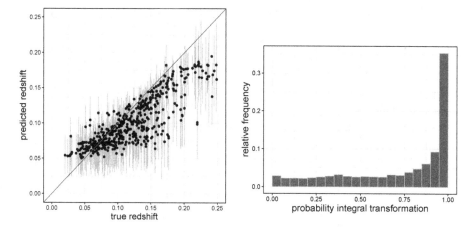

Fig. 5. True versus predicted redshifts for 500 random test points. Median predicted redshifts are denoted by black dots, and vertical grey bars show 90% prediction intervals on the left. A probability integral transformation (PIT) histogram plot is on the right.

(6); the samples were then used in a kernel density estimator to construct the predicted density function. As expected, the distributions show asymmetric, multimodal behaviors that would be impossible to capture by a single GP model. On the other hand, unlike in a traditional setting, comparing mean prediction with the true response would not be appropriate here and would result in low prediction accuracy. In other cases, degeneracies in the prediction are resolved by considering the highest mode among all mixture components or the mean from the mixture component with the highest probability [9]. Figure 5 shows the prediction performance of 500 random test (out-of-sample) examples. For each example, we obtained the median prediction (denoted by black dots) and 90% high-density probability region (denoted by vertical grey bars). Because of the selection effect of observing cosmological objects in a restricted region, imbalance arises with respect to cosmological scale [9]. Objects with higher redshifts are usually less represented in the full dataset in consideration. This is also evident from the probability integral transformation (PIT) plot in Fig. 5. The PIT plot is used to assess the quality of the probabilistic predictions—a perfect uniform distribution suggests perfect accuracy. In our case it is far from uniform, because of the shift in prediction at higher redshifts.

Scalability. Our framework allows for fairly straightforward parallel computations; each stage is amenable to parallelism. Once the data is partitioned according to any partitioning scheme, the ensemble members can be obtained in an embarrassingly parallel way. Each ensemble member consists of sampling data from the partitions and training the GP model, independently from other ensemble members. Almost linear speedup can be achieved when each ensemble member is run in parallel by using multiple cores in a single computing unit.

5   Discussion and Conclusion

In the SDSS dataset considered here, redshift ranges from 0 to 0.3, with a high concentration around 0.1. To make sense of the Gaussianity assumption for each GP model, we modeled logged redshifts instead of raw ones. Input values were transformed to $[0, 1]$ to have consistent length-scale parameter estimates in the covariance. These transformations were handled at the beginning of the data-preprocessing step, before splitting the data into training and testing sets. Such an approach ensured that no extrapolation was being done in predicting the redshifts for the testing examples. Selection of $N_{\min}$ and $N_{\max}$ for balanced hyperrectangle partitioning is conceived to be a sequential updating process. In most applications, the choices are highly influenced by the intended size of the training set for each individual model in the ensemble. Prior knowledge of the expected smoothness of the global response surface can assist in choosing the size of the data subsample. For example, estimating a smooth surface using a GP with Gaussian covariance would require a substantially small number of training examples. We used $N_{\min} = 50$ and $N_{\max} = 150$ based on prior experience. Construction of the graph induced by partitions is simple and intuitive. One could argue for considering two hyperrectangles $\mathcal{C}_{k_i}$ and $\mathcal{C}_{k_j}$ to be neighbors when their boundaries intersect only at a corner, namely, $|cl(\mathcal{C}_{k_i}) \cap cl(\mathcal{C}_{k_j})| = 1$. In fact, neighbor relationships on the partitions can be as arbitrary as an application desires. Considering corner-sharing partitions as neighbors will produce a graph with many more edges, which in turn requires more compute time in the subsequent sampling and modeling stages.

Here we have presented a novel approach to regression modeling for large data that leverages efficient data sampling and an advanced statistical model such as GP modeling. The problem of estimating the full predictive distribution of the photometric redshift provides a base case for demonstrating novel aspects of our proposed methodology. Redshift estimation based on photometric surveys is an important problem. Almost all recent estimation protocols emphasize being able to use large photometry data to train their models [3,9,20]: since the new generation of telescopes promises a much larger survey area of the sky, models need to ingest this huge amount of data and produce forecasts for new observations in a reasonable amount of time. One way to meet this need is by using a combination of new algorithms and more powerful supercomputers. New modeling paradigms also need to be developed that find a sweet spot between careful choices of efficient algorithms, good models, and opportunity to scale. Indeed, the present work is motivated by this pursuit.

Another aspect of our work involves prediction targets, which are different from simply estimating the mean or median under parametric uncertainty assumptions. Best-case scenarios would involve nonparametric approximation of the unknown predictive distribution, an example of which can be found in Dirichlet process-based models [12]. Full Bayesian estimation here is rarely feasible with large data. Combining several simple models is a step toward the same goal but at a fraction of the cost. Successful ensemble techniques [14,15,22] and ML models [8,21] achieve superior accuracy by combining locally learned structures in the response surface. While we draw motivation from such endeavors, our approach differs from them in a significant way. Each individual model in our ensemble does not target local structures but, rather, learns a global response surface every time a new model is trained on a sample of the data, allowing for a broad range of distribution to be covered. To the best of our knowledge, this work is the first of its kind that draws local inference by combining global information. Evaluating the accuracy of different combinations of partitioning schemes, statistical models, and ensemble sizes is a promising direction for future work.

Acknowledgments. This material was based upon work supported by the U.S. Department of Energy, Office of Science, Office of Advanced Scientific Computing Research, Applied Mathematics and SciDAC programs under Contract No. DE-AC02-06CH11357.

References

1. Banerjee, S., Fuentes, M.: Bayesian modeling for large spatial datasets. Wiley Interdiscip. Rev. Comput. Stat. **4**(1), 59–66 (2012)
2. Baum, W.A.: Photoelectric determinations of redshifts beyond 0.2 c. Astronomical J. **62**, 6–7 (1957)
3. Beck, R., Dobos, L., Budavári, T., Szalay, A.S., Csabai, I.: Photometric redshifts for the SDSS Data Release 12. Mon. Not. R. Astron. Soc. **460**(2), 1371–1381 (2016)
4. Benitez, N.: Bayesian photometric redshift estimation. Astrophys. J. **536**(2), 571 (2000)
5. Binois, M., Gramacy, R.B.: hetGP: heteroskedastic gaussian process modeling and design under replication (2019). https://CRAN.R-project.org/package=hetGP, r package version 1.1.2
6. Brahim-Belhouari, S., Bermak, A.: Gaussian process for nonstationary time series prediction. Comput. Stat. Data Anal. **47**(4), 705–712 (2004)
7. Cavuoti, S., Brescia, M., Longo, G., Mercurio, A.: Photometric redshifts with the quasi Newton algorithm (MLPQNA) Results in the PHAT1 contest. Astron. Astrophys. **546**, A13 (2012)
8. Dietterich, T.G., et al.: Ensemble learning. Handb. Brain Theory Neural Netw. **2**, 110–125 (2002)
9. D'Isanto, A., Polsterer, K.L.: Photometric redshift estimation via deep learning-generalized and pre-classification-less, image based, fully probabilistic redshifts. Astron. Astrophys. **609**, A111 (2018)
10. Fernández-Soto, A., Lanzetta, K.M., Yahil, A.: A new catalog of photometric redshifts in the Hubble deep field. Astrophys. J. **513**, 34–50 (1999)
11. Firth, A.E., Lahav, O., Somerville, R.S.: Estimating photometric redshifts with artificial neural networks. Mon. Not. R. Astron. Soc. **339**(4), 1195–1202 (2003)
12. Gelfand, A.E., Kottas, A., MacEachern, S.N.: Bayesian nonparametric spatial modeling with Dirichlet process mixing. J. Am. Stat. Assoc. **100**(471), 1021–1035 (2005)
13. Gramacy, R.B.: Surrogates: Gaussian Process Modeling. Design and Optimization for the Applied Sciences. Chapman Hall/CRC, Boca Raton, Florida (2020)
14. Gramacy, R.B., Apley, D.W.: Local Gaussian process approximation for large computer experiments. J. Comput. Graph. Stat. **24**(2), 561–578 (2015)

15. Hastie, T., Rosset, S., Zhu, J., Zou, H.: Multi-class AdaBoost. Stat. Interface **2**(3), 349–360 (2009)
16. Hastie, T., Tibshirani, R., Friedman, J.: Random forests. The Elements of Statistical Learning. SSS, pp. 587–604. Springer, New York (2009). https://doi.org/10. 1007/978-0-387-84858-7_15
17. Higdon, D.: Space and space-time modeling using process convolutions. In: Anderson C.W., Barnett V., Chatwin P.C., El-Shaarawi A.H. (eds.) Quantitative Methods for Current Environmental Issues. Springer, London (2002). https://doi.org/ 10.1007/978-1-4471-0657-9_2
18. Hu, Y.H., Hwang, J.N.: Handbook of neural network signal processing. Acoustical Society of America (2002)
19. Ilbert, O., et al.: COSMOS photometric redshifts with 30-bands for 2-deg2. Astrophys. J. **690**(2), 1236–1249 (2009)
20. Kaufman, C.G., Bingham, D., Habib, S., Heitmann, K., Frieman, J.A., et al.: Efficient emulators of computer experiments using compactly supported correlation functions, with an application to cosmology. Ann. Appl. Stat. **5**(4), 2470–2492 (2011)
21. Lawrence, S., Giles, C.L., Tsoi, A.C., Back, A.D.: Face recognition: a convolutional neural-network approach. IEEE Trans. Neural Netw. **8**(1), 98–113 (1997)
22. Liaw, A., Wiener, M., et al.: Classification and regression by randomForest. R News **2**(3), 18–22 (2002)
23. Neal, R.M.: Regression and classification using Gaussian process priors. In: Bernardo, J.M., Berger, J.O., Dawid, A., Smith, A.F.M., et al. (eds.) Bayesian Stat., vol. 6, pp. 476–501. Oxford University Press, Oxford (1998)
24. Neal, R.M.: Annealed importance sampling. Stat. Comput. **11**(2), 125–139 (2001)
25. Puschell, J.J., Owen, F.N., Laing, R.A.: Near-infrared photometry of distant radio galaxies - Spectral flux distributions and redshift estimates. Astrophys. J. Lett. **257**, L57–L61 (1982)
26. R Core Team: R: A Language and Environment for Statistical Computing (2020). https://www.R-project.org
27. Rasmussen, C.E., Ghahramani, Z.: Infinite mixtures of Gaussian process experts. Adv. Neural Inf. Process. Syst. **14**, 881–888 (2001)
28. Rasmussen, C.E., Williams, C.K.I.: Gaussian Processes for Machine Learning. MIT Press, Cambridge, MA (2005)
29. Sacks, J., Welch, W.J., Mitchell, T.J., Wynn, H.P.: Design and analysis of computer experiments. Stat. Sci. **4**, 409–423 (1989)
30. Snelson, E., Ghahramani, Z.: Sparse Gaussian processes using pseudo-inputs. Adv. Neural Inf. Process. Syst. **18**, 1257–1264 (2005)
31. Wang, J., Hertzmann, A., Fleet, D.J.: Gaussian process dynamical models. Adv. Neural Inf. Process. Syst. **18**, 1441–1448 (2005)
32. Weinberg, D.H., Mortonson, M.J., Eisenstein, D.J., Hirata, C., Riess, A.G., Rozo, E.: Observational probes of cosmic acceleration. Phys. Reports **530**, 87–255 (2013)
33. York, D.G., et al.: The Sloan digital sky survey: technical summary. Astron. J. **120**(3), 1579 (2000)

Timeseries Based Deep Hybrid Transfer Learning Frameworks: A Case Study of Electric Vehicle Energy Prediction

Paul Banda$^{(\boxtimes)}$, Muhammed A. Bhuiyan, Kazi N. Hasan, Kevin Zhang, and Andy Song

RMIT University, School of Engineering, Melbourne 3000, Australia

Abstract. The problem of limited labelled data availability causes under-fitting, which negatively affects the development of accurate time series based prediction models. Two-hybrid deep neural network architectures, namely the CNN-BiLSTM and the Conv-BiLSTM, are proposed for time series based transductive transfer learning and compared to the baseline CNN model. The automatic feature extraction abilities of the encoder CNN module combined with the superior recall of both short and long term sequences by the decoder LSTM module have shown to be advantageous in transfer learning tasks. The extra ability to process in both forward and backward directions by the proposed models shows promising results to aiding transfer learning. The most consistent transfer learning strategy involved freezing both the CNN and BiLSTM modules while retraining only the fully connected layers. These proposed hybrid transfer learning models were compared to the baseline CNN transfer learning model and newly created hybrid models using the R^2, MAE and RMSE metrics. Three electrical vehicle data-sets were used to test the proposed transfer frameworks. The results favour the hybrid architectures for better transfer learning abilities relative to utilising the baseline CNN transfer learning model. This study offers guidance to enhance time series-based transfer learning by using available data sources.

Keywords: Hybrid deep learning · Electric vehicle load prediction · Transfer learning

1 Introduction

Due to the environment-friendly policies and emission reduction schemes, the transportation sector is expected to go through a significant transformation with the adoption of many electric vehicles (EVs) into the existing vehicle fleet [8]. The usage of EVs is still in the early stages; thus, the EV charging demand data is scarce. Hence, it is challenging to perform the EV impact studies with the EV data that accurately represents the EV charging profiles [9]. Moreover, the deployment of more public EV charging stations with a large capacity charging requirement would pose a new challenge to the secure operation of the power

© Springer Nature Switzerland AG 2021
M. Paszynski et al. (Eds.): ICCS 2021, LNCS 12746, pp. 259–272, 2021.
https://doi.org/10.1007/978-3-030-77977-1_20

grid [21]. In this perspective, adequately capturing the EV charging demand and accurately predict the near future charging demand is critical for the electrical power grid's secure operation.

The implementation of forecasting techniques for EV charging demand prediction has been conducted by using fuzzy clustering and back-propagation neural network models in [22], by employing evolutionary optimisation techniques in [14] and by examining various data mining techniques [20]. An unsupervised algorithm was proposed in [13] for non-intrusive EV load extraction. A comparison of the traditional time series methods and machine learning methods for EV load prediction was presented in [4]. Machine learning methods were concatenated to give an ensemble method that performed better than individual machine learning methods in [1] for residential EV load prediction. A comparative assessment of the deep learning methods for EV charging demand prediction has been performed in [23]. The sub-hourly and hourly EV charging load prediction models have been developed using a hybrid lion algorithm consist of convolutional neural network (CNN) and long-short-term memory (LSTM) inspired models in [12]. The above studies have demonstrated effective EV prediction methods; however, the only drawback is that they require independent and identical distribution to exist in a data-set and that they must be enough training data to learn a good model. To counter these limitations, we introduce transfer learning for EV load prediction, which is not bound by the mentioned limitations.

1.1 Transfer Learning Models Review

Transfer learning using deep learning models has received great attention in various application domains such as image processing [18], time series classification, [7], natural language processing tasks [2,17] and building energy forecasting [5]. The CNN model has been the dominant model facilitating transfer learning in most studies.

In non-transfer learning studies [10,15,19], hybrid deep learning models have been shown to outperform the conventional CNN model as they leverage on the advantages of the encoder and decoder modules during operations. In the same line of thought, the authors propose using hybrid deep learning model architectures, namely, the CNN-BiLSTM and Conv-BiLSTM models for transfer learning and compare their performance to the commonly used CNN transfer learning model. Furthermore, most transfer learning studies seem to report transfer methods is isolation, without showing a comparative assessment to determine which transfer learning method shows superior results, an observation which is addressed in this study. The Conv-BiLSTM model can capture salient spatial features using a convolution operator within the LSTM cell on multiple-dimensional data and has the extra ability to process in both forward and backward directions (bidirectional) is helpful for transfer learning. The CNN-BiLSTM model can leverage the automatic feature extraction advantages of the encoder CNN module and the superior recall of both short and long term sequences by the decoder LSTM module, which processes both forward and backward directions.

The study proposes the implementation of CNN-BiLSTM and Conv-BiLSTM hybrid deep learning architectures for time series transfer learning. Also, the study introduces transductive transfer learning for electrical vehicle load prediction to enhance prediction efforts in limited labelled EV load data situations.

The rest of the paper is organised as follows: Sect. 2 briefly introduces transfer learning and the proposed hybrid transfer learning models, Sect. 3 describes the three case study data-sets, Sect. 4 discusses the transfer learning results and discussion, and Sect. 5 highlights the conclusions and future works.

2 Methodology

This section provides a brief description of the application of transfer learning for EV load prediction. It also discusses the proposed hybrid architectures and how they are modified for implementation in transfer learning.

2.1 Transfer Learning

Given a source domain EV data-set (data-rich), a target domain (limited labelled data) EV data-set and a learning task, transfer learning seeks to improve the learning of the target predictive function in the target domain EV data-sets (slow and fast commercial EV charging stations (CEVCS)) using the knowledge learnt from the source domain (residential EVCS) data-set. The formal expression to define transfer learning is given as; A domain comprises feature space X and label space Y, thus given a source domain

$$D_s = \{x_S^i, y_S^i\}_{i=1}^{N_S} \tag{1}$$

and a target domain

$$D_T = \{x_T^i, y_T^i\}_{i=1}^{N_T} \tag{2}$$

where $N_S > N_T$ and N is the labelled data size. It is challenging for a model to learn well using little data in the target domain. Since the source and target domains have different data distributions, it is unlikely for a model trained on the source domain to predict the target domain's test data-set accurately. Instead of creating two separate models for the source and target domain, as usually done in traditional machine learning, transfer learning seeks to utilise the knowledge learnt on the source domain to help predict the data domain.

The proposed hybrid transfer learning workflow illustrated in Fig. 1 applies to both the CNN-BiLSTM and Conv-BiLSTM hybrid models.

Following the above formal transfer learning definition, the transfer learning procedure implementation in this study is thus summarised as below;

(i) Pre-process data and develop the hybrid deep neural networks for source domain (residential EVCS) prediction with full data complement, (ii) Train the pre-trained models (fine-tuning) with limited data from the target domain (slow and fast CEVCS) (iii) Develop new neural networks models from scratch for the target stations with limited data (same data size as used in step II), (iv) Compare the pre-trained model's (ii) results to the new target models developed from scratch for both slow and fast CEVCS.

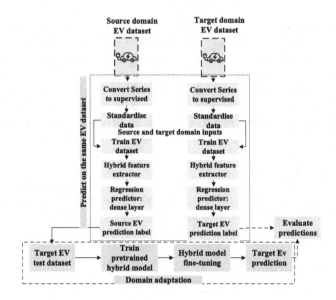

Fig. 1. An illustration of the implemented hybrid deep transfer learning workflow.

2.2 Hybrid Deep Learning Models for Transfer Learning

This section describes the proposed models for transfer learning using the available three EV data-sets. Hybrid deep learning models are designed by the fusion of conventional deep learning models that combine multiple models by diversifying the input features and varying the initialisation of the neural network's weights.

2.3 Baseline Convolutional Neural Networks (CNN)

Originally designed to handle image data, the CNN model has achieved a state of art results in image classification and object recognition, among other tasks. The CNN model can extract useful features automatically from time series data by treating a sequence of observations as a one-dimensional image. The CNN model is relatively easy and faster to train because the weights are less than those of a fully connected artificial neural network architecture. A predetermined number of historical EV energy consumption observations are fed into the convolutional layers, which perform a one-dimensional convolution operation on this data, whose output is then passed to the fully connected layers for final processing. A CNN operation is described in detail in [11];

2.4 Hybrid Convolutional Bidirectional Long Short-Term Memory (Conv-BiLSTM)

The hybrid Conv-BiLSTM does a convolution operation within the LSTM cells. The Conv-BiLSTM layer is a recurrent layer (same as LSTM) that replaces the usual matrix multiplication operation with a convolutional process. The convolution operator is applied directly to read input into the BiLSTM cells, that is, during the input-to-state transitions and during the state-to-state transitions [16]. The Conv-BiLSTM compresses the EV sequence into a hidden state tensor decodable by an LSTM layer that processes this input in both forward and backward directions (bidirectional), forwarding its output to the fully connected layer for final prediction. The critical equations of the Conv-BiLSTM cell gates are given in Eqs. (3–7) below;

$$I_t = \sigma(W_{XI} * X_t + W_{HI} * H_{t-1} + W_{CI} \circ C_{T-1} + b_I) \tag{3}$$

$$O_t = \sigma(W_{XO} * X_t + W_{HO} * H_{T-1} + W_{CO} \circ C_t + b_0) \tag{4}$$

$$F_t = \sigma(W_{XF} * X_t + W_{HF} * H_{t-1} + W_{CF} \circ C_{t-1} + b_F) \tag{5}$$

$$C_t = F \circ C + i_t \circ (W_{XC} * x_t + W_{HC} * h_{t-1} + b_C) \tag{6}$$

$$H_t = O \circ tanh(C_t) \tag{7}$$

The convolutional product and element-wise multiplication operations are denoted by "$*$" and "$\circ$" respectively. In Eqs. (3–5), I_t, F_t, and O_t are the input, forget, and output gate, respectively. W represents the weight matrix, x_t is the current input data, h_{t-1} is the previously hidden output, and C_t denotes the cell state at timestep t. The traditional LSTM equations use convolution operation ($*$), in comparison, the Conv-BiLSTM uses matrix multiplication between W and X_t ,h_{t-1} for every gate. This matrix multiplication replaces the fully connected layer with a convolutional layer, leading to a reduced number of weight parameters in the model. The Hybrid Conv-BiLSTM model is implemented to capture the advantages of the CNN and BiLSTM techniques to improve the overall prediction accuracy. The expected input shape into the Conv-BiLSTM model must be of the form [samples, timesteps, rows, columns, channels]. When using the Conv-BiLSTM, the previous EV timeseries of 47 (lags) is split such that it has one row of 47 timesteps, and the Conv-BiLSTM performs the convolutional operation on this particular row. This sequence design operation results in a 5D input tensor with shape [s, 1, 1, 47, 9] denoting sample, timestep, rows, columns and channels, respectively, as shown in Fig. 4(a). Finally, the hidden layer of the Conv-BiLSTM encoder is defined and flattened in readiness for decoding using the BiLSTM operation. The last layer is made up of the fully connected (dense layer) with 200 neurons for processing the output from the BiLSTM operation.

2.5 Hybrid Bidirectional Deep Convolutional Neural Network Long Short-Term Memory (CNN-BiLSTM)

The CNN-BiLSTM is a hybrid model that combines a 1D-CNN with a BiLSTM model to solve the vanishing gradient problem. This hybrid system consequently captures the advantages of both models. In this hybrid system, the 1D-CNN acts as an encoder responsible for interpreting the input sequence from the EV time series. The CNN encoder model then outputs a sequence that is passed on to the bidirectional BiLSTM model (decoder) for interpretation. The encoder CNN model does the convolutional operation and outputs a sequence. Naturally, the LSTM has an inherent strong ability to remember the structure of short and long term sequences; thus, by combining the BiLSTM with the CNN, which automatically learns features from sequence data, the hybrid CNN-BiLSTM model offers improved accuracy. The CNN and BiLSTM model hybrid structure expect the input EV demand data to have a 3-dimensional form (sample, input and time-step). Figure 4(b) presents an illustration of hybrid CNN-BiLSTM workflow. In this implementation, the historical EV time series of 47 h is input into the CNN encoder architecture for reading. The first convolutional layer reads across this input sequence using a filter of size three timesteps (1 × 3) and then projects its output onto 32 feature maps. The second convolutional layer reads the previous layer's output and uses the max-pooling layer to simplify the feature maps sizes by preserving the maximum possible amount of information (signal). The final extracted features map from the max-pooling layer is then flattened for use with the BiLSTM decoding module, which processes the cell state in both forward and backward directions before passing the output to the fully connected layer prediction.

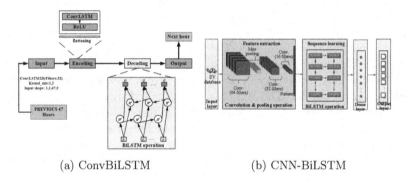

(a) ConvBiLSTM   (b) CNN-BiLSTM

Fig. 2. Proposed Hybrid models for EV charging load transfer learning implementation

3   Data-Sets Description

The EV datasets (https://data.dundeecity.gov.uk/dataset/ev-charging-data), collected over an entire year from 01/09/2017 to 31/08/2018 equates to 8760 data points, representing hourly EV charging power (in kilo Watt, kW) of a year. As single residential charging stations have small data sizes with highly random data, collective residential EV charging profiles were created by grouping data from 100 single residential profiles located in relative proximity to each other. This was done to ascertain the EV charging behavior at distribution sub-station level. Additionally, slow commercial EV charging station profile represents data from a Health Care Centre, and fast commercial EV charging station profile represents data from a multi-story car park, which are referred throughout this report as slow commercial EV charging station (slow CEVCS) and fast commercial EV charging station (fast CEVCS), respectively. These commercial profiles differ in charging capacities; slow charging (6−10 kW) and fast charging (22−70 kW). The EV energy consumption profiles for the three types of charging stations, namely residential EVCS, slow CEVCS and fast CEVCS, are illustrated in Fig. 3.

The residential EVCS consumption patterns are similar throughout the seven days of the week, steadily rising from 6 am up until 7 pm, remaining low in the night when users are sleeping. The slow commercial CEVCS demand has two distinct consumption patterns within the week: weekdays and weekends. As seen in Fig. 3, the EV charging demand is higher during the weekdays and lower during the weekends. During the weekends it is low throughout the whole day because of less demand. The fast CEVCS offer quick charging (about 30mins) for users, resulting in the irregular demand with no distinct pattern. Predicting the EV demand for the fast charging station is expected to be challenging, given the irregular consumption patterns. Consequently, transfer learning can be expected to be problematic for fast CEVCS relative to the slow CEVCS.

3.1 Input Processing

The EV time series data must be transformed into supervised learning to allow reading by the model. The EV sequence is reframed into pairs of input and output variables using the sliding window method. The inputs comprise values of the current EV observation t and previous EV observations at times $t − 1$, $t − 2....t − n$ which predict the output EV observation at time $t + 1$. The length of the sliding window is defined based on [3,6].

3.2 Data Standardisation

The differences in numerical ranges between the input and output values require that the data-set be standardised. By standardisation, each feature is scaled to have a distribution that is centred around 0, having a standard deviation of 1. Standardisation allows for comparability among input data-sets, and it also enhances the training efficiency of models since the numerical condition of optimisation is improved. The mean and standard deviation for each feature is calculated, then the feature is scaled using Eq. 8.

$$z = (x_i − \mu)/\sigma \tag{8}$$

where, z represents the standardized value, x_i is the observed energy consumption value, μ is the mean, and σ is the standard deviation of the EV energy data-sets.

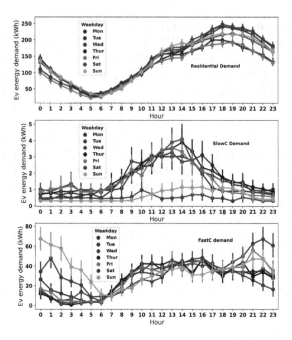

Fig. 3. EV energy consumption weekly profile by the hour.

3.3 Evaluation Metrics

An evaluation of the models' skill is done using the mean absolute error MAE, root-mean-square error RMSE, and R-squared R^2 metrics. The R^2 metric describes the proportion of variance of the input variable that the regression model explains. The MAE metric calculates the positive mean error value for the test data. The root-mean-square error (RMSE), which calculates the square root of the means square error, is a strict metric that penalises significant errors; thus, it is a reliable metric. One must look at these metrics concurrently when selecting the superior model. These performance evaluation metrics are calculated using Eqs. (9–11) below

$$R^2 = 1 - \frac{\sum(y - y')^2}{\sum(y - \bar{y}')^2}, \tag{9}$$

$$RMSE = \sqrt{\Sigma \frac{(y' - y)^2}{N}}, \tag{10}$$

$$\text{MAE} = \frac{1}{n} \sum_{i=1}^{n} |y' - y| \tag{11}$$

where y is the measured EV energy consumption value, y' is the predicted EV energy consumption value, and N represents the number of data pairs considered.

4 Results and Discussion

This study investigates the possibility of knowledge transfer from; (1) residential EVCS to slow CEVCS and (2) residential EVCS to fast CEVCS to enhance prediction in the target datasets with limited labelled data. The baseline transfer learning-based model (CNN-T) is compared against the proposed hybrid transfer learning-based models (CNN-BiLSTM-T and Conv-BiLSTM-T) to determine which model is more superior. The newly created (CNN-N), CNN-BiLSTM-N and Conv-BiLSTM-N non-transfer learning models are also compared against their transfer learning-based counterpart models. If a new model without transfer learning can outperform the transfer learning-based models, we can conclude that transfer learning is not practical. Thus, we may need to fine-tune the transfer learning model or abandon it altogether. However, if the transfer learning-based models outperform the new model (which is the expected outcome), we can conclude that transfer learning is beneficial.

Table 1. Comparing tested models for residential commercial charging load prediction

MODEL	RMSE (kWh)	MAE (kWh)	R^2
CNN	27.87	22.13	0.86
CNN-BiLSTM	27.17	21.25	0.87
Conv-BiLSTM	26.96	21.12	0.87

4.1 Training with Full Data Complement

The results of the initial training of the three models on residential EVCS before transfer learning are shown in Table 1. Lag 47 provided the best EV prediction results compared to other tested lags. The results in Table 1 are in reference to lag 47; that is, the previous 47 h of EV energy consumption are used as input for predicting the next hour of EV energy consumption. The tested models demonstrated almost similar performance in predicting the next hour charging load for the residential EVCS (source domain), with the CNN, CNN-BiLSTM and Conv-BiLSTM models having comparable RMSE scores of 27.87 kWh, 27.17 kWh and 26.96 kWh, respectively.

4.2 Comparing Transfer Learning and Non-transfer Learning-Based Methods

This section compares the three transfer learning-based models' performance against their newly created counterparts for residential EVCS to slow CEVCS and residential EVCS to fast CEVCS transfer learning. The results from this part of the investigation are shown in Tables 2 and 3 and illustrated in Fig. 4(a) and (b).

Table 2. Comparing RMSE values for transfer and non-transfer learning-based model for residential EVCS to slow CEVCS

Data size	CNN-T		CNN-N		CNN-BiLSTM-T		CNN-BiLSTM-N		Conv-BiLSTM-T		Conv-BiLSTM-N	
	rmse	mae	rmse	mae	rmse	mae	rmse	mae	rmse	mae	rmse	mae
12	44.87	39.23	0.12	0.10	0.08	0.07	0.06	0.05	0.15	0.14	0.04	0.04
24	0.26	0.25	0.65	0.65	0.39	0.39	0.61	0.61	0.47	0.47	0.63	0.63
36	1.05	1.04	0.12	0.11	0.42	0.41	0.03	0.03	0.04	0.03	0.08	0.07
48	2.56	2.00	1.20	1.08	0.82	0.04	0.96	0.84	0.82	0.48	0.86	0.66
60	0.27	0.26	1.36	1.31	0.24	0.19	0.54	0.53	0.13	0.11	0.16	0.15
72	1.69	1.50	1.91	1.58	1.43	1.11	1.60	1.29	1.49	1.24	1.65	1.33
96	2.17	1.21	2.55	1.67	2.32	1.32	2.77	1.70	2.32	1.27	2.72	1.78
120	2.37	1.58	3.65	3.43	2.27	1.62	2.36	2.09	2.61	1.64	2.79	2.38
1200	1.25	0.65	1.16	0.69	1.40	0.62	2.14	1.14	1.48	0.81	1.79	0.9
6961	1.99	1.17	2.03	1.13	1.38	0.98	1.77	1.14	1.36	0.98	1.46	0.88

The results from this part of the investigation indicate the potential of transferring learning in aiding model prediction in cases of limited labelled data. In the residential EVCS to slow CEVCS, the table and charts show competitive performance by all tested models. Only at data instance 36 do the newly-created CNN-N, and CNN-BiLSTM-N models seem to present superior results than the transfer based hybrid models, with the latter presenting equal MAE scores with the Conv-BiLSTM-T model. In the rest of the tested data instances, the proposed transfer learning-based hybrid models, particularly the Conv-BiLSTM-T model presenting superior results.

In the residential EVCS to fast CEVCS transfer scheme, hybrid transfer models continue to show dominance over newly created counterpart models. The CNN-N and the CNN-BiLSTM-N supersede counterpart models at data instance 48. Besides this instance, transfer learning models continue to dominate in both the limited data instances and beyond.

4.3 Comparing Transfer Learning-Based Models

In another part of the investigation, transfer learning-based models were compared to determine which methods are most effective during transfer learning. It is observed that hybrid-based transfer learning models outperform the baseline CNN-T model at most tested data sizes, which point to the hybrid architectures' effectiveness during transfer learning.

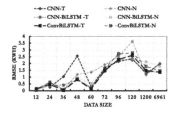
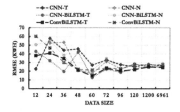

(a) Residential to slow CEVCS   (b) Residential to fast CEVCS

Fig. 4. An RMSE comparison of the implemented non transfer and transfer learning-based models

Table 3. A comparison of the transfer learning and non-transfer learning-based model for residential EVCS to fast CEVCS

Data size	CNN-T		CNN-N		CNN-BiLSTM-T		CNN-BiLSTM-N		Conv-BiLSTM-T		Conv-BiLSTM-N	
	rmse	mae	rmse	mae	rmse	mae	rmse	mae	rmse	mae	rmse	mae
12	22.64	18.54	143.64	136.50	42.90	36.33	50.63	46.08	37.97	33.40	60.27	56.44
24	57.77	51.74	52.48	47.50	32.29	29.58	54.90	50.27	40.98	38.10	46.82	41.29
36	44.24	36.35	53.35	50.54	19.70	18.70	38.36	34.05	34.73	30.32	30.04	34.63
48	45.77	39.07	21.23	16.31	42.14	36.02	20.64	16.30	21.34	17.88	22.32	17.00
60	27.13	19.05	17.63	16.71	12.82	10.21	21.08	19.89	14.70	12.43	21.35	20.10
72	33.30	26.81	31.46	25.67	22.51	17.92	35.03	29.23	23.52	18.39	23.84	19.06
96	27.67	20.95	22.46	18.75	18.73	15.24	22.77	18.02	20.33	17.27	22.80	18.75
120	25.18	20.06	28.82	24.68	22.95	18.01	27.56	23.95	21.85	18.11	28.02	24.55
1200	28.03	21.38	25.13	19.60	25.90	20.28	27.42	20.69	24.92	18.41	27.20	20.68
6961	26.20	20.13	25.50	19.34	24.58	19.61	29.08	21.79	24.30	19.27	27.68	20.78

As seen in Table 4, there is dominance by the proposed hybrid transfer learning-based models over the baseline CNN-T model. As clearly seen in the illustration given in Fig. 5, the CNN-T model at all instances recorded higher RMSE values in both tested transfer learning schemes, that is, from residential EVCS to slow CEVCS and from residential EVCS to fast CEVCS. When considering the hybrid transfer learning-based models, it is observed that in small data settings, the Conv-BiLSTM-T model tend to show superior performance over its counterpart CNN-BiLSTM-T hybrid model.

4.4   Summary of Findings

Transfer learning models dominate the newly created (CNN-N) between data sizes of 24 and 60 h and beyond, except at isolated instances (12, 24 and 48 h). A 12-h sized sample did not provide enough information to enhance transfer learning in most tested cases, explaining the relatively high error values recorded at this testing point by transfer learning models. Both hybrid deep transfer models

Table 4. Comparing the implemented transfer learning-based models from residential EVCS to slow CEVCS and to fast CEVCS

Data size	CNN-T		CNN-BiLSTM-T		Conv-BiLSTM-T		CNN-T		CNN-BiLSTM-T		Conv-BiLSTM-T	
	rmse	mae	rmse	mae	rmse	mae	rmse	mae	rmse	mae	rmse	mae
12	44.87	39.23	0.08	0.07	0.15	0.14	22.64	18.54	42.90	36.33	37.97	33.40
24	0.26	0.25	0.39	0.39	0.47	0.47	57.77	51.74	32.29	29.58	40.98	38.10
36	1.05	1.04	0.42	0.41	0.04	0.03	44.24	36.35	19.70	18.70	34.73	30.32
48	2.56	2.00	0.82	0.04	0.82	0.48	45.77	39.07	42.14	36.02	21.34	17.88
60	0.27	0.26	0.24	0.19	0.13	0.11	27.13	19.05	12.82	10.21	14.70	12.43
72	1.69	1.50	1.43	1.11	1.49	1.24	33.30	26.81	22.51	17.92	23.52	18.39
96	2.17	1.21	2.32	1.32	2.32	1.27	27.67	20.95	18.73	15.24	20.33	17.27
120	2.37	1.58	2.27	1.62	2.61	1.64	25.18	20.06	22.95	18.01	21.85	18.11
1200	1.25	0.65	1.40	0.62	1.48	0.81	28.03	21.38	25.90	20.28	24.92	18.41
6961	1.99	1.17	1.38	0.98	1.36	0.98	26.20	20.13	24.58	19.61	24.30	19.27

prove robust models, with consistent superior performance over the CNN-N and CNN-T models for both residential EVCS to slow CEVCS transfer and residential EVCS to fast CEVCS transfer learning tasks. The hybrid CNN-BiLSTM-T and the Conv-BiLSTM-T show interchangeable performance between themselves in tested transfer learning tasks. That is, no outright dominance by either model in the critical transfer window and beyond.

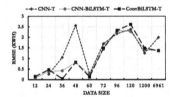 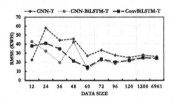

(a) Residential to slow CEVCS   (b) Residential to fast CEVCS

Fig. 5. An RMSE comparison of the implemented transfer learning-based models

The above observations confirm the benefits of transfer learning for electrical vehicle knowledge transfer. In cases of little available labelled data at either the slow or fast CEVCS, a model trained at the residential EVCS can improve prediction efforts at these target data-sets.

5   Conclusions and Future Works

The CNN model is a standard model for transfer learning due to its enabling properties. Most research efforts study the effective strategies of improving transfer learning using the CNN model. However, little is known about the performance of hybrid deep learning models as mediums for transfer learning. Experiments were set up using electric vehicle data-sets to determine the performance of hybrid CNN-BiLSTM and Conv-BiLSTM models against the commonly used CNN-based model in the transfer learning tasks. The experimental results confirmed the superiority of hybrid deep learning transfer learning-based models over the conventional CNN transfer learning model. These results show that the hybrid structure of the implemented models is beneficial for the transfer learning tasks. As such, it can be concluded that the use of the hybrid CNN-BiLSTM and Conv-BiLSTM models for time series data-sets can improve the performance of the models in transfer learning settings. This study is also a pioneer transfer learning study to electric vehicles prediction literature. Future works would involve pre-processing input data to improve the transfer learning performance, such as the weighting of the samples selected for transfer learning; that way, only important data pairs are chosen, negating instances of negative transfer learning. Source domain models can be improved by considering more input data to enhance their generalisation capacity.

References

1. Ai, S., Chakravorty, A., Rong, C.: Household EV charging demand prediction using machine and ensemble learning. In: 2018 IEEE International Conference on Energy Internet (ICEI), pp. 163–168. IEEE (2018)
2. Bahdanau, D., Cho, K., Bengio, Y.: Neural machine translation by jointly learning to align and translate. arXiv preprint arXiv:1409.0473 (2014)
3. Brownlee, J.: Deep learning for time series forecasting: predict the future with MLPs. CNNs and LSTMs in Python, Machine Learning Mastery (2018)
4. Buzna, L., De Falco, P., Khormali, S., Proto, D., Straka, M.: Electric vehicle load forecasting: A comparison between time series and machine learning approaches. In: 2019 1st International Conference on Energy Transition in the Mediterranean Area (SyNERGY MED), pp. 1–5. IEEE (2019)
5. Fan, C., Sun, Y., Xiao, F., Ma, J., Lee, D., Wang, J., Tseng, Y.C.: Statistical investigations of transfer learning-based methodology for short-term building energy predictions. Appl. Energy **262**, 114499 (2020)
6. Fan, C., Wang, J., Gang, W., Li, S.: Assessment of deep recurrent neural network-based strategies for short-term building energy predictions. Appl. energy **236**, 700–710 (2019)
7. Fawaz, H.I., Forestier, G., Weber, J., Idoumghar, L., Muller, P.A.: Transfer learning for time series classification. In: 2018 IEEE international conference on big data (Big Data), pp. 1367–1376. IEEE (2018)
8. Grubb, M., Vrolijk, C., Brack, D.: Routledge Revivals: Kyoto Protocol (1999): A Guide and Assessment. Routledge (2018)
9. Hilson, D.: Managing the impacts of renewably powered electric vehicles on distribution networks. Tech. rep., Technical Report
10. Kim, T.Y., Cho, S.B.: Predicting residential energy consumption using CNN-LSTM neural networks. Energy **182**, 72–81 (2019)
11. Krizhevsky, A., Sutskever, I., Hinton, G.E.: ImageNet classification with deep convolutional neural networks. Commun. ACM **60**(6), 84–90 (2017)

12. Li, Y., Huang, Y., Zhang, M.: Short-term load forecasting for electric vehicle charging station based on niche immunity lion algorithm and convolutional neural network. Energies **11**(5), 1253 (2018)
13. Munshi, A.A., Mohamed, Y.A.R.I.: Unsupervised nonintrusive extraction of electrical vehicle charging load patterns. IEEE Trans. Ind. Inform. **15**(1), 266–279 (2018)
14. Niu, D., Ma, T., Wang, H., Liu, H., Huang, Y.: Short-term load forecasting of electric vehicle charging station based on KPCA and CNN parameters optimized by NSGA. In: Electric Power Construction, p. 03 (2017)
15. Sajjad, M., et al.: A novel CNN-GRU-based hybrid approach for short-term residential load forecasting. IEEE Access **8**, 143759–143768 (2020)
16. Schuster, M., Paliwal, K.K.: Bidirectional recurrent neural networks. IEEE Trans. Signal Process. **45**(11), 2673–2681 (1997)
17. Sutskever, I., Vinyals, O., Le, Q.V.: Sequence to sequence learning with neural networks. arXiv preprint arXiv:1409.3215 (2014)
18. Szegedy, C., et al.: Going deeper with convolutions. In: Proceedings of the IEEE conference on computer vision and pattern recognition, pp. 1–9 (2015)
19. Ullah, F.U.M., Ullah, A., Haq, I.U., Rho, S., Baik, S.W.: Short-term prediction of residential power energy consumption via CNN and multi-layer bi-directional lstm networks. IEEE Access **8**, 123369–123380 (2019)
20. Xydas, S., Marmaras, C., Cipcigan, L.M., Hassan, A., Jenkins, N.: Electric vehicle load forecasting using data mining methods (2013)
21. Zhang, C., Yang, Z., Li, K.: Modeling of electric vehicle batteries using RBF neural networks. In: 2014 International Conference on Computing, Management and Telecommunications (ComManTel), pp. 116–121. IEEE (2014)
22. Zhang, W.G., Xie, F.X., Huang, M., Li, J., Li, Y.F.: Research on short-term load forecasting methods of electric buses charging station. Power Syst. Protec. Control **41**(4), 61–66 (2013)
23. Zhu, J., et al.: Electric vehicle charging load forecasting: a comparative study of deep learning approaches. Energies **12**(14), 2692 (2019)

Hybrid Machine Learning for Time-Series Energy Data for Enhancing Energy Efficiency in Buildings

Ngoc-Tri Ngo[1]([✉]) [ID], Anh-Duc Pham[1] [ID], Ngoc-Son Truong[1], Thi Thu Ha Truong[2], and Nhat-To Huynh[1]

[1] The University of Danang - University of Science and Technology, Danang, Vietnam
{trinn,paduc,tnson,hnto}@dut.udn.vn
[2] The University of Danang – University of Technology and Education, Danang, Vietnam
tttha@ute.udn.vn

Abstract. Buildings consume about 40% of the world's energy use. Energy efficiency in buildings is an increasing concern for the building owners. A reliable energy use prediction model is crucial for decision-makers. This study proposed a hybrid machine learning model for predicting one-day-ahead time-series electricity use data in buildings. The proposed SAMFOR model combined support vector regression (SVR) and firefly algorithm (FA) with conventional time-series seasonal autoregressive integrated moving average (SARIMA) forecasting model. Large datasets of electricity use in office buildings in Vietnam were used to develop the forecasting model. Results show that the proposed SAMFOR model was more effective than the baselines machine learning models. The proposed model has the lowest errors, which yielded 0.90 kWh in RMSE, 0.96 kWh in MAE, 9.04% in MAPE, 0.904 in R in the test phase. The prediction results provide building managers with useful information to enhance energy-saving solutions.

Keywords: Energy consumption data · Machine learning · Data analytics · Prediction model

1 Introduction

Buildings consumes about 40% of the world's energy use and 30% of carbon dioxide generation [1, 2]. Saving electricity consumption in buildings is valuable [3]. Energy-saving solutions in buildings have been attracted concerns of various researches [4]. Forecasting future energy consumption can provide a reference for users or building managers to save their energy use. Future energy data prediction is a method of projecting future data based on historical time-series data.

Building energy data is recognized as time-series data that vary along with hourly or daily - based timestamps. Statistics-based methods and machine learning (ML) methods have been developed for predicting time-series data. An autoregressive integrated moving average (ARIMA) is an example of powerful statistical methods [5]. However, ARIMA models are suitable for modeling the linear relationship between inputs and outputs.

© Springer Nature Switzerland AG 2021
M. Paszynski et al. (Eds.): ICCS 2021, LNCS 12746, pp. 273–285, 2021.
https://doi.org/10.1007/978-3-030-77977-1_21

Energy consumption prediction is difficult because it is affected uncertainly by occupant's behaviors [6]. Because the nature of the energy use exhibits the complex and seasonal pattern, the unreliable forecast may result in an additional production or waste of resources [7]. Meanwhile, machine learning (ML) have been increasingly used in various domains [8]. Prediction models were developed based on a single machine learning model, ensemble ML models such as XGboost, the feedforward deep networks (FDN) [9], and hybrid ML models [10].

Artificial neural networks (ANNs) models were used to forecast building electricity consumption [11]. The integration of ANNs and ARIMA models was proposed for predicting time-series data [12]. Support vector regression (SVR) was used to forecast the hourly cooling energy demand in office buildings [13]. The SVR was combined with the genetic algorithm (GA) to forecast energy use [14]. However, the SVR is relatively slow in dealing with huge data [15] and a high computational burden [16]. The least-squares support vector regression (LSSVR) [17] is also widely used for prediction problems because it can reduce the computational effort [18].

Pham et al. (2020) proposed the random forests (RF)–based ML model for forecasting short-term electricity use patterns in buildings [4]. Ngo (2019) has investigated the effectiveness of various single and ensemble approaches for building energy simulation and prediction [19]. Although ML models have been used to develop prediction models in previous works, few studies have used a hybrid approach. This study proposed a hybrid machine learning model to predicting one-day-ahead energy use in office buildings. The proposed model combined the SARIMA model, the LSSVR model, and the firefly algorithm (FA). The proposed hybrid ML model can learn the linear and nonlinear patterns in energy data. The model can involve the temporal data and weather data, and historical energy data as the inputs. Energy consumption data in Vietnam was used to evaluate the proposed model.

2 Hybrid Machine Learning Model

2.1 SARIMA Model

Seasonal AR and MA terms in the SARIMA model predict energy consumption in building y_t by using data values and errors at previous periods with lags that are multiples of the seasonality length S. The SARIMA$(p, d, q) \times (P, D, Q)_S$, is a multiplicative model that consists of nonseasonal and seasonal elements. Equation (1) presents the mathematical expression of the SARIMA model as described in [20, 21]. The terms of the model are expressed in Eqs. (2)–(5) [5].

$$\theta_p(B)\Theta_P(B^S)(1 - B)^d(1 - B^S)^D y_t = w_q(B)W_Q(B^S)\alpha_t \tag{1}$$

$$w_q(B) = 1 - w_1 B - w_2 B_2 - w_3 B^3 - \dots - w_q B^q \tag{2}$$

$$\theta_p(B) = 1 - \theta_1 B - \theta_2 B^2 - \theta_3 B^3 - \dots - \theta_p B^p \tag{3}$$

$$\Theta_P(B^S) = 1 - \Theta_1(B^S) - \Theta_2(B^{2S}) - \Theta_3(B^{3S}) - \dots - \Theta_P(B^{PS}) \tag{4}$$

$$W_Q(B^S) = 1 - W_1(B^S) - W_2(B^{2S}) - W_3(B^{3S}) - \ldots - W_Q(B^{QS}) \tag{5}$$

where p is the nonseasonal AR; d is nonseasonal differencing; q is the nonseasonal MA; P is the seasonal AR; D is seasonal differencing; Q is the seasonal MA order; S is the season length; B is the backward shift operator; $w_q(B)$, $\theta_p(B)$, $\Theta_P(B^S)$, and $W_Q(B^S)$ are polynomials in B; y_t is the actual value at the time t; α_t is the estimated residual at the time t; d, q, P, D, Q are integers.

2.2 Support Vector Regression Model

LSSVR models were developed [17] to deal with the large data sets. During the training phase, the least squares cost function was used to get a linear set of equations in a dual space. Then, the conjugate gradient method is used to derive a solution by efficiently solving a set of linear equations [22]. Given a training data set $\{x_k, y_k\}_{k=1}^N$, function estimation using LSSVR is formulated as an optimization problem, as expressed in the Eq. (6):

$$\min_{\omega, b, e} J(\omega, e) = \frac{1}{2}\|\omega\|^2 + \frac{1}{2}C\sum_{k=1}^N e_k^2; \text{ subject to } y_k = \langle \omega, \varphi(x_k) \rangle + b + e_k, \ k = 1, \ldots N \tag{6}$$

where $J(\omega, e)$ denotes the optimization function, ω denotes the linear approximator parameter, $e_k \in R$ denote error variables, $C \geq 0$ denotes a regularization constant specifying the constant representing the trade-off between empirical error and function flatness, x_k denotes input patterns, y_k denotes prediction outputs, and N denotes the sample size.

The resulting LSSVR model for function estimation is expressed as Eq. (7).

$$f(x) = \sum_{k=1}^N \alpha_k K(x, x_k) + b \tag{7}$$

where α_k, b denote the Lagrange multipliers and the bias term, respectively, and $K(x, x_k)$ denotes the kernel function. The Gaussian radial basis function (RBF) in the LSSVR model is expressed mathematically in Eq. (8)

$$K(x, x_k) = \exp(-\|x - x_k\|^2 / 2\sigma^2) \tag{8}$$

where σ is the RBF width (Fig. 1).

2.3 Firefly – Based Optimization Algorithm

A firefly algorithm is a nature-inspired metaheuristic algorithm [23] which is potential to identify the global solution and local solution. The FA operation is based on three main principles: a firefly is attracted to other fireflies; the brightness of fireflies impacts its attractiveness regarding the distance among fireflies, and the brightness is affected

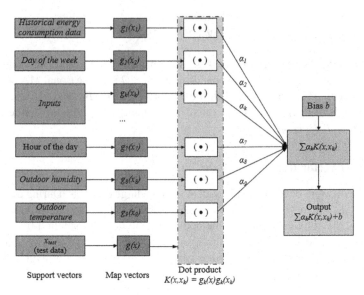

Fig. 1. Structure of SVR model for regression.

by the search space of the optimization problems. The attractiveness of a firefly β can be expressed as Eq. (9). The distance between fireflies i and j is calculated as Eq. (10):

$$\beta = \beta_0 e^{-\gamma r^2} \tag{9}$$

$$r_{ij} = \|x_i - x_j\| = \sqrt{\sum_{k=1}^{d} (x_{i,k} - x_{j,k})^2} \tag{10}$$

where β is the attractiveness of a firefly; β_0 is the attractiveness at $r = 0$; r is the distance between the firefly and other fireflies; e is a constant coefficient; γ is the absorption coefficient; r_{ij} is the distance between any fireflies i and j at x_i and x_j, respectively; $x_{i,k}$ is the kth component of the spatial coordinate x_i of the ith firefly; $x_{j,k}$ is the kth component of the spatial coordinate x_j of the jth firefly; and d is the number of dimensions in the search space.

An optimal solution is affected by the movement of fireflies during the optimization process. The movement of a firefly is expressed as Eq. (11)

$$x_i^{t+1} = x_i^t + \beta_0 e^{-\gamma r_{ij}^2}(x_j^t - x_i^t) + \alpha^t \theta_i^t \tag{11}$$

where x_i^{t+1} is the position of the ith firefly; x_i^t is the position of the ith firefly; x_j^t is the position of the jth firefly; α^t is a randomization parameter; and θ_i^t is random numbers.

To improve the performance of the FA, this study adopted the modified version of FA that was developed by Chou and Ngo [24]. A Gauss/mouse map was applied to change an attractiveness parameter while a logistic map in the modified FA generates a diverse population of fireflies. The adaptive inertia weight (AIW) was adopted to vary the randomization parameter α, which can improve the local exploitation and the global exploration during the progress of the optimization process. Moreover, Lévy flights facilitate local exploitation. Figure 2 reveals the pseudocode of the modified FA.

```
Begin
Define objective function f(x), x = (x1, ..., xd)^T
Set search space and iterations number
Fireflies population is generated by logistic chaotic map xi (i = 1, 2,..., n)
Light intensity Ii at xi is determined by f(xi)
Define the light absorption coefficient
Initial generation, t = 0
while (t ≤maximum iteration) do
  Vary value of α using AIW
  Tune value of β using Gauss/mouse chaotic map
       for i = 1: n
               for j = 1: n
if (light intensity j> light intensity i)
  Firefly i moves to firefly j using Eq. (11) with the addition of Lévy flight;
end if
                       Calculate attractiveness with distance r via exp[-γ*r]
                       Evaluate new solutions and update light intensity
end for j
end for i
       Rank and confirm the current optimum
end while
Export optimal solutions
End
```

Fig. 2. Pseudocode of modified firefly algorithm.

2.4 Proposed Hybrid Machine Learning Model

Figure 3 depicts the two-stage flowchart of the proposed model in predicting time-series energy consumption in buildings. The energy data consists of linear and nonlinear parts, as shown in Eq. (12). In the first stage, the historical energy data was input to the linear SARIMA model to infer the linear building energy consumption data. In the second stage, the nonlinear FA-SVR was used to forecast the nonlinear building energy consumption.

$$Y_t = L_t + N_t \tag{12}$$

where Y_t represents the building energy consumption data, L_t and N_t represent the linear part and the nonlinear part in building energy consumption data, respectively.

Equation (13) depicts the predictive results obtained by the SARIMA model in which the linear part in building energy consumption data is modeled as the predicted building energy consumption ($\hat{L}_t$) and residual values o (R_t). As shown in Fig. 3, the inputs in the 1st stage are only historical building energy consumption data.

$$L_t = \hat{L}_t + R_t \tag{13}$$

where $\hat{L}_t$ are the forecasted values by the SARIMA model and R_t are the residual values.

The final prediction results of future building energy consumption were performed in the 2nd stage by the FA-SVR model. Inputs for this stage consists of the forecasted values $\hat{L}_t$, historical building energy consumption, temporal data (i.e., day of the week – DoW and hour of the day – HoD), and weather data (i.e., outdoor temperature and humidity data). Therefore, the forecasted results of building energy consumption were presented as Eq. (14)

$$Y_t = (DoW_t, HoD_t, T_t, H_t, \hat{L}_t, Y_{t-1}, Y_{t-2}, ..., Y_{t-lag}) \tag{14}$$

where DoW_t is the day of the week; HoD_t is the hour of the day; T_t is outdoor temperature; H_t is outdoor humidity data; Y_{t-1} is building energy consumption value at the time $t-1$; Y_{t-lag} is the time $(t-lag)$.

The nonlinear time-series prediction model was built based on the integration of the SVR model and the FA optimization algorithm (SVR-FA). This integration can significantly improve the predictive performance of the proposed model because the configuration of the SVR model was optimized automatically to fit with data patterns.

The proposed model was experienced the learning phase and test phase using various data sets from real-world buildings. Particularly, the proposed model was learned and tested multiple times. During an evaluation, the learning data were to build the time-series prediction model for building energy consumption in the learning phase.

The SARIMA projected the predicted linear building energy consumption in the 1st stage based on the learning data. At the 2nd stage, the proportion of learning data (i.e., 70% of the total size of the learning data) was applied to train the SVR model while the remaining proportion of the learning data (i.e., 30%) was used to optimize the predictive accuracy of the proposed model via the optimization process by the FA. The FA optimized the optimal hyperparameters of the SVR in the search space via the objective function. In this study, the root-mean-square error (RMSE) was used as the OF for the optimization problem. The RMSE was calculated upon the collected actual building energy consumption data and predicted building energy consumption data. After the learning phase, the learned predicted model was produced. The accuracy of the learned model was tested in the test phase. The test data include the 24-h building energy consumption data.

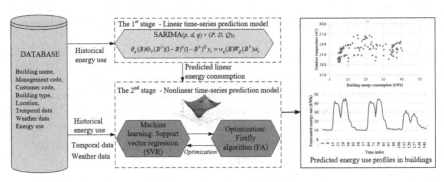

Fig. 3. Flowchart of the proposed SAMFOR model.

3   Dataset and Model Evaluation Results

3.1   Dataset

30-min energy consumption data and weather data were collected from three office buildings in Vietnam. In years of 2018 and 2019. Each dataset consists of 35,040 data points. Three buildings were selected to test model accuracy. Their energy use profiles for the years of 2018 and 2019 were plotted in Fig. 4. Table 1 summaries the descriptive analysis of these data. These buildings are office buildings. Building 1 is a software development center while building 2 is for a logistic company. Building 3 is working space for a construction company in Danang. The energy use of these buildings is mainly from the air-conditioning system, lighting, and electric appliances.

Table 1. Data description of case studies.

Dataset	30-minutely energy consumption Y (kWh)				Outdoor temperature T (°C)			
	Min	Ave.	Max	Std. dev.	Min	Ave.	Max	Std. dev.
Office building 1	1.74	35.08	144.79	31.49	15.5	27.1	39.3	3.9
Office building 2	0.00	7.25	21.23	2.63	15.5	27.1	39.3	3.9
Office building 3	0.06	4.15	30.10	4.30	15.5	27.1	39.3	3.9

3.2   Model Evaluation Results

The SARIMA model was set as SARIMA $(1, 0, 1) \times (48, 0, 48)_{48}$. The seasonal length was set as 48 which consists of a recorded number of data points in a day. The search space for C and σ were set in the range of $[10^{-3}\ 10^{12}]$. The firefly's population and maximum iteration were set at 50 and 25, respectively. The proposed model was evaluated 12 times using learning data and test data as shown in Table 2.

Table 2. Data settings for evaluations of all buildings.

Evaluation	Learning data (4-month historical data)	Test data (one-day-ahead data)
1	January 8–May 7, 2018	May 8, 2018 (Tuesday)
2	June 15–October 14, 2018	October 15, 2018 (Monday)
3	April 1–July 31, 2019	August 1, 2019 (Thursday)
4	August 27–Dec. 26, 2019	December 27, 2019 (Friday)

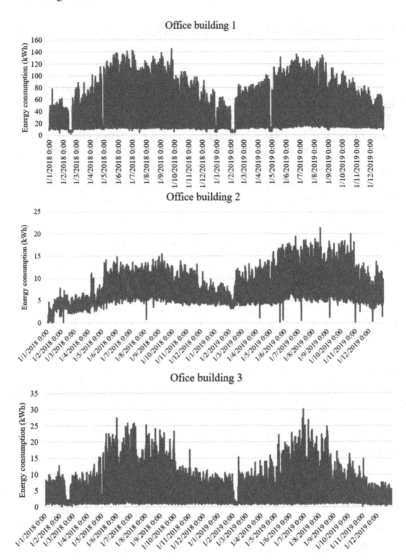

Fig. 4. Energy consumption in the buildings in the years of 2018 and 2019.

Table 3 depicts the accuracy achieved by the proposed hybrid SAMFOR model. In the learning phase, the average accuracy in three office buildings were 0.89 kWh in mean-square-error (RMSE), 0.91 kWh in mean absolute error (MAE), 10.28% in the mean absolute percentage error (MAPE), 0.975 in the correlation coefficient (R). In the test phase, the hybrid machine learning model yielded 0.90 kWh in RMSE, 0.96 kWh in MAE, 9.04% in MAPE, 0.904 in R. Figure 5 visualizes the actual energy data and predicted energy data obtained by the SAMFOR. The results revealed that the proposed model was effective in predicting 30-hourly consumed energy data in office buildings. The SAMFOR model was performed in a desktop with the Intel (R) Core (TM) i7-9750H

CPU and the RAM of 8.00 GB. The running CPU time was about 1 min. Running time of the SVR model was 26 s while that of SARIMA was 20 s.

Table 3. Performance results by SAMFOR model for three office buildings in the learning phase and test phase.

Dataset	Evaluation	Accuracy by SAMFOR in learning phase				Performance by SAMFOR in test phase			
		RMSE (kWh)	MAE (kWh)	MAPE (%)	R	RMSE (kWh)	MAE (kWh)	MAPE (%)	R
Office building 1	1	1.24	1.54	6.77	0.993	1.69	2.86	5.95	0.996
	2	1.54	2.36	6.63	0.995	1.40	1.97	5.20	0.996
	3	1.39	1.94	5.76	0.996	1.44	2.08	4.45	0.996
	4	1.27	1.61	5.22	0.995	1.11	1.23	3.65	0.995
	Average	1.36	1.86	6.10	0.995	1.41	2.04	4.81	0.996
Office building 2	1	0.51	0.26	24.19	0.960	0.71	0.50	6.43	0.922
	2	0.64	0.41	5.17	0.933	0.55	0.30	4.87	0.904
	3	0.71	0.50	8.48	0.953	0.71	0.50	7.25	0.782
	4	0.70	0.49	7.09	0.958	0.48	0.23	4.19	0.955
	Average	0.64	0.42	11.23	0.951	0.61	0.38	5.69	0.891
Office building 3	1	0.59	0.34	13.73	0.975	0.88	0.77	14.71	0.986
	2	0.76	0.58	13.57	0.983	0.54	0.29	28.92	0.407
	3	0.74	0.55	14.03	0.983	0.79	0.62	13.19	0.966
	4	0.57	0.33	12.71	0.975	0.47	0.22	9.65	0.946
	Average	0.67	0.45	13.51	0.979	0.67	0.48	16.62	0.827
Overall average		0.89	0.91	10.28	0.975	0.90	0.96	9.04	0.904
Std. dev		0.36	0.74	5.62	0.020	0.42	0.88	7.20	0.168

In Table 4, performance of the proposed SAMFOR model was compared against the SARIMA and SVR models. Scatter plots of actual and predicted energy data produced by the SAMFOR, SARIMA, and SVR models were presented in Fig. 6. For predicting energy consumption in office buildings, the SARIMA model obtained 44.08 kWh in RMSE, 36.94 kWh in MAE, 59.19% in MAPE, and 0.806 in R. The SVR model was better than the SARIMA model, which yielded 11.70 kWh in RMSE, 5.78 kWh in MAE, 9.89% in MAPE, and 0.909 in R. Comparison results show that the SAMFOR model was more effective than the SARIMA and SVR models in forecasting 30-min energy consumption in office buildings. The proposed model has the lowest errors with 0.90 kWh in the RMSE, 0.96 kWh in the MAE, and 9.04% in the MAPE. The hybrid machine learning model enhanced significantly the predictive accuracy compared to other investigated models. Therefore, SAMFOR model was suggested as a forecasting model in predicting energy consumption in office buildings.

Table 4. Performance comparison among base models and proposed model.

Prediction model	Accuracy measures			
	RMSE (kWh)	MAE (kWh)	MAPE (%)	R
SARIMA	44.08	36.94	58.19	0.806
SVR	11.70	5.78	9.89	0.909
Proposed SAMFOR	0.90	0.96	9.04	0.904

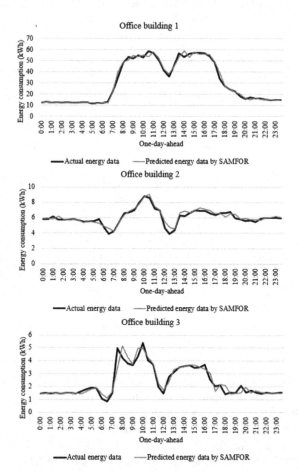

Fig. 5. Actual and predicted energy data by the SAMFOR for office buildings.

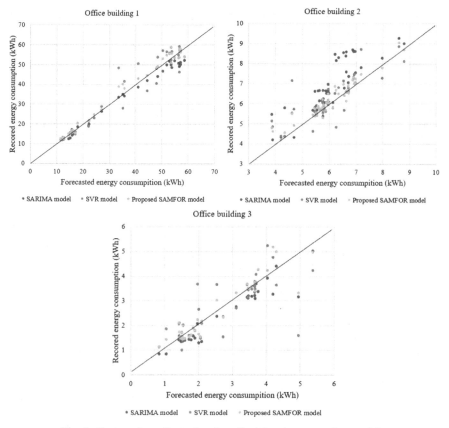

Fig. 6. Scatter plots of actual and predited data by comparing models.

4 Conclusions

This study proposed a hybrid machine learning forecasting model for forecasting energy consumption in office buildings. The proposed SAMFOR model was developed based on the SARIMA, SVR, and FA. A large dataset of 30-min energy consumption from three office buildings in Vietnam was applied to develop and evaluate the proposed model. the hybrid machine learning model yielded 0.90 kWh in RMSE, 0.96 kWh in MAE, 9.04% in MAPE, 0.904 in R in the test phase.

Performance of the proposed SAMFOR model was compared against the SARIMA and SVR models. Comparison results show that the SAMFOR model was more effective than the SARIMA and SVR models in forecasting 30-min energy consumption in office buildings. Therefore, SAMFOR model was suggested as a forecasting model in predicting energy consumption in office buildings.

The contribution of the study was the effective prediction model in forecasting the one-day-ahead energy data in office buildings. The power of the hybrid approach comes from taking advantages of a linear model and a nonlinear model, in which an optimization algorithm was applied to fine tune the configuration of the proposed model.

The predicted results provide building managers and users with useful information to enhance effectiveness of energy use in office buildings. Future studies may perform a comparative analysis between the proposed model and other deep learning models such as a convolutional neural network. Besides, sensitivity analysis on the model parameters should be studied in the future to provide more convincible prediction results.

Acknowledgements. This work was funded by Gia Lam Urban Development and Investment Company Limited, Vingroup and supported by Vingroup Innovation Foundation (VINIF) under project code VINIF.2019.DA05.

References

1. Klein, L., Kwak, J.-Y., Kavulya, G., Jazizadeh, F., Becerik-Gerber, B., Varakantham, P., et al.: Coordinating occupant behavior for building energy and comfort management using multi-agent systems. Autom. Constr. **22**, 525–536 (2012)
2. Allouhi, A., El Fouih, Y., Kousksou, T., Jamil, A., Zeraouli, Y., Mourad, Y.: Energy consumption and efficiency in buildings: current status and future trends. J. Clean. Prod. **109**, 118–130 (2015)
3. Mousavi, A., Vyatkin, V.: Energy efficient agent function block: a semantic agent approach to IEC 61499 function blocks in energy efficient building automation systems. Autom. Constr. **54**, 127–142 (2015)
4. Pham, A.-D., Ngo, N.-T., Ha Truong, T.T., Huynh, N.-T., Truong, N.-S.: Predicting energy consumption in multiple buildings using machine learning for improving energy efficiency and sustainability. J. Clean. Prod. **260**, 121082 (2020)
5. Box, G.E.P., Jenkins, G.M.: Time Series Analysis: Forecasting and Control, 3rd edn. Holden-Day, San Francisco (1970)
6. Shen, M., Lu, Y., Wei, K.H., Cui, Q.: Prediction of household electricity consumption and effectiveness of concerted intervention strategies based on occupant behaviour and personality traits. Renew. Sustain. Energy Rev. **127**, 109839 (2020)
7. Li, R., Jiang, P., Yang, H., Li, C.: A novel hybrid forecasting scheme for electricity demand time series. Sustain. Urban Areas **55**, 102036 (2020)
8. Eligüzel, N., Çetinkaya, C., Dereli, T.: Comparison of different machine learning techniques on location extraction by utilizing geo-tagged tweets: a case study. Adv. Eng. Inform. **46**, 101151 (2020)
9. Chen, K., Jiang, J., Zheng, F., Chen, K.: A novel data-driven approach for residential electricity consumption prediction based on ensemble learning. Energy **150**, 49–60 (2018)
10. Nguyen, T.-D., Tran, T.-H., Hoang, N.-D.: Prediction of interface yield stress and plastic viscosity of fresh concrete using a hybrid machine learning approach. Adv. Eng. Inform. **44**, 101057 (2020)
11. Kalogirou, S.A., Bojic, M.: Artificial neural networks for the prediction of the energy consumption of a passive solar building. Energy **25**, 479–491 (2000)
12. Khashei, M., Bijari, M.: A novel hybridization of artificial neural networks and ARIMA models for time series forecasting. Appl. Soft Comput. **11**, 2664–2675 (2011)
13. Li, Q., Meng, Q., Cai, J., Yoshino, H., Mochida, A.: Applying support vector machine to predict hourly cooling load in the building. Appl. Energy **86**, 2249–2256 (2009)
14. Jung, H.C., Kim, J.S., Heo, H.: Prediction of building energy consumption using an improved real coded genetic algorithm based least squares support vector machine approach. Energy Build. **90**, 76–84 (2015)

15. Su, S., Zhang, W., Zhao, S.: Fault prediction for nonlinear system using sliding ARMA combined with online LS-SVR. Math. Probl. Eng. **2014**, 9 (2014)

16. Wang, H., Hu, D.: Comparison of SVM and LS-SVM for regression. In: International Conference on Neural Networks and Brain, pp. 279–283. IEEE (2005)

17. Suykens, J.A.K., Gestel, T.V., Brabanter, J.D., Moor, B.D., Vandewalle, J.: Least Squares Support Vector Machines. World Scientific, Singapore (2002)

18. Chou, J.-S., Ngo, N.-T., Pham, A.-D.: Shear strength prediction in reinforced concrete deep beams using nature-inspired metaheuristic support vector regression. J. Comput. Civ. Eng. **30**, 04015002 (2015)

19. Ngo, N.-T.: Early predicting cooling loads for energy-efficient design in office buildings by machine learning. Energy Build. **182**, 264–273 (2019)

20. Tseng, F.-M., Tzeng, G.-H.: A fuzzy seasonal ARIMA model for forecasting. Fuzzy Sets Syst. **126**, 367–376 (2002)

21. Wang, Y., Wang, J., Zhao, G., Dong, Y.: Application of residual modification approach in seasonal ARIMA for electricity demand forecasting: a case study of China. Energy Policy **48**, 284–294 (2012)

22. Shamshirband, S., Mohammadi, K., Yee, P.L., Petković, D., Mostafaeipour, A.: A comparative evaluation for identifying the suitability of extreme learning machine to predict horizontal global solar radiation. Renew. Sustain. Energy Rev. **52**, 1031–1042 (2015)

23. Yang, X.-S.: Firefly algorithm. Luniver Press, Bristol, UK (2008)

24. Chou, J.-S., Ngo, N.-T.: Modified firefly algorithm for multidimensional optimization in structural design problems. Struct. Multidiscip. Optim. **55**(6), 2013–2028 (2016). https://doi.org/10.1007/s00158-016-1624-x

I-80 Closures: An Autonomous Machine Learning Approach

Clay Carper[1], Aaron McClellan[1], and Craig C. Douglas[2(✉)]

[1] Department of Computer Science, University of Wyoming,
Laramie, WY 82071, USA
{ccarper2,amcclel2}@uwyo.edu
[2] School of Energy Resources, Department of Mathematics and Statistics,
University of Wyoming, Laramie, WY 82072-3036, USA

Abstract. Road closures due to adverse and severe weather continue to affect Wyoming due to hazardous driving conditions and temporarily suspending interstate commerce. The mountain ranges and elevation in Wyoming makes generating accurate predictions challenging, both from a meteorological and machine learning stand point. In a continuation of prior research, we investigate the 80 km stretch of Interstate-80 between Laramie and Cheyenne using autonomous machine learning to create an improved model that yields a 10% increase in closure prediction accuracy. We explore both serial and parallel implementations run on a supercomputer. We apply auto-sklearn, a popular and well documented autonomous machine learning toolkit, to generate a model utilizing ensemble learning. In the previous study, we applied a linear support vector machine with ensemble learning. We will compare our new found results to previous results.

Keywords: Machine learning · Autonomous machine learning · Road closure

1 Introduction

In 2018, the first author commuted between Cheyenne and the University of Wyoming in Laramie, Wyoming. During this nine month period it became apparent while driving between the two towns there were inconsistent classifications concerning road closures. Road conditions sometimes were severe and dangerous while the roadway was classified as open. Road conditions sometimes were clearly safe while the roadway was classified as closed. The experience overall was one that is shared with anyone who regularly travels the corridor between these two towns.

This research was supported in part by National Science Foundation grant DMS-1722621.

M. Paszynski et al. (Eds.): ICCS 2021, LNCS 12746, pp. 286–291, 2021.
https://doi.org/10.1007/978-3-030-77977-1_22

Featuring a maximum elevation differential of nearly 800 m between Cheyenne and Laramie (1,848 and 2,184 m above sea level, respectively), this section of Interstate-80 (I-80) is a predictive nightmare. The misclassification many Wyoming residents and frequent visitors have grown accustomed to leads to needless loss of life, accidents, higher insurances rates for Wyoming residents, and delays in interstate commerce. Prior research [5,6] has shown that applying machine learning to locally obtained sensor data provides a viable solution to this misclassification problem.

In this paper we will review previous work and results, the details of our application of autonomous using the University of Wyoming's Advanced Research Computing Center (ARCC) [1], our results and comparisons to prior results, and the implication of our new results and future work in Sects. 2, 3, 4, and 5, respectively.

2   Prior Work

This project begin as a class project in fall 2018. To date, two documents have been published: an international conference paper [5] and a master's thesis [6].

The first publication was in the Proceedings of the 2019 18th International Symposium on Distributed Computing and Applications for Business Engineering and Science (DCABES), Wuhan, China. In this work, we first obtained raw sensor data on the roadway between Laramie and Cheyenne, Wyoming from MesoWest [10]. This data consisted of 29 individual quantities measured at each sensor at irregular time steps and was reduced to six parameters: air temperature, relative humidity, wind speed, wind bust, visibility, and dew point. This was done due to sparsity of parameters, usability, and principle component analysis. Sparse parameters were omitted due to skewing concerns. While replacing missing parameters with the median is standard practice, doing so on millions of missing entries will undoubtedly skew the data in an undesirable fashion.

The weather condition was a feature among the cut parameters. Although this parameter was dense enough to be viable, the datatype presented issues. Since the weather condition parameter contained plaintext values such as *snow squall*, *thunderstorm*, and *mostly cloudy*, we were unable to assign meaningful and consistent numerical weights. Such an endeavor would require insight into meteorology beyond the scope of our work. Applying a one-hot encoding of these parameters may have been possible but would have drastically increased the number of features and each one would be very sparse.

There are many was to ensure selected features maintain predictive power. In this case, we ensured the selected features maintained at least 90% of the energy in the system using principle compenent analysis. To supplement the sensor data, we obtained closure status data from the Wyoming Department of Transportation [12]. This data consisted of binary classifications of weather related roadway closures and ad hoc roadway closures. For consistency, we only used the weather related roadway closures. Using the scikit-learn machine learning framework, we trained a linear support vector machine (LSVM) on a subset of the aforementioned data. Such a model yielded a maximum accuracy of 71% [5].

Further work occurred for the first author's master's thesis. Ensemble learning was the major addition to the project [2,4,9]. Using scikit-learn's implementation of AdaBoost, we obtained a 10% boost in accuracy for a maximum of 81% [6]. Boosting is a common form of ensemble learning, allowing for additional copies of the original classifier to be applied to the same subset of data. This yielded a 13% improvement over a single linear support vector machine model. Additionally, we performed confusion matrix analysis for the standard and boosted models.

Rationale behind all choices can be found in prior publications [5,6], while results from the previous publications can be found in Table 1. Note throughout both analyses, cross validation was the metric of choice.

Table 1. Prior models, cross validation confusion matrix

Model	True negative	False positive	False negative	True positive
Standard	59.05%	10.42%	18.56%	11.97%
Boosted	64.49%	4.98%	14.57%	15.95%

3   Methods

In this section, we present the tools used in this research and their versions, our reasons for choosing these tools, and the methods that led to an improved model. We strive to provide research that is verifiable and reproducible, thus everything needed to recreate our results is provided. Not only do we describe our tools, we include the versions of all software used, the versions of all tools' immediate (but not transient) dependencies, our code and data, and sample execution scripts for use on supercomputers.

The code was written in Python, the most prevalent programming language in academic data science research. This choice was necessitated by the fact that our other tools only provide Python APIs, leaving little choice for other languages.

Our most important tool is auto-sklearn [8]. It is an autonomous machine learning (AML) framework, which attempts to find the best machine learning model automatically with little or no guidance from the researcher. The tool auto-sklearn has undergone stringent comparisons with other AML frameworks and compares favorably in most categories [8]. In order to automatically find optimal models, auto-sklearn creates a parameter space that models can be selected from. Then it algorithmically searches through the parameter space until a termination condition is met. For our purposes, there are two termination conditions: auto-sklearn exhausts its allocated execution time or an optimal model is found (for a user-specified definition of optimal). Several optimality metrics are included with the framework.

The University of Wyoming's supercomputer, ARCC [1], is our execution platform of choice. The ability to run our code in an environment where compute resources were effectively unconstrained allowed for shorter iteration times and the discovery of higher quality models.

As is common for supercomputers, ARCC uses a software scheduler to allocate compute resources. We provide sample scripts to run our code for use with ARCC's scheduler, Slurm [1]. Since many supercomputers schedule with Slurm, we hope our scripts allow researchers to verify and reproduce our results on their supercomputer of choice.

Five individual iterations were run, using one, two, four, eight, and sixteen CPUs. The reason for this is partially because scikit-learn is built on top of Dask [7] and partially due to the nature of auto-sklearn. Dask allows Python programs to be scalable across multiple nodes by generating a Dask cluster. In doing so, it organizes the workers and handles the management of the cluster. Further, in allocating additional CPUs with a fixed amount of time we are simply running additional models. This does not change running times but yields greater accuracy by checking additional types and variations of models.

Since software is constantly changing, we provide a Conda environment containing all the versions of all software described and their dependencies. At this time, the code and data is available upon request.

4 Results

To benchmark an algorithm using auto-sklearn, we use a fixed subset of data that has previously been tested on. This data is available upon request. For our testing, we used the `AutoSklearnClassifier` [8] with randomized training and test sets. scikit-learn [3,11] provides a function called `sklearn.metrics.accuracy_score`, which we used to determine the accuracy of the model. Doing so yields an average accuracy of nearly 91%, an improvement of 22% over the base LSVM model [5] and an improvement of 11% over the ADABoosted LSVM [6]. It is important to note that auto-sklearn automatically applies ensemble learning, allowing for a linearly-boosted model to be one of the possible parameters [8]. Accuracy for each of the five variations can be found in Table 2. The number of models checked are the number of models that ran successfully. Instances of failed models are either models that crash, exceed the time limit, or exceed the memory limit. Memory usage is the amount of memory reported by Slurm. It is worth noting that the memory allocated for each experiment was 80–100 GB. We are yet to identify why this is necessary, however, without allocating a significant amount of additional memory the multi-core instances will fail to produce output.

The models auto-sklearn found is an ensemble consisting of nine individual models, each with their own weights. Of the models found, 70% of the predictive power is encapsulated in the first three models. Those models are two different ADABoosted linear models and an extra trees model, respectively. The remaining models are a collection of tree-based algorithms and k-nearest neighbors.

Table 2. Accuracy for auto-sklearn models and number of models checked

CPUs	Accuracy	Models evaluated	Memory usage (GB)
1	90.69%	72	2.68
2	90.83%	101	4.99
4	90.96%	161	8.16
8	90.85%	191	10.13
16	90.76%	255	14.82

More specifically, they are three random forest models, another ADABoosted linear model, a quadratic discriminant analysis, and k-nearest neighbors.

Another typical metric for analyzing machine learning algorithms is confusion matrix analysis. We elected to do so in this research, the results of which are given in Table 3. Please note that in this research, negative refers to a prediction of a nonclosure and positive refers to a closure.

Table 3. Confusion matrices for auto-sklearn models

CPUs	True negative	False positive	False negative	True positive
1	67.55%	1.93%	3.10%	27.43%
2	67.72%	1.75%	3.22%	27.31%
4	67.65%	1.82%	3.13%	27.39%
8	67.74%	1.74%	3.18%	27.34%
16	67.67%	1.81%	3.15%	27.38%

Interestingly, an average of 35% of the incorrect classifications are related to predicting the roadway being closed when it in fact was open and on average 65% of the incorrect classifications are related to predicting the roadway being open when it was in fact closed. This is nearly identical to previous confusion matrix values of 35.96% and 64.04%, respectively. It is worth noting that the variations in the confusion matrices is within variational norms; adding models doesn't necessarily imply an increase in accuracy of predictions.

5   Conclusions and Future Work

As a continuation of previous work, an improvement of 11% in accuracy is substantial, with an average accuracy of 91%. Similar results in the confusion matrix lead us to conclude that the application of auto-sklearn is a viable and meaningful next step for classifying roadway closures. Future directions for this project include dataset subset analysis and validation and automation.

References

1. Advanced Research Computing Center: Teton Computing Environment. Intel x86_64 cluster (2020). https://doi.org/10.15786/M2FY47
2. Aurélien, G.: Hands-on Machine Learning with Scikit-Learn and TensorFlow: Concepts, Tools, and Techniques to Build Intelligent Systems. OReilly Media, Sebastopol, CA (2017)
3. Buitinck, L., et al.: API design for machine learning software: experiences from the scikit-learn project. In: ECML PKDD Workshop: Languages for Data Mining and Machine Learning, pp. 108–122 (2013)
4. Burkov, A.: The Hundred-Page Machine Learning Book. Andriy Burkov (2019)
5. Carper, C., Douglas, C.C.: I-80 closures: a support vector machine model. In: Yucheng, G. (eds.) Proceedings of DCABES 2019. Wuhan, China, 8–10 November, IEEE Computer Society Press, Los Alamitos, CA, pp 199–202 (2019)
6. Carper, C.: A support vector machine model for predicting closures on interstate 80. Master's thesis, University of Wyoming, Laramie, WY (2020)
7. Dask: https://dask.org/. Accessed 18 Feb 2021
8. Feurer, M., Klein, A., Eggensperger, K., Springenberg, J., Blum, M., Hutter, F.: Efficient and robust automated machine learning. In: Cortes, C., Lawrence, N.D., Lee, D.D., Sugiyama, M., Garnett, R. (eds.) Advances in Neural Information Processing Systems 28, pp. 2962–2970. Curran Associates, Inc. (2015). http://papers.nips.cc/paper/5872-efficient-and-robust-automated-machine-learning.pdf
9. Leskovec, J., Rajaraman, A., Ullman, J.D.: Mining of Massive Datasets, 3rd edn. Cambridge University Press, New York, NY (2020)
10. MesoWest Data: https://mesowest.utah.edu/. Accessed 10 Dec 2020
11. Pedregosa, F., et al.: Scikit-learn: machine learning in Python. J. Mach. Learn. Res. **12**, 2825–2830 (2011)
12. Wyoming Department of Transportation: http://www.dot.state.wy.us/home.html. Accessed 10 Dec 2020

Energy Consumption Prediction for Multi-functional Buildings Using Convolutional Bidirectional Recurrent Neural Networks

Paul Banda[✉], Muhammed A. Bhuiyan, Kevin Zhang, and Andy Song

RMIT University, Melbourne 3000, Australia
paul.banda@student.rmit.edu.au

Abstract. In this paper, a Conv-BiLSTM hybrid architecture is proposed to improve building energy consumption reconstruction of a new multi-functional building type. Experiments indicate that using the proposed hybrid architecture results in improved prediction accuracy for two case multi-functional buildings in ultra-short-term to short term energy use modelling, with R^2 score ranging between 0.81 to 0.94. The proposed model architecture comprising the CNN, dropout, bidirectional and dense layer modules superseded the performance of the commonly used baseline deep learning models tested in the investigation, demonstrating the effectiveness of the proposed architectural structure. The proposed model is satisfactorily applicable to modelling multi-functional building energy consumption.

Keywords: Deep learning · Energy use prediction · Leisure centre

1 Introduction

As the world awakens to the need to protect and conserve the natural environment, building energy efficiency continues to take centre stage in global sustainability issues. The commonly used techniques for building energy consumption prediction (ECP) are classified under engineering methods, statistical methods and artificial intelligence-based methods [2]. Statistical methods for building ECP insist on assumptions such as stationarity of historical data, while engineering methods tend to be time-consuming as they try to account for every building parameter. Machine learning methods may fail to identify intricate building energy patterns buried in the ultra-short-term timeseries data. Deep learning methods have been adopted for building ECP with successful results, mainly because of their automatic feature extraction and higher information abstraction capabilities.

Research on building energy consumption prediction (ECP) has been underway for decades. Building ECP has been extensively researched for many building types such as office buildings, schools, hotels, commercial and residential

© Springer Nature Switzerland AG 2021
M. Paszynski et al. (Eds.): ICCS 2021, LNCS 12746, pp. 292–305, 2021.
https://doi.org/10.1007/978-3-030-77977-1_23

buildings, as highlighted in the review paper [21]. However, multi-functional building types have been marginally represented in building energy literature. Data-driven techniques have been marginally applied in sports and recreation facilities' energy consumption prediction studies. A toolkit was proposed for preliminary estimation of power and energy requirements in sports centres [5]. Artificial neural networks were designed using simulated data to predict and optimise the energy consumption of an indoor swimming pool [23].

1.1 Deep Learning Models Review

The deep learning group of techniques provide a practical approach to building energy consumption prediction. The monthly energy consumption of customers of an energy company was forecasted in [6] using three deep learning models. The fully connected MLP, Long Short-Term Memory neural networks and convolutional neural networks were tested on their prediction skill based on the MAE, RMSE, MSE and (R^2) metrics. The LSTM model was reported as the superior model, while both the MLP and CNN did not show significant differences. Deep recurrent networks and MLP were studied for residential and commercial buildings for medium to long term predictions [17]. The findings highlighted the better performances of the recurrent networks. Deep highway networks and ensemble-based tree methods were used in short-term building energy prediction, with the former reported being superior to its counterpart models in [1]. Day-ahead multi-step forecasting of commercial buildings energy consumption was studied in [9] using CNN, recurrent neural network and seasonal ARIMA models. In this work, the temperature-based CNN model was reported as the superior model using one-year-long historical data, while the SARIMA model was noted as the inferior model. At the district level, [3] demonstrated the superior performance of the deep learning models for short-term load forecasting in commercial buildings; however, a deterioration in performance as forecasting horizon increases was noted. An effective genetic algorithm-LSTM hybrid was developed in [4] for buildings ECP. Aggregated power load and photo-voltaic power output were successfully predicted using a recurrent neural network and an LSTM hybrid model [22]. The aggregate residential building active power use estimation was tested using conditional restricted Boltzmann machines (FCRMB) and factored restricted Boltzmann machine, artificial neural network, recurrent neural network and support vector machines by [16] for different prediction horizons, and the authors reported the superiority of FCRMB over other tested models. The IHEPC dataset on the UCI Machine Learning Repository spans, which over almost four years, was used to develop a CNN-LSTM model which outperformed the prior FCRMB model in [13]. In another work, a CNN and multilayer bidirectional LSTM model was proposed [20] and outperformed other rival techniques in excessive power consumption prediction. A CNN-GRU hybrid model was proposed [18] for short-term residential energy consumption prediction. In this study, two benchmark datasets, namely the AEP and IHEPC, were considered, and the proposed CNN-GRU architecture outperformed the rival machine learning and deep learning models. More recently, attention mechanism was incorporated on a CNN-BiLSTM [15] in daily load forecasting and proved effective.

The authors propose a Conv-BiLSTM model that can capture salient spatial features using a convolution operator within the LSTM cell on multiple-dimensional data. After careful examination of the input timeseries features, the architectures of the conventional CNN and LSTM models separately, a hybrid model Conv-BiLSTM is proposed. In the hybrid Conv-BiLSTM model, both the input-to-state and state-to-state cells have convolutional structures; that is, convolutions on the input timeseries are directly inputted into each BiLSTM unit which processes the cell states in both forward and backward directions. The proposed hybrid model shows superior accuracy for building ECP because of its robust architectural structure. Most of the above-cited works using hybrid models relate mostly to residential buildings ECP. However, in this study, we consider a unique multi-functional building type marginalised used in building energy studies.

The rest of the paper is organised as follows: Sect. 2 introduces the proposed hybrid Conv-BiLSTM model, the rest of the forecasting methods used in this study are described in Sect. 3, Sect. 4 gives a detailed description of the two case study datasets, Sect. 5 gives the results and discussion, and Sect. 6 highlights the conclusions and future works.

2 Proposed Hybrid Convolutional Bidirectional Long Short-Term Memory (Conv-BiLSTM)

Hybrid deep learning models are designed by the fusion of conventional deep learning models that combine multiple models by diversifying the input features and varying the initialisation of the weights of the neural network. The Hybrid Conv-BiLSTM model is implemented to capture the advantages of the CNN and BiLSTM techniques to improve the overall prediction accuracy. The Conv-BiLSTM is a different hybrid variant of the CNN-BiLSTM model, which does a convolution operation within the BiLSTM cells. The Conv-BiLSTM layer is a recurrent layer that replaces the usual matrix multiplication by a convolution operation. The convolution operator is applied directly to read input into the LSTM cells, that is, during the input-to-state transitions and during the state-to-state transitions [19]. The Conv-BiLSTM compresses the building energy consumption sequence into a hidden state tensor that is then decoded by an LSTM layer which processes this input in both the forward and backward directions to give the final prediction. The critical equations of the Conv-BiLSTM cell gates are given in Eqs. 1–5;

$$I_t = \sigma(W_{XI} * X_t + W_{HI} * H_{t-1} + W_{CI} \circ C_{T-1} + b_I) \tag{1}$$

$$O_t = \sigma(W_{XO} * X_t + W_{HO} * H_{T-1} + W_{CO} \circ C_t + b_O) \tag{2}$$

$$F_t = \sigma(W_{XF} * X_t + W_{HF} * H_{t-1} + W_{CF} \circ C_{t-1} + b_F) \tag{3}$$

$$C_t = F{\circ}C + i_t \circ (W_{XC} * x_t + W_{HC} * h_{t-1} + b_C) \tag{4}$$

$$H_t = O \circ tanh(C_t) \tag{5}$$

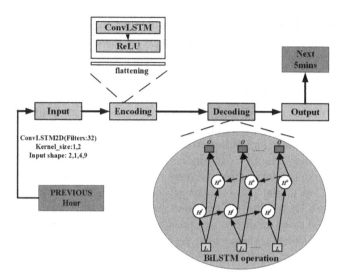

Fig. 1. An illustration of the implemented hybrid Conv-BiLSTM operational workflow.

The convolutional product and element-wise multiplication operations are denoted by "$*$" and "$\circ$" respectively. In (1–5), I_t, F_t, and O_t are the input, forget, and output gate, respectively. W represents the weight matrix, x_t is the current input data, h_{t-1} is the previously hidden output, and C_t denotes the cell state at a given timestep t. While the traditional LSTM equations use convolution operation ($*$), the Conv-BiLSTM instead uses matrix multiplication between W and X_t, h_{t-1} for every gate. This matrix multiplication replaces the fully connected layer with a convolutional layer which leads to a reduced number of weight parameters in the model. The default expected input shape must be of the form [samples, timesteps, rows, columns, channels], which is similar to the form used with the image data. However, the building energy consumption timeseries data is in sequence form, which is one dimensional (1D). As such, the data is read as a row with columns. In this case, since data is in the 5-min interval, 60 min at a 5-min interval gives 12 columns per row. When using the Conv-BiLSTM, the previous BEC timeseries of 60 min is split into two subsequences of thirty minutes each (that is, two rows with six columns each), and the Conv-BiLSTM can then perform the convolutional operation on each of the two subsequences. This sequence-splitting operation results in a 5D input tensor with shape [s, 2, 1, 4, 9] denoting sample, timestep, rows, columns and channels, respectively, as shown in Fig. 1. Finally, the hidden layer of the Conv-BiLSTM encoder is defined and flattened in readiness for decoding using the traditional BiLSTM operation discussed below. The last layer is made up of the fully connected layer (dense layer), which processes the output from the BiLSTM operation.

3   Forecasting Methods

This section briefly discusses the forecasting methods adopted in this study for multi-functional buildings ECP, namely, the linear regression model, CNN model, LSTM model and BiLSTM model.

3.1   Baseline Convolutional Neural Networks (CNN)

This type of neural network was originally designed to handle image data and has achieved the state of the art results in the field of computer vision on tasks such as image classification, object recognition, among other tasks. Through representation learning, the CNN model can extract useful features automatically from timeseries data by treating a sequence of observations as a one-dimensional image. The CNN model [14] is relatively easy and faster to train because the weights are less than a fully connected architecture. In performing the ECP, a predetermined number of energy consumption historical observations are fed into the convolutional layers, which perform a one-dimensional convolution of this data. The output from the convolution operation is then passed to the fully connected layers for final processing.

3.2   Baseline Long Short-Term Memory (LSTM)

The LSTM model reads a one-time step of the sequence at a time and then builds up an internal state representation. During learning a mapping function between inputs and outputs, LSTMs, unlike MLPs and CNNs, can remember the observations seen previously and can deduce their importance to the prediction task, and since the relevant context of inputs changes dynamically, LSTM can adapt and respond appropriately [12].

3.3   Bidirectional LSTM (BiLSTM)

The LSTM is unidirectional, which has one group of hidden layers for the energy consumption sequence in the positive time directions. However, the bidirectional LSTM (BiLSTM) [19] maintains two groups by adding energy consumption input sequence in the negative time direction. The two groups of hidden layers are independent of each other, and their outputs are concatenated and linked to the same output layer. The mathematical representations describing the BiLSTM architecture are the same as that of the unidirectional LSTM except that there exist two hidden states at timestep t, that is H_t^f and H_t^f representing the forward and backward hidden states, respectively. The final hidden states representation H_t results from merging H_t^f and H_t^f is as follows [19]:

$$H_t = H_t^f + H_t^f \tag{6}$$

4   Datasets Description

This section summaries the datasets used for developing and evaluating the different forecasting methods for ECP.

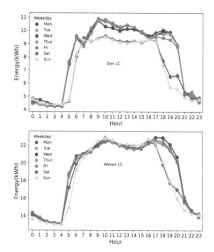

Fig. 2. An illustration of the weekday energy consumption profiles at the two leisure centres.

The datasets comprise two and half years of 5-min resolution energy consumption observations for two leisure centres precisely, from May 2017 to December 2019. Waves leisure centre's (WLC) aggregate energy use profile is made up of 14 individual meters while Don Tatnell leisure centre's (DTLC) energy use's timeseries is a total of 5 electrical meters, all from various sections within the individual leisure centres.

Figure 2 shows the 24-h weekly energy consumption profiles of the two leisure centres. There is substantial similarity in the 24-h energy use profiles between the two centres. Energy consumption rises around the 4th hour and begins dropping off at about 9 pm for most days of the week. However, on weekends, particularly on Sundays, energy consumption drops off a little earlier, between 5 pm and 6 pm for both centres.

The 5-min resolution energy consumption profiles for the two buildings demand dataset was resampled to 15 min, hourly datasets to test the model performances on forecasting the next step of energy consumption and facilitate merging with climatic data. The resampling was achieved through downsampling the 5 min energy consumption timeseries into 15 min and hourly bins and taking the sum values of the timestamps falling into a bin.

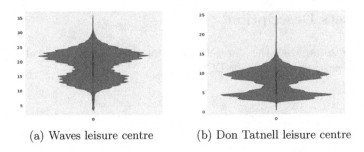

(a) Waves leisure centre   (b) Don Tatnell leisure centre

Fig. 3. Violin plots showing the energy consumption distribution at the two centres

The violin plots highlighted in Fig. 3 shows the observed density distribution of different electrical energy use values at the two centres. The box-plot component (inner section) with a thick black line in the centre represents the inter-quartile ranges of the electrical energy profiles (between 15 kWh and 25 kWh) for WLC and between 4 kWh and 10 kWh for DTLC. The little white dot points at the centre represent the median electrical energy use values of 21 kWh and 9 kWh for WLC and DTLC, respectively. The long tapered top end sections show the existence of few very high energy use values at both centres. These high values were outliers from system recording errors. Any values per 5-min interval above 35 kWh for WLC and above 19 kWh for DTLC were deemed outliers; as such, they were removed and replaced by the observation occurring immediately after (backfilling).

On the other hand, the broader section of the violin plots shows high frequency (occurrence) of electrical energy use-values. The violin plots provide a vital visual perspective that assists in data anomaly detection and designing a robust modelling process for electrical energy forecasting.

4.1   Input Processing

The building energy time-series data needs to be transformed into supervised learning to allow reading by the model. The sequence data is reframed into pairs of input and output variables using the sliding window method. The inputs comprise values of the current BEC observation t and previous BEC observations at times $t-1, t-2....t-n$ which predict the output BEC observation at time $t+1$. The length of the sliding window is defined based on [8,11].

4.2   Data Standardisation

The differences in numerical ranges between the input and output values require that the dataset be standardised. Standardisation scales each feature to have a distribution that is centred around 0, with a standard deviation of 1. Standardisation allows for comparability among input datasets, and it also enhances the training efficiency of models since the numerical condition of optimisation is improved. The mean and standard deviation for each feature is calculated, then the feature is scaled using Eq. 7.

$$z = (x_i - \mu)/\sigma \qquad (7)$$

where z is the standardized value, x_i is the observed energy consumption value, μ is the mean, and σ is the standard deviation of the electrical energy datasets.

4.3 Evaluation Metrics

An evaluation of the models' accuracy is done using the mean absolute error MAE, root-mean-square error RMSE, and R-squared (R^2) metrics. The (R^2) metric defines the proportion of variance of the dependent variable that is explained by the regression model. A low (R^2) value closer to 0 highlights a low correlation level, while an (R^2) closer to 1 means a strong correlation exists between considered variables. The MAE metric calculates the positive mean error value for the test data. Calculating the root of the mean square error gives a metric called root-mean-square error RMSE. The RMSE penalises significant errors, which makes it a strict and reliable metric. One cannot look at these metrics in isolation in sizing up the model; rather, all the metrics must be considered at the same time. These performance evaluation metrics are calculated using Eqs. (8–10)

$$R^2 = 1 - \frac{\sum(y - y')^2}{\sum(y - \bar{y}')^2}, \qquad (8)$$

$$RMSE = \sqrt{\sum \frac{(y' - y)^2}{N}}, \qquad (9)$$

$$\text{MAE} = \frac{1}{n} \sum_{i=1}^{n} |y' - y| \qquad (10)$$

where y is the measured energy consumption value, y' is the predicted energy consumption value, and N represents the number of data pairs considered.

5   Results and Discussion

This section describes the results for the next 5 min (mins), 15 mins and hour for multi-functional buildings ECP, given the previous energy consumption observations and the calendar timestamps. The ultra-short-term to short-term forecasts are crucial for developing strategies that facilitate the safe, economical and efficient operation of multi-functional buildings. Table 1 and Table 2 show the testing phase results using the considered models for the next step energy consumption prediction for DTLC and WLC, respectively. The results of the most influential previous energy consumption values (lags) are discussed here.

5.1   Implementation Details

The data was split into 70% training, 10% validation and 20% testing while maintaining the temporal order of the timeseries data. Temporal order maintenance, unlike random sampling, helps avoid information 'leaking' into the training set; that is, it ensures no future values infiltrates the training set.

All learning algorithms were implemented in the Python programming language on google collaboratory [7] online platform. The deep learning models were built using Keras library with Tensorflow backend [10]. All development and experiments were conducted on a macOS Mojave (2.90 GHz Intel Core i9 16 GB 2400 MHz DDR4) machine.

5.2   Prediction Using 5-Min Resolution

Prediction using five-minute resolution has been chosen because it is useful in near real-time market activities such as resources dispatch and anomaly detection. The next 5 min of energy consumption has been performed using the previous 25-min of historical energy consumption data and calendar data inputs, namely, the hour of the day, weekday or weekend, the week of the year, the month of the year, the quarter of the year, the day of the month. Figure 4 shows extracted snapshots of the ground truth and the next 5-min predictions performance of the tested models at DTLC and WLC. According to the charts, the tested models do capture the trend and patterns of the ground truth line but struggle with the abrupt changes in the consumption profiles. DTLC was relatively easy to predict at this resolution, with all the tested models performing relatively well, with the R^2 score metric reaching as high as 0.87 for the deep learning models and 0.82 for the linear regression model. The proposed Conv-BiLSTM model demonstrated superior performance with RMSE, MAE and R^2 scores of 1.01 kWh, 0.76 kWh and 0.87, respectively.

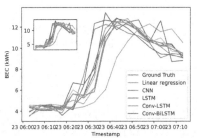
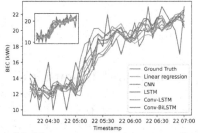

(a) Don Tatnell prediction results.   (b) Waves centre prediction results.

Fig. 4. An snapshort Illustration of the observed and predicted values for Don Tatnell and Waves leisure centres in the testing phase. (Color figure online)

Predicting the next 5-min of energy consumption at WLC was a relatively challenging task for the tested models with the RMSE score ranging between 2.04 kWh and 2.17 kWh and R^2 scores between 0.79 and 0.81 for the LR and the Conv-BiLSTM models, respectively. The hour of the day was influential in predicting the next 5-min of energy consumption at both leisure centres. Figure 5 shows the correlations between the predicted and actual BEC values for both centers. Don Tatnell leisure centre presents are more closer fit between observed and predicted values. The existence of a large observed BEC value (45 kWh) in the test set may be responsible for degrading the correlation value between observed and predicted at this centre. Waves centre does show relative higher differences between the observed and actual values.

Table 1. Generalisation capabilities of the tested models in the testing phase for Waves leisure centre

Model	Lag	RMSE (kWh)	MAE (kWh)	R^2	Lag	RMSE (kWh)	MAE (kWh)	R^2	Lag	RMSE (kWh)	MAE (kWh)	R^2
	5min				**15min**				**1hour**			
CNN	5	2.06	1.57	0.81	4	4.34	3.21	0.9	4	14.17	9.59	0.92
LSTM	5	2.10	1.61	0.81	4	3.68	2.79	0.92	4	15.09	10.65	0.92
ConvLSTM	2,4	2.08	1.59	0.81	2,4	3.88	2.94	0.92	2,4	13.49	9.13	0.93
ConvBiLSTM	2,4	**2.05**	**1.56**	**0.81**	2,4	**3.80**	**2.86**	**0.92**	2,4	**13.00**	**8.99**	**0.94**
LR	5	2.17	1.66	0.79	4	4.62	3.29	0.88	4	18.94	12.83	0.86

5.3   Prediction Using 15-Min Resolution

Prediction of the next 15-min of energy consumption is crucial for planning effective network utilisation by energy suppliers and monitoring energy market prices. Additional features in the form of climatic variables (temperature, dew point, relative humidity, mean wind velocity and wind direction) at 15-min intervals were added to determine their effect on building energy consumption at the two case buildings. Prediction at 15-min resolution resulted in an improved prediction performance for both case buildings. The deep learning suite of models scored R^2 scores of up to 0.92, representing an improvement of up to 15% for WLC from the 5-min prediction scores. However, DTLC saw a marginal increase (2%) in prediction performance by the deep learning models. The proposed Conv-BiLSTM again showed superior performance by outperforming other models in both case studies with R^2 scores of 0.91 and 0.92 for DTLC and WLC, respectively. Most building energy performance studies highlight improvement in model prediction performance by use of climatic variables; however, for the two case study buildings considered in this study, the climatic variables' addition did not improve models' performances at 15-min resolution.

Table 2. Generalisation capabilities of the tested models in the testing phase for Don Tatnell leisure centre

Model	Lag	RMSE (kWh)	MAE (kWh)	R^2	Lag	RMSE (kWh)	MAE (kWh)	R^2	Lag	RMSE (kWh)	MAE (kWh)	R^2
	5min				15min				1hour			
CNN	4	1.05	0.80	0.86	2	2.71	2.04	0.89	4	12.74	9.58	0.84
LSTM	4	1.06	0.80	0.86	2	2.88	2.23	0.88	4	10.96	7.96	0.88
ConvLSTM	2,4	1.02	0.78	0.87	2,4	2.57	1.93	0.90	2,2	12.38	9.31	0.84
ConvBiLSTM	2,4	**1.01**	**0.76**	**0.87**	2,4	**2.54**	**1.92**	**0.91**	2,2	**10.34**	**7.56**	**0.88**
LR	4	1.18	0.89	0.82	2	3.07	2.0	0.86	4	14.05	9.93	0.79

5.4   Prediction Using Hourly Resolution

This resolution represents the common choices of energy consumption prediction in building energy performance literature. All tested models did well at predicting the next hour of energy consumption at WLC, with the LR models scoring RMSE, MAE and R^2 values of 18.73 kWh, 12.62 kWh and 0.87, respectively. The proposed hybrid Conv-BiLSTM model continues to dominate with RMSE, MAE and R^2 scores of 13.00 kWh, 8.99 kWh and 0.94, respectively, outperforming all the other models for the WLC dataset. However, at DTLC, predicting the next hour of energy consumption was relatively challenging with superior performance from among the deep learning models scoring RMSE, MAE and R^2 values of 12.37 kWh, 9.31 kWh and 0.84, respectively. The Conv-BiLSTM model showed superior performance outperforming the popular LSTM by 6% and 5% margins in the RMSE and MAE scores, respectively.

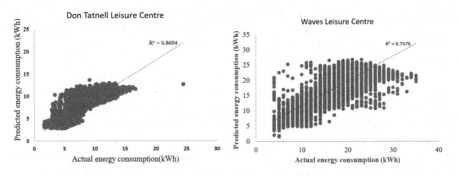

Fig. 5. An Illustration showing the correlations between the observed and predicted BEC values for the leisure centres in the testing phase.

5.5   Model Parameters

The hybrid Conv-BiLSTM architectures comprising the layer name, output shape and number of parameters for the next 5-min ECP at both WLC and DTLC are presented in Table 3. The proposed model had a total of 186 405 and 147 721 trainable parameters for DTLC and WLC, respectively. A dropout

Table 3. Architecture of the best performing model using 5-min resolution

DTLC Parameters	WLC Parameters
Layer (type) Output Shape Param # (ConvLSTM2D_2) (None, 1, 3, 32) 10624	
(Dropout) (None, 1, 3, 32) 0	**Layer (type) Output Shape Param #** (ConvLSTM2D_3) (None, 1, 1, 32) 21120
(Flatten_2) (None, 96) 0	(Flatten_3) (None, 32) 0
(RepeatVector_2 (None, 1, 96) 0	(RepeatVector_3) (None, 1, 32) 0
(Bidirectional_2) (None, 1, 200) 157600	(Bidirectional_3) (None, 1, 200) 106400
(Dropout_2) (None, 1, 200) 0	(TimeDistributed_6) (None, 1, 100) 20100
(TimeDistributed_4) (None, 1, 90) 18090	(TimeDistributed_7 (None, 1, 1) 101
(TimeDistributed_5 (None, 1, 1) 91	Total params: 147,721 Trainable params: 147,721 Non-trainable params: 0
Total params: 186,405 Trainable params: 186,405 Non-trainable params: 0	

(20%) module was only beneficial for DTLC, while no dropout was necessary WLC as it did not improve model performance.

At both DTLC and WLC, the proposed Conv-BiLSTM hybrid technique showed superior non-linear mapping generalisation abilities, outperforming the baseline deep learning-based techniques. It can be concluded that the proposed hybrid Conv-BiLSTM model is effective in learning complex decision boundaries for near real-time to short term aggregate energy consumption prediction at the considered multi-functional buildings. The hybrid Conv-BiLSTM model showed satisfactory performance capturing the trends and seasonality present in the energy consumption observations as shown in Fig. 4. The proposed hybrid Conv-BiLSTM followed the electricity consumption patterns closely and performed better in predicting lower and higher electricity consumption values. This is highlighted by the closeness to the ground truth line (blue line) by the Conv-BiLSTM model line.

Predicting the ultra-short-term (5 min) energy consumption patterns has been particularly challenging for the tested models for both case buildings studied, with the R^2 scores ranging between 0.79–0.87. The fine resolution has tendencies to bury the energy use patterns in the noise, thus making the energy use patterns invisible. However, as the resolution becomes coarser, up to the hourly resolution, the predictive performance of models increases with the proposed superior model (Conv-BiLSTM), attaining an R^2 score of 0.94 and 0.89 for WLC and DTLC, respectively. The improvement in performance as the granularity changes from 5-min to an hour can be attributed to the smoothening effect of the energy consumption profile, with the hourly and 15-min resolution providing the best results for WLC and DTLC, respectively. This study demonstrates that the proposed Conv-BiLSTM is a valuable computational intelligence technique to predict energy consumption for unique multi-functional buildings.

6   Conclusion and Future Works

This paper proposes a hybrid Conv-BiLSTM model for energy consumption prediction of multi-functional leisure centres. In pursuit of accurate and robust forecasting models, the hybrid Conv-BiLSTM model was tested against a suite of baseline competitive deep learning models for the next 5-min, 15-min and hour of energy consumption prediction. The results of the experiments reveal that the proposed hybrid. Conv-BiLSTM outperformed its counterpart models for multi-functional building energy consumption prediction. By directly reading the input into the LSTM cells and processing the cell states in both the forward and backward directions, the hybrid Conv-BiLSTM model was able to effectively reconstruct the energy consumption patterns at the two tested multi-functional buildings.

Multi-functional leisure centres have high and irregular energy consumption patterns than most studied building types. The study determined that for the considered building types, the previous energy consumption observations and the calendar inputs, particularly the hour of the day, had a significant effect on the energy consumption in both case buildings, and therefore the study recommends their adoption as primary inputs. The study showed that the hybrid Conv-BiLSTM model could be used for aggregate energy consumption prediction at these new building types. Ongoing work with the proposed model involves its applicability for transfer learning in which the developed model for one leisure centre (source building) will be configured to predict the other centre (target building) in an effort to curb data shortages issues in the latter.

References

1. Ahmad, M.W., Mouraud, A., Rezgui, Y., Mourshed, M.: Deep highway networks and tree-based ensemble for predicting short-term building energy consumption. Energies **11**(12), 3408 (2018)
2. Ahmad, M.W., Mourshed, M., Yuce, B., Rezgui, Y.: Computational intelligence techniques for HVAC systems: a review. Build. Simul. **9**, 359–398 (2016). https://doi.org/10.1007/s12273-016-0285-4
3. Ahmad, T., Chen, H., Huang, Y.: Short-term energy prediction for district-level load management using machine learning based approaches. Energy Procedia **158**, 3331–3338 (2019)
4. Almalaq, A., Zhang, J.J.: Evolutionary deep learning-based energy consumption prediction for buildings. IEEE Access **7**, 1520–1531 (2018)
5. Artuso, P., Santiangeli, A.: Energy solutions for sports facilities. International J. Hydrogen Energy **33**(12), 3182–3187 (2008)
6. Berriel, R.F., Lopes, A.T., Rodrigues, A., Varejao, F.M., Oliveira-Santos, T.: Monthly energy consumption forecast: a deep learning approach. In: 2017 International Joint Conference on Neural Networks (IJCNN), pp. 4283–4290. IEEE (2017)

7. Bisong, E.: Google Colaboratory. In: Building Machine Learning and Deep Learning Models on Google Cloud Platform, pp. 59–64. Apress, Berkeley (2019). https://doi.org/10.1007/978-1-4842-4470-8_7
8. Brownlee, J.: Deep learning for time series forecasting: predict the future with MLPs. CNNs and LSTMs in Python, Machine Learning Mastery (2018)
9. Cai, M., Pipattanasomporn, M., Rahman, S.: Day-ahead building-level load forecasts using deep learning vs. traditional time-series techniques. Appl. Energy **236**, 1078–1088 (2019)
10. Chollet, F., et al.: Keras documentation. keras. io, vol. 33 (2015)
11. Fan, C., Wang, J., Gang, W., Li, S.: Assessment of deep recurrent neural network-based strategies for short-term building energy predictions. Appl. Energy **236**, 700–710 (2019)
12. Gers, F.A., Schmidhuber, J.: Recurrent nets that time and count. In: Proceedings of the IEEE-INNS-ENNS International Joint Conference on Neural Networks. IJCNN 2000. Neural Computing: New Challenges and Perspectives for the New Millennium, vol. 3, pp. 189–194. IEEE (2000)
13. Kim, T.Y., Cho, S.B.: Predicting residential energy consumption using CNN-LSTM neural networks. Energy **182**, 72–81 (2019)
14. Krizhevsky, A., Sutskever, I., Hinton, G.E.: ImageNet classification with deep convolutional neural networks. Commun. ACM **60**(6), 84–90 (2017)
15. Miao, K., Hua, Q., Shi, H.: Short-Term load forecasting based on CNN-BiLSTM with Bayesian optimization and attention mechanism. In: Zhang, Y., Xu, Y., Tian, H. (eds.) PDCAT 2020. LNCS, vol. 12606, pp. 116–128. Springer, Cham (2021). https://doi.org/10.1007/978-3-030-69244-5_10
16. Mocanu, E., Nguyen, P.H., Kling, W.L., Gibescu, M.: Unsupervised energy prediction in a smart grid context using reinforcement cross-building transfer learning. Energy Build. **116**, 646–655 (2016)
17. Rahman, A., Srikumar, V., Smith, A.D.: Predicting electricity consumption for commercial and residential buildings using deep recurrent neural networks. Appl. Energy **212**, 372–385 (2018)
18. Sajjad, M., et al.: A novel CNN-GRU-based hybrid approach for short-term residential load forecasting. IEEE Access **8**, 143759–143768 (2020)
19. Schuster, M., Paliwal, K.K.: Bidirectional recurrent neural networks. IEEE Trans. Signal Process. **45**(11), 2673–2681 (1997)
20. Ullah, F.U.M., Ullah, A., Haq, I.U., Rho, S., Baik, S.W.: Short-term prediction of residential power energy consumption via CNN and multi-layer bi-directional LSTM networks. IEEE Access **8**, 123369–123380 (2019)
21. Wei, Y., et al.: A review of data-driven approaches for prediction and classification of building energy consumption. Renew. Sustain. Energy Rev. **82**, 1027–1047 (2018)
22. Wen, L., Zhou, K., Yang, S., Lu, X.: Optimal load dispatch of community microgrid with deep learning based solar power and load forecasting. Energy **171**, 1053–1065 (2019)
23. Yuce, B., Li, H., Rezgui, Y., Petri, I., Jayan, B., Yang, C.: Utilizing artificial neural network to predict energy consumption and thermal comfort level: an indoor swimming pool case study. Energy Build. **80**, 45–56 (2014)

Machine Learning and Data
Assimilation for Dynamical Systems

Deep Learning for Solar Irradiance Nowcasting: A Comparison of a Recurrent Neural Network and Two Traditional Methods

Dennis Knol[1]([✉]), Fons de Leeuw[2], Jan Fokke Meirink[3], and Valeria V. Krzhizhanovskaya[1,4]

[1] University of Amsterdam, Amsterdam, The Netherlands
[2] Dexter Energy Services, Amsterdam, The Netherlands
[3] Royal Netherlands Meteorological Institute (KNMI), De Bilt, The Netherlands
[4] ITMO University, Saint Petersburg, Russia

Abstract. This paper aims to improve short-term forecasting of clouds to accelerate the usability of solar energy. It compares the Convolutional Gated Recurrent Unit (ConvGRU) model to an optical flow baseline and the Numerical Weather Prediction (NWP) Weather Research and Forecast (WRF) model. The models are evaluated over 75 days in the summer of 2019 for an area covering the Netherlands, and it is studied under what circumstance the models perform best. The ConvGRU model proved to outperform both extrapolation-based methods and an operational NWP system in the precipitation domain. For our study, the model trains on sequences containing irradiance data from the Meteosat Second Generation Cloud Physical Properties (MSG-CPP) dataset. Additionally, we design an extension to the model, enabling the model also to exploit geographical data. The experimental results show that the ConvGRU outperforms the other methods in all weather conditions and improves the optical flow benchmark by 9% in terms of Mean Absolute Error (MAE). However, the ConvGRU prediction samples demonstrate that the model suffers from a blurry image problem, which causes cloud structures to smooth out over time. The optical flow model is better at representing cloud fields throughout the forecast. The WRF model performs best on clear days in terms of the Structural Similarity Index Metric (SSIM) but suffers from the simulation's short-range.

Keywords: Nowcasting · Solar irradiance · Deep learning · Convolutional GRU · Numerical weather prediction · WRF · Optical flow

1 Introduction

To help constrain global warming, the energy produced by fossil fuels is being replaced more and more by renewable energy. This transition is accelerated

© Springer Nature Switzerland AG 2021
M. Paszynski et al. (Eds.): ICCS 2021, LNCS 12746, pp. 309–322, 2021.
https://doi.org/10.1007/978-3-030-77977-1_24

because wind and solar energy are now cheaper than energy from traditional resources [1,2]. However, in terms of usability, wind and solar are not yet fully competitive due to their variable nature. The way the energy market works dictates that more accurate weather forecasts are essential to increase solar usability. The energy supply must continuously match demand, keeping the electricity grid in balance. As renewable resources replace conventional power plants, balancing the grid is becoming increasingly complex [3,4].

One of the critical challenges is forecasting solar irradiance at the Earth's surface. When it comes to solar irradiance forecasting, clouds are the most critical driver and notoriously challenging to predict [5]. More traditional forecasting methods to forecast clouds divide roughly into two classes: image-based extrapolation methods (e.g., optical flow) and Numerical Weather Prediction (NWP) based methods. Extrapolation based methods perform well on a small temporal scale, but accuracy decreases markedly for increasing temporal scales as these methods do not take into account the evolution of clouds. NWP models generally have a better forecast accuracy for larger temporal scales, particularly for clear sky conditions [6]. However, performance decreases for cloudy conditions as NWP models do not explicitly resolve sub-grid cloud processes. Unresolved processes are parameterised and add a source of uncertainty to the model.

More recent studies propose to address these limitations from a machine learning perspective, taking advantage of the vast amount of weather data available. These studies formulate nowcasting as a spatiotemporal sequence forecasting problem and specifically focus on nowcasting precipitation, using Recurrent Neural Networks (RNN). Shi et al. [7] propose a novel Long Short Term Memory (LSTM) with convolutional layers to capture spatial correlation, which outperforms the optical flow based ROVER algorithm. A follow-up study proposes a Convolutional Gated Recurrent Unit (ConvGRU) [8], an architecture which is less complicated and more efficient to train. Google's MetNet is the first deep learning model to produce more accurate precipitation nowcasts than NWP [9].

This paper describes the implementation of a ConvGRU model for the nowcasting of solar irradiance trained on satellite observations. The ConvGRU model's performance is evaluated using two more traditional models as the benchmark: an optical flow algorithm and the Weather and Research Forecast (WRF) NWP model. We implement parameterisations specifically designed to meet the growing demand for specialised numerical forecast products for solar power applications for the latter. We also study the strengths and weaknesses of each method. A method's performance depends on the weather conditions (e.g., clear sky vs cloudy) and different models perform well on different temporal scales. This study aims to assess under what particular circumstances and spatiotemporal scales the models perform best.

2   Methods

2.1   Deep Neural Networks in the Nowcasting Domain

One of the most recent advances in the space of nowcasting is the use of Deep Neural Networks (DNN). Contrary to Optical Flow and NWP, deep neural networks can exploit the large amount of data collected continuously from ground-based cameras, radars, weather stations, and satellites.

Most progress is made in nowcasting precipitation. One of the early studies tackling the nowcasting problem from a deep learning perspective is [7]. In this study, researchers introduce an end-to-end trainable model, where both input and output are spatiotemporal sequences. The proposed model architecture is a Convolutional Long Short-Term Memory (ConvLSTM), which captures temporal dependencies using the LSTM cells and spatial dependencies with the convolutional layers. The model is trained using 812 days of radar data and outperformed a state-of-the-art operational optical flow algorithm called ROVER.

The follow-up study [8], led to the emergence of two more accurate model architectures: Convolutional Gated Recurrent Unit (ConvGRU) and Trajectory Gated Recurrent Unit (TrajGRU). The first updates the ConvLSTM by replacing the LSTM cells with Gated Recurrent Unit (GRU) cells. The latter differs from ConvLSTM and ConvGRU. In the TrajGRU model, convolutions generate flow fields, and this enables the model to learn the location-variant structures and capture spatiotemporal correlations more efficiently.

More recently, Google's researchers introduced MetNet and showed that the model outperforms the current operational NWP by the National Oceanic and Atmospheric Administration (NOAA), High-Resolution Rapid Refresh (HRRR), at predictions up to 8 h. MetNet is a neural network model trained on both radar and satellite data. The data are first processed by a convolutional LSTM, second by axial attention layers. The attention mechanisms allow the model to ignore some parts of the data and focus on others and enable the model to learn long-term dependencies.

To our knowledge, solar nowcasting based on satellite data and deep learning techniques has not been covered by literature yet. The published literature on nowcasting irradiance with DNNs is based on sky-images retrieved from ground-based camera's [10,11]. This approach limits the forecast's lead time because the ground-based cameras cover only a small geographic area. Consequently, models trained on such data can only consider the clouds' possible motion over a very short period. Using satellite data, we can model larger geographical areas and generate forecasts further ahead in time.

2.2 Sequence-to-Sequence Model

Similar as in [7,8], this paper considers irradiance nowcasting a spatiotemporal sequence forecasting problem. A definition is presented in (1), where the input is a sequence of length J containing previous observations and the forecast a sequence of K frames ahead:

$$\tilde{I}_{t+1}, \ldots, \tilde{I}_{t+K} = \underset{I_{t+1}, \ldots, I_{t+K}}{\arg\max} \; p\left(I_{t+1}, \ldots, I_{t+K} \mid \hat{I}_{t-J+1}, \hat{I}_{t-J+2}, \ldots, \hat{I}_t\right) \quad (1)$$

Here, the input and predictions are a sequence containing tensors, more formally $I \in R^{C \times H \times W}$. The spatial dimensions are H and W, the temporal dimension C. The input sequence contains the current observation, I_t. The model learns by minimising the forecast error through back-propagation and learns spatiotemporal relations without explicit assumptions about atmospheric physics.

For the problem, we adapt the encoder-decoder sequence-to-sequence structure proposed by [8]. This structure maps the input sequence with a fixed length on an output sequence with a fixed length and allows the input and output to have different lengths. An example is visualised in Fig. 1. First, the encoder processes the elements in the spatiotemporal input sequence and creates a smaller and higher dimensional representation. Subsequently, the decoder learns to generate the predictions from the hidden state through multiple layers.

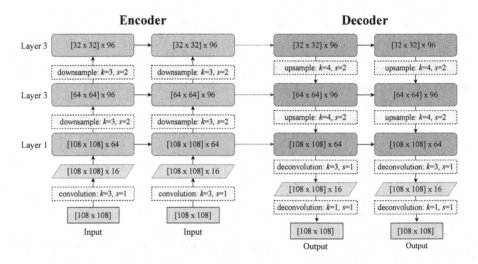

Fig. 1. Example of the sequence-to-sequence structure with three RNN layers, where $J = 2$ and $K = 2$. In the example, the model predicts the next two images based on the input of two images. Convolutions with stride, in between RNN layers, downsample the input in the encoder. Deconvolutional layers upsample the hidden representation in the decoder. We implement ConvGRUs as the RNN units, represented by the coloured cells. Figure inspired by [12].

2.3   Convolutional Gated Recurrent Unit

The ConvGRU convolutional layers assign importance to specific objects in the image, learn what parts of an input frame are important, and differentiate between different objects. The Gated Recurrent Unit (GRU), represented by the formulas in (2), captures the temporal dependencies by learning what previous information in a sequence is important to predict the future.

The GRU takes two sources of information: the memory state holding information of the previous units denoted with H_{t-1} and the information new to the model. The latter is denoted as X_t and is a single element of an inputted sequence. The first step in a GRU is to determine what information to pass on into the future using the update gate Z_t. Subsequently, the reset gate R_t determines what information to forget. With the output of the reset gate, the unit computes the new memory content H'_t through the activation function f. As the last step, the unit computes the memory state H_t passed on to the next unit in the network. The corresponding formulas are as follows:

$$
\begin{aligned}
Z_t &= \sigma \left(W_{xz} * X_t + W_{hz} * H_{t-1} \right) \\
R_t &= \sigma \left(W_{xr} * X_t + W_{hr} * H_{t-1} \right) \\
H'_t &= f \left(W_{xh} * X_t + R_t \circ \left(W_{hh} * H_{t-1} \right) \right) \\
H_t &= (1 - Z_t) \circ H'_t + Z_t \circ H_{t-1}
\end{aligned}
\tag{2}
$$

where $X_t \in \mathbb{R}^{C_i \times H \times W}$ and $H_t, R_t, Z_t, H'_t \in \mathbb{R}^{C_h \times H \times W}$. Here C is the number of input channels, H the height of the input image, and W the width. In the formulas, the convolutional operation is denoted as $*$, the Hadamard product as $\circ$ and f is the Leaky Rectified Linear Unit (ReLU) activation function with a negative slope of 0.2. The sigmoid activation function is applied to the update and reset gates.

2.4   Model Extension

Recent work in the medical field introduced model architectures that can exploit both static and dynamical medical data features, such as gender and patient visits [13]. The proposed architectures combine RNNs to process dynamic data with an independent Feed Forward Neural Networks processing the static data. The weather is also driven by a combination of dynamic and static features. Elevating terrain height, for example, causes orographic lifting and stimulates the formation of clouds. The ConvGRU model discussed above is specifically designed to work with sequence data and not exploit such data types. For the static input data, we develop an independent CNN with ReLU activation functions and batch normalisation, and we adapt the encoder-decoder structure. The encoder part of the network runs parallel to the encoder of the RNN and outputs the hidden state representation of the static information. Both encoders' hidden state is concatenated mid-way through the decoder and provide this information to the last layers, which output the irradiance prediction.

2.5   Optical Flow Baseline

The optical flow baseline used for this paper is an ensemble model of three optical flow algorithms available in the RainyMotion library [14]. The ensemble model, introduced by [15], combines the predictions of the Farneback, DeepFlow and Dense Inverse Search algorithms and provided more accurate irradiance forecast than the individual algorithms. The model is computed by taking the mean of the prediction of N optical flow algorithms:

$$
\bar{e}_t(x, y) = \frac{1}{N} \sum_{i=1}^{N} p_t(x, y)
\tag{3}
$$

where i is the summation index, $\bar{e}_t(x,y)$ is the forecast of the ensemble model at lead time t and at coordinates x, y. The prediction of the optical flow algorithms is represented by p_t. We initialise the model on an input sequence of 5 frames containing the satellite-based irradiance parameter (see Sect. 2.7).

2.6   The WRF Baseline Model

Jimenez et al. [6] introduced an augmentation to the standard WRF model that makes the model appropriate for solar power forecasting. One of the studies that implemented the WRF model augmentation compares the model to the forecasts provided by the Global Forecasting System (GFS) and finds that the forecasting error is lower than of the GFS baseline [16]. However, this study does not consider Data Assimilation (DA), meaning that the WRF model suffers from incomplete information on the initial atmospheric state.

This limitation is addressed by a follow-up study [17]. This work incorporates satellite and ground-based observations into the WRF model's initial conditions using 3DVAR or 4DVAR initialisation. Both approaches improve the next-day forecast accuracy and the use of the latter results in the most accurate irradiance forecasts at all lead times. However, this comes at a computational expense. The 4DVAR initialisation takes 2 h to compute on a high-performance machine. Given the steep increase in computational requirements and the studied forecasting window of 0 to 6 h lead time, we did not apply DA in our study.

To set a baseline for this study, we use version 4.2 of the Advanced Research WRF. The 00:00UTC release of the Global Forecast System (GFS) provides the initial conditions of for the WRF model and updates the lateral boundary conditions hourly throughout the forecast. We set up the WRF model with a nested domain and implement the two-way nest option. The parent domain has a grid size of 60×60 and a horizontal resolution of 27 km, the middle point is at latitude 52.3702 and longitude 4.8952. The nested domain's resolution is 9 km and spans 70×70 grid cells, and we use the Lambert conformal conic projection. We summarise the important parameterisations in Table 1.

Table 1. The key physics settings used for the baseline WRF model.

Setting	Implemented scheme
Microphysics	Aerosol Aware Thomson micro-physics scheme [18]
SW Radiation	Rapid Radiative Transfer Model (RRTMG) scheme [19]
LW Radiation	Rapid Radiative Transfer Model (RRTMG) scheme [19]
Shallow Cumulus	Deng cumulus scheme [20]
Cumulus	Updated Kain-Fritsch cumulus scheme [21]
Land Surface model	Noah Land Surface Model [22]

2.7   Irradiance Data

We chose the irradiance data from the Meteosat Second Generation Cloud Physical Properties (MSG-CPP) algorithm developed by the KNMI to derive cloud parameters and solar radiation at the surface from the SEVIRI instrument onboard the Meteosat Second Generation satellite [23]. The data is available per 15 min, and the spatial resolution is $3 \times 3 \, \mathrm{km}^2$. The MSG-CPP algorithm retrieves cloud properties, such as cloud optical thickness, thermodynamic phase and particle size, from the SEVIRI measurements of radiation reflected and emitted by the Earth and the atmosphere. Based on the computed cloud properties, the algorithm derives the direct and diffuse surface irradiance components in $\mathrm{W/m}^2$ [24]. We use the MSG-CPP irradiance data to initialise the optical flow and ConvGRU models and to evaluate the performance of all models.

A limitation of using this particular dataset is that the range of the processed data is somewhat shorter than the range of actual sunlight. The algorithm can only derive the cloud properties when the solar zenith angle (SZA) is not too high, as the estimations otherwise become very inaccurate. In the current version, the cloud properties are computed when the SZA is less than 78°, consequently missing a part of the day.

The data is retrieved from a geostationary satellite and spans an area much larger than the studied domain. For the preprocessing, we reproject the data to the Lambert Conformal Conic projection. Subsequently, we cut out the model domain presented in Sect. 3.

3   Experimental Setup

We next describe the overall setup for this research. We first compare different versions of the ConvGRU model. After that, we compare the best performing ConvGRU model to the baseline models. We consider a forecasting window of 0 to 6 h and assess the forecasts at 15-min intervals. We evaluate the models over 75 days in 2019, from July 18 until October 31.

This research's case study area is a geographical area corresponding to the Netherlands and is centred at latitude 52.4° and longitude 4.9°. For the WRF model, we rely on a nested domain setup. The optical flow and ConvGRU models' domain correspond to the outer domain in Fig. 2. This domain provides a spatial context of approximately 300 km in all direction to the domain of interest. The spatial context is essential as it provides the information to model incoming clouds. For all models, we evaluate the forecast in the inner domain.

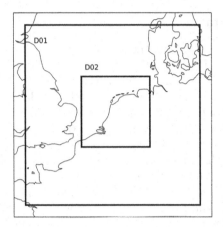

Fig. 2. Domain decomposition. The WRF domains have a spatial resolution of 9 and 27 km. The ConvGRU and optical flow models' resolution is 3 km.

3.1 Training Set

The training data for the ConvGRU model comprises frames containing the MSG-CPP irradiance data, which have a resolution of 323×323 pixels. To reduce the computational requirements needed to train the models, we downsample the images to a resolution of 108×108. The ConvGRU model receives a four-dimensional tensor of size $[t, c, h, w]$ with dimensions time, number of channels, height and width. The time dimension of the input patch is 5, ranging from 08:00–09:00UTC as one frame is provided every 15 min. The output tensor contains 24 predicted frames, corresponding to 6 h lead time. For training the models, we used 486 days of irradiance data (2018-03-15–2018-09-30, 2019-03-15–2019-07-17 and 2020-03-15, 2020-08-25) and the test set contains 75 days (2019-07-18–2019-09-30). By the time of the experiments, these were all the available days of MSG-CPP data with at least seven hours of irradiance data on the studied domain. We train all models using the Adam optimiser with a learning rate of 10^{-4}, 3000 epochs and a batch size of 6. The loss function we optimise is the smooth L1 loss function.

3.2 Forecast Evaluation

We calculate the Mean Bias Error (MBE) and the Mean Absolute Error (MAE) to determine the forecast accuracy of the models and quantify the strength of the error signal over the whole image. To complement the error-based metrics, we also compute the Structural Similarity Index Metric (SSIM) [25]. This metric accounts for patterns and textures in the images and can be used to assess the quality of the predicted images as it measures the loss of structural information by comparing local patterns in images. We compute the metrics based on normalised data to account for varying irradiance strengths throughout the day. We normalise the data with the clear sky irradiance from the MSG-CPP dataset.

Following [16], we assess the models' performance under different weather conditions. We compute Clear Sky Index (CSI) and categorise days into sunny, partly cloudy and very cloudy. We compute the CSI by dividing the average irradiance over a day by the average clear sky irradiance. We define days on which the average CSI is higher than 0.85 as sunny days and days with a lower CSI than 0.6 as mostly cloudy days. Days on which the CSI is between those values are defined as partly cloudy days.

4   Results

4.1   The ConvGRU Experiments

With the ConvGRU models, we conduct three experiments. The first experiment is based on the model introduced by [8] and trained on only a sequence of irradiance data. As we have a limited amount of data, we transpose the training data for the second experiment, doubling the number of images in the training set and increasing diversity. We thus train the ConvGRU model on twice as much data. The third experiment is based on the model we introduced in Sect. 2.4. In this model, we input the sequence data into the ConvGRU and model the static data in the neural network specifically designed to model such features. The static features we add are terrain height, and land mask [26]. The frames containing these features cover the same spatial area as the WRF parent domain and have a 60×60 pixels resolution.

The model that performs best overall is the ConvGRU trained on only sequence data. Figure 3 shows an exemplary forecast, demonstrating that the model learned the advection of clouds. Additionally, we find that in all ConvGRU model experiments, image quality degrades rapidly and that structures are not preserved as a consequence of blurriness. The blurring effect, also visible in Fig. 3, can be explained by the optimisation of a global-evaluating loss function and could be a result of the relatively small training set.

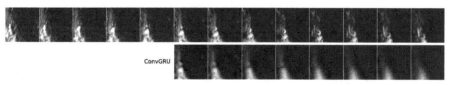

08:00 UTC   08:15 UTC   08:30 UTC   08:45 UTC   09:00 UTC   09:15 UTC   09:30 UTC   09:45 UTC   10:00 UTC   10:15 UTC   10:30 UTC   10:45 UTC   11:00 UTC

Fig. 3. Two-hour prediction sample of the ConvGRU model trained on only sequence data on 2019-07-24. The top row shows MSG-CPP data, of which the five left images are input to the model and the other images are the expected output. The second row shows the model's predictions. The normalised irradiance ranges from 0 (white) to 1 (black). One is the clear sky irradiance.

4.2   Comparison to the Baseline Models

Figure 4 contains prediction samples from each method. The optical flow algorithm generates the top prediction sample. In this prediction, clouds closely resemble the clouds in the expected output, and throughout the first two hours of the simulation, we note that the structures from the initialisation frames are preserved. This can be explained by the model's primary assumption that pixel values do not change over time. When comparing the methods, we note a clear difference at 09:15 UTC between the WRF prediction and the two others. This is because WRF is initialised on GFS data, while the ConvGRU and optical flow models are both initialised on the MSG-CPP data, and therefore, the output at the first lead time is very similar. After that, both generate increasingly different forecasts and portray the behaviour specific to both methods.

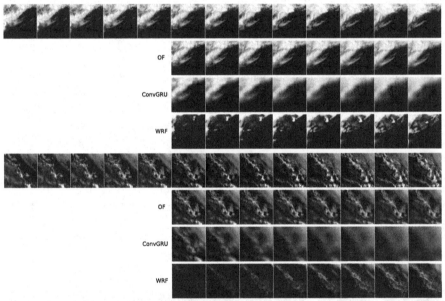

08:00UTC 08:15UTC 08:30UTC 08:45UTC 09:00UTC 09:15UTC 09:30UTC 09:45UTC 10:00UTC 10:15UTC 10:30UTC 10:45UTC 11:00UTC

Fig. 4. Examples of two-hour irradiance predictions by the three different methods on 2019-07-24 (top) and 2019-08-20 (bottom). For each date, the images in the top row are the input and expected output, and the images in the second, third and fourth row are the predictions by the optical flow, ConvGRU and WRF model, respectively.

In the MSG-CPP images on the top rows of the figure, the shape of the clouds evolves over the forecast. In the optical flow predictions, the clouds found in the input sequence move across the images and are stretched out in the figure. The ConvGRU model blurs the images to optimise the global error metrics and as a result, causes cloud-like shapes to disappear over time. The WRF model is the only forecasting method that can simulate the formation, growth and dissipation of clouds over time, but it suffers from inadequate initialisation and requires a spin-up time before more realistic cloud fields are simulated.

Figure 5 demonstrates that in terms of MBE and MAE, the WRF model is outperformed by the other models. In the first couple of model-steps, the optical flow and ConvGRU model performance are very similar, whereas the WRF performance is particularly poor due to the spin-up problem. After about 45 min, ConvGRU outperforms optical flow in terms of MAE and, on average, improves the MAE by 9%. In terms of SSIM, the optical flow model is best during the first 3.5 h, and the WRF model has the highest score after that. All models tend to overestimate irradiance (positive MBE) for nearly the entire forecast range in the study period.

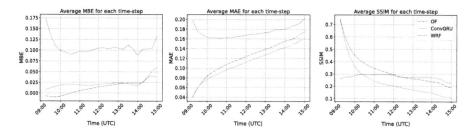

Fig. 5. The MBE, MAE and SSIM for the three methods. The metrics are averaged over the model domain and over all days from 2019-07-18 to 2019-09-30. Time is in UTC and the metrics are computed after normalisation of the data.

4.3   Weather Dependent Analysis

We next examine the metrics in different weather conditions. In terms of MAE, it is clear that the ConvGRU outperforms the other methods in all weather conditions and from Fig. 6, we see that the difference is largest on cloudy and very cloudy days. On such days, the ConvGRU model improves the optical flow MAE by 12% and 10%, respectively. On days with a clear sky index higher than 0.85, the improvement is 4%. The ConvGRU model, however, degrades more rapidly than the Optical Flow model in terms of SSIM. Especially on sunny days, optical flow is much better at predicting the irradiance than the ConvGRU model in terms of structural similarity. On average, the optical flow SSIM is 78% higher when compared to the ConvGRU SSIM.

Overall, the WRF model's performance is affected more strongly by the presence of clouds than the other models' performance. The MAE, for example, on partly cloudy days is more than double when compared to the same metrics for sunny days and increases by 42% for very cloudy days. Similarly, the MBE of the WRF forecast increases under more cloudy conditions (not shown), meaning that the model overestimates the irradiance more under such conditions. This demonstrates that the model consistently underestimates the irradiance absorbed and reflected (back to space) by clouds. The ConvGRU and optical flow predictions are less dependent on the weather than the WRF model.

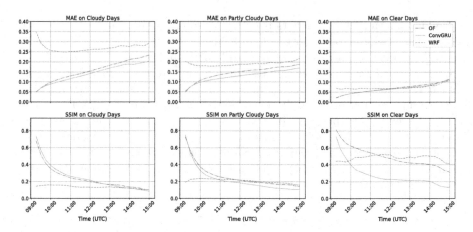

Fig. 6. The MAE (top) and SSIM (bottom) for all models grouped per weather type.

5   Conclusion

This research shows that the ConvGRU model trained on only sequences of MSG-CPP data provides the most accurate solar irradiance forecasts and improves the optical flow predictions by 9% in terms of MAE. The ConvGRU samples demonstrate that the model learned to represent clouds' advection, but it does show a blurring effect. The latter can be explained by the fact that the ConvGRU model is guided by a global-evaluating loss function and by the relatively small dataset the model is trained on. The blurring effect is the model's primary shortcoming as it causes the model not to represent cloud-like structures.

The cloud structures are better represented by the optical flow algorithm, especially on mostly clear days. The optical flow forecasts result in a 36% higher SSIM when compared to the ConvGRU predictions. Because of the assumption that pixel values stay constant from one frame to the next, optical flow mostly preserves the cloud-like structures from the initialisation data and advects the clouds over time. This enables the model to provide accurate predictions on a very short-term, but accuracy decreases for larger temporal scales as the optical flow model fails to represent the growth or dissipation of clouds.

One of the main strengths of trained DNNs and optical flow based methods is that models are easily initialised on the latest available weather data. Initialising WRF on the latest cloud data through data assimilation is a complex process, introduces uncertainty into the model and significantly increases the computational requirements. Because WRF was not initialised on the latest weather in our study, we cannot perform a reasonable comparison with the optical flow and ConvGRU models in the simulation's short range.

Which model is best depends on the forecast horizon and use cases. We also found that the performance of the different models depends on weather conditions. For that reason, operational forecasts are often provided by a forecasting system composed of several models. At lead times of 0–1 h, the predictions by the optical flow model are best. On clear days and when cloud fields do not quickly dissipate or grow, the model performs well for a longer time. When compared to the ConvGRU model, optical flow is better at preserving the cloud field structure. This is a significant benefit when producing very short-term predictions of the PV output in a specific area, i.e. a solar farm. However, at longer lead times, predicting clearly defined cloud structures that are slightly different from the real clouds or in a different location can result in a very imprecise estimation of the future PV output of a specific solar farm. In such cases, the blurry output of the ConvGRU model might be preferred.

For future work, we recommend modifying the DNN model to optimise for a locally-oriented loss function to solve the blurry image problem of the ConvGRU model. Recent literature on precipitation nowcasting replaced the globally-oriented L1 loss function with a loss function based on the SSIM and showed that this approach improves the quality of the predictions and significantly reduces blurry image issue [12]. Furthermore, we recommend training the ConvGRU model on more data. This increases variability and generally improves the performance of RNNs and can be achieved in various ways.

References

1. Gimon, E., O'Boyle, M., Clack, C.T.: The Coal Cost Crossover: Economic Viability of Existing Coal Compared to New Local Wind and Solar Resources. Vibrant Clean Energy, LLC (2019)
2. Brunekreeft, G., Buchmann, M., Meyer, R.: New developments in electricity markets following large-scale integration of renewable energy. In: The Routledge Companion to Network Industries (2015)
3. Bird, L., Milligan, M., Lew, D.: Integrating variable renewable energy: challenges and solutions (2013)
4. Lund, P.D., Lindgren, J., Mikkola, J., Salpakari, J.: Review of energy system flexibility measures to enable high levels of variable renewable electricity. Renew. Sustain. Energy Rev. **45**, 785–807 (2015)
5. Bauer, P., Thorpe, A., Brunet, G.: The quiet revolution of numerical weather prediction. Nature **525**, 47–55 (2015)
6. Jimenez, P.A., et al.: WRF-Solar: description and clear-sky assessment of an augmented NWP model for solar power prediction. Bull. Am. Meteorol. Soc. **97**, 1249–1264 (2016)
7. Shi, X., Chen, Z., Wang, H., Yeung, D.-Y., Wong, W.-K., Woo, W.: Convolutional LSTM network: a machine learning approach for precipitation nowcasting. arXiv preprint arXiv:1506.04214 (2015)
8. Shi, X., et al.: Deep learning for precipitation nowcasting: a benchmark and a new model. arXiv preprint arXiv:1706.03458 (2017)
9. Sonderby, C.K., et al.: MetNet: a neural weather model for precipitation forecasting. arXiv preprint arXiv:2003.12140 (2020)
10. Siddiqui, T.A., Bharadwaj, S., Kalyanaraman, S.: A deep learning approach to solar-irradiance forecasting in sky-videos. In: 2019 IEEE Winter Conference on Applications of Computer Vision (WACV), pp. 2166–2174 (2019)
11. Zhang, J., Verschae, R., Nobuhara, S., Lalonde, J.-F.: Deep photovoltaic nowcasting. Sol. Energy **176**, 267–276 (2018)

12. Tran, Q.-K., Song, S.: Computer vision in precipitation nowcasting: applying image quality assessment metrics for training deep neural networks. Atmosphere **10**, 244 (2019)
13. Esteban, C., Staeck, O., Baier, S., Yang, Y., Tresp, V.: Predicting clinical events by combining static and dynamic information using recurrent neural networks (2016)
14. Ayzel, G., Heistermann, M., Winterrath, T.: Optical flow models as an open benchmark for radar-based precipitation nowcasting (rainymotion v0.1). Geosci. Model Dev. **12**, 1387–1402 (2019)
15. Kellerhals, S.: Cloud and solar radiation nowcasting using optical flow and convolutional gated recurrent units. MSc Thesis. University of Amsterdam (2020)
16. Verbois, H., Huva, R., Rusydi, A., Walsh, W.: Solar irradiance forecasting in the tropics using numerical weather prediction and statistical learning. Sol. Energy **162**, 265–277 (2018)
17. Huva, R., Verbois, H., Walsh, W.: Comparisons of next-day solar forecasting for Singapore using 3DVAR and 4DVAR data assimilation approaches with the WRF model. Renew. Energy **147**, 663–671 (2020)
18. Thompson, G., Eidhammer, T.: A study of aerosol impacts on clouds and precipitation development in a large winter cyclone. J. Atmos. Sci. **71**, 3636–3658 (2014)
19. Iacono, M.J., Delamere, J.S., Mlawer, E.J., Shephard, M.W., Clough, S.A., Collins, W.D.: Radiative forcing by long-lived greenhouse gases: calculations with the AER radiative transfer models. J. Geophys. Res. **113** (2008)
20. Deng, A., Gaudet, B., Dudhia, J., Alapaty, K.: Implementation and evaluation of a new shallow convection scheme in WRF. In: 26th Conference on Weather Analysis and Forecasting/22nd Conference on Numerical Weather Prediction (2014)
21. Kain, J.S.: The Kain-Fritsch convective parameterization: an update. J. Appl. Meteorol. **43**, 170–181 (2004)
22. Chen, F., Dudhia, J.: Coupling an advanced land surface-hydrology model with the Penn State-NCAR MM5 modeling system. Part I: model implementation and sensitivity. Mon. Weather Rev. **129**, 569–585 (2001)
23. Roebeling, R.A., Feijt, A.J., Stammes, P.: Cloud property retrievals for climate monitoring: implications of differences between Spinning Enhanced Visible and Infrared Imager (SEVIRI) on METEOSAT-8 and Advanced Very High Resolution Radiometer (AVHRR) on NOAA-17. J. Geophys. Res. Atmos. **111** (2006)
24. Greuell, W., Meirink, J.F., Wang, P.: Retrieval and validation of global, direct, and diffuse irradiance derived from SEVIRI satellite observations. J. Geophys. Res. Atmos. **118**, 2340–2361 (2013)
25. Wang, Z., Bovik, A.C., Sheikh, H.R., Simoncelli, E.P.: Image quality assessment: from error visibility to structural similarity. IEEE Trans. Image Process. **13**, 600–612 (2004)
26. WRF Users Page: WPS V4 Geographical Static Data Downloads Page. https://www2.mmm.ucar.edu/wrf/users/download/get_sources_wps_geog.html. Accessed 02 Apr 2021

Automatic-differentiated Physics-Informed Echo State Network (API-ESN)

Alberto Racca[1] and Luca Magri[1,2,3(✉)]

[1] Department of Engineering, University of Cambridge, Cambridge, UK
lm547@cam.ac.uk
[2] The Alan Turing Institute, London, UK
[3] Aeronautics Department, Imperial College London, London, UK

Abstract. We propose the Automatic-differentiated Physics-Informed Echo State Network (API-ESN). The network is constrained by the physical equations through the reservoir's exact time-derivative, which is computed by automatic differentiation. As compared to the original Physics-Informed Echo State Network, the accuracy of the time-derivative is increased by up to seven orders of magnitude. This increased accuracy is key in chaotic dynamical systems, where errors grow exponentially in time. The network is showcased in the reconstruction of unmeasured (hidden) states of a chaotic system. The API-ESN eliminates a source of error, which is present in existing physics-informed echo state networks, in the computation of the time-derivative. This opens up new possibilities for an accurate reconstruction of chaotic dynamical states.

Keywords: Reservoir computing · Automatic differentiation · Physics-informed echo state network

1 Introduction

In fluid mechanics, we only rarely have experimental measurements on the entire state of the system because of technological/budget constraints on the number and placement of sensors. In fact, we typically measure only a subset of the state, the *observed* states, but we do not have data on the remaining variables, the *hidden* states. In recent years, machine learning techniques have been proposed to

A. Racca is supported by the EPSRC-DTP and the Cambridge Commonwealth, European & International Trust under a Cambridge European Scholarship. L. Magri is supported by the Royal Academy of Engineering Research Fellowship scheme and the visiting fellowship at the Technical University of Munich – Institute for Advanced Study, funded by the German Excellence Initiative and the European Union Seventh Framework Programme under grant agreement n. 291763.
L. Magri—(visiting) Institute for Advanced Study, Technical University of Munich, Germany

© Springer Nature Switzerland AG 2021
M. Paszynski et al. (Eds.): ICCS 2021, LNCS 12746, pp. 323–329, 2021.
https://doi.org/10.1007/978-3-030-77977-1_25

infer hidden variables, which is also known as *reconstruction*. A fully-data driven approach to reconstruction assumes that data for the hidden states is available only for a limited time interval, which is used to train the network [8]. On the other hand, a physics-informed approach to reconstruction employs the governing equations [4,10]. In this work, we use automatic differentiation to eliminate the source of error in the Physics-Informed Echo State Network, which originates from approximating the time-derivative of the network [4]. Automatic differentiation records the elementary operations of the model and evaluates the derivative by applying the chain rule to the derivatives of these operations [2]. With automatic differentiation, we compute exactly the time-derivative, thereby extending the network's ability to reconstruct hidden states in chaotic systems. In Sect. 2, we present the proposed network: the Automatic-differentiated Physics-Informed Echo State Network (API-ESN). In Sect. 3, we discuss the results. We summarize the work and present future developments in Sect. 4. The code for the API-ESN can be found in the openly-available GitLab repository https://gitlab.com/ar994/automatic-differentiated-pi-esn.

2 Automatic-Differentiated Physics-Informed Echo State Network (API-ESN)

We study the nonlinear dynamical system

$$\dot{\mathbf{y}} = \mathbf{f}(\mathbf{y}), \tag{1}$$

where $\mathbf{y} \in \mathbb{R}^{N_y}$ is the state of the physical system, $\mathbf{f}$ is a nonlinear operator, and $\dot{}$ is the time-derivative. We consider a case where $\mathbf{y}$ consists of an observed state, $\mathbf{x} \in \mathbb{R}^{N_x}$, and a hidden state, $\mathbf{h} \in \mathbb{R}^{N_h}$: $\mathbf{y} = [\mathbf{x}; \mathbf{h}]$, where $[\cdot\,;\cdot]$ indicates vertical concatenation and $N_y = N_x + N_h$. We assume we have non-noisy data on $\mathbf{x}$, and its derivative, $\dot{\mathbf{x}}$, which can be computed offline. We wish to reconstruct $\mathbf{h}$ given the data. The $N_t + 1$ training data points for the observed states are $\mathbf{x}(t_i)$ for $i = 0, 1, 2, \ldots, N_t$, taken from a time series that ranges from $t_0 = 0$ to $t_{N_t} = N_t \Delta t$, where Δt is the constant time step. We introduce the Automatic-differentiated Physics-Informed Echo State Network (API-ESN) to reconstruct $\mathbf{h}$ at the same time instants, by constraining the time-derivative of the network through the governing equations. The network is based on the Physics-Informed Echo State Network (PI-ESN) [3,4], which, in turn, is based on the fully data-driven ESN [5,9]. In the PI-ESN, the time-derivative of the network is approximated by a first-order forward Euler numerical scheme. In this work, we compute the derivative at machine precision through automatic differentiation [1,2].

In the API-ESN, the data for the observed state, $\mathbf{x}$, updates the state of the high-dimensional reservoir, $\mathbf{r} \in \mathbb{R}^{N_r}$, which acts as the memory of the network. At the i-th time step, $\mathbf{r}(t_i)$ is a function of its previous value, $\mathbf{r}(t_{i-1})$, and the current input, $\mathbf{x}(t_i)$. The output is the predicted state at the next time step: $\hat{\mathbf{y}}(t_i) = [\hat{\mathbf{x}}(t_{i+1}); \hat{\mathbf{h}}(t_{i+1})] \in \mathbb{R}^{N_y}$. It is the linear combination of $\mathbf{r}(t_i)$ and $\mathbf{x}(t_i)$

$$\mathbf{r}(t_i) = \tanh\left(\mathbf{W}_{\text{in}}[\mathbf{x}(t_i); b_{\text{in}}] + \mathbf{W}\mathbf{r}(t_{i-1})\right); \quad \hat{\mathbf{y}}(t_i) = \mathbf{W}_{\text{out}}[\mathbf{r}(t_i); \mathbf{x}(t_i); 1] \tag{2}$$

where $\mathbf{W} \in \mathbb{R}^{N_r \times N_r}$ is the state matrix, $\mathbf{W}_{\text{in}} \in \mathbb{R}^{N_r \times (N_x+1)}$ is the input matrix, $\mathbf{W}_{\text{out}} \in \mathbb{R}^{N_y \times (N_r+N_x+1)}$ is the output matrix and b_{in} is the input bias. The input matrix, $\mathbf{W}_{\text{in}}$, and state matrix, $\mathbf{W}$, are sparse, randomly generated and fixed. These are constructed in order for the network to satisfy the echo state property [9]. The input matrix, $\mathbf{W}_{\text{in}}$, has only one element different from zero per row, which is sampled from a uniform distribution in $[-\sigma_{\text{in}}, \sigma_{\text{in}}]$, where σ_{in} is the input scaling. The state matrix, $\mathbf{W}$, is an Erdős-Renyi matrix with average connectivity $\langle d \rangle$. This means that each neuron (each row of $\mathbf{W}$) has on average only $\langle d \rangle$ connections (non-zero elements). The value of the non-zero elements is obtained by sampling from an uniform distribution in $[-1, 1]$; the entire matrix is then scaled by a multiplication factor to set its spectral radius, ρ. The only trainable weights are those in the output matrix, $\mathbf{W}_{\text{out}}$. The first N_x rows of the output matrix, $\mathbf{W}_{\text{out}}^{(x)}$, are computed through ridge regression on the available data for the observed state by solving the linear system

$$\left(\mathbf{R}\mathbf{R}^T + \gamma \mathbf{I}\right) \mathbf{W}_{\text{out}}^{(x)^T} = \mathbf{R}\mathbf{X}^T, \tag{3}$$

where $\mathbf{X} \in \mathbb{R}^{N_x \times N_t}$ and $\mathbf{R} \in \mathbb{R}^{(N_r+N_x+1) \times N_t}$ are the horizontal concatenation of the observed states, $\mathbf{x}$, and associated reservoir states, $[\mathbf{r}; \mathbf{x}; 1]$, respectively; γ is the Tikhonov regularization factor and $\mathbf{I}$ is the identity matrix [9]. The last N_h rows of the output matrix, $\mathbf{W}_{\text{out}}^{(h)}$, are initialized by solving (3), where we embed prior knowledge of the physics by substituting $\mathbf{X}$ with $\mathbf{H} \in \mathbb{R}^{N_h \times N_t}$, whose rows are constants and equal to the components of the estimate of the mean of the hidden state, $\overline{\mathbf{h}} \in \mathbb{R}^{N_h}$. To train $\mathbf{W}_{\text{out}}^{(h)}$ only, we minimize the loss function, $\mathcal{L}_{\text{Phys}}$

$$\mathcal{L}_{\text{Phys}} = \frac{1}{N_t N_y} \sum_{j=1}^{N_t} ||\dot{\hat{\mathbf{y}}}(t_j) - \mathbf{f}(\hat{\mathbf{y}}(t_j))||^2, \tag{4}$$

where $|| \cdot ||$ is the L_2 norm; $\mathcal{L}_{\text{Phys}}$ is the Mean Squared Error between the time-derivative of the output, $\dot{\hat{\mathbf{y}}}$, and the right-hand side of the governing equations evaluated at the output, $\hat{\mathbf{y}}$. To compute $\dot{\hat{\mathbf{y}}}$, we need to differentiate $\hat{\mathbf{y}}$ with respect to $\mathbf{x}$, because the time dependence of the network is implicit in the input, $\mathbf{x}(t)$, i.e. $\frac{d\hat{\mathbf{y}}}{dt} = \frac{\partial \hat{\mathbf{y}}}{\partial \mathbf{x}} \frac{d\mathbf{x}}{dt}$. The fact that $\dot{\hat{\mathbf{y}}}$ is a function of $\dot{\mathbf{x}}$ means that the accuracy of the derivative of the output is limited by the accuracy of the derivative of the input. In this work, we compute $\dot{\mathbf{x}}$ exactly using the entire state to evaluate $\mathbf{f}(\mathbf{y})$ in (1). Furthermore, $\hat{\mathbf{y}}$ depends on all the inputs up to the current input due to the recurrent connections between the neurons. In Echo State Networks, we have the recursive dependence of the reservoir state, $\mathbf{r}$, with its previous values, i.e., omitting the input bias for brevity

$$\mathbf{r}(t_i) = \tanh\left(\mathbf{W}_{\text{in}}\mathbf{x}(t_i) + \mathbf{W}\tanh\left(\mathbf{W}_{\text{in}}\mathbf{x}(t_{i-1}) + \mathbf{W}\tanh(\mathbf{W}_{\text{in}}\mathbf{x}(t_{i-2}) + \dots)\right)\right). \tag{5}$$

Because of this, the time-derivative of the current output has to be computed with respect to all the previous inputs in the training set

$$\dot{\hat{\mathbf{y}}}(t_i) = \frac{d\hat{\mathbf{y}}}{dt}\bigg|_{t_i} = \frac{\partial \hat{\mathbf{y}}(t_i)}{\partial \mathbf{x}(t_i)} \frac{d\mathbf{x}}{dt}\bigg|_{t_i} + \frac{\partial \hat{\mathbf{y}}(t_i)}{\partial \mathbf{x}(t_{i-1})} \frac{d\mathbf{x}}{dt}\bigg|_{t_{i-1}} + \frac{\partial \hat{\mathbf{y}}(t_i)}{\partial \mathbf{x}(t_{i-2})} \frac{d\mathbf{x}}{dt}\bigg|_{t_{i-2}} + \dots \tag{6}$$

This is computationally cumbersome. To circumvent this extra computational cost, we compute $\dot{\hat{y}}$ through the derivative of the reservoir's state, $\dot{r} = \frac{\partial r}{\partial x} \frac{dx}{dt}$. By differentiating (2) with respect to time, we obtain $\dot{\hat{y}} = W_{out}[\dot{r}; \dot{x}; 0]$. Because $\dot{r}$ is independent of W_{out}, $\dot{r}$ is fixed during training. Hence, the automatic differentiation of the network is performed only once during initialization. We compute $\dot{r}$ as a function of the current input and previous state as the network evolves

$$\dot{r}(t_i) = \frac{dr}{dt}\bigg|_{t_i} = \frac{\partial r(t_i)}{\partial x(t_i)} \frac{dx}{dt}\bigg|_{t_i} + \frac{\partial r(t_i)}{\partial r(t_{i-1})} \frac{dr}{dt}\bigg|_{t_{i-1}}, \qquad (7)$$

in which we initialize $\dot{r}(t_0) = 0$ and $r(t_0) = 0$ at the beginning of the washout interval (the washout interval is the initial transient of the network, during which we feed the inputs without recording the outputs in order for the state of the network to be uniquely defined by the sequence of the inputs [9]).

3   Reconstruction of Hidden States in a Chaotic System

We study the Lorenz system [7], which is a prototypical chaotic system that models Rayleigh–Bénard convection

$$\dot{\phi}_1 = \sigma_L(\phi_2 - \phi_1), \qquad \dot{\phi}_2 = \phi_1(\rho_L - \phi_3) - \phi_2, \qquad \dot{\phi}_3 = \phi_1\phi_2 - \beta_L\phi_3, \qquad (8)$$

where the parameters are $[\sigma_L, \beta_L, \rho_L] = [10, 8/3, 28]$. To obtain the data, we integrate the equation through the implicit adaptive step scheme of the function odeint in the scipy library. The training set consists of $N_t = 10000$ points with step $\Delta t = 0.01$LTs, where a Lyapunov Time (LT) is the inverse of the leading Lyapunov exponent Λ of the system, which, in turn, is the exponential rate at which arbitrarily close trajectories diverge. In the Lorenz system, $\Lambda = LT^{-1} \simeq 0.906$. We use networks of variable sizes from $N_r = 100$ to $N_r = 1000$, with parameters $\langle d \rangle = 20$, $\gamma = 10^{-6}$ and $\rho = 0.9$ [8]. We set $\sigma_{in} = 0.1$, $b_{in} = 10$, and $\bar{h} = 10$, to take into account the order of magnitude of the inputs ($\sim 10^1$). We train the networks using the Adam optimizer [6] with initial learning rate $l = 0.1$, which we decrease during optimization to prevent the training from not converging to the optimal weights due to large steps of the gradient descent.

In Fig. 1, we compare the accuracy of the Automatic Differentiation (AD) derivative of the API-ESN with respect to the first-order Forward Euler (FE) approximation of the PI-ESN [4]. Here, we study the case where the entire state is known, $y = x$, to be able to compute the true derivative of the predicted state, $f(\hat{y})$, (1). In plot (a), we show $f(\hat{y})$ in an interval of the training set for $N_r = 100$. In plot (b), we show in the same interval the squared norm of the error for FE, $\mathcal{L}_{FE}$, and AD, $\mathcal{L}_{AD}$, with respect to $f(\hat{y})$. In addition, we show the squared norm of the error, $\mathcal{L}_Y$, of the output, $\hat{y}$, with respect to the data, y. Because FE and AD share the same $\hat{y}$, $\mathcal{L}_Y$ is the same for the two networks. In plot (c), we show the time average of the squared norms, indicated by the overline, as a function of the size of the reservoir. In the case of $\overline{\mathcal{L}}_{FE}$ and $\overline{\mathcal{L}}_{AD}$, they coincide with $\mathcal{L}_{Phys}$ (4). AD is four to seven orders of magnitude more accurate than FE. The error

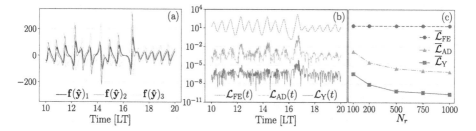

Fig. 1. Time-derivative of the output for $N_r = 100$ in an interval of the training set, (a). Squared norm of the error of the output derivative, $\mathcal{L}_{\text{FE}}$ and $\mathcal{L}_{\text{AD}}$, and output, $\mathcal{L}_{\text{Y}}$, in the same interval, (b). Mean Squared Error of the output derivative, $\overline{\mathcal{L}}_{\text{FE}}$ and $\overline{\mathcal{L}}_{\text{AD}}$, and output, $\overline{\mathcal{L}}_{\text{Y}}$, as a function of the reservoir size, (c).

of the FE numerical approximation is dominant from $N_r = 100$. This prevents its accuracy from increasing in larger networks.

We use the API-ESN to reconstruct the hidden states in three testcases: (i) reconstruction of $\mathbf{h} = [\phi_2]$ given $\mathbf{x} = [\phi_1; \phi_3]$; (ii) reconstruction of $\mathbf{h} = [\phi_3]$ given $\mathbf{x} = [\phi_1; \phi_2]$; (iii) reconstruction of $\mathbf{h} = [\phi_2; \phi_3]$ given $\mathbf{x} = [\phi_1]$. We choose (i) and (ii) to highlight the difference in performance when we reconstruct different variables, and (iii) to compare the reconstruction of the states ϕ_2 and ϕ_3 when fewer observed states are available. We reconstruct the hidden states in the training set and in a 10000 points test set subsequent to the training set. The reconstructed states in an interval of the training set for networks of size $N_r = 1000$ are shown in plots (a, d) in Fig. 2. The network is able to reconstruct satisfactorily the hidden states. The accuracy deteriorates only in the large amplitude oscillations of ϕ_3, (d). To visualize the global performance over both the training set and the test set, we plot the Probability Density Functions (PDF) in (b, e), and (c, f), respectively. The PDFs are reconstructed with similar accuracy between the two sets. Interestingly, the increased difficulty in reconstructing ϕ_3 is due to the dynamical the system's equations (rather than the network's ability to learn). Indeed, ϕ_3 appears in only two of the three equations (8), whereas ϕ_2 appears in all the equations and it is a linear function in the first equation. In other words, we can extract more information for ϕ_2 than for ϕ_3 from the constraint of the physical equations, which means that the network can better reconstruct the dynamics of ϕ_2 vs. ϕ_3. In the lower part of large amplitude oscillations of ϕ_3 in particular, we have small values for the derivatives, so that the error in the governing equations is small.

To quantitatively assess the reconstruction, we compute for each component, h_i, of the hidden state, $\mathbf{h}$, the Normalized Root Mean Squared Error: $\text{NRMSE}(h_i) = \sqrt{N^{-1}\sum_{j}^{N}(\hat{h}_i(t_j) - h_i(t_j))^2}/(\max(h_i) - \min(h_i))$; where $(\max(h_i) - \min(h_i))$ is the range of h_i. In Fig. 3, we show the values of the NRMSE for different sizes of the reservoir, N_r, in the training (continuous lines) and test (dash-dotted lines) sets for FE, (a, c), and AD, (b, d). The error of the FE approximation dominates the reconstruction of ϕ_2, (a), while AD produces

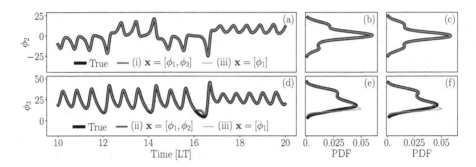

Fig. 2. Reconstruction of the hidden variables for $N_r = 1000$ in an interval of the training set, (a) and (d), Probability Density Function (PDF) in the training, (b, e), and test set, (c, f). Reconstruction of $\mathbf{h} = [\phi_2]$ (top row) and $\mathbf{h} = [\phi_3]$ (bottom row): testcase (i) is shown in (a–c), testcase (ii) is shown in (d–f), and testcase (iii) is shown in (a–f).

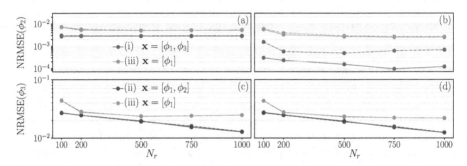

Fig. 3. NRMSE in the training (continuous lines) and test (dash-dotted lines) sets as a function of the size of the reservoir. The reconstruction is performed through forward Euler, (a, c), and automatic differentiation, (b, d). Reconstruction of $\mathbf{h} = [\phi_2]$, (top row) and $\mathbf{h} = [\phi_3]$ (bottom row): testcase (i) is shown in (a, b), testcase (ii) is shown in (c, d), and testcase (iii) is shown in (a–d).

NRMSEs up to one order of magnitude smaller, (b). In the reconstruction of ϕ_3, (c, d), AD and FE perform similarly because of the dominant error related to the mathematical structure of the dynamical system's equations, as argued for Fig. 2(e, f). In general, the values of the NRMSE are similar between the training and test sets, which indicates that the reconstruction also works on unseen data. There is a difference between the two sets only in the reconstruction of ϕ_2 given $\mathbf{x} = [\phi_1; \phi_3]$, (b). In this case, the NRMSE in the test set is smaller than the NRMSE we obtain when using data on the hidden state to train the network, through the network described in [8] (results not shown). In addition, we compare the reconstruction of the states when fewer observed states are available. When only ϕ_1 is available ($\mathbf{x} = [\phi_1]$), the larger error on ϕ_3 limits the accuracy on the reconstruction on ϕ_2. This results in a reconstruction of ϕ_3, (d), with similar accuracy to the case where $\mathbf{x} = [\phi_1; \phi_2]$, while there is a larger error in the reconstruction of ϕ_2, (b), with respect to the case where $\mathbf{x} = [\phi_1; \phi_3]$.

4   Conclusions and Future Directions

We propose the Automatic-differentiated Physics-Informed Echo State Network (API-ESN) to leverage the knowledge of the governing equations in an echo state network. We use automatic differentiation to compute the exact time-derivative of the output, which is shown to be a function of all the network's previous inputs through the recursive time dependence intrinsic in the neurons' recurrent connections. Albeit this long time-dependence in the past would make the computation of the time-derivative computationally cumbersome, we eliminate this cost by computing the derivative of the reservoir's state. We apply the API-ESN to a prototypical chaotic system to reconstruct the hidden states in different datasets both in the training points and on unseen data. We compare the API-ESN to the forward Euler approximation of the Physics-Informed Echo State Network. We obtain a Normalized Mean Squared Error up to one order of magnitude smaller in the reconstruction of the hidden state. Future work will focus on using the API-ESN to reconstruct and predict (by letting the network evolve autonomously) hidden states from experimental data.

References

1. Abadi, M., et al.: TensorFlow: large-scale machine learning on heterogeneous systems (2015). https://www.tensorflow.org/
2. Baydin, A.G., Pearlmutter, B.A., Radul, A.A., Siskind, J.M.: Automatic differentiation in machine learning: a survey. J. Mach. Learn. Res. **18**(153), 1–43 (2018). http://jmlr.org/papers/v18/17-468.html
3. Doan, N.A.K., Polifke, W., Magri, L.: Physics-informed echo state networks for chaotic systems forecasting. In: Rodrigues, J.M.F., et al. (eds.) ICCS 2019. LNCS, vol. 11539, pp. 192–198. Springer, Cham (2019). https://doi.org/10.1007/978-3-030-22747-0_15
4. Doan, N.A.K., Polifke, W., Magri, L.: Learning hidden states in a chaotic system: a physics-informed echo state network approach. In: Krzhizhanovskaya, V.V., et al. (eds.) ICCS 2020. LNCS, vol. 12142, pp. 117–123. Springer, Cham (2020). https://doi.org/10.1007/978-3-030-50433-5_9
5. Jaeger, H., Haas, H.: Harnessing nonlinearity: predicting chaotic systems and saving energy in wireless communication. Science **304**(5667), 78–80 (2004)
6. Kingma, D.P., Ba, J.: Adam: a method for stochastic gradient descent. In: ICLR: International Conference on Learning Representations, pp. 1–15 (2015)
7. Lorenz, E.N.: Deterministic Nonperiodic Flow. J. Atmos. Sci. **20**(2), 130–141 (1963)
8. Lu, Z., Pathak, J., Hunt, B., Girvan, M., Brockett, R., Ott, E.: Reservoir observers: model-free inference of unmeasured variables in chaotic systems. Chaos Interdisc. J. Nonlinear Sci. **27**(4), 041102 (2017)
9. Lukoševičius, M.: A practical guide to applying echo state networks. In: Montavon, G., Orr, G.B., Müller, K.-R. (eds.) Neural Networks: Tricks of the Trade. LNCS, vol. 7700, pp. 659–686. Springer, Heidelberg (2012). https://doi.org/10.1007/978-3-642-35289-8_36
10. Raissi, M., Yazdani, A., Karniadakis, G.E.: Hidden fluid mechanics: learning velocity and pressure fields from flow visualizations. Science **367**(6481), 1026–1030 (2020)

A Machine Learning Method
for Parameter Estimation and Sensitivity
Analysis

Marcella Torres[(✉)] [iD]

University of Richmond, Richmond, VA 23173, USA
mtorres@richmond.edu

Abstract. We discuss the application of a supervised machine learning method, random forest algorithm (RF), to perform parameter space
exploration and sensitivity analysis on ordinary differential equation
models. Decision trees can provide complex decision boundaries and can
help visualize decision rules in an easily digested format that can aid
in understanding the predictive structure of a dynamic model and the
relationship between input parameters and model output. We study a
simplified process for model parameter tuning and sensitivity analysis
that can be used in the early stages of model development.

Keywords: Parameter estimation · Machine learning · Sensitivity
analysis · Ordinary differential equations · Random forest

1 Introduction

Machine learning, a sub-field of artificial intelligence, is most commonly used
to produce an accurate classifier given data. For example, given a data set containing various predictors such as blood pressure and age, a machine learning
model could be used to predict whether a given patient has diabetes. This type
of application requires adequate experimental data to produce a good classifier, and then the function of such a machine learning model is to accurately
classify a new case given predictors. Dynamical systems models, in contrast,
offer greater flexibility and more information about the time course of a system rather than just an outcome. Yet these models contain parameters that can
be hard to estimate given limited data and model complexity, and the effectiveness of a model depends on the quality of the connection between model
parameters and output. In deterministic models such as the ordinary differential
equations model presented here, the only model inputs are the model parameters
and initial conditions and so these are the sources of any uncertainty in model
prediction. Before such a model can be deemed useful for prediction, this uncertainty must be quantified by performing sensitivity analysis. This is a critical
step in the model building process, since knowledge of influential parameters will
guide experimental design, data assimilation, parameter estimation, and model
refinement in the form of complexity reduction.

© Springer Nature Switzerland AG 2021
M. Paszynski et al. (Eds.): ICCS 2021, LNCS 12746, pp. 330–343, 2021.
https://doi.org/10.1007/978-3-030-77977-1_26

Machine learning techniques have previously been applied to the problem of parameter estimation in ordinary differential equations, and to identifying multicollinearity among parameters. For example, support vector clustering (a combination of support vector machine and clustering) has been used for model parameter estimation [24], and clustering (an unsupervised learning method) has also been used to reduce the number of parameters that need to be estimated by identifying collinearities in pairwise groupings of parameters [6,23]. Neural networks have also been combined with differential equations to produce models with large data sets [18,19], and in the case of limited data, sparse identification of nonlinear dynamical systems has been used to discover model structure [5]. Since machine learning uses pattern recognition methods to construct classifiers for data sets, it can also be used to uncover the predictive structure of a model, i.e. the strength of the connection between inputs and outputs. If the goal of developing a good dynamical model is that the model can produce an expected range of biologically reasonable outcomes (epidemic or endemic in SIR models, for example, or survival versus extinction in a predator prey model), then we already have some idea of how output could be classified. When we know the class of each observation in the data set, we can use a supervised learning method to classify future observations. Decision tree algorithms, also called classification and regression tree algorithms (CART), developed by Breiman [3], are one example of many available supervised learning methods, and this is the method we apply here. Decision trees are ideal for parameter space exploration and global sensitivity analysis because the complex relationships between parameters can be easily visualized and decision rules can guide parameter subset selection.

By combining uncertainty analysis (UA) using Latin Hypercube Sampling, as developed by Marino et al. [12], and sensitivity analysis using decision trees, intuition can be gained about how partitions in parameter space produce different model behaviors. In this way we can gain early insight into what behaviors the model is capable of producing and which parameters are driving outcomes. This preliminary exploration can also serve several practical purposes:

- **Visual communication of the significance of model parameters to non-mathematicians.** The tree format allows interactions between multiple parameters and the associated output to be represented simultaneously. In this way it can be made clear that the same behavior can result from different combinations of parameters. Most other learning-based methods of parameter exploration are not easily visualized or are restricted to pairwise parameter plots.
- **Sensitivity analysis.** The first step in estimating model parameters is to identify sensitive parameters - the parameters that impact model output [23]. This is easily obtained from decision tree algorithms as a feature importance measure. In general, this a very quick way of getting a first look at the key

parameters that drive model behavior. We can change class definitions dependant on the behavior for which we want to identify these important parameters; for example, instead of endemic versus epidemic we can consider stable versus unstable.

- **Decision rules can be used to find representative sets for simulation.** Collections of parameters that serve as input to a model can represent different hypothetical individuals and sub-populations in a biological model. To model these individuals or groups specifically, we can choose parameter sets from value ranges restricted to those suggested by a decision tree.
- **Decision rules can be used to set bounds on parameter space for fitting.** Given experimental data, we know the ultimate class of the observation. Setting bounds on the parameter space and choosing initial parameter values for data fitting is nontrivial and can be time consuming. Generation of a decision tree from simulated data is relatively fast and straightforward. We can then refer to the decision rules on parameters that result in the same class the observation is in to initiate parameter estimation.
- **Decision rules can be used to restrict parameters to those that produce biologically reasonable behavior.** When representative subsets of parameters found using decision rules are investigated and found to produce non-physiological behavior, these parameter sets can be excluded or constraints can be identified. This could also be done by labeling output as either biologically viable or nonviable in the data set as the class label.

To perform sensitivity analysis using decision tree algorithms, we first perform uncertainty analysis by sampling model output over a range of parameter values. Next, a data set is created by appending associated model output to the matrix containing all sampled parameter sets. Each parameter set, together with its output, is then given a binary class label that characterizes the behavior of interest which will serve as the class in the decision tree classifier. Since model output is continuously defined, this is done by defining a criteria on output values which transforms each output value to one of the two class labels. The decision tree is then trained, pruned, and tested on the data. Given acceptable classification accuracy, the tree, or an ensemble of trees, can then be used to analyze parameter importance and identify subsets of parameters that produce particular model outcomes.

2   Methods

First, we perform uncertainty analysis using Latin Hypercube Sampling (LHS), the most efficient of the Monte Carlo methods. LHS is a stratified sampling without replacement method developed by McKay et al. [8,10,12,14] in which each parameter is independently sampled from a statistical distribution in order to efficiently create an unbiased collection of parameter sets that can each be used to generate model output, thus simulating a variety of responses. The choice of statistical distribution from which to sample will be determined by knowledge of the modeled phenomena or an examination of available data. In the absence of information about the underlying distribution, the usual choice is to sample from a uniform distribution, with maximum and minimum values for each parameter determined by physiological constraints or through experimentation.

Next, we combine the matrix of parameter vectors with a vector (or vectors) of model output sampled at a particular time or times where we expect behavior to differ for different outcomes. Each row is labeled with a binary classifier according to a criterion on model output values which transforms each output value to a class label. A simplified table showing the representation of data is given in Fig. 1, where the classes are labeled "0" or "1".

In statistical learning, measurements are first made on an observation. The goal is to predict which class the observation is in based on the measurements. In this case, the measurements are the parameters $\vec{\theta}$ sampled using LHS which serve as input to the model for simulation and generation of model output. Each observation is a vector of m measurements $\vec{\theta}^{(i)} = \left[\theta_1^{(i)}, \theta_2^{(i)}, \cdots, \theta_m^{(i)} \right]$ which are the parameter sets that are used in combination as model input, and Ω is the parameter space of all possible parameter vectors which is defined by the LHS sample space.

Next, we employ a feature importance calculation with random forest algorithm (RF) to determine the sensitive parameters. A RF is composed of a large number of individual decision trees which sample from the data set with replacement [1,4,9]. Finally, we examine an individual decision tree to obtain the rules on parameters which lead to each of the states of interest. Using these rules, we obtain subsets of parameters that may exhibit different behavior.

3   Simple Benchmark Model Example

Here we illustrate the process of using decision trees to identify parameter subsets and significant parameters, decision tree sensitivity analysis). Ultimately, we will apply this method to more complex, large-scale systems which can benefit from its simplicity, but here we aim to compare the method to more traditional approaches, including solution of the sensitivity equations, so a simple model was chosen. We test the method by comparing results to global sensitivity analysis performed using PRCC as well as the magnitude of computed local sensitivities integrated over time for a well studied model of HIV infection.

3.1   Decision Tree Sensitivity Analysis Example: An HIV Infection Model

Perelson et al. developed a simplified model of HIV interaction with T cells consisting of four differential equations with eight parameters that model concentration of uninfected (T), latently infected (T^*), and actively infected (T^{**}) CD4$^+$ cells and free infectious virus particles [17]:

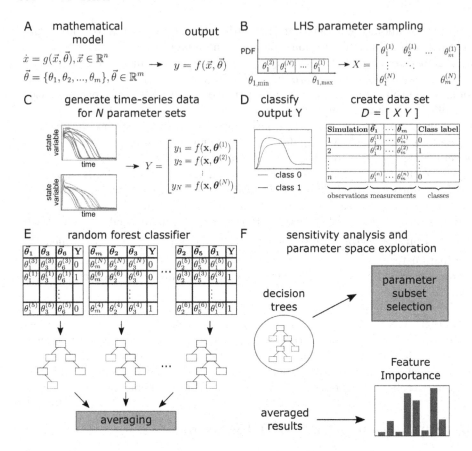

Fig. 1. Process overview. (A) An example mathematical model, an n-dimensional system of ordinary differential equations, where $\vec{x}$ represents a vector of state variables and $\vec{\theta}$ is the parameter vector with m model parameters. Model output y depends on state variables and parameters. (B) LHS sampling is performed over all non-fixed parameters, by random sampling of each parameter without replacement from a specified probability density function (a uniform distribution is pictured) over N equally-size bins. This creates N parameter sets, each containing all m model parameters, to form LHS matrix X. (C) N sets of time-series data is produced using all parameter sets. A vector Y of model output sampled at a specified time is selected as the response variable. (D) Assign class label to each output y and create data set D, the set of parameter sets and their associated labeled output. (E) Train random forest classifier with bootstrapping. (F) Obtain feature or permutation importance measure on averaged results as a parameter sensitivity measure. Examine decision trees individually to obtain parameter subsets for parameter space exploration.

$$\frac{dT}{dt} = s - \mu_T T + rT \left(1 - \frac{T + T^* + T^{**}}{T_{\max}}\right) - k_1 VT \tag{1}$$

$$\frac{dT^*}{dt} = k_1 VT - \mu_T T^* - k_2 T^* \tag{2}$$

$$\frac{dT^{**}}{dt} = k_2 T^* - \mu_b T^{**} \tag{3}$$

$$\frac{dV}{dt} = N\mu_b T^{**} - k_1 VT - \mu_V V. \tag{4}$$

Parameters and constants that Perelson et al. defined are in Table 1. Initial conditions used in the model are $T(0) = 1000\,\text{mm}^{-3}$, $T^*(0) = T^{**}(0) = 0$, and $V(0) = 10^{-3}\,\text{mm}^{-3}$. Marino et al. performed uncertainty analysis using LHS and sensitivity analysis using PRCCs, and the range used for sampling is also given in Table 1. We will use the same ranges on parameters for LHS sampling to create a data set for constructing decision trees, for comparison to PRCC results.

Two steady states are possible in this model: an uninfected state E_B with no virus, and an endemically infected state E_P with a constant level of the virus [12,17].

Table 1. Parameters and constants from Perelson et al. [17] and parameter ranges used for LHS sampling by Marino et al. [12].

Parameter	Description [17]	Default [17]	LHS range [12]
s	Rate of supply of CD4$^+$ T cells from precursors	10day^{-1} mm^{-3}	$[10^{-2}, 50]$
r	Rate of growth of CD4$^+$	0.03day^{-1}	$[10^{-4}, 0.2]$
μ_T	Death rate of infected and latently infected CD4$^+$ cells	0.02day^{-1}	$[10^{-2}, 50]$
$T_{\max}$	Maximum CD4$^+$ population level	1500 mm^{-3}	1500
k_1	Rate at which CD4$^+$ cells become infected	2.4×10^{-5} mm^3 day^{-1}	$[10^{-7}, 10^{-3}]$
k_2	Rate at which latently infected CD4$^+$ cells become actively infected	3×10^{-3} day^{-1}	$[10^{-5}, 10^{-2}]$
μ_b	Death rate of actively infected CD4$^+$ cells	0.24day^{-1}	$[10^{-1}, 0.4]$
N	Number of free virus produced by lysing a CD4$^+$ cell	Not fixed	$[1, 2^3]$
μ_V	Death rate of free virus	2.4day^{-1}	$[10^{-1}, 10]$

Perelson et al. discovered that a criterion for achieving the uninfected steady state E_B is

$$N < \frac{(k_2 + \mu_T)\mu_V + k_1 T_0}{k_1 k_2 T_0}, \tag{5}$$

and showed further showed analytically that parameters contained in this critical value for N, defined as N_{crit}, are bifurcation parameters.

Generating the Data Set. LHS was performed in MATLAB with sample size $N = 1000$, using code provided by Marino et al. [13]. A data matrix was created such that each row of the matrix contained a parameter set consisting of parameters s, r, μ_T, k_1, k_2, μ_b, μ_V, and N was sampled from a uniform distribution. The model given in Equations (1)–(4) tracks four variables that are possible model outputs to define as a classifier in decision tree sensitivity analysis. We chose concentration of free virus particles, V, as a binary classifier, sampled at time $t = 4000$ days, since this output should characterize the two steady states of biological interest: uninfected (E_B) or endemically infected (E_P). We labeled model output as uninfected if its value was near zero ($V < 1 \times 10^{-6}$), and endemically infected otherwise. Finally, the data set was balanced such that each class was equally represented, resulting in a data set of size $N = 434$ in the form given in Fig. 1.

Random Forest for the Model of HIV Infection. In the RF algorithm, individual decision trees in the forest are built using bootstrap samples selected with replacement from the total data set, and the predictions of all trees are averaged in the final step. In this way, variance is reduced. For each tree, observations not used for fitting (out-of-bag or OOB observations) can then be used for validation using each of the individual decision tree models for which it was not contained in the bootstrap sample. The out-of-bag score (OOB score) is then the number of correctly predicted observations from the out of bag sample.

There are many RF implementations available. In our example, we use RandomForestClassifier in the Python *scikit-learn* machine learning package [16]. A brief overview of the general RF algorithm, with reference to important arguments that can be used to tune the model, is:

1. Draw a same-size bootstrap sample from the original data set (argument: number of samples to draw, `max_sample`). Observations left out of the sample (OOB data) is used as an unbiased accuracy measure (out-of-bag score) and in computing feature importance.
2. Train a tree on the bootstrapped sample by randomly selecting features without replacement at each node (argument: number of features to consider at each split, `max_features`), then splitting according to the feature choice that optimizes the objective function (argument: splitting criterion, `criterion`).
3. Repeat (argument: number of trees in forest, `n_estimators`).
4. Aggregate the predictions of all trees by majority voting. The OOB score of the RF is then computed as the averaged correct predictions of OOB observations using trees for which they are OOB.
5. The feature importance, used here as a measure of parameter sensitivity, is averaged over all trees.

In Fig. 2, we show the impact of tuning several of the RF algorithm parameters on prediction accuracy for a RF trained on the LHS-generated data for the HIV dynamical system, with the eight model input parameters defined in Table 1 as features and model output variable free virus concentration after 4000 days defined as class. In this example, limiting the number of model input parameters to consider as candidate predictors at each split does not impact model performance. A minimum number of trees in the forest to achieve a high accuracy as measured by OOB score could be as low as 50 trees, however for a data set of this size a larger forest is also a reasonable choice given the negligible difference in computation time. Large decision trees can suffer from overfitting, so it is best to choose the smallest tree size with sufficient classification accuracy. Here it appears that a maximum tree depth of four, or even three, is sufficient. Boulesteix et al. [2] contains an excellent practical guide to RF tuning parameters important in computational biology applications.

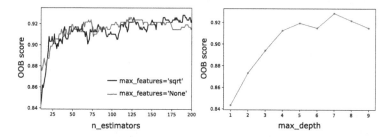

Fig. 2. Dependence of OOB score on RF tuning parameters. On the left, out-of-bag accuracy score is computed for varying number of trees in the forest (n_estimators), for two different values of number of features (model parameters) to consider at each split (max_features). On the right, out-of-bag accuracy score is computed for varying tree size (max_depth).

Feature Importance: Comparison to PRCC and Local Sensitivity Analysis. A comparison between RF feature importance, PRCCs, and sensitivities as measures of parameter importance is shown in Fig. 3, with sensitive parameters as determined by each method appearing with (*). PRCCs (Partial Rank Correlation Coefficients) uses partial rank correlation to first rank transform the vectors containing sample parameters and associated outputs, and then calculates the correlation between each parameter and output after discounting the effects of the remaining parameters [7,12]. Thus this global method apportions variability in model output to variability in parameters, allowing us to determine how each parameter effects model output (sensitivity analysis). Traditional, local sensitivity analysis methods, unlike RF feature importance or PRCC methods, neglect relationships between parameters and consider only the impact of individual parameters on model output by holding all others fixed.

These values are obtained by integrating the partial differential equations for each parameter with respect to free virus V over the whole time course, something that is computationally intensive or impossible for a larger model.

In comparison, feature importance is calculated as part of the RF classification algorithm in *scikit-learn* [16]; this is the total amount that the selection criterion decreases with each split on the given feature (the selection criterion used here is Gini impurity). Since we are splitting on parameters with model output defined as class, this is a parameter importance measure that indicates which parameters contribute most to determining model outcomes which implicitly takes into account interactions between model parameters (unlike traditional, local methods) [2]. Unlike PRCCs, these are values that range from 0 to 1 and are normalized to sum to 1. Averaged results for a random forest of 100 decision trees are shown in Fig. 3, and we expect these results to be unbiased since parameters are continuous variables [21]. With a feature importance cutoff of 5%, we identify the same important parameters identified analytically by Perelson et al. as bifurcation parameters, which in turn are the same parameters identified by PRCC and traditional methods.

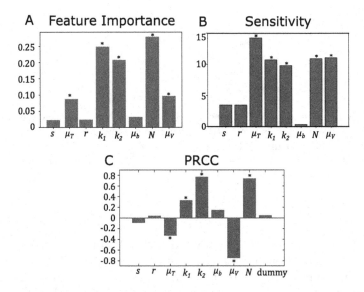

Fig. 3. Comparison of parameter importance measures. Parameters denoted sensitive by each measure are denoted with (*). (A) Averaged feature importance estimates for a random forest of 100 decision trees. The parameter importance cutoff was set to 5%, with parameters contributing more than 5% to reduction in the selection criterion considered sensitive. (B) Magnitude of the computed sensitivities of free virus V to parameters integrated over time, obtained by solving the partial differential sensitivity equations. (C) PRCC results from Marino et al. [12] computed by sampling HIV model output at time $t = 4000$ days with sample size of $N = 200$. Additional PRCC values, including time $t = 2000$ days and varying sample sizes, are given in Supplement D, Table D.1 in [12].

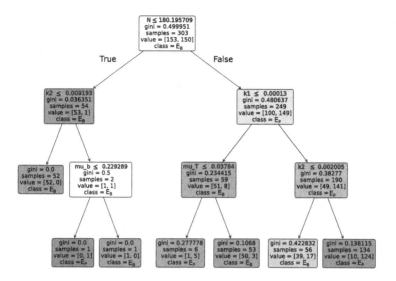

Fig. 4. Decision tree for the HIV model. Classes are labeled as uninfected (E_B) or endemically infected (E_P). In the tree, each node has a decision rule based on a parameter that splits the data. If evaluation of the rule has a true result, the data will be sorted to the left in the split. A class is designated at each node depending on which class is most represented (the number of data points in each class is given in "value".) For example, from the leftmost branch of the tree, we see if $117.9 \leq N \leq 180.2$ and $k_2 \leq 0.009$, the predicted model steady state is uninfected.

Parameter Subset Selection. A unique benefit of decision tree classification algorithms is visualization of complex relationships among features. While we used a random forest of trees to check accuracy and determine parameter importance, for visualization and interpretation of decision rules individual trees are examined.

For example, the decision tree shown in Fig. 4 can be interpreted as collections of parameter sets that are input into the model that produce particular steady states. Classes are labeled as uninfected (E_B, no virus remains in the long term) or endemically infected (E_P, a constant level of virus remains in the long term). In the tree, each node has a decision rule based on one of the parameters that splits the data. If evaluation of the rule has a true result, the data will be sorted to the left in the split. A class is designated at each node depending on which class is most represented (the number of data points in each class is given in "value".) From the leftmost branch of the tree, we see if $117.9 \leq N \leq 180.2$ and $k_2 \leq 0.009$, where N is the number of free virus produced by lysing a CD4$^+$ cell and k_2 is the transition rate of CD4$^+$ cells from latently to actively infected, the predicted model steady state is uninfected unless $\mu_b \leq 0.229$.

A decision tree can exported in text format that is easily translated to code for setting parameter ranges in running model simulation. For example, using the `export_text` function in Python *scikit-learn* for the decision tree in Fig. 4 results in the output given in Listing 1.1 [16].

Listing 1.1. Text output of decision rules from `export_text` function in Python *sklearn* for the decision tree in Fig. 4

```
|--- N <= 180.195709
|    |--- k2 <= 0.009193
|    |    |--- class: 0
|    |--- k2 >  0.009193
|    |    |--- mu_b <= 0.229289
|    |    |    |--- class: 1
|    |    |--- mu_b >  0.229289
|    |    |    |--- class: 0
|--- N >  180.195709
|    |--- k1 <= 0.000130
|    |    |--- mu_T <= 0.037840
|    |    |    |--- class: 1
|    |    |--- mu_T >  0.037840
|    |    |    |--- class: 0
|    |--- k1 >  0.000130
|    |    |--- k2 <= 0.002005
|    |    |    |--- class: 0
|    |    |--- k2 >  0.002005
|    |    |    |--- class: 1
```

These rules on parameters can be used to guide simulation and parameter fitting to get a sense of the different paths to the same outcomes. For example, given the tree in Fig. 4 and the decision rules in Listing 1.1, we can choose parameter sets that satisfy decision rules

1. $N > 180.19571, k_1 \leq 0.00013, \mu_T \leq 0.03784$ (parameter inputs sampled from these ranges pictured in Fig. 5A)
2. $N > 180.19571, k_1 > 0.00013, k_2 > 0.00201$ (parameter inputs sampled from these ranges pictured in Fig. 5B)
3. $N \leq 180.19571, k_2 > 0.00919, \mu_b \leq 0.22929$ (parameter inputs sampled from these ranges pictured in Fig. 5C).

The rules allow us to select collections of parameters that, when input in the model, ultimately lead to the endemically infected steady state yet may exhibit different behavior. Figure 5 depicts free virus versus time, with parameters sampled from three regions of parameter space determined by each of the three decision rules discovered with decision tree sensitivity analysis.

Parameter sets sampled from ranges set by decision rules produce different behavior across groups even though most of trajectories end up in the endemically infected state as predicted. As compared to Fig. 5A, the twenty simulations run using the Rule 2 inputs shown in Fig. 5B have an earlier peak, and some exhibit an early spike in free virus concentration. In contrast to simulations run

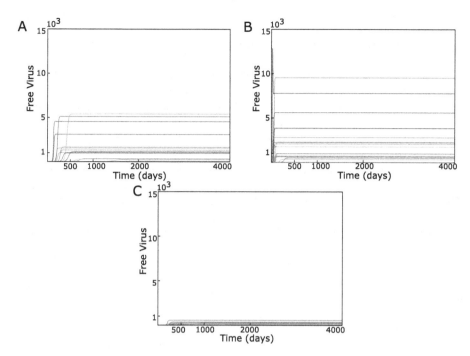

Fig. 5. Free virus (model output) versus time in days. The three decision tree paths that lead to the endemically infected steady state are used to set parameter ranges for sampling. In (A), parameter inputs to the model are sampled from ranges set by Rule $1 = N > 180.19571, k_1 \leq 0.00013, \mu_T \leq 0.03784$. In (B), parameter inputs to the model are sampled from ranges set by Rule $2 = N > 180.19571, k_1 > 0.00013, k_2 > 0.00201$. In (C), parameter inputs to the model are sampled from ranges set by Rule $3 = N \leq 180.19571, k_2 > 0.00919, \mu_b \leq 0.22929$.

using Rule 1 and Rule 2, simulations run using Rule 3 shown in Fig. 5C all have much lower magnitude of free virus concentration over the entire time course of infection. This is not meant to be an exhaustive demonstration of how decision rules could be used and is intended to show that we gain insight from this simple approach if what are interested in is primarily an exploration of parameter space. We could also consider fixing all parameters that were not found sensitive by the feature importance metric before allowing only the sensitive parameters to vary within boundaries set by decision rules, for example.

4   Discussion and Conclusion

We have demonstrated a process for qualitative analysis of dynamical model behavior using decision trees. The main advantages of this method are the clarity in the connection between the model outcomes of interest and the parameters, its ease of implementation, the simultaneous attainment of both important parameters (sensitivity analysis) and subsets of parameters linked to particular outcomes (parameter subset selection), and the visualization of parameter relationships.

This method employs simulated data, which is a limitation that allows early investigation of model behavior that can be expanded into a larger investigation of parameter space as experimental data becomes available. In addition, as with most sampling-based global sensitivity methods, experimentation is required to find an adequately sized data set to produce a good classifier. A benefit of this method is that RF is known to implicitly account for interactions between features, an important consideration where features are dynamical systems model parameters [2, 20]; however, no information about the nature of interactions is provided that can be obtained without sorting through tree decision rules. Methods for systematically analyzing RF results to obtain more information about interactions have, however, been proposed in bioinformatics applications [11, 15, 22, 25].

In future work, we propose to further investigating how interactions among parameters could be analyzed by comparing splitting attributes on parameters across decision trees in a random forest, for more complex models with known interactions. We will apply this process to such a larger-scale ordinary differential equations model with many parameters next, as well as to an agent-based model, for which traditional methods of sensitivity analysis are more difficult to conduct and results are hard to interpret.

References

1. Amit, Y., Geman, D.: Shape quantization and recognition with randomized trees. Neural Comput. **9**(7), 1545–1588 (1997)
2. Boulesteix, A.L., Janitza, S., Kruppa, J., König, I.R.: Overview of random forest methodology and practical guidance with emphasis on computational biology and bioinformatics. Wiley Interdisc. Rev. Data Min. Knowl. Discov. **2**(6), 493–507 (2012)
3. Breiman, L.: Classification and Regression Trees. Wadsworth Statistics/probability Series. Wadsworth International Group (1984). https://books.google.com/books?id=uxPvAAAAMAAJ
4. Breiman, L.: Random forests. Mach. Learn. **45**(1), 5–32 (2001). https://doi.org/10.1023/A:1010933404324
5. Brunton, S.L., Proctor, J.L., Kutz, J.N.: Discovering governing equations from data by sparse identification of nonlinear dynamical systems. Proc. Natl. Acad. Sci. **113**(15), 3932–3937 (2016)
6. Chu, Y., Hahn, J.: Parameter set selection via clustering of parameters into pairwise indistinguishable groups of parameters. Ind. Eng. Chem. Res. **48**(13), 6000–6009 (2009)
7. Conover, W.J., Iman, R.L.: Rank transformations as a bridge between parametric and nonparametric statistics. Am. Stat. **35**(3), 124–129 (1981)
8. Helton, J.C., Davis, F.J.: Latin hypercube sampling and the propagation of uncertainty in analyses of complex systems. Reliab. Eng. Syst. Saf. **81**(1), 23–69 (2003)
9. Ho, T.K.: The random subspace method for constructing decision forests. IEEE Trans. Pattern Anal. Mach. Intell. **20**(8), 832–844 (1998)

10. Iman, R.L., Conover, W.J.: The use of the rank transform in regression. Techno-metrics **21**(4), 499–509 (1979)
11. Jiang, R., Tang, W., Wu, X., Fu, W.: A random forest approach to the detection of epistatic interactions in case-control studies. BMC Bioinform. **10**(1), 1–12 (2009)
12. Marino, S., Hogue, I.B., Ray, C.J., Kirschner, D.E.: A methodology for perform-ing global uncertainty and sensitivity analysis in systems biology. J. Theor. Biol. **254**(1), 178–196 (2008). https://doi.org/10.1016/j.jtbi.2008.04.011
13. Marino, S., Hogue, I.B., Ray, C.J., Kirschner, D.E.: Uncertainty and sensitivity functions and implementation (Matlab functions for PRCC and eFAST). http://malthus.micro.med.umich.edu/lab/usanalysis.html
14. McKay, M.: Latin hypercube sampling as a tool in uncertainty analysis of computer models. In: Proceedings of the 1992 Winter Simulation Conference (1992). https://doi.org/10.1145/167293.167637
15. Meng, Y., Yang, Q., Cuenco, K.T., Cupples, L.A., DeStefano, A.L., Lunetta, K.L.: Two-stage approach for identifying single-nucleotide polymorphisms associated with rheumatoid arthritis using random forests and Bayesian networks. In: BMC Proceedings, vol. 1, pp. 1–6. BioMed Central (2007)
16. Pedregosa, F., et al.: Scikit-learn: machine learning in Python. J. Mach. Learn. Res. **12**, 2825–2830 (2011)
17. Perelson, A.S., Kirschner, D.E., De Boer, R.: Dynamics of HIV infection of CD4+ T cells. Math. Biosci. **114**(1), 81–125 (1993)
18. Rackauckas, C., et al.: Universal differential equations for scientific machine learn-ing. arXiv preprint arXiv:2001.04385 (2020)
19. Raissi, M., Perdikaris, P., Karniadakis, G.E.: Physics-informed neural networks: a deep learning framework for solving forward and inverse problems involving non-linear partial differential equations. J. Comput. Phys. **378**, 686–707 (2019)
20. Rodenburg, W., et al.: A framework to identify physiological responses in microarray-based gene expression studies: selection and interpretation of biolog-ically relevant genes. Physiol. Genomics **33**(1), 78–90 (2008)
21. Strobl, C., Boulesteix, A.L., Zeileis, A., Hothorn, T.: Bias in random forest variable importance measures: illustrations, sources and a solution. BMC Bioinform. **8**(1), 1–21 (2007)
22. Tang, R., Sinnwell, J.P., Li, J., Rider, D.N., de Andrade, M., Biernacka, J.M.: Iden-tification of genes and haplotypes that predict rheumatoid arthritis using random forests. In: BMC Proceedings, vol. 3, pp. 1–5. BioMed Central (2009)
23. Torres, M., Wang, J., Yannie, P.J., Ghosh, S., Segal, R.A., Reynolds, A.M.: Iden-tifying important parameters in the inflammatory process with a mathematical model of immune cell influx and macrophage polarization. PLoS Comput. Biol. **15**(7), e1007172 (2019)
24. Yılmaz, Ö., Achenie, L.E., Srivastava, R.: Systematic tuning of parameters in sup-port vector clustering. Math. Biosci. **205**(2), 252–270 (2007)
25. Yoshida, M., Koike, A.: SNPInterForest: a new method for detecting epistatic interactions. BMC Bioinform. **12**(1), 1–10 (2011). https://doi.org/10.1186/1471-2105-12-469

Auto-Encoded Reservoir Computing
for Turbulence Learning

Nguyen Anh Khoa Doan[1,2(✉)], Wolfgang Polifke[2], and Luca Magri[3,4,5,6]

[1] Faculty of Aerospace Engineering, Delft University of Technology,
Delft, Netherlands
n.a.k.doan@tudelft.nl
[2] Department of Mechanical Engineering, Technical University of Munich,
Garching, Germany
[3] Aeronautics Department, Imperial College London, London, UK
[4] The Alan Turing Institute, London, UK
[5] Institute for Advanced Study, Technical University of Munich, Garching, Germany
[6] Department of Engineering, University of Cambridge, Cambridge, UK

Abstract. We present an Auto-Encoded Reservoir-Computing (AE-RC) approach to learn the dynamics of a 2D turbulent flow. The AE-RC consists of an Autoencoder, which discovers an efficient manifold representation of the flow state, and an Echo State Network, which learns the time evolution of the flow in the manifold. The AE-RC is able to both learn the time-accurate dynamics of the flow and predict its first-order statistical moments. The AE-RC approach opens up new possibilities for the spatio-temporal prediction of turbulence with machine learning.

Keywords: Echo state network · Autoencoder · Turbulence

1 Introduction

The spatio-temporal prediction of turbulence is challenging because of the extreme sensitivity of chaotic flows to perturbations, the nonlinear interactions between turbulent structures of different scales, and the unpredictable nature of energy/dissipation bursts. Despite these intricate characteristics of turbulence, many advances have been achieved in its understanding with, for example, the energy cascade concept that provides a statistical description of the energy transfer between different scales in turbulent flows [5]. Additionally, the existence of coherent structures, such as vortices, which evolve in a deterministic way, provides a basis for understanding turbulence [12]: Within the chaotic dynamics of turbulence, there exist identifiable patterns that can help us predict the evolution of turbulent flows. To discover such patterns, recent works have relied on

The authors acknowledge the support of TUM-IAS, funded by the German Excellence Initiative and the EU 7th Framework Programme (grant no. 291763) and PRACE for awarding access to ARIS at GRNET, Greece. L.M. also acknowledges the RAEng Research Fellowship Scheme. L. Magri—Visiting.

© Springer Nature Switzerland AG 2021
M. Paszynski et al. (Eds.): ICCS 2021, LNCS 12746, pp. 344–351, 2021.
https://doi.org/10.1007/978-3-030-77977-1_27

machine learning [1]. In particular, the dynamics of models of turbulent flows have been learned by recurrent neural networks (RNNs) such as the Long Short-Term Memory units [10,11] or a physics-informed reservoir computing (RC) approach, based on Echo State Networks (ESN) [2]. Because RNNs are generally limited to low-dimensional datasets due to the complexity of training, past studies have been restricted to fairly low-dimensional systems. To deal with high dimensional fluid mechanical systems, recent approaches based on convolutional neural networks (CNNs), and in particular Autoencoders (AE), have shown great potential in discovering coherent structures in turbulent flows and reducing the dimensionality of flows [1,7], more efficiently than linear reduced-order modelling approaches (for a review of reduced-order models in fluids refer to [9]).

In this paper, we propose the Auto-Encoded Reservoir Computing framework (AE-RC). This combines an ESN and an AE with the objective of learning the spatio-temporal dynamics of a 2D turbulent flow governed by the Navier-Stokes equations (the Kolmogorov flow). The flow is discussed in Sect. 2. The AE-RC framework is presented in Sect. 3 and results are discussed in Sect. 4. The final section summarizes the results and outlines avenues for future work.

2   Turbulent Flow

We investigate 2D turbulence governed by the incompressible Navier-Stokes equations

$$\nabla \cdot \boldsymbol{u} = 0 \tag{1}$$

$$\partial_t \boldsymbol{u} + \boldsymbol{u} \cdot \nabla \boldsymbol{u} = -\nabla p + \frac{1}{\mathrm{Re}} \Delta \boldsymbol{u} + \boldsymbol{f} \tag{2}$$

where $\boldsymbol{u} = (u, v)$ is the velocity field, p is the pressure, Re is the Reynolds number, and $\boldsymbol{f}$ is a harmonic volume force defined as $\boldsymbol{f} = (\sin(k_f y), 0)$ in cartesian coordinates. The Navier-Stokes equations are solved on a domain $\Omega \equiv [0, 2\pi] \times [0, 2\pi]$ with periodic boundary conditions. (The solution of this problem is also known as the 2D Kolmogorov flow.) The flow has a laminar solution $u = \mathrm{Re} k_f^{-2} \sin(k_f y), v = 0$, which is unstable for sufficiently large Reynolds numbers and wave numbers k_f [8]. Here, we take $k_f = 4$ and Re $= 30$ to guarantee the development of a turbulent solution [11]. The set of Eqs. (1) and (2) is solved on a uniform $N \times N$ grid, with $N = 24$, using a pseudo-spectral code with explicit Euler in time [11] with a timestep, $\Delta t = 0.01$, to ensure numerical stability. Snapshots of the velocity and vorticity, ω, fields are shown in Fig. 1, in which the complexity and chaotic pattern of the turbulent flow can be observed. Figures 1d and 1e show the time evolution of the kinetic energy, k, and dissipation, D, which are calculated as $k(\boldsymbol{u}) = (2\pi)^{-2} \int_\Omega \frac{1}{2} |\boldsymbol{u}|^2 d\Omega$ and $D(\boldsymbol{u}) = \mathrm{Re}^{-1} (2\pi)^{-2} \int_\Omega |\nabla \boldsymbol{u}|^2 d\Omega$, respectively. The solution is turbulent.

3   Auto-Encoded Reservoir Computing

The proposed Auto-Encoded Reservoir-Computing (AE-RC) framework is shown in Fig. 2a. The AE-RC is composed of two parts: (i) an Autoencoder (AE), which is composed of an encoder and a decoder; and (ii) an echo state network, which is a form of reservoir computing [6]. The role of the AE is to discover an efficient reduced-order representation of the original data, $u \in \mathbb{R}^{N \times N \times 2 = N_u}$. The encoder reduces the dimension of the data to a code, $c \in \mathbb{R}^{N_c}$, where $N_c < N_u$, while the decoder reconstructs the data from the code, c, by minimizing the error between the reconstructed solution, $\hat{u}$, and the data. Here, the AE consists of a series of CNNs, which identify patterns within images through kernel operations [3]. The details of the AE are shown in Fig. 2b. On the downsampling side, the encoder is composed of multiple blocks of successive 2D CNNs, max pooling and dropout layers. Dropout layers prevent overfitting, while max pooling layers decrease the dimension of the input data. The dropout rate is 0.001 and was chosen during the training of the AE to have mean-squared errors of the same order of magnitude (and as small as possible) on both training and validation datasets. (The dropout rate is rather small because the AE-RC has a small number of trainable weights with respect to the size of the dataset, which reduces the risk of overfitting). After the last layer of the encoder, a dense feedforward neural network is used to combine the information from the previous layer and compress the data into the final code of dimension 192, compared to the original data of dimension $24 \times 24 \times 2 = 1152$. On the upsampling side, the architecture of the decoder mirrors that of the encoder, but the dimension of the code is progressively increased using bilinear upsampling layers to recover the original data [7].

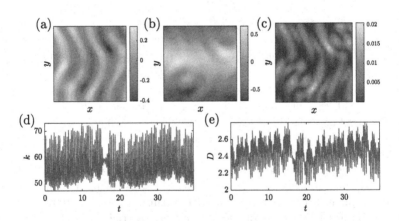

Fig. 1. Isocontours of (a) u, (b) v and (c) vorticity for the 2D turbulent flow at time $t = 0$. Time evolution of (d) kinetic energy, k, and (e) dissipation, D.

To learn the temporal dynamics of the reduced representation obtained with the AE, an Echo State Network (ESN) [6] is employed as ESNs are accurate learners of chaotic dynamics and flows, e.g., [2,6]. The ESN receives the code as an input at a time n, $c(n)$, and approximates the code at the subsequent time step, $c(n+1)$, as an output. An ESN is composed of three parts: (i) a randomized

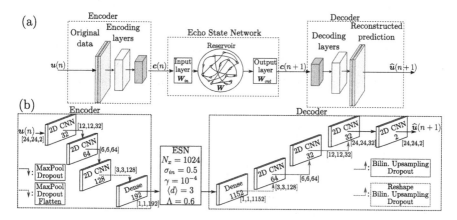

Fig. 2. (a) Schematic and (b) details of the AE-RC. Number in boxes indicate the number of filters (for CNN) or neurons (for Dense layer). [·]: dimension of the signal. tanh activation is used for all layers except the last layer in which a linear activation is used to reconstruct the velocity in the full range of real numbers. All MaxPool and upsampling layers have a window of $(2, 2)$. All CNNs layers have a kernel of $(3, 3)$.

high dimensional dynamical system, called the reservoir, whose states of neurons (or units) at time n are represented by a vector, $\boldsymbol{x}(n) \in \mathbb{R}^{N_x}$, N_x being the number of neurons; (ii) an input matrix, $\boldsymbol{W}_{in} \in \mathbb{R}^{N_x \times N_c}$, and (iii) an output matrix, $\boldsymbol{W}_{out} \in \mathbb{R}^{N_y \times N_x}$ where N_y is the dimension of the output of the ESN. The output of the ESN, $\widehat{\boldsymbol{y}}$, is a linear combination of the reservoir states, $\widehat{\boldsymbol{y}}(n) = \boldsymbol{W}_{out}\boldsymbol{x}(n)$. The evolution of the neurons' states is given by the discrete nonlinear law

$$\boldsymbol{x}(n) = \tanh\left(\boldsymbol{W}_{in}\boldsymbol{c}(n) + \boldsymbol{W}\boldsymbol{x}(n-1)\right) \tag{3}$$

Because the aim is to predict the dynamics of the reduced-order representation, the output of the ESN is the predicted subsequent state of the reduced-order representation, i.e., $\widehat{\boldsymbol{y}}(n) \approx \boldsymbol{c}(n + 1)$. In the ESN approach, $\boldsymbol{W}_{in}$ and $\boldsymbol{W}$ are randomly initialized once and are not trained. Only $\boldsymbol{W}_{out}$ is trained. The sparse matrices $\boldsymbol{W}_{in}$ and $\boldsymbol{W}$ are constructed to satisfy the Echo State Property. Following [2], $\boldsymbol{W}_{in}$ is generated such that each row of the matrix has only one randomly chosen nonzero element, which is independently taken from a uniform distribution in the interval $[-\sigma_{in}, \sigma_{in}]$. Matrix $\boldsymbol{W}$ is constructed with an average connectivity $\langle d \rangle$, and the non-zero elements are taken from a uniform distribution over the interval $[-1, 1]$. All the coefficients of $\boldsymbol{W}$ are then multiplied by a constant coefficient for the largest absolute eigenvalue of $\boldsymbol{W}$, i.e. the spectral radius, to be equal to a value Λ, which is typically smaller than (or equal to) unity. The exact parameters of the ESN used here are provided in Fig. 2b. The training procedure to train the AE-RC is provided in the grey box below.

AE-RC TRAINING PROCEDURE

1. **Pre-train the AE** with the 2D velocity field as input/output. The reconstruction error, $E = \frac{1}{N_t} \sum_{n=1}^{N_t} ||u(n) - \widehat{u}(n)||^2$ where N_t is the number of samples, is minimized. The AE learns a nonlinear reduced-order representation, c, of u.

2. **Compute the reduced representation, $c(n)$,** of the original dataset, $u(n)$, using the encoder part of the pre-trained AE.

3. **Pre-train the ESN** using the dataset $c(n)$ and ridge regression, $W_{out} = YX^T (XX^T + \gamma I)^{-1}$, where Y and X are the horizontal concatenations of the target data, $c(n)$, and the associated ESN states $x(n)$, respectively. γ is the Tikhonov regularization factor [6].

4. **Train the combined AE-RC** for further fine-tuning. The AE-RC receives $u(n)$ as an input and predicts $\widehat{u}(n+1)$. The training minimizes $L = \frac{1}{N_t} \sum_{n=1}^{N_t} ||u(n+1) - \widehat{u}(n+1)||^2$, where $\widehat{u}(n+1)$ is the prediction of the AE-RC at the next timestep, given an input $u(n)$.

Steps 1 to 3 are used to obtain an initial AE-RC, which is the initial guess for the training of the entire AE-RC in Step 4. This accelerates the overall training of the AE-RC by taking advantage of the fast training of the ESN with ridge regression compared to a random initialization of the AE-RC. The ADAM optimizer [4] is used for Steps 1 and 4 with a learning rate of 0.0001.

4   Results

The AE-RC framework presented in Sect. 3 is applied to learning the dynamics of a 2D turbulent flow. The training dataset corresponds to the first 80% of the time-evolution shown in Fig. 1 and the last 20% are used for validation. The AE-RC receives the 2D velocity field at a given timestep, as the input, and predicts the velocity field at the next timestep, as the output. The predictions of k and D during training (quantities noted with $\hat{}$) are shown in Fig. 3 with their errors. The AE-RC accurately reproduces the evolution in the training data. To assess the extrapolation capability, the output of the AE-RC is looped back as an input so that the AE-RC evolves autonomously. The learned extrapolated time-series of k and D are shown in Fig. 4 (the insets of the vorticity fields are shown for different time instants). The AE-RC reproduces the spatio-temporal evolution of k and D, which is in agreement with the physical evolution of the turbulent flow. The phase difference between the AE-RC solution and the benchmark solution may be due to the spatio-temporally chaotic nature of the flow, in which small errors in the initial conditions are amplified exponentially in a short time. This is why, in turbulent flows, the statistics are typically compared to assess the accuracy of a solution. Figure 5 shows the time-averaged velocity profiles respectively, computed over the duration shown in Fig. 4. Because the error is small (the average absolute error of u and v normalized by their respective maximum values is less than 6% and 4% respectively), it is concluded that the

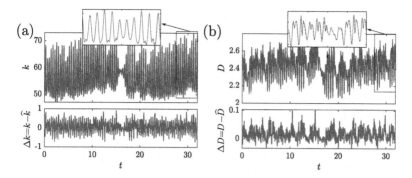

Fig. 3. Time evolution of (a) k and the error, Δk, and (b) D and the error, ΔD, for the 2D turbulent flow during the training stage. The AE-RC is "teacher-forced" (the input is provided by the training data). Blue lines: benchmark evolution. Dashed red lines: AE-RC. (Color figure online)

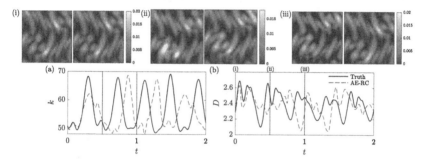

Fig. 4. Time evolution of (a) k and (b) D for the 2D turbulent flow. Vertical dotted lines indicates the time-instants for the snapshots (i) to (iii) of vorticity. For each top panel: (left) benchmark evolution, (right) AE-RC prediction.

AE-RC has learned the dynamics of the Kolmogorov flow also in a statistical sense (for the first moment). The standard deviations of the velocity profile were also computed and found to be of similar accuracy as those of time-averaged velocity (not shown here).

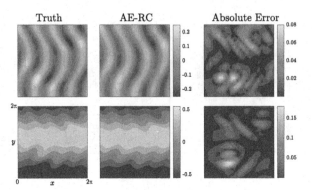

Fig. 5. Prediction of time-averaged u (top row) and v (bottom row). Right column: absolute error between AE-RC prediction and benchmark solution.

5   Conclusions and Future Directions

We propose the Auto-Encoded Reservoir-Computing framework (AE-RC) to learn the dynamics of high-dimensional turbulent flows, which are both spatially and temporally chaotic. This framework consists of an Autoencoder, which learns an efficient reduced-order representation of the spatial dynamics, and an Echo State Network, which learns the temporal dynamics of the reduced-order representation. With these two components, the AE-RC is able to learn both the instantaneous and average dynamics of the two-dimensional turbulent flow governed by the incompressible Navier-Stokes equations. This framework is being assessed on flow conditions that also exhibit bursts of kinetic energy. In future work, the effect of the code dimension on the accuracy of the AE-RC will be analysed. A comparative study of the AE-RC performance with respect to existing non-intrusive linear reduced-order models, such as the Proper Orthogonal Decomposition with Galerkin projection, and with respect to Long-Short-Term Memory units for the time prediction, is scope for further research.

References

1. Brunton, S.L., Noack, B.R., Koumoutsakos, P.: Machine learning for fluid mechanics. Ann. Rev. Fluid Mech. **52**(1), 477–508 (2020)
2. Doan, N.A.K., Polifke, W., Magri, L.: Physics-informed echo state networks. J. Comput. Sci. **47**, 101237 (2020)
3. Goodfellow, I., Bengio, Y., Courville, A.: Deep Learning. MIT Press, Cambridge (2016)
4. Kingma, D.P., Ba, J.L.: Adam: a method for stochastic optimization. In: 3rd International Conference on Learning Representations, ICLR 2015 - Conference Track Proceedings, pp. 1–15 (2015)
5. Kolmogorov, A.N.: Dissipation of energy in locally isotropic turbulence. Doklady Akademiia Nauk SSSR **32**, 19–21 (1941)

6. Lukoševičius, M., Jaeger, H.: Reservoir computing approaches to recurrent neural network training. Comput. Sci. Rev. **3**(3), 127–149 (2009)
7. Murata, T., Fukami, K., Fukagata, K.: Nonlinear mode decomposition with convolutional neural networks for fluid dynamics. J. Fluid Mech. **882**, A13 (2019)
8. Platt, N., Sirovich, L., Fitzmaurice, N.: An investigation of chaotic Kolmogorov flows. Phys. Fluids A **3**(4), 681–696 (1991)
9. Rowley, C.W., Dawson, S.T.: Model reduction for flow analysis and control. Ann. Rev. Fluid Mech. **49**, 387–417 (2017)
10. Srinivasan, P.A., Guastoni, L., Azizpour, H., Schlatter, P., Vinuesa, R.: Predictions of turbulent shear flows using deep neural networks. Phys. Rev. Fluids **4**, 054603 (2019)
11. Wan, Z.Y., Vlachas, P., Koumoutsakos, P., Sapsis, T.P.: Data-assisted reduced-order modeling of extreme events in complex dynamical systems. PLoS ONE **13**(5), 1–22 (2018)
12. Yao, J., Hussain, F.: A physical model of turbulence cascade via vortex reconnection sequence and avalanche. J. Fluid Mech. **883**, A51 (2020)

Low-Dimensional Decompositions for Nonlinear Finite Impulse Response Modeling

Maciej Filiński[1]($\boxtimes$)(iD), Paweł Wachel[1](iD), and Koen Tiels[2](iD)

[1] Wrocław University of Science and Technology, 50-556 Wrocław, Poland
{Maciej.Filinski,Pawel.Wachel}@pwr.edu.pl
[2] Eindhoven University of Technology, 5600 Eindhoven, MB, The Netherlands
K.Tiels@tue.nl

Abstract. This paper proposes a new decomposition technique for the general class of Non-linear Finite Impulse Response (NFIR) systems. Based on the estimates of projection operators, we construct a set of coefficients, sensitive to the separated internal system components with short-term memory, both linear and nonlinear. The proposed technique allows for the internal structure inference in the presence of unknown additive disturbance on the system output and for a class of arbitrary but bounded nonlinear characteristics.

The results of numerical experiments, shown and discussed in the paper, indicate applicability of the method for different types of nonlinear characteristics in the system.

Keywords: NFIR systems · Short-term memory · Structure detection

1 Introduction

Continuous development of technology results in new challenges in various engineering fields, but some basic fundamental problems remain invariably relevant and, to some extent, open. Among them, one can indicate modelling of nonlinear phenomena, [10], often encountered in various types of cyber-physical systems [9], technical processes, etc. In this paper, we focus on a modelling task for a wide class of stationary nonlinear dynamical systems with almost unknown internal structure. Assuming only that the system at hand has a finite memory and bounded non-linearity, we discuss a family of Non-linear Finite Impulse Response (NFIR) objects [5,6,11]. However, rather than direct identification, our goal is to reveal (if it exists) their hidden internal structure, *cf.* [1,4], leading to the decomposition of the system into smaller, short-term memory linear or non-linear, elements. Such an approach leads to at least two benefits. Firstly, it potentially simplifies system identification through the reduction of problem dimensionality. Secondly, it supports inference about the role and properties of

Supported by Wrocław University of Science and Technology.

the particular short-term memory substructures. Our main motivation is that estimation of the separability of the system can be helpful in choosing a suitable model structure before estimating the system. The proposed method is based on the projection operators technique [7,8] and allows to infer about systems with almost arbitrary non-linearity in the presence of additive output disturbance.

The paper is organized as follows. Section 2 formally introduces the class of NFIR systems and defines the general decomposition problem. Sections 3 and 4 introduce the proposed methodology, its motivation, and dedicated algorithms. Sections 5 and 6 present results of numerical experiments and final remarks.

2   Problem Statement

We consider single-input single-output (SISO), time-invariant, Non-linear Finite Impulse Response (NFIR) systems with memory length d, described by

$$y_n = g(v_n) + e_n \tag{1}$$
$$v_n = [u_n, u_{n-1}, \ldots, u_{n-d}], \tag{2}$$

where $g : \mathbb{R}^{d+1} \to \mathbb{R}^1$ is an unknown nonlinear characteristic, u_n, y_n are the input and output signals, respectively, and e_n is an additive output disturbance; *cf.* Fig. 1. Regarding the system (1)–(2), input signal and output noise, we assume:

A1. The system has a finite memory of known length $d < \infty$, and unknown but bounded nonlinear characteristic $g : \mathbb{R}^{d+1} \to \mathbb{R}^1$.
A2. Input u_n is an *i.i.d.* sequence, uniformly distributed on the interval $[0, 1]$.
A3. The noise e_n is a zero mean *i.i.d.* sequence with a finite variance $\sigma_e^2 < \infty$ and is independent of u_n.

The requirements above are general in this sense that (for large enough d) the considered NFIR systems well approximate a general class of fading memory nonlinear objects, *cf.* [2]. We admit that assumption **A2** is rather restrictive and, in practice, allows to apply the proposed method if the system at hand can be actively excited by the user defined inputs, *cf.* [3]. Assumption **A3** is a standard assumption in literature.

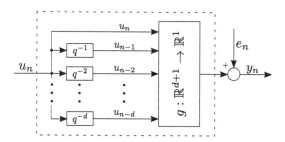

Fig. 1. The system under consideration. The non-linearity $g(\cdot)$ is shown as a mono-block, as its internal structure is unknown; q is the time-shift operator

Based on the set of measurements of the input and output of the system, $\{(u_1, y_1), (u_2, y_2), \ldots, (u_N, y_N)\}$, captured in a steady state, the aim is to investigate a potential separability of characteristic $g(\cdot)$ for its alternative representation, composed of additive, Short-Term Memory Nonlinear blocks (STMN), cf. exemplary decompositions in Fig. 2b, 2c. Here, short-term memory refers to splitting nonlinear characteristic $g(\cdot)$, which has memory length d, into a sum of two or more nonlinear characteristics g_{ξ_i}, each having a memory length shorter than d. A more formal description is provided in the next sections.

3   Separation Algorithm

We begin with a simple observation: due to assumptions **A1** and **A2**, $g(\cdot)$ is square integrable in a domain of system input. Let $\{1, 2, \ldots, d+1\}$ be a set denoting indices of the consecutive arguments of $g(\cdot)$, and $\xi_1 \subseteq \{1, 2, \ldots, d+1\}$ be its arbitrary subset. Let $\xi_2 = \{1, 2, \ldots, d+1\} \setminus \xi_1$ be a complement of ξ_1. We say that $g(\cdot)$ is separable with respect to (wrt) $\{\xi_1, \xi_2\}$ if $g \equiv g_{\xi_1} + g_{\xi_2}$, where g_{ξ_j} is a function depending *only* on the variables with indices from ξ_j. Recursive extension of this rule defines separability of $g(\cdot)$ for extended separation schemes $\{\xi_1, \xi_2, \ldots, \xi_\mu; \mu \leq d+1\}$. We note that $g(\cdot)$ in (1)–(2) is separable wrt $\{\xi_1, \xi_2, \ldots, \xi_\mu\}$ iff $S(\xi_1, \xi_2, \ldots, \xi_\mu) := S_1 - S_2$ equals zero [7, Th. 2] (cf. [8]), with

$$S_1 = \int_{[0,1]^{(d+1)}} \int_{[0,1]^{(d+1)}} g(\mathbf{s})(g(\mathbf{s}) + (\mu - 1)g(\mathbf{t})) ds dt \tag{3}$$

$$S_2 = \int_{[0,1]^{(d+1)}} \int_{[0,1]^{(d+1)}} g(\mathbf{s}) \sum_{j=1}^{\mu} g(\mathbf{s}_{\xi_j}, \mathbf{t}_{\{1,2,\ldots,d+1\}\setminus\xi_j}) ds dt, \tag{4}$$

and with $g(\mathbf{s}_{\xi_j}, \mathbf{t}_{\{1,2,\ldots,d+1\}\setminus\xi_j})$ denoting the value of $g(\cdot)$ for the argument composed of ξ_j–indexed entries of $\mathbf{s}$ and $(\{1, 2, \ldots, d+1\} \setminus \xi_j)$–indexed entries of $\mathbf{t}$.

Clearly, the above integrals cannot be evaluated analytically without system knowledge. Yet, based on their stochastic interpretation as expectations wrt the uniform probability distribution, one can estimate S_1 using input-output observations of the system (cf. assumption **A2**). The corresponding estimator is

$$\hat{S}_1 = \frac{1}{|\mathbf{I}|} \sum_{i \in \mathbf{I}} y_i(y_i + (\mu - 1)y_{i+c}), \tag{5}$$

where $\mathbf{I} = \{n : n = (d+1), 2(d+1), 3(d+1), \ldots; d+1 \leq n \leq c(d+1)\}$, $c = \lfloor N/(2(d+1)) \rfloor$ and $|\mathbf{I}|$ is the cardinality of set $\mathbf{I}$.

Slightly more effort is needed to estimate integral S_2, since due to argument $(\mathbf{s}_{\xi_j}, \mathbf{t}_{\{1,2,\ldots,d+1\}\setminus\xi_j})$ of non-linearity g, the system has to be excited with a properly designed input, determined by the actually probed separation scheme ξ_j. We, therefore, design a supplementary input sequence U according to Algorithm 1, excite the system with U and collect the corresponding output, denoted as Y.

Algorithm 1. Data Generation For Active Experiment

1: **Input:** $\{u_1, u_2, \ldots, u_N\}$, candidate separation scheme $\{\xi_1, \xi_2, \ldots, \xi_\mu\}$
2: $c := \lfloor N/2 \rfloor$
3: **for** $i = 1$ **to** μ **do**
4:     **for** $j = 1$ **to** c **do**
5:         **for** $k = 1$ **to** $d + 1$ **do**
6:             **if** $k \in \xi_i$ **then** $U_{(i-1)(d+1)c+(j-1)(d+1)+k} := u_{c(d+1)+k+(d+1)(j-1)}$
7:             **else** $U_{(i-1)(d+1)c+(j-1)(d+1)+d+1-(k-1)} := u_{d+1-(k-1)+(d+1)(j-1)}$
8:             **end if**
9:         **end for**
10:     **end for**
11: **end for**
12: **Output:** Supplementary input sequence U.

Based on the active experiment outcome, the following estimate of integral $\hat{S}_2$ can be defined

$$\hat{S}_2 = \frac{1}{|\mathbf{I}|} \sum_{i \in \mathbf{I}} \sum_{j=1}^{\mu} y_i Y_{c(j-1)(d+1)+i}. \tag{6}$$

Finally, as a resulting estimate of separability coefficient $S(\xi_1, \xi_2, \ldots, \xi_\mu)$ we take $\hat{S}(\xi_1, \xi_2, \ldots, \xi_\mu) = \hat{S}_1 - \hat{S}_2$.

Remark 1. Theoretical analysis of estimates $\hat{S}_1$, $\hat{S}_2$ is out of scope of the paper. Here, we only note that $\hat{S}_1$ is in fact a *biased* estimate of S_1 (with a bias equal to the variance of the output noise, σ_e^2). In effect, $\hat{S}$ is a *biased* estimate of S, with bias$\{\hat{S}\} = \sigma_e^2$. This is exploited in the numerical experiments in Sect. 5.

4   Short-Term Memory Separation Searching

We are now about to apply empirical coefficient $\hat{S}$ in decomposition of NFIR systems (with total memory length d) into a parallel connection of *Short-Term Memory Nonlinear* blocks (STMN); cf. Fig. 2b, 2c. Assuming its existence, such a representation is equivalent to the requirement that the genuine separation scheme, $\{\xi_1, \xi_2, \ldots, \xi_\mu; \mu \leq d+1\}$, is an *ordered* set of indices. For instance, for $d = 2$, $g(\cdot)$ could be separable with respect to $\xi_1 = \{1, 2\}$, $\xi_2 = \{3\}$, but not with respect to $\xi_1 = \{1, 3\}$ and $\xi_2 = \{2\}$. Hence, the resulting representation of the system, if exists, is composed of the Short-Term Memory blocks. In general, observe that if $g(\cdot)$ is separable with respect to $\xi_p = \{1, 2, \ldots, p\}$ and $\xi_q = \{1, 2, \ldots, q\}$, some $p < q$, then it is also separable with respect to $\{1, 2, \ldots, p\}$ and $\{p+1, p+2, \ldots, q\}$. Therefore, the proposed separation procedure estimates all the coefficients $\hat{S}$ for $\xi_p = \{1, \ldots, p\}$, vs. its complement for $p = 1, 2, \ldots, d+1$, and the outcome is next used as a recommendation for the final inference about the considered NFIR system separation scheme.

Algorithm 2. Short-Term Memory Separation Searching

1: **Input:** $\{(u_1, y_1), (u_2, y_2), \ldots, (u_N, y_N)\}$
2: **for** $p = 1$ **to** $d + 1$ **do**
3:     Design supplementary input U: apply Algorithm 1. for $\xi_p = \{1, \ldots, p\}$.
4:     Excite the system with U and measure the corresponding output sequence Y.
5:     Compute $\hat{S} = \hat{S}_1 - \hat{S}_2$ according to (5)–(6).
6: **end for**
7: **Output:** $\hat{S}(\xi_1), \hat{S}(\xi_2), \ldots, \hat{S}(\xi_{d+1})$.

Remark 2. Note that $\hat{S}(\xi_{d+1})$ with $\xi_{d+1} = \{1, \ldots, d+1\}$ is an estimate of σ_e^2. Indeed, for the above-mentioned separation scheme, characteristic $g(\cdot)$ is 'separable', and therefore, due to biasness of $\hat{S}$ (see Remark 1), it converges to σ_e^2.

Clearly, any system (1)–(2) is 'separable' with respect to the full set of indices $\xi_{d+1} = \{1, 2, \ldots, d+1\}$, and therefore, based on the observation in Remark 2, we use $\hat{S}(\xi_{d+1})$ as a reference for the relative assessing of $\hat{S}(\xi_1), \hat{S}(\xi_2), \ldots, \hat{S}(\xi_d)$. If $g(\cdot)$ is not separable for some subset of indices, the corresponding value of $\hat{S}$ is high (for large enough N) with respect to $\hat{S}(\xi_{d+1})$. Hence, for the interpretational purposes, all the values $\hat{S}(\xi_1), \hat{S}(\xi_2), \ldots$ are scaled according to the formula $\bar{S}(\xi_p) = |\hat{S}(\xi_p)| / \sum_{i=1}^{d+1} |\hat{S}(\xi_i)|$. Finally, the comparison of $\bar{S}(\xi_{d+1})$ with respect to $\bar{S}(\xi_p)$ is used as the indicator of separability. Although the above approach is not yet theoretically justified, the results of numerical experiments are promising, as we show in the next section.

Remark 3. The proposed method has a simple construction and linear time complexity with respect to N, although the computing time strongly depends on the total system memory length d.

5   Numerical Experiments

In this section, we present selected results of numerical experiments, performed for the NFIR systems with various types of admissible separation schemes and memory length $d = 9$. The following types of systems are considered: (A) non-separable system with non-linearity $g_A(v_n) = \sum_{i=1}^{d}(100(u_{n-i}^2 - u_{n-i+1})^2 + (u_{n-i} - 1)^2)$, see Fig. 2a, (B) partially separated system with non-linearity $g_B(v_n) = \sum_{i=0}^{\lfloor (d+1)/2 \rfloor}(100(u_{n-2i+1} - u_{n-2i})^2)$, see Fig. 2b, and (C) fully-separable system $g_C(v_n) = \sum_{i=0}^{d}(u_{n-i}^2 - 10\cos(2\pi u_{n-i}) + 10)$, see Fig. 2c. All the systems, A, B, C, are driven with uniformly distributed signal $\mathcal{U}[0, 1]$, scaled internally to $\mathcal{U}[-2, 2]$, $\mathcal{U}[-3, 3]$, $\mathcal{U}[-5.12, 5.12]$, respectively, *cf.* [7]. We ran the simulations for the three different numbers of samples N and three levels of output noise, as indicated in Table 1. According to the results (see Table 1), one can infer which sets of indices indicate separability. Note, that for $\bar{S}(\xi_{10})$ in the case of systems A and B the results are very small (but in fact $\bar{S}(\xi_{10}) > 0$). This is also the case for indices revealing separability. However, different type of the results are visible for system C, where most of the values are high compared to $\bar{S}(\xi_{10})$.

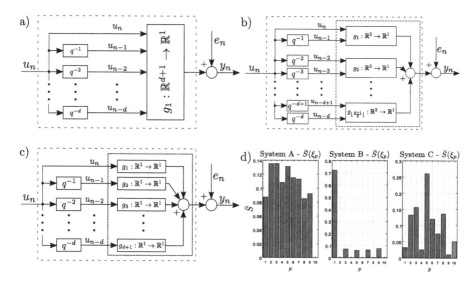

Fig. 2. Internal structure of a) a non-separable system A, b) a partially separable system B, and c) a fully separable system C. d) algorithm outcomes for the systems A, B, and C. Notice the small relative values of ξ_{10} for systems A, B and a higher one (with respect to $\xi_1, \ldots \xi_9$) for system C (in red). (Color figure online)

6   Conclusions and Future Work

A new separation method was introduced for NFIR systems, representing a wide class of non-linear models with finite memory. The proposed method allows for the recovering of hidden *short-term memory structures* in the system under mild requirements regarding the form of non-linearity in the system. The method was investigated numerically for the systems with various levels of potential separability. According to the results shown in Table 1, the method correctly indicates separability patterns in the considered cases.

Future work includes a theoretical analysis in which separability is determined based on different threshold levels. Furthermore, the convergence rates of the proposed estimates, as well as the computational complexity of the method with respect to the system memory length will be thoroughly investigated.

Table 1. The results of Algorithm 2 for non-separable system A, partially separable system B and fully-separable system C. The gray colored rows indicate true negative outcomes (lack of separability), whereas white rows represent true positive indications – possible separation of the system

		$N = 10^4$			$N = 10^5$			$N = 10^6$		
	σ_e^2	0	0.1	1	0	0.1	1	0	0.1	1
System A	ξ_1	0.028	0.009	0.003	0.068	0.145	0.154	0.099	0.068	0.087
	ξ_2	0.140	0.057	0.045	0.138	0.083	0.122	0.109	0.090	0.135
	ξ_3	0.227	0.039	0.204	0.108	0.035	0.079	0.130	0.136	0.135
	ξ_4	0.065	0.215	0.042	0.209	0.091	0.179	0.116	0.119	0.108
	ξ_5	0.139	0.205	0.066	0.039	0.122	0.100	0.096	0.133	0.131
	ξ_6	0.138	0.095	0.216	0.095	0.140	0.099	0.122	0.100	0.115
	ξ_7	0.160	0.161	0.249	0.099	0.182	0.080	0.086	0.108	0.113
	ξ_8	0.038	0.025	0.163	0.165	0.138	0.129	0.136	0.096	0.085
	ξ_9	0.065	0.195	0.011	0.080	0.064	0.058	0.107	0.150	0.092
	ξ_{10}	0.000	0.000	0.000	0.000	0.000	0.000	0.000	0.000	0.000
System B	ξ_1	0.869	0.773	0.681	0.736	0.676	0.688	0.732	0.728	0.719
	ξ_2	0.000	0.003	0.001	0.000	0.000	0.001	0.000	0.001	0.002
	ξ_3	0.018	0.010	0.039	0.058	0.113	0.077	0.069	0.071	0.071
	ξ_4	0.000	0.002	0.001	0.000	0.000	0.003	0.000	0.000	0.001
	ξ_5	0.024	0.097	0.094	0.060	0.082	0.076	0.065	0.063	0.061
	ξ_6	0.000	0.002	0.001	0.000	0.000	0.006	0.000	0.000	0.001
	ξ_7	0.052	0.050	0.058	0.066	0.052	0.055	0.070	0.069	0.065
	ξ_8	0.000	0.001	0.001	0.000	0.000	0.000	0.000	0.000	0.002
	ξ_9	0.038	0.062	0.114	0.081	0.076	0.093	0.064	0.068	0.076
	ξ_{10}	0.000	0.002	0.011	0.000	0.000	0.000	0.000	0.000	0.002
System C	ξ_1	0.125	0.105	0.036	0.100	0.069	0.192	0.111	0.002	0.032
	ξ_2	0.000	0.058	0.208	0.100	0.229	0.086	0.111	0.073	0.133
	ξ_3	0.125	0.012	0.103	0.100	0.068	0.024	0.111	0.161	0.156
	ξ_4	0.125	0.323	0.070	0.100	0.011	0.128	0.111	0.000	0.025
	ξ_5	0.125	0.134	0.045	0.100	0.028	0.129	0.000	0.236	0.260
	ξ_6	0.000	0.015	0.057	0.100	0.119	0.108	0.000	0.007	0.120
	ξ_7	0.125	0.073	0.024	0.100	0.068	0.055	0.111	0.232	0.075
	ξ_8	0.125	0.137	0.065	0.100	0.087	0.037	0.111	0.110	0.137
	ξ_9	0.125	0.017	0.236	0.100	0.212	0.095	0.222	0.023	0.011
	ξ_{10}	0.125	0.127	0.156	0.100	0.110	0.147	0.111	0.155	0.051

References

1. Bai, E.W., Cerone, V., Regruto, D.: Separable inputs for the identification of block-oriented nonlinear systems. In: 2007 American Control Conference, pp. 1548–1553. IEEE (2007)
2. Boyd, S., Chua, L.: Fading memory and the problem of approximating nonlinear operators with Volterra series. IEEE Trans. Circuits Syst. **32**(11), 1150–1161 (1985)
3. De Cock, A., Gevers, M., Schoukens, J.: D-optimal input design for nonlinear fir-type systems: a dispersion-based approach. Automatica **73**, 88–100 (2016)

4. Decuyper, J., Tiels, K., Runacres, M.C., Schoukens, J.: Retrieving highly structured models starting from black-box nonlinear state-space models using polynomial decoupling. Mech. Syst. Signal Process. **146**, 106966 (2019)
5. Enqvist, M., Ljung, L.: Linear models of nonlinear FIR systems with Gaussian inputs. IFAC Proc. Volumes **36**(16), 1873–1878 (2003)
6. Enqvist, M., Ljung, L.: Linear approximations of nonlinear FIR systems for separable input processes. Automatica **41**(3), 459–473 (2005)
7. Goda, T.: On the separability of multivariate functions. Math. Comput. Simul. **159**, 210–219 (2019)
8. Kuo, F., Sloan, I., Wasilkowski, G., Woźniakowski, H.: On decompositions of multivariate functions. Math. Comput. **79**(270), 953–966 (2010)
9. Lee, E.A.: Cyber physical systems: design challenges. In: 2008 11th IEEE International Symposium on Object and Component-Oriented Real-Time Distributed Computing (ISORC), pp. 363–369. IEEE (2008)
10. Schoukens, J., Ljung, L.: Nonlinear system identification: a user-oriented road map. IEEE Control Syst. Mag. **39**(6), 28–99 (2019)
11. Śliwiński, P., Marconato, A., Wachel, P., Birpoutsoukis, G.: Non-linear system modelling based on constrained Volterra series estimates. IET Control Theory Appl. **11**(15), 2623–2629 (2017)

Latent GAN: Using a Latent Space-Based GAN for Rapid Forecasting of CFD Models

Jamal Afzali, César Quilodrán Casas, and Rossella Arcucci[✉]

Data Science Institute, Imperial College London, London, UK
`r.arcucci@imperial.ac.uk`

Abstract. The focus of this study is to simulate realistic fluid flow, through Machine Learning techniques that could be utilised in real-time forecasting of urban air pollution. We propose a novel *Latent GAN* architecture which looks at combining an AutoEncoder with a Generative Adversarial Network to predict fluid flow at the proceeding timestep of a given input, whilst keeping computational costs low. This architecture is applied to tracer flows and velocity fields around an urban city. We present a pair of AutoEncoders capable of dimensionality reduction of 3 orders of magnitude. Further, we present a pair of Generator models capable of performing real-time forecasting of tracer flows and velocity fields. We demonstrate that the models, as well as the latent spaces generated, learn and retain meaningful physical features of the domain. Despite the domain of this project being that of computational fluid dynamics, the *Latent GAN* architecture is designed to be generalisable such that it can be applied to other dynamical systems.

Keywords: Generative adversarial networks · Reduced order models · Urban air pollution

1 Introduction

Computational Fluid Dynamics (CFD) concerns itself with using applied mathematics and/or computational software to resolve fluid flows in a domain. A crucial component of CFD are the Navier-Stokes (NS) equations; these describe the motions of incompressible fluid flow. A major drawback of CFD is the extreme difficulty in solving the NS equations due to their non-linearity. Another is the high dimensionality involved. Although the introduction of computers has allowed researchers to automate the calculations involved, with current hardware limitations, the computational complexity involved is far too great to solve in a reasonable amount of time. The revolution of Machine Learning (ML) in recent years has been invaluable for the field of CFD [6]. Previous data-driven studies tend to represent the fluids using linear basis functions such as principal component analysis (PCA). However, CFD is a non-linear problem and so more complex methods are required.

© Springer Nature Switzerland AG 2021
M. Paszynski et al. (Eds.): ICCS 2021, LNCS 12746, pp. 360–372, 2021.
https://doi.org/10.1007/978-3-030-77977-1_29

The aim of this study is the development of a neural network that could be utilised in real-time forecasting of urban air pollution in London. Accurate and fast forecasts of air pollution simulations have tremendous potential for healthcare, especially to explore the impact of exposure of individuals to air pollution. Generalising to the wider field of CFD, historically velocity fields and turbulent flows have been extremely difficult to resolve. The non-linearity of the problem leads to complex solutions that are far too difficult to solve analytically, with numerical solutions being too slow to do within a reasonable amount of time. A wider objective of this project is to train a surrogate generative model capable of predicting velocity fields in real-time.

In this study, we propose a novel *Latent GAN* architecture that looks at combining a Convolutional AutoEncoder (CAE) with a Generative Adversarial Network (GAN) [7] to produce surrogate generative models capable of predicting fluid flows such as tracers and velocity fields, whilst keeping computational costs low. The CAE focuses on reducing the dimensionality of given data samples before passing it to the GAN, which attempts to predict outputs that are a single timestep ahead. Despite the domain for this study being that of CFD, the network architecture is designed to be general such that it could be applied to many other dynamical domains that aren't necessarily related to CFD.

One study [8] of interest looks at combining an autoencoder (AE) with a generative model to predict fluid flows, which shows promising results. Another model of interest is the variational autoencoder (VAE)/GAN [9] which looks at generations of random faces using a combination of VAEs with GANs. Further, the Structural and Denoising Generative Adversarial Network (SD-GAN) [5] makes use of a piece-wise loss function to guide its GAN generations. This is useful as we also want to restrict the generations of our GANs; namely, to predict at the proceeding timestep. Other studies that have successfully combined fluid predictions with machine learning include: an application to a realistic case in China [17], application to unsteady flows [15], prediction of flow fields with deep convolutional recurrent autoencoders [16], and for data assimilation [3] with ROMs indoors [2] and outdoors [10,13]. Furthermore, [12] used adversarial training to improve the divergence of the data-driven forecast prediction over time and achieve better compression from full-space to latent space of urban air pollution CFD simulations.

In summary, in this paper we

- propose a novel *Latent GAN* architecture that combines a CAE along with a GAN to produce a model that is capable of predicting fluid flow. The Generator is restricted to generate data at the proceeding timestep, given an input data sample. The architecture has been designed to be generalisable such that it can be applied to any dynamical system.

– demonstrate the ability to encode fluid data on unstructured meshes down
to a latent space with several orders of reduction through the use of a CAE,
an architecture typically used on structured domains that do not have real
physical meanings, such as images.
– present a pair of Generator models, trained via the *Latent GAN* architecture,
that are capable of predicting tracer flows and velocity fields at the proceeding
timestep of a given input.

2   Latent GAN

The *Latent GAN* architecture looks at combing a Convolutional AutoEncoder
with a Generative Adversarial Network to predict the proceeding timestep, given
an input data sample. A high level overview of the model architecture proposed is
displayed in Fig. 1. The AutoEncoder comprised of an Encoder and a Generator,
as shown. The Encoder model reduces dimensionality of the data down to a
reduced latent space. This is done to reduce the computational complexity when
training the GAN and to attempt to reduce the instability that is inherit when
training GANs.

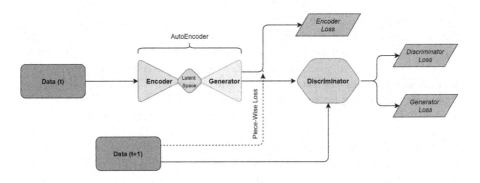

Fig. 1. High level overview of *Latent GAN* architecture.

Initially, we attempted to train all three models in a single training process.
Whilst in theory this is expected to work, in practice the training process showed
extreme instability across all three models. This is understandable as we were
essentially trying to find three solutions simultaneously with targets that were
constantly moving as each of the three networks were updated. A further, less
obvious, issue with this technique is that an unintuitive latent space is being
generated. As the Encoder is encoding w.r.t timestep t and the Generator is
decoding w.r.t timestep $t + 1$, this meant the latent space became a strange
cross between both timesteps. Whilst this may not have been an issue for scope
of this particular project, it meant the AE could not be used alone as it provided

no useful latent space. Instead, we opted to train the AE and GANs separately. The AE training process is performed, as standard, with a Mean Squared Error (MSE) loss being calculated. Once a stable AE is trained, the Decoder is dropped, and the Encoder is implemented as a fixed standalone pre-trained model into the *Latent GAN* architecture displayed previously in Fig. 1.

The input of the *Latent GAN* network is a data sample at time t. This is reduced into a latent space representation via the pre-trained Encoder. The output is then passed through the Generator which attempts to reconstruct the data back to its original size, but also incrementing the time by one. The generated outputs, along with real data at time $t+1$ are then fed into the Discriminator for classification. As per standard GAN training [7], Binary Cross Entropy (BCE) losses were calculated for the Discriminator and Generator, based on whether the generated outputs had successfully "tricked" the Discriminator. In addition, a separate MSE is calculated for the Generator between its generations and the ground truths at $t+1$. This is done to guide our Generator to produce a specific output, and not one that only appeared to be plausibly from the dataset as per regular GANs.

To summarise, the losses for each network were as follows:

$$L_{AE}(\boldsymbol{x}_t) = \min\left[\frac{1}{N}\sum_{i=1}^{N}[AE(\boldsymbol{x}_t) - \boldsymbol{x}_t]^2\right] \tag{1}$$

$$L_D(\boldsymbol{x}_t) = \max\left[\log D(\boldsymbol{x}_{t+1}) + \log\left(1 - D(G(E(\boldsymbol{x}_t)))\right)\right] \tag{2}$$

$$L_G(\boldsymbol{x}_t) = \max\left[\log D(G(E(\boldsymbol{x}_t)))\right] + \min\left[\frac{1}{N}\sum_{i=1}^{N}[G(E(\boldsymbol{x}_t)) - \boldsymbol{x}_{t+1}]^2\right] \tag{3}$$

where AE represents a forward pass through both Encoder and Decoder architectures, E represents the Encoder alone, G represents the Generator and D represents the Discriminator. A full Algorithm of both training processes can be seen in Sect. 3.1.

3 Models

The full codebase constructed as part of this study can be found at the following repository:
https://github.com/DL-WG/LatentGAN

3.1 Algorithms

For the AutoEncoder, training is as described in Algorithm 1. The inputs to this were unstructured tracer/velocity field data arrays at time t.

Algorithm 1: AutoEncoder Training

---

1 Instantiate Encoder (**E**) and Decoder (**D**) architectures;
2 Initialise **E** and **D** weights from normal distribution;
3 Instantiate DataLoader;
4 **for** *epoch* ← *0* **to** *max_epoch* **do**
      /* iterate over all batches in DataLoader   */
5   **for** *data* **in** *DataLoader* **do**
6   Zero gradients in Encoder and Decoder;
7   output ← **E**(data) ; // Forward pass through Encoder
8   output ← **D**(output) ; // Forward pass through Decoder
9   loss ← MSE(data, output) ; // Calculate loss
10   Calculate gradients;
11   Update weights using Optimiser step;
12   **end**
13 **end**
14 Save models;

For the GAN, training is as described in Algorithm 2. The inputs to this were unstructured tracer/velocity fields at time t, and $t + 1$ for the Generator loss.

It is worth noting that in Algorithm 2 Line 23, the order of magnitude of the two components of the Generator's error are starkly different. Typical values for the BCE loss are of 1 order of magnitude, whereas for the MSE error, we'd hope for this to be as small as possible. From our experimentations, we noticed that some MSE errors reaches as small as 10^{-7}. Therefore, an α coefficient is included with the MSE to try to balance the scales of the errors to be of the same order. If we were to perform the standard addition of both components ($\alpha = 1$), the MSE error would become irrelevant and ignored when calculating the gradients for gradient descent. For example, say $errBCE = 2.3$ and $errMSE = 0.0000006$, then $errG = 2.3 + 0.0000006 = 2.3000006 \approx 2.3$. This then means the Generator is just learning to become a standard GAN Generator, and ignores information about $t + 1$.

Algorithm 2: GAN Training

---

1 Instantiate Encoder (**E**), Generator (**G**) and Discriminator (**D**) architectures;
2 Load pretrained **E** weights;
3 Initialise **G** and **D** weights from normal distribution;
4 Instantiate DataLoader;

5 *real* ← 1 ;
6 *fake* ← 0 ;

7 **for** *epoch* ← *0* **to** *max_epoch* **do**
    /* iterate over all batches in DataLoader   */
8   **for** *data* **in** *DataLoader* **do**
9   data_incr ← (data samples incremented by 1);
10   Zero gradients in **D** and **G**;
       /* Discriminator Training   */
       /* Pass all-real batch   */
11   outputD_real ← **D**(data) ;
12   errD_real ← BCE(outputD_real, real) ;
       /* Pass all-fake batch   */
13   outputE ← **E**(data) ; // Do not track gradients here
14   outputG ← **G**(outputE) ;
15   outputD_fake ← **D**(outputG) ;
16   errD_fake ← BCE(outputD_fake, fake) ;

17   errD ← errD_real + errD_fake ;
18   Calculate gradients;
19   Update **D** weights using Optimiser step;

       /* Generator Training   */
20   output ← **D**(outputG) ; // Pass outputG through updated D
21   errBCE ← BCE(output, real) ; // Use real labels as suggested by Goodfellow
22   errMSE ← MSE(data_incr, outputG) ;

23   errG ← errBCE + α errMSE ; // See comment below
24   Calculate gradients;
25   Update **G** weights using Optimiser step;
26   **end**
27 **end**
28 Save models;

3.2 Model Architectures

Note that all LeakyReLU layers had a *negative_slope* value of 0.2 and *in_place* set to True. Tables 1, 2, and 3 show the final hyperparameters used. The visualisation of the networks are shown in Figs. 2 and 3.

Table 1. Model architecture for Encoder.

Layer name	Kernel size	Stride	Padding
Conv1D	4	2	1
LeakyReLU			
BatchNorm1D			
Conv1D	4	2	1
LeakyReLU			
BatchNorm1D			
Conv1D	4	2	1
LeakyReLU			
BatchNorm1D			
Conv1D	4	2	1
LeakyReLU			
BatchNorm1D			
Conv1D	4	2	1

Table 2. Model architecture for Decoder and Generator.

Layer name	Kernel size	Stride	Padding	Output padding
ConvTranspose1D	4	2	1	0
BatchNorm1D				
LeakyReLU				
ConvTranspose1D	4	2	1	1
BatchNorm1D				
LeakyReLU				
ConvTranspose1D	4	2	1	0
BatchNorm1D				
LeakyReLU				
ConvTranspose1D	4	2	1	0
BatchNorm1D				
LeakyReLU				
ConvTranspose1D	4	2	1	0
Tanh				

Table 3. Model architecture for Discriminator.

Layer name	Kernel size	Stride	Padding
Conv1D	4	8	1
LeakyReLU			
Conv1D	4	8	1
BatchNorm1D			
LeakyReLU			
Conv1D	4	8	1
BatchNorm1D			
LeakyReLU			
Conv1D	4	8	1
BatchNorm1D			
LeakyReLU			
Conv1D	4	4	1
BatchNorm1D			
LeakyReLU			
Conv1D	4	4	1
BatchNorm1D			
LeakyReLU			
Conv1D	1	2	1
Sigmoid			

4   Testing and Evaluation on a Real Test Case

The dataset used to train the network is that of an urban city comprising of tracers with dimensionality $100,040$ and velocity fields with dimensionality $3 \times 100,040$. Note that all screenshots displayed showcase a 2D slice of the 3D domain, using ParaView [1].

Using DCGAN [14] as a basis for the GAN with modifications made to fit the specific dataset, the Decoder is a duplicate of the Generator and the Encoder is designed to be an inverse of this. All networks were run for 1000 epochs, taking around 24 h for the AutoEncoders and around 27 h for the GANs. We saved the networks every 200 epochs, and the best of these were chosen.

The final tracer AE, trained for 600 epochs, is capable of reducing the dimensionality down to 256, whilst achieving an average MSE loss of 7.68×10^{-7}. Showcased in Fig. 4 is an AE reconstruction of timestep $t = 3000$ from our test set along with its ground truth. We see similar reconstructions, with flow shapes and intensities maintained. Its worth noting that the final timestep used to train the tracer models is $t = 988$, demonstrating that the AE network has learnt meaningful physical features of the dataset and is generalisable to unseen data. The final tracer *Latent GAN* is trained for 400 epochs, achieving an average

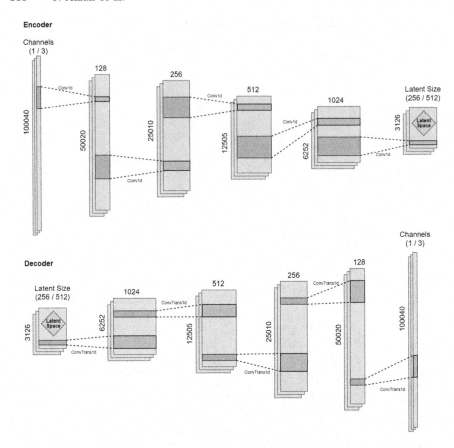

Fig. 2. Encoder and Decoder model visualisation.

MSE loss of 7.14×10^{-6} as shown in Fig. 8. An example of interpolation results can be viewed in Fig. 5. The network is fed in timestep $t = 3000$ and a prediction is made for timestep $t = 3001$. We see that the shapes of the tracer are correctly predicted along with the intensities. Despite this however, we notice that some artifacts are created around the domain with a somewhat patchy prediction of the actual flow. This is perhaps a side effect caused by the unstructured meshes used, leading to the convolutional layers incorrectly learning features in the data arrays that are not present in the actual flow.

The final velocity field AE, trained for 1000 epochs, reduces the dimensionality down to 512, whilst achieving an average MSE loss of 1.49×10^{-2}. We immediately notice the degraded performance compared to that of the tracers. This is, however, to be expected as the velocity fields are far more complex, with 3 channels instead of 1 for the tracers, as well as having flows occurring across the entire domain instead of just the centre. Figure 6 displays AE reconstructions for $t = 980$, where the final timestep to train the model is $t = 899$. We find that the reconstructions appear to be very similar, maintaining shapes and

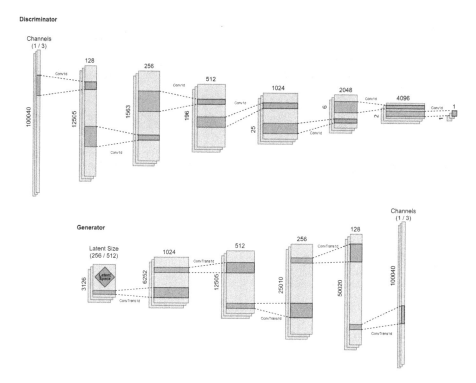

Fig. 3. Discriminator and Generator model visualisation.

intensities. The final velocity field *Latent GAN* is also trained for 1000 epochs, achieving an average MSE loss of 2.31×10^{-2} as shown in Fig. 9. Note that the initial 25 timesteps have been removed for better visualisation. Interpolated results of these can be viewed in Fig. 7, where the network is fed in $t = 980$ and a prediction of $t = 981$ is generated. Similarly to the tracers, we find that the prediction correctly maintained intensities as well as flow shapes. We do note however, that the predictions here also suffer from patchy generations.

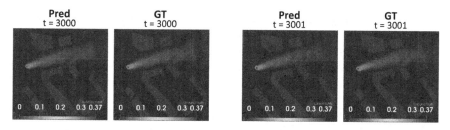

Fig. 4. Tracer AutoEncoder reconstruction.

Fig. 5. Tracer *Latent GAN* prediction given input $t = 3000$

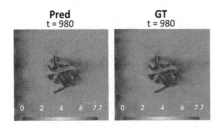

Fig. 6. Velocity field AutoEncoder reconstruction

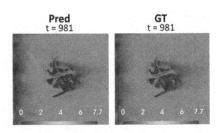

Fig. 7. Velocity field *Latent GAN* prediction given input $t = 980$

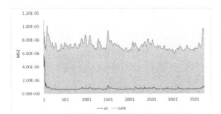

Fig. 8. Tracer MSE losses

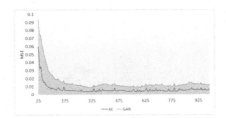

Fig. 9. Velocity field MSE losses

Generating real timesteps, where the actual physical equations were solved, took around 172 s using an *E5-2650 v3 CPU* with 250 GB RAM. Predictions using the *Latent GAN* models took only 1.1 s and 1.3 s for tracers and velocities respectively using a single core of an *i7-4790k* with 16 GB RAM. Unfortunately, due to limitations we were unable to reproduce these on the same hardware, although its worth noting that the i7 has around a 15% better single core performance. Running on a NVIDIA CUDA [11] enabled GPU, a GTX 970 with 4 GB of vRAM, took 0.25 and 0.3 s for tracers/velocities respectively.

5   Summary and Future Work

This paper introduced *Latent GAN*. We showed that the latent spaces generated, learnt and retained meaningful physical features of the domain. Despite the domain of this project being that of CFD, *Latent GAN* is generalisable such that it can be applied to other dynamical systems.

As future work, exploiting other kinds of CAE using 3D data arrays would allow us to incorporate known real physics into the network, in the form of physics informed loss functions. A different possible route to take would be to look at the Wasserstein GAN [4]. This is a modification to the existing GAN that are generally more robust and stable, and so may allow us to train the entire *Latent GAN* in a single training process. Further directions could be to possibly extend the *Latent GAN* architecture by combining a Long-Short Term

Memory (LSTM) network. Currently, only a single timestep is fed as input to the network; it would be interesting to see results where multiple timesteps are fed in instead.

Acknowledgements. This work is supported by the EPSRC Grand Challenge grant Managing Air for Green Inner Cities (MAGIC) EP/N010221/1, the EP/T003189/1 Health assessment across biological length scales for personal pollution exposure and its mitigation (INHALE), the EP/T000414/1 PREdictive Modelling with QuantIfication of UncERtainty for MultiphasE Systems (PREMIERE) and the Leonardo Centre for Sustainable Business at Imperial College London.

References

1. Paraview (2018)
2. Amendola, M., et al.: Data assimilation in the latent space of a neural network. arXiv preprint arXiv:2012.12056 (2020)
3. Arcucci, R., Zhu, J., Hu, S., Guo, Y.K.: Deep data assimilation: integrating deep learning with data assimilation. Appl. Sci. **11**(3), 1114 (2021)
4. Arjovsky, M., Chintala, S., Bottou, L.: Wasserstein GAN. arXiv preprint arXiv:1701.07875 (2017)
5. Banerjee, S., Das, S.: SD-GAN: structural and denoising GAN reveals facial parts under occlusion. arXiv preprint arXiv:2002.08448 (2020)
6. Brunton, S.L., Noack, B.R., Koumoutsakos, P.: Annual review of fluid mechanics: machine learning for fluid mechanics. Ann. Rev. **52**, 477–508 (2020). https://doi.org/10.1146/annurev-fluid-010719-060214
7. Goodfellow, I., et al.: Generative adversarial nets. In: Advances in Neural Information Processing Systems, pp. 2672–2680 (2014)
8. Kim, B., Azevedo, V.C., Thuerey, N., Kim, T., Gross, M., Solenthaler, B.: Deep fluids: a generative network for parameterized fluid simulations. In: Computer Graphics Forum. vol. 38, pp. 59–70. Wiley Online Library (2019)
9. Larsen, A.B.L., Sønderby, S.K., Larochelle, H., Winther, O.: Autoencoding beyond pixels using a learned similarity metric. arXiv preprint arXiv:1512.09300 (2015)
10. Mack, J., Arcucci, R., Molina-Solana, M., Guo, Y.K.: Attention-based convolutional autoencoders for 3d-variational data assimilation. Comput. Methods Appl. Mech. Eng. **372**, 113291 (2020)
11. NVIDIA: Cuda. https://developer.nvidia.com/cuda-zone
12. Quilodrán-Casas, C., Arcucci, R., Pain, C., Guo, Y.: Adversarially trained LSTMs on reduced order models of urban air pollution simulations. arXiv preprint arXiv:2101.01568 (2021)
13. Quilodrán-Casas, C., Arcucci, R., Wu, P., Pain, C., Guo, Y.K.: A reduced order deep data assimilation model. Physica D **412**, 132615 (2020)
14. Radford, A., Metz, L., Chintala, S.: Unsupervised representation learning with deep convolutional generative adversarial networks. arXiv preprint arXiv:1511.06434 (2015)
15. Reddy, S.B., et al.: Reduced order model for unsteady fluid flows via recurrent neural networks. In: International Conference on Offshore Mechanics and Arctic Engineering, vol. 58776, p. V002T08A007. American Society of Mechanical Engineers (2019)

16. Reddy Bukka, S., Magee, A.R., Jaiman, R.K.: Deep convolutional recurrent autoencoders for flow field prediction. arXiv pp. arXiv-2003 (2020)
17. Xiao, D., Fang, F., Zheng, J., Pain, C., Navon, I.: Machine learning-based rapid response tools for regional air pollution modelling. Atmos. Environ. **199**, 463–473 (2019)

Data Assimilation in the Latent Space
of a Convolutional Autoencoder

Maddalena Amendola[1], Rossella Arcucci[1(✉)], Laetitia Mottet[2],
César Quilodrán Casas[1], Shiwei Fan[3], Christopher Pain[1,2], Paul Linden[4],
and Yi-Ke Guo[1,5]

[1] Data Science Institute, Department of Computing, Imperial College London,
London, UK
r.arcucci@imperial.ac.uk
[2] Department of Earth Science and Engineering, Imperial College London,
London, UK
[3] Department of Chemistry, University of Cambridge, Cambridge, UK
[4] Department of Applied Mathematics and Theoretical Physics,
University of Cambridge, Cambridge, UK
[5] Department of Computer Science, Hong Kong Baptist University,
Kowloon, Hong Kong

Abstract. Data Assimilation (DA) is a Bayesian inference that combines the state of a dynamical system with real data collected by instruments at a given time. The goal of DA is to improve the accuracy of the dynamic system making its result as real as possible. One of the most popular technique for DA is the Kalman Filter (KF). When the dynamic system refers to a real world application, the representation of the state of a physical system usually leads to a big data problem. For these problems, KF results computationally too expensive and mandates to use of reduced order modeling techniques. In this paper we proposed a new methodology we called Latent Assimilation (LA). It consists in performing the KF in the latent space obtained by an Autoencoder with non-linear encoder functions and non-linear decoder functions. In the latent space, the dynamic system is represented by a surrogate model built by a Recurrent Neural Network. In particular, an Long Short Term Memory (LSTM) network is used to train a function which emulates the dynamic system in the latent space. The data from the dynamic model and the real data coming from the instruments are both processed through the Autoencoder. We apply the methodology to a real test case and we show that the LA has a good performance both in accuracy and in efficiency.

Keywords: Data assimilation · Machine learning · Neural network ·
Convolutional autoencoder · Long short term memory

1 Introduction and Motivation

Data Assimilation (DA) is an approach for fusing data (observations) with prior knowledge (e.g., mathematical representations of physical laws; model output) to

© Springer Nature Switzerland AG 2021
M. Paszynski et al. (Eds.): ICCS 2021, LNCS 12746, pp. 373–386, 2021.
https://doi.org/10.1007/978-3-030-77977-1_30

obtain an estimate of the distribution of the true state of a process [24]. In order to perform DA, one needs observations (i.e., a data or measurement model), a background (i.e., a priori state or process model) and information about the distribution of the errors on these two. In real world applications, DA is usually used to improve the accuracy of dynamic systems which represent the evolution of some physical fields in time (e.g. weather, climate, ocean, air pollution). For those applications, the background is defined in big computational grids which lead to a big data problem sometimes impossible to handle without introducing approximations or space reductions. To overcome this problem, Reduced Order Modeling (ROM) techniques are used [1,2]. ROM allows to speed up the dynamic model and the DA process. Popular approaches to reduce the domain are the Principal Component Analysis (PCA) and the Empirical Orthogonal Functions (EOF) technique both based on a truncated singular value decomposition (TSVD) analysis [17]. The simplicity and the analytic derivation of those approaches are the main reasons behind their popularity in atmospheric and ocean science. However, the physical interpretability of the obtained patterns is a matter of controversy because of the constraints satisfied by these approaches, e.g. orthogonality in both space and time. Also, despite those are powerful approaches, the accuracy of the obtained solution exhibits a severe sensibility to the variation of the value of the truncation parameters. This issue introduces a severe drawback to the reliability of these approaches, hence their usability in operative software in different scenarios [16]. An approach to reduce the dimensionality maintaining information of the data is the Neural Network (NN), precisely the Autoencoders. NNs have the ability to fit functions and they can fit almost any unknown function theoretically. That is the ability which makes it possible for neural networks to face complex problems. Autoencoders (AE) with non-linear encoder functions and non-linear decoder functions can thus learn a more powerful non-linear generalization of methods based on TSVD. The co-domain of an encoder function is named latent space. The latent space is also the domain of the decoder function. In the latent space, the evolution of the transformed state variables defined in time, is still a dynamical system. Thanks to what the Universal Approximation Theorem [14,21] claims, we can assume that any non-linear dynamical system can be approximated to any accuracy by a Recurrent Neural Network (RNN), with no restrictions on the compactness of the state space, provided that the network has enough sigmoidal hidden units. In the present work, we propose a new methodology we called Latent Assimilation (LA). It consists in reducing the dimensionality with an AE and perform both prediction through a surrogate dynamic model and DA directly in the latent space. In the latent space, the surrogate dynamical system is built by a RNN which learns the dynamic of the transformed variable. We apply LA to improve the prediction of air flows and indoor pollution transport. In fluid dynamics problems such as the propagation of air pollution, the data represents physical variables that are spatial distributed and contains information about geographical position (e.g. the sensor placement). We have to take into account those variables in our process. On the other hand, usually, such problems are defined in high dimensional spaces. To reduce the complexity of the problem we use the Autoencoder. The AE performs a non-linear transformation on the input. We process the system states and the observations coming from sensors using the same AE. In this way, we have a latent version of the model states and observations transforming the physical variables in the same way.

1.1   Related Works and Contribution of the Present Work

The use of machine learning in correcting model forecasts is promising in several geophysics applications [10,15]. In some operational centres, data driven models have been already introduced to support CFD simulations [9,13,20]. The future challenges of numerical weather prediction (NWP) include more accurate initial conditions that take advantage of the increasing volume of real-time observations, and improved post-processing of model outputs, among others [9]. Neural networks for correction of error in forecasting have been previously studied in [5–7]. However, in this literature, the error correction by NN does not have a direct relation with the updated model system in each step and the training is not on the results of the assimilation process. In [3,25] the authors describe a framework for integration of NN with physical models by DA algorithms. In this approach, the NNs are iteratively trained when observed data are updated. However, this approach presents a limit due to the time complexity of the numerical models involved, which limits the use of the forecast model for large data problems. In [11], the authors presented an approach for employing artificial neural networks (NNs) to emulate the local ensemble transform Kalman filter (LETKF) as a method of data assimilation. In [19], the authors combined Deep Learning and Data Assimilation to predict the production of gas from mature gas wells. They used a modified deep Long Short Term Memory (LSTM) model as their prediction model in the EnKF framework for parameter estimation. In [8] and [23], modified versions of KF based on NNs are applied to simulated pendulum and other four visual tasks (an agent in a plane with obstacles, a visual version of the classic inverted pendulum swing-up task, balancing a cart-pole system, and control of a three-link arm with larger images), respectively.

In the present work, we focus on the problem of assimilating data to improve the prediction of air flows and indoor pollution transport [22]. We propose a methodology we called Latent Assimilation (LA) which consists in:

- **Dimensionality reduction:** The dimensionality of both background of the dynamic model and observations (real data coming from the instruments) is reduced by Autoencoder;
- **Surrogate Model:** A data driven version of the dynamic model we call "surrogate dynamic model" is build using RNN. In particular, an LSTM is used to train a function which emulates the dynamic system in the latent space;
- **Data Assimilation:** In the latent space, DA is performed by a modification of a standard KF;
- **Physical space:** The results of the DA in the latent space are then reported in the physical space through the Decoder.

In this paper, we apply Latent Assimilation to improve the prediction of air flows and indoor pollution transport. However, the technology and the model are very general and can be applied to other kinds of computational fluid dynamic systems which simulate other dynamical systems. Both prediction and DA show a speed up in the reduced space. We apply the methodology to a real test case and we show that the LA has a good performance both in accuracy and in time.

This paper is structured as follows. In the next section the Kalman Filter is described. Then, it follows a section where the Latent Assimilation is introduced. Next, experimental results are provided. Finally, in the last section we summarize conclusions and future works.

2   Kalman Filter

DA merges the estimated state $x_t \in \mathbb{R}^n$ of a discrete-time dynamic process at time t:

$$x_{t+1} = M_{t+1}x_t + w_t \tag{1}$$

with an observation $y_t \in \mathbb{R}^m$:

$$y_t = H_t x_t + v_t \tag{2}$$

where M_t is a dynamics linear operator and H_t is called linear observation operator. The vectors w_t and v_t represent the process and observation errors, respectively. They are usually assumed to be independent, white-noise processes with Gaussian probability distributions

$$w_t \sim \mathcal{N}(0, Q_t), \quad v_t \sim \mathcal{N}(0, R_t)$$

where Q_t and R_t are called errors covariance matrices of the model and observation respectively.

DA is a Bayesian inference that combines the state x_t with y_t at each given time. The Bayes theorem conducts to the estimation of x_t^a which maximise a probability density function given the observation y_t and a prior from x_t. This approach is implemented in one of the most popular DA methods which is the Kalman filter (KF) [18]. The goal of the KF is to compute an optimal a posteriori estimate, x_t^a, that is a linear combination of an a priori estimate, x_t, and a weighted difference between the actual measurement, y_t, and the measurement prediction, $H_t x_t$ as described in Eq. (5) where the quantity $d_t = y_t - H_t x_t$ is called misfit and represents the distance between the actual and the predicted measurements at time t. For big data problems, KF is usually implemented in a simplified version as an Optimal Interpolation method [4] which consists of fix the covariance matrix $Q_t = Q$, for each time step t. It mainly consists of two steps: a prediction and a correction step:

1. **Prediction:**

$$x_{t+1} = M_{t+1}x_t^a \tag{3}$$

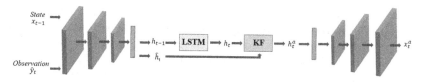

Fig. 1. The work flow of the Latent Assimilation model. Let us assume that we want to predict the state of the system at time t and we assume that the LSTM needs one observation back to predict the next time-step. The input of the system is the state x_{t-1}. We encode x_{t-1} producing its encoded version h_{t-1}. From h_{t-1} we compute h_t through LSTM. To perform the Kalman Filter, we need the observation y_t at time-step t. We encode y_t and we combine the result, $\hat{h}_t$, with the prediction h_t through the KF. The result h_t^a is the updated prediction. We report the updated prediction in its physical space through the decoder, producing x_t^a.

2. **Correction:**

$$K_{t+1} = QH^T(HQH^T + R_{t+1})^{-1} \tag{4}$$

$$x_{t+1}^a = x_{t+1} + K_{t+1}(y_{t+1} - Hx_{t+1}) \tag{5}$$

The prediction-correction cycle is complex and time-consuming and it mandates the introduction of simplifications, approximations or data reductions techniques. In the next section, we present the Latent Assimilation approach which consists in performing KF in the latent space of an Autoencoder with nonlinear encoder and nonlinear decoder functions. In the latent space, the dynamic system in (3) is replaced by a surrogate model built with a RNN.

3   Latent Assimilation

Latent Assimilation is a model that implements the new idea of assimilating real data in the Latent Space. Instead of using PCA or others mathematical approaches to reduce the space, we decide to experiment the reduction with non-linear transformations using Deep NNs. Specifically, we choose to use Convolutional Autoencoder to reduce the space.

Figure 1 is a graphical representation of the Latent Assimilation model. The model is structured in four main parts:

Dimensionality Reduction: The dimensionality reduction is implemented by a Convolutional Autoencoder which produces a representation of the state vector $x_t \in \mathbb{R}^n$ in (1) in a "latent" state vector $h_t \in \mathbb{R}^p$ defined in a Latent Space where $p < n$. We denote with $f : \mathbb{R}^n \to \mathbb{R}^p$ the Encoder function

$$h_t = f(x_t) \tag{6}$$

which transforms the state x_t in a latent variable h_t.

Surrogate Model: In the latent space we perform a regression through a LSTM function $l : \mathbb{R}^{p \times q} \to \mathbb{R}^p$

$$h_{t+1} = l(\boldsymbol{h}_{t,q}) \tag{7}$$

where $\boldsymbol{h}_{t,q} = \{h_i\}_{i=t,\dots,t-q}$ is a sequence of q encoded time-steps up to time t.

Data Assimilation: The assimilation is performed in the latent space. In order to merge the observations in (2) with the "latent" state vector h_t, the observations are processed by the Encoder in the same way as the state vector. As $y_t \in \mathbb{R}^m$ where usually $m \leq n$, i.e. the observations are usually held or measured in just few point in space, the observations vector y_t is interpolated in the state space $\mathbb{R}^n$ obtaining $\hat{y}_t \in \mathbb{R}^n$. The observations $\hat{y}_t$ are then processed in the same way as the state vector trough f:

$$\hat{h}_t = f(\hat{y}_t) \tag{8}$$

The "latent" observations $\hat{h}_t$, transformed by the Encoder in the latent space, are then assimilated by the prediction-correction steps as described in Eqs. (9)–(11):

1. **Prediction:**

$$h_{t+1} = l(\boldsymbol{h_{t,q}}) \tag{9}$$

2. **Correction:**

$$\hat{K}_{t+1} = \hat{Q}\hat{H}^T(\hat{H}\hat{Q}\hat{H}^T + \hat{R}_{t+1})^{-1} \tag{10}$$

$$h_{t+1}^a = h_{t+1} + \hat{K}_{t+1}(\hat{h}_{t+1} - \hat{H}h_{t+1}) \tag{11}$$

where l in (9) is the surrogate model defined in (7) computed by the LSTM, $\hat{Q}$ and $\hat{R}$ are the errors covariance matrices of the transformed background h_t and observations $\hat{h}_t$ respectively, they are computed directly in the latent space. $\hat{H}$ is the observation operator: if $m = n$, it is the identity function, otherwise it is an interpolation function. The background covariance matrix $\hat{Q}$ is computed with a sample of s model state forecasts $\boldsymbol{h}$ that we set aside as background such that:

$$\boldsymbol{h} = [h_1, ..., h_s] \in \mathbb{R}^{p \times s}, \quad V = (\boldsymbol{h} - \bar{h}) \in \mathbb{R}^{p \times s} \tag{12}$$

where $\bar{h}$ is the mean of the sample of background states, then $\hat{Q} = VV^T$. The observations errors covariance matrix $\hat{R}$ can be computed with the same process in (12) replacing h_t with $\hat{h}_t$, $\forall t$ or, it can be estimated by evaluations of measurements (instruments) errors. $\hat{K}$ is the Kalman Gain matrix defined in the latent space and $\hat{H}$ is the observation operator.

Physical Space: The results of the DA in the latent space are then reported in the physical space through the Decoder, applying the function $g : \mathbb{R}^p \rightarrow \mathbb{R}^n$ to compute

$$x_{t+1}^a = g(h_{t+1}^a). \tag{13}$$

The Decoder is almost a mirror of the Encoder: it is composed of a Fully Connected Layer followed by some Convolutional Layers.

In the next Section we apply Latent Assimilation to the problem of assimilating data to improve the prediction of air flows and indoor pollution transport in a real scenario [22]. We show the performance of the model step by step and we compare results with a standard DA performed in the physical space.

4   Experimental Results

Latent Assimilation is here applied to merge indoor air quality measurements from sensors with an indoor air quality simulation made by a computational fluid dynamic (CFD) software named Fluidity.

CFD Data: The domain is an office room in the Clarence Centre in the Borough of Southwark, London [22]. The CFD simulation constitutes a time series composed of 3500 time-steps. For each time step, the CFD simulation provides a matrix of dimension 180×250 where each value of the matrix represents the concentration of CO_2 expressed in PPM in a specific location of the room.

The time series we use is composed of 2500 time-steps, then the size of the data set is $\mathcal{O}(10^8)$ just considering the data related to the CFD simulation. The CFD data represents the real simulation of a flow and it doesn't change much between two consecutive steps. For this reason, we decided to divide the data in train, validation and test set making small jumps. We consider two consecutive time-steps for train and we make a jump. Every position not considered yet (the ones we jump) is assigned to validation and test set alternately.

Considering a jump $= 1$, the series is divided as in Fig. 2:

Fig. 2. Train, validation and test split

Observed Data from Sensors: For the observations, we have measurements from 7 sensors providing information for 10 time-steps. The sensors are spatially distributed in the room. In fluid dynamics problems, the data contains physical variables that are spatial distributed and contains information about geographical position. It's important to maintain those information during the data assimilation process. We preserve them processing both CFD and observed data with the same AE. We first bring the observation in the same space of the CFD data using an interpolation function. Then we process the system states and the observations coming from sensors using the same AE. In this way, we have a latent version of the model states and observations transforming the physical variables in the same way. The measurements from sensors are extended in the radius of 30 cm in the room. Then the values are linearly interpolated using the Scipy library obtaining matrices of the same dimension of the data from the CFD, i.e. 180×250. The interpolation is repeated for each time-step. All the data are normalized in the range [0, 1]. We normalized the dataset with Min-Max Normalization considering the global minimum and maximum PPM values found in both dataset and observations.

The LA code and the pre-processed data can be downloaded using the link https://github.com/DL-WG/LatentAssimilation.

4.1 Dimensionality Reduction

The Autoencoder we implement is a Convolutional Autoencoder. In particular, the Encoder is composed of some Convolutional layers followed by a Fully connected layer that determines the shape of the Latent space. The Decoder is nearly the mirror of the Encoder.

The construction of the Autoencoder is divided in two steps: the search of the structure and the Grid Search of hyper-parameters. All the experiments were performed with 4 GPUs K80.

Structure: The search of the structure is conducted progressively. We start comparing few autoencoders (that, for example, differ on the number of hidden convolutional layers). We pick the best structure and we create new configurations to compare. The model used to create new configurations, is here called *baseline*. We use the Mean Squared Error[1] (MSE) as metrics to evaluate the model. For each configuration, we compute a 5-Cross Fold Validation[2] (CV) and we choose the model with lower mean and standard deviation of the MSE values (Mean-MSE and Std-MSE respectively). A low standard deviation tells us that the model is stable and it does not depend on the data we use to train and validate it. We fix the following parameters: number of filters 32, activation function ReLu[3], Kernel size 3, latent space 7, optimizer Adam, epochs 300 and batch size 32. We use the MSE as loss function. This choice of value for the latent space is the result of an analysis of accuracy and efficiency.

We shuffle the data before to start to make the neural network independent from the order of the data. We first check the good number of hidden Convolutional layers. We try three different configurations:

1. Encoder with 3 Convolutional layers and Decoder with 4 Convolutional layers
2. Encoder with 4 Convolutional layers and Decoder with 5 Convolutional layers
3. Encoder with 5 Convolutional layers and Decoder with 6 Convolutional layers

Comparing in Table 1 configurations 1, 2 and 3 for which only the number of convolutional layers is changing, configuration 2 is the one highlighting the best performance in terms of both Mean-MSE and Mean-MAE with MSE two order of magnitude lower than configurations 1 and 3. Moreover, configuration 2 is the most stable regarding the standard deviations, reflecting well that this CAE network architecture does not depend on the data used to train and validate it. In addition, the execution time of configuration 2 is relatively acceptable to answer real-time problems. Hence, in the following, the number of layers is taken as the same than configuration 2: 4 for the encoder and 5 for the decoder.

---

[1] https://www.tensorflow.org/api_docs/python/tf/keras/losses/MeanSquaredError.

[2] https://scikit-learn.org/stable/modules/cross_validation.html.

[3] https://www.tensorflow.org/api_docs/python/tf/keras/layers/ReLU.

Table 1. Convolutional AutoEncoder performance. N denotes the configuration number as listed in the main text. Time is given in seconds.

N	Mean-MSE	Std-MSE	Mean-time	Std-time
1	1.324e−02	2.072e−02	775.665	5.606e+00
2	**2.435e−04**	**3.851e−05**	**812.293**	**1.436e+00**
3	2.293e−02	2.763e−02	828.328	1.574e+00

Configuration 2 is then our *baseline*. We substitute the Convolutional layers with the Convolutional Transpose layers in the Decoder and we will call this the configuration number 4. As we can see from Table 2, the accuracy (Mean-MSE) and the stability (Std-MSE) are slightly better, while the execution time is slightly longer, when using convolutional layers (config. 2) rather than transpose convolutional layers in the decoder. As no major improvements in terms of MSE is observed when switching from convolutional (config. 2) to transpose convolutional layers in the decoder, convolutional layers are used for the decoder.

Table 2. Performance of the baseline (config. 2) and the baseline with Convolutional Transpose layers (config. 5).

N	Mean-MSE	Std-MSE	Mean-time	Std-time
2	**2.435e−04**	**3.851e−05**	**812.293**	**1.436e+00**
4	2.587e−04	4.114e−05	746.804	3.089e+00

Finally, we discard the Convolutional Transpose layers and we change the kernel size increasing it a little bit. We build the model with Kernel size equal to 5×5 everywhere, defining the configuration number 5. Table 3 shows that this choice works well but not better than the *baseline*.

Table 3. Performance of the baseline (config. 2) and the baseline with kernel size equal to 5 (config. 5)

N	Mean-MSE	Std-MSE	Mean-time	Std-time
2	**2.435e−04**	**3.851e−05**	**812.293**	**1.436e+00**
5	1.055e−02	2.099e−02	1222.251	6.627e+00

Grid Search: We make a grid search varying (i) the number of filters $\in \{16, 32, 64\}$, (ii) the activation function $\in \{ReLu, Elu\}$, (iii) the number of epochs $\in \{250, 300, 400\}$ and (iv) the batch size $\in \{16, 32, 64\}$. Table 4 shows the optimal hyper-parameters found.

Table 4. Grid search results of the autoencoder.

Filters	Activation	Epochs	Batch size
64	ReLu	400	32

The performance of the *baseline* with the hyper-parameters found are reported in Table 5. Because we shuffle the data making the neural network independent from the order of the data, we can say that the model is not dependent on the set of input we choose to train it. This is important in our case because we will use the encoder to reduce the data from the observations too.

Table 5. Autoencoder performance with the chosen hyper-parameters.

Mean-MSE	Std-MSE	Mean-time	Std-time
8.509e−05	1.577e−05	1887.612	6.845e+00

4.2 Surrogate Model

The surrogate model is built implementing an LSTM on the results of the Autoencoder. All data are encoded with the Autoencoder: each sample is a vector of 7 scalar. We followed the same strategy as for the Autoencoder: we define the structure of the model and then we compute the grid search. To this purpose, we encode the train and validation sets defined in Fig. 2. We split the data in small sequences based on the number of time-steps we look back and the number of steps we want to predict. In this case, we predict one step forward. We do not perform the CV but we repeated the fitting and the validation of the model 5 times. We fix the following parameters: neurons 30, activation function ReLu, number of time-steps 3, optimizer Adam, epochs 300, batch size 32. We use the MSE as loss function. LSTMs are stacked from 1 to 5 times in order to see if the model gains in accuracy, stability and efficiency: the results are shown in Table 6.

The single layer LSTM is the one highlighting the best accuracy with the lowest Mean-MSE value. Indeed, the input of the LSTM consists of a 7×1 vector and adding more LSTM layer introduces overfitting bias. In addition, the standard deviation, reflecting the stability, of the single layer LSTM are about one order of magnitude lower than the other tested LSTM. The single layer LSTM is also the most efficient in term of computation cost.

We compute the grid search changing the hyper-parameters: (i) number of neurons $\in \{30, 50, 70\}$, (ii) activation function $\in \{ReLu, Elu\}$, (iii) number of steps $\in \{3, 5, 7\}$, (iv) number of epochs $\in \{200, 300, 400\}$ and (v) batch size $\in \{16, 32, 64\}$.

Table 7 shows the result of the Grid Search considering a single LSTM, while Table 8 shows the performance of the LSTM with the chosen hyper-parameters.

Table 6. LSTM performance evaluation for 5 network architectures. N denotes the number of stacked LSTMs. Time is given in seconds.

N	Mean-MSE	Std-MSE	Mean-time	Std-time
1	**1.634e−02**	**8.553e−03**	**230.355**	**1.050e+00**
2	2.822e−02	7.244e−03	360.877	6.618e−01
3	4.619e−02	1.942e−02	494.254	2.258e+00
4	5.020e−02	1.675e−02	658.039	2.632e+00
5	4.742e−02	1.183e−02	806.001	5.921e+00

Table 7. Grid search results of the single LSTM.

Neurons	Activation	Steps	Epochs	Batch size
30	Elu	3	400	16

Table 8. Single LSTM performance with the chosen hyper-parameters.

Mean-MSE	Std-MSE	Mean-time	Std-time
1.233e−02	1.398e−03	949.328	7.508e+00

4.3   Data Assimilation

In this phase, we encoded both the states and the observations. The assimilation is performed in the latent space. From the Test set of the CFD data, we select the sequences that predict the time-steps where we have measurements from sensors. We make the prediction through the LSTM and we update the prediction using the corresponding observation with the Kalman Filter. In the KF, the error covariance matrix $\hat{Q}$ is computed as $\hat{Q} = VV^T$ where V is computed as described in (12). Since both predictions of the model and observations are values of CO_2, i.e. the observations don't have to be transformed, the operator $\hat{H}$ is an identity matrix.

We studied how KF improves the accuracy of the prediction testing different values of $\hat{R}$ computed by the procedure in (12) or, fixed as $\hat{R} = 0.01\ I, 0.001\ I, 0.0001\ I$ where $I \in \mathbb{R}^{p \times p}$ denotes the identity matrix. This last assumption is usually made to give higher fidelity and trust to the observations.

The MSE of the background data in the latent space, without performing data assimilation is MSE $= 7.220e-01$. Tables 9 shows values of MSE after the assimilation in the latent space. It also reports values of execution time. As expected, we can observe an improvement in the execution time in assuming $\hat{R}$ as a diagonal matrix instead of a full matrix.

Table 9. Value of MSE of h_t^a and execution time of the Latent Assimilation for different values of the observations errors covariance matrix $\hat{R}$.

$\hat{R}$	Cov	0.01 I	0.001 I	0.0001 I
MSE	3.215e−01	1.250e−02	1.787e−03	3.722e−05
Time	2.053e−03	3.541e−04	2.761e−04	2.618e−04

4.4   Physical Space

After performing DA in the latent space, the results h_t^a are reported in the physical space through the Decoder which gives x_t^a. Table 10 and Table 11 show values of MSE after the assimilation in the physical space for LA and for a standard DA respectively. They also report values of execution time. The MSE in the physical space without the assimilation is MSE $= 6.491e{-}02$. Tables 10 and 11 show that both LA and DA improve the accuracy of the forecasting. Comparing the tables we can also observe that LA performs better both in terms of execution time and accuracy with respect to a Standard DA where the assimilation works directly with big matrices becoming very slow.

Table 10. Values of MSE of x_t^a in the physical space for different values of the observations errors covariance matrix $\hat{R}$.

$\hat{R}$	Cov	0.01 I	0.001 I	0.0001 I
MSE	3.356e−02	6.933e−04	1.211e−04	2.691e−06
Time	3.191e+00	2.899e+00	2.896e+00	2.896e+00

Table 11. Standard assimilation in the physical space performed by a KF (see Eqs. (3)–(5)). Here $R \in \mathbb{R}^{n \times n}$ is defined in the physical space.

R	Cov	0.01 I	0.001 I	0.0001 I
MSE	5.179e−02	6.928e−03	6.928e−03	6.997e−03
Time	2.231e+03	2.148e+03	2.186e+03	2.159e+03

5   Conclusion and Future Works

In this paper we proposed a new methodology we called Latent Assimilation (LA) to efficiently and accurately perform DA. LA consists in performing the KF in the latent space obtained by an Autoencoder with non-linear encoder functions and non-linear decoder functions. In the latent space, the dynamic system is represented by a surrogate model built by an LSTM network to train a function which emulates the dynamic system in the latent space. The data from the dynamic model and the real data coming from the instruments are both processed through the Autoencoder. We apply the methodology to a real test case and we show that the LA performs better than a standard DA in both accuracy and efficiency. An implementation of LA to emulate variational DA [4] will be developed as future work. In particular, we will focus on a 4D variational (4DVar) method. 4DVar is a computational expensive method as it is developed to assimilate several observations (distributed in time) for each time step of the forecasting model. We will develop an extended version of LA able to assimilate set of distributed observations for each time step and, then, able to perform a 4DVar.

Acknowledgements. This work is supported by the EPSRC Grand Challenge grant Managing Air for Green Inner Cities (MAGIC) EP/N010221/1, the EP/T003189/1 Health assessment across biological length scales for personal pollution exposure and its mitigation (INHALE), the EP/T000414/1 PREdictive Modelling with QuantIfication of UncERtainty for MultiphasE Systems (PREMIERE) and the Leonardo Centre for Sustainable Business at Imperial College London.

References

1. Arcucci, R., Pain, C., Guo, Y.: Effective variational data assimilation in air-pollution prediction. Big Data Min. Anal. **1**(4), 297–307 (2018)
2. Arcucci, R., Mottet, L., Pain, C., Guo, Y.: Optimal reduced space for variational data assimilation. J. Comput. Phys. **379**, 51–69 (2019)
3. Arcucci, R., Zhu, J., Hu, S., Guo, Y.: Deep data assimilation: integrating deep learning with data assimilation. Appl. Sci. **11**(3), 11–14 (2021). Multidisciplinary Digital Publishing Institute
4. Asch, M., Bocquet, M., Nodet, M.: Data Assimilation: Methods, Algorithms, and Applications. SIAM, Fundamentals of Algorithms, Philadelphia (2016)
5. Babovic, V., Keijzer, M., Bundzel, M.: From global to local modelling: a case study in error correction of deterministic models. In: Proceedings of Fourth International Conference on Hydroinformatics (2000)
6. Babovic, V., Cañizares, R., Jensen, H., Klinting, A.: Neural networks as routine for error updating of numerical models. J. Hydraul. Eng. **127**(3), 181–193 (2001). American Society of Civil Engineers
7. Babovic, V., Fuhrman, D.: Data assimilation of local model error forecasts in a deterministic model. Int. J. Numer. Methods Fluids **39**(10), 887–918 (2002). Wiley Online Library
8. Becker, P., Pandya, H., Gebhardt, G., Zhao, C., Taylor, J., Neumann, G.: Recurrent Kalman networks: factorized inference in high-dimensional deep feature spaces. arXiv preprint arXiv:1905.07357 (2019)
9. Boukabara, S., Krasnopolsky, V., Stewart, J., Maddy, E., Shahroudi, N., Hoffman, R.: Leveraging modern artificial intelligence for remote sensing and NWP: benefits and challenges. Bull. Am. Meteorol. Soc. **100**(12), ES473–ES491 (2019)
10. Campos, R., Krasnopolsky, V., Alves, J., Penny, S.: Nonlinear wave ensemble averaging in the Gulf of Mexico using neural networks. J. Atmos. Oceanic Technol. **36**(1), 113–127 (2019)

11. Cintra, R., Campos Velho, H.: Data assimilation by artificial neural networks for an atmospheric general circulation model. In: Advanced Applications for Artificial Neural Networks, p. 265. BoD-Books on Demand (2018)
12. Dozat, T.: Incorporating nesterov momentum into adam (2016)
13. Dueben, P., Bauer, P.: Challenges and design choices for global weather and climate models based on machine learning. Geosci. Model Dev. 11(10), 3999–4009 (2018). Copernicus GmbH
14. Funahashi, K., Nakamura, Y.: Approximation of dynamical systems by continuous time recurrent neural networks. Neural Netw. 6(6), 801–806 (1993). Elsevier
15. Gagne, D., McGovern, A., Haupt, S., Sobash, R., Williams, J., Xue, M.: Storm-based probabilistic hail forecasting with machine learning applied to convection-allowing ensembles. Weather Forecast. 32(5), 1819–1840 (2017)
16. Hannachi, A.: A primer for EOF analysis of climate data. Department of Meteorology University of Reading, pp. 1–33 (2004)
17. Hansen, P., Nagy, J., O'leary, D.: Deblurring Images: Matrices, Spectra, and Filtering, vol. 3. SIAM, Philadelphia (2006)
18. Kalman, R.: A new approach to linear filtering and prediction problems. J. Basic Eng. 82(1), 35–45 (1960). American Society of Mechanical Engineers
19. Loh, K., Omrani, P.S., van der Linden, R.: Deep learning and data assimilation for real-time production prediction in natural gas wells. arXiv preprint arXiv:1802.05141 (2018)
20. Rasp, S., Dueben, P., Scher, S., Weyn, J., Mouatadid, S., Thuerey, N.: Weather-Bench: a benchmark dataset for data-driven weather forecasting. arXiv preprint arXiv:2002.00469 (2020)
21. Schäfer, A.M., Zimmermann, H.G.: Recurrent neural networks are universal approximators. In: Kollias, S.D., Stafylopatis, A., Duch, W., Oja, E. (eds.) ICANN 2006. LNCS, vol. 4131, pp. 632–640. Springer, Heidelberg (2006). https://doi.org/10.1007/11840817_66
22. Song, J., et al.: Natural ventilation in cities: the implications of fluid mechanics. Build. Res. Inform. 46(8), 809–828 (2018). Taylor & Francis
23. Watter, M., Springenberg, J., Boedecker, J., Riedmiller, M.: Embed to control: a locally linear latent dynamics model for control from raw images. In: Advances in Neural Information Processing Systems, pp. 2746–2754 (2015)
24. Wikle, C., Berliner, M.: A Bayesian tutorial for data assimilation. Phys. D Nonlinear Phenom. 230(1–2), 1–16 (2007). Elsevier
25. Zhu, J., Hu, S., Arcucci, R., Xu, C., Zhu, J., Guo, Y.: Model error correction in data assimilation by integrating neural networks. Big Data Min. Anal. 2(2), 83–91 (2019). TUP

Higher-Order Hierarchical Spectral Clustering for Multidimensional Data

Giuseppe Brandi[1](✉) and Tiziana Di Matteo[1,2,3]

[1] Department of Mathematics, King's College London,
The Strand, London WC2R 2LS, UK
giuseppe.brandi@kcl.ac.uk
[2] Complexity Science Hub Vienna, Josefstaedter Strasse 39, A, 1080 Vienna, Austria
[3] Centro Ricerche Enrico Fermi, Via Panisperna 89 A, 00184 Rome, Italy

Abstract. Understanding the community structure of countries in the international food network is of great importance for policymakers. Indeed, clusters might be the key for the understanding of the geopolitical and economic interconnectedness between countries. Their detection and analysis might lead to a bona fide evaluation of the impact of spillover effects between countries in situations of distress. In this paper, we introduce a clustering methodology that we name Higher-order Hierarchical Spectral Clustering (HHSC), which combines a higher-order tensor factorization and a hierarchical clustering algorithm. We apply this methodology to a multidimensional system of countries and products involved in the import-export trade network (FAO dataset). We find a structural proxy of countries interconnectedness that is not only valid for a specific product but for the whole trade system. We retrieve clusters that are linked to economic activity and geographical proximity.

Keywords: Tensor decomposition · Factor analysis · Clustering analysis · Spectral clustering · Multidimensional data · Multilayer networks · Food networks

1 Introduction

In this paper, we propose a clustering methodology that can be used by policymakers to analyse and extrapolate community structures in multidimensional datasets, such as the FAO data network. Food networks have been widely studied in the literature [10–12, 17, 20, 21, 26, 31]. The interest on the subject comes from different perspectives, such as the study of trade networks [1, 12, 17], the fragility of the food network [20, 31], health-related shocks which can flow from contaminated food through the trade network [10] and the connection between the food import-export and economic development of countries [21]. Other studies tried to detect common pattern of countries in relation to specific products through clustering techniques [11, 26]. However, all these papers analyse the food networks product by product without exploiting the multidimensionality of the data. In this paper, we instead introduce a methodology that is able to produce

© Springer Nature Switzerland AG 2021
M. Paszynski et al. (Eds.): ICCS 2021, LNCS 12746, pp. 387–400, 2021.
https://doi.org/10.1007/978-3-030-77977-1_31

a synthetic proxy of geopolitical and economic interconnectedness by applying a multidimensional data consistent approach.

Several dimensionality reduction techniques have been introduced in the literature to synthesize datasets. In particular, factor analysis is a dimensionality reduction technique which has been extensively applied in time series and cross-sectional data [3,23]. The main objective of factor analysis is to reduce the full dataset to a set of few relevant factors which explain most of the information contained in the original dataset [3,24]. These models have been also extensively employed in multivariate analysis to inspect latent features of the data [23,25] and has then been later extended to the analysis of multidimensional data, i.e. tensors [13,27]. Tensors, also known as multiway or multidimensional data [15,16] arise in several research fields, e.g. economics, 3D tomographic images, psychometrics, and factor analysis applied to these systems is commonly known as tensor decomposition [15]. In this paper we use the higher order Tucker decomposition [27], which is an extension of the bilinear factor analysis to multidimensional data. With respect to community detection, several clustering algorithms have been proposed in the literature. In this paper, we apply the Directed Bubble Hierarchical Tree (DBHT), which is a hierarchical filtering algorithm able to retrieve clusters, without the need of choosing the number of clusters a priori or a threshold parameter. This clustering is an unsupervised learning technique that can be applied to multivariate data and in particular, when applied to bilateral economic networks, can be employed to tailor economic and political interventions [11,26] or to create synthetic factors combining intra-cluster components [28]. Often when applied to multidimensional data, clustering techniques are commonly employed on single slices of data, e.g. layers in a multiplex resulting in a computationally intensive procedure and do not synthesize the dataset, resulting in different clustering outputs for each layer. In this paper we propose a methodology which overcomes this issue by implementing a tensor decomposition analysis in combination with a hierarchical clustering technique. The application of this methodology to the Food and Agriculture Organization of the United Nations (FAO) network is able to produce a proxy granted by geographic and economic interpretation. The paper is structured as follows. In Sect. 2 we describe the methodology, in Sect. 3 we report the results of the application of this methodology to the network of the FAO while Sect. 4 concludes.

2   Clustering Multidimensional Data

The higher-order hierarchical spectral clustering method is based on the combination of tensor decomposition [15,27] and the DBHT clustering tool [22,28] by means of a 2-steps approach. In the first step, we decompose the multidimensional dataset using the Tucker decomposition [15,27] from which we obtain a set of factor loadings matrices that projects the higher dimensional dataset in a low-dimensional space which compresses the information on common factors. Then, the DBHT algorithm [22,28] is performed on such matrices to obtain clusters of specific dimension, e.g. countries or product. In the next subsections, we provide a brief review of the methods applied in the 2-steps approach.

2.1   Tensor Decomposition

Tensors are a generalization of vectors and matrices and are ideal instruments to study multidimensional data. Tensors, like matrices, can be decomposed into smaller (in terms of rank) objects [15]. Among several tensor decomposition methods, we employ the Tucker decomposition, which is mainly used for factor analysis or dimensionality reduction and extends the bilinear factor analysis to the higher dimensional case [4,6,7,27]. Throughout this work, the notation follows the standard convention introduced in [15]: x is a scalar, $\mathbf{x}$ is a vector, $\mathbf{X}$ is a matrix and $\mathcal{X}$ is a tensor. Take an n-th order tensor $\mathcal{X} \in \mathbb{R}^{I_1 \times I_2 \cdots \times I_N}$, the Tucker decomposition of $\mathcal{X}$ can be written as a n-mode product, i.e.:

$$\mathcal{X} \approx \mathcal{F} \times_1 \Lambda^{(1)} \times_2 \Lambda^{(2)} \cdots \times_N \Lambda^{(N)} = \mathcal{F} \times \{\Lambda^{(n)}\}, \tag{1}$$

where $\Lambda^{(n)}$ are the factor loading matrices and $\mathcal{F}$ is the *core* tensor and it is usually of smaller dimension than $\mathcal{X}$.

2.2   DBHT Clustering Algorithm

Directed Bubble Hierarchical Tree (DBHT) is a machine learning hierarchical clustering method that exploits the topological property of the Planar Maximally Filtered Graph (PMFG) in order to find the clusters [22,28]. The PMFG is a generalization of the Minimum Spanning Tree (MST), that allows for loops and more edges by preserving all hierarchical properties of the MST. This is constructed following the same procedure of the MST, except that the non-loop condition is replaced with the weaker condition of planarity (i.e. each added link must not cut a pre-existent link). Thanks to this more relaxed topological constraint, the PMFG is able to retain a larger number of edges, hence of information. In particular, it can be shown that each PMFG contains exactly $3(N - 2)$ edges for a system of N nodes. The key elements of a PMFG are the three-cliques elements, subgraphs made of three nodes all reciprocally connected (i.e., triangles). The DBHT exploits this topological structure, and in particular the distinction between separating and non-separating three-cliques, to identify a clustering partition of all nodes in the PMFG. A complete hierarchical structure is then obtained for both inter-clusters and intra-clusters by following a traditional agglomerative clustering procedure. The algorithm requires as inputs a distance matrix D and a similarity matrix S.

2.3   Higher-Order Hierarchical Spectral Clustering (HHSC)

After presenting the 2-steps approach, we here introduce the HHSC used to extract the clusters from the multidimensional dataset. In the first step, HHSC extracts a set of factor matrices $\Lambda^{(n)}$ by means of Eq. 1 and then it computes a distance and a similarity matrix between the factor loadings corresponding to each element, i.e. country or product. In the second step, by inputting the two matrices in the DBHT algorithm, we identify the clusters.[1] This approach

---

[1] The algorithm's time complexity is described in Appendix A.

follows the same spirit of spectral clustering [30] by not directly performing the clustering procedure on the original dataset but rather, on the dimensionally reduced system. This avoids the over-dispersion of information in the original dataset, which can make the analysis very noisy. Some papers in the tensor literature use the values of the factor loadings matrices to directly cluster the data (nodes in the case of tensor networks) by identifying to which factors they are more related to [2]. Even if this is a sensible approach, it neglects the distribution of the factor loadings by focusing only on the maximum value for each node. Conversely, the use of the DBHT algorithm in our procedure, takes into account the full distribution of the factor loadings. Despite it is clear that the maximum loading of each node will weight more, they will not be the only drivers of the community detection in our procedure because all factor loadings are considered. Indeed, the DBHT has been proved to outperform standard factor model analyses [28].

3   FAO Trade Network

In this section, we apply the HHSC described in Sect. 2 to an economic network, i.e. the FAO trade network. In particular, we show how HHSC can be used to extrapolate relevant structures from a multidimensional dataset and synthesize the information in geographically and economically meaningful clusters.

3.1   Data

The dataset is collected from the Food and Agriculture Organization of the United Nations (FAO) website.[2] The FAO trade matrix is an economic network in which nodes correspond to importing and exporting countries, layers represent the products and the last dimension is related to the time. Edges at each layer represent the trade relationships of a specific product between countries in a specific time period. We have collected yearly data between 1986 and 2018 for 128 countries and 137 products. We represent this data by a 4-th order tensor $\mathcal{Y}_t$ of dimension $128 \times 128 \times 137 \times 33$.[3] In order to mitigate the difference in magnitude between the data and avoid the model to only fit high data values, we apply the log transformation which is commonly used in the literature for bilateral trades, i.e. $\bar{\mathcal{Y}}_t = log(1 + \mathcal{Y}_t)$. Finally, to ensure data stationary, we use the following first-order difference of the log-transformed trade tensor, i.e.:

$$\mathcal{X}_t = \bar{\mathcal{Y}}_t - \bar{\mathcal{Y}}_{t-1},$$

where $\mathcal{X}_t$ is a tensor of dimension $128 \times 128 \times 137 \times 32$ and t is the time index. This transformation represents the rate of change of the original dataset.

---

[2] http://www.fao.org/faostat/en/data/TM.

[3] We filtered out some data from the full dataset. We report the data filtering methodology in Appendix B.

3.2 Step-by-Step HHSC Methodology

In the multivariate time series and panel data literature, the dynamic factor model [3,24] starts from a function of the data of the following form:

$$\mathbf{X}_t = \mathbf{\Lambda}\mathbf{F}_t + \mathbf{E}_t, \tag{2}$$

where $\mathbf{X}_t$ is the data, $\mathbf{\Lambda}$ is the factor loading matrix, $\mathbf{F}_t$ is the factor matrix and $\mathbf{E}_t$ is the error term. By assuming Gaussianity of $\mathbf{F}_t$ and $\mathbf{E}_t$, $\mathrm{Cov}(\mathbf{X}_t) = \mathbf{\Lambda}\mathbf{\Lambda}' + \mathbf{\Sigma}^2$, where $\mathbf{\Sigma}^2$ is a diagonal matrix representing the idiosyncratic error of each component of $\mathbf{X}_t$ incorporated in $\mathbf{E}_t$ while $\mathbf{\Lambda}\mathbf{\Lambda}'$ is the factor covariance matrix with rank equal to the number of factors in $\mathbf{\Lambda}$. This is equivalent to the n-mode product formulation of the Tucker decomposition in Eq. 1, i.e.:

$$\mathbf{X}_t = \mathbf{F}_t \times_1 \mathbf{\Lambda} + \mathbf{E}_t. \tag{3}$$

This model can be easily extended to the multidimensional case by using the higher-order Tucker representation [14] as:

$$\mathcal{X}_t = \mathcal{F}_t \times_1 \mathbf{\Lambda}^{(1)} \times_2 \mathbf{\Lambda}^{(2)} \cdots \times_{N-1} \mathbf{\Lambda}^{(N-1)} + \mathcal{E}_t, \tag{4}$$

where now $\mathcal{X}_t$ is a tensor representing the multidimensional dataset at time t, $\mathcal{F}_t$ is the dynamic core factor tensor while each $\mathbf{\Lambda}^{(i)}$ is a factor loading matrix for the i-th mode. This formulation corresponds to the '$N-1$' Tucker decomposition where the time dimension is not factorized but rather used to estimate the factor components of the other modes. As for the multivariate case, each mode covariance matrix is assumed to be of the form $\mathrm{Cov}(\mathcal{X}_t^{(n)}) = \mathbf{\Lambda}^{(n)}\mathbf{\Lambda}'^{(n)} + \mathbf{\Sigma}^{(n)}\mathbf{\Sigma}'^{(n)}$, where $\mathbf{\Lambda}^{(n)}$ is estimated through the Higher Order Singular Value Decomposition (HOSVD) [9] for the Tucker model, and $\mathbf{\Sigma}^{(n)}$ is estimated through the flip-flop algorithm applied to the residuals of the model [6,14]. This algorithm is based on the assumption that the residuals follow the array Normal distribution [14] and iteratively estimate the covariance matrix of each mode considering the others as fixed. The Algorithm 1 is reported below:

Algorithm 1. Flip-flop algorithm for covariance estimation

1: **Initialize the algorithm to some** $\mathbf{\Sigma}^{(1)} \dots \mathbf{\Sigma}^{(N)}$

2: Compute $\mathcal{E}_t = \mathcal{X}_t - \widehat{\mathcal{F}}_t \times_1 \widehat{\mathbf{\Lambda}}^{(1)} \times_2 \widehat{\mathbf{\Lambda}}^{(2)} \cdots \times_N \widehat{\mathbf{\Lambda}}^{(N)}$

3: **for** $n = 1, \dots, N$

4: Compute $\mathcal{E}_t^{(n)} = \mathcal{E}_t \times_1 \mathbf{\Sigma}^{(1)} \cdots \times_{n-1} \mathbf{\Sigma}^{(n-1)} \times_n \mathbf{I}^{(n)} \times_{n+1} \mathbf{\Sigma}^{(n+1)} \cdots \times_N \mathbf{\Sigma}^{(N)}$

5: Compute $\widehat{\mathbf{\Sigma}}^{(n)}\widehat{\mathbf{\Sigma}}'^{(n)} = \mathbb{E}[\mathbf{E}_{(n)}\mathbf{E}_{(n)}^T]$

6: **Return** $\dfrac{\widehat{\mathbf{\Sigma}}^{(1)}\widehat{\mathbf{\Sigma}}'^{(1)}}{Tr(\widehat{\mathbf{\Sigma}}^{(1)}\widehat{\mathbf{\Sigma}}'^{(1)})} \cdots \dfrac{\widehat{\mathbf{\Sigma}}^{(N)}\widehat{\mathbf{\Sigma}}'^{(N)}}{Tr(\widehat{\mathbf{\Sigma}}^{(N)}\widehat{\mathbf{\Sigma}}'^{(N)})}$

It is important to notice that $\widehat{\boldsymbol{\Sigma}}^{(n)}\widehat{\boldsymbol{\Sigma}}'^{(n)}$ are not identifiable because by multiplying one of the covariance matrices for a scalar w and another covariance matrix for the inverse value w^{-1}, the optimization logarithm reaches the same value. For this reason, and to have covariance matrices which are comparable in magnitude, we estimate the covariance matrices normalized by their trace, that is the sum of the elements of the main diagonal. In this paper, we do not assume any structure of the error term and compute the covariance matrices only as post-modeling diagnostic to check if any residual information is present in the error term, i.e. strongly non-diagonal covariance matrices, and to check if the autocovariance matrix (the fourth mode covariance matrix) exhibits any sort of dynamics that can be exploited in a forecasting setting. Yet, the latter analysis is beyond the scope of the present paper since our main focus is on the factor loadings matrices. From each factor loading matrix, we compute a distance and a similarity matrix which are then used as inputs in the DBHT algorithm. For the distance matrix, we use the Euclidean distance[4], i.e.:

$$D_{a,b}^{(i)} = \left\| \boldsymbol{\Lambda}_a^{(i)} - \boldsymbol{\Lambda}_b^{(i)} \right\| \tag{5}$$

where $\boldsymbol{\Lambda}_j^{(i)}$ is the j-th element (country or product) of the i-th factor loadings matrix. The distance synthesizes the dissimilarity between two items' factor loadings. For the similarity matrix, we use the Gaussian kernel [22, 30], i.e.:

$$S_{a,b}^{(i)} = e^{\frac{-\left\| \Lambda_a^{(i)} - \Lambda_b^{(i)} \right\|^2}{2\sigma^2}}, \tag{6}$$

where σ^2 is the variance of the set of distances in D. This matrix weights more pairwise distances $D_{a,b}^{(i)}$ near to 0 and less values with higher distances. These two matrices suffice for the DBHT to cluster the data as the algorithm first extracts the PMFG and then uses its three-cliques elements to hierarchically cluster the nodes. In the next section, we present the results of our analysis of the FAO international trade network using the HHSC method.

3.3 Application of HHSC to FAO Data

The number of factors introduced in Sect. 3.2 used in the Higher-order Tucker decomposition can be chosen in two ways: either by using a theoretically or economically motivated number of factors or by using some data-driven methods [29] which heuristically choose the best model compared to its complexity. However, standard information criteria cannot be used in this context because of the strong imbalance between the huge multidimensional data and the number of parameters. To select the number of components, we fit various models with increasing number of factors, starting from the [1 1 1] specification (one-factor model) to the [50 50 50] specification and search for the elbow in a scree type of plot [8] in which we compare the log of the Explained Sum of Squares (ESS) and

---

[4] We row normalize the factor loadings before computing the distance and similarity measures in order to harmonize the different items, i.e. countries or products.

the number of parameters in the model. Figure 1 shows that the right combination of the explained variance and the number of parameters is [28 28 17], which corresponds to the elbow of the curve. Using this rank specification, we obtain three factor loading matrices, i.e. the import matrix $\mathbf{\Lambda}^{(1)} \in \mathbb{R}^{128 \times 28}$, the export matrix $\mathbf{\Lambda}^{(2)} \in \mathbb{R}^{128 \times 28}$, and the product matrix $\mathbf{\Lambda}^{(3)} \in \mathbb{R}^{137 \times 17}$. These matrices represent the information related to each mode of the tensor filtered by the effect of the other modes. They can be used both as outputs to directly analyze or as inputs for a clustering algorithm. In fact, for each row of the matrix (country or product) we identify a set of factor loadings and these provide the information on which factor (hub) the country or product is more related to. Countries or products with similar factor loadings in terms of distribution and magnitude are expected to have a strong similarity and to be allocated to the same cluster. However, by the use of a clustering algorithm, the analysis is made more robust as the full distribution of factor loadings for each country or product will be used instead of taking only the maximum value. By applying the DBHT algorithm on the set of latent factor matrices, we obtain 12 clusters for the import mode, 13 clusters for the export mode, and 9 clusters for the products mode. Results are graphically reported in Figs. 2, 3 and 4.[5] From Fig. 2, we can observe that clusters are mainly explained by geographical proximity and economic growth. Indeed, we detect the European clusters highlighted in blue, cyan, yellow, and orange on the center left of the plot. However, even if geographically close to each others, they have different trading patterns for some products which make them to fall in different clusters. There is a second block of European countries on the opposite side of the plot, highlighted in magenta, violet and light blue. It is important to observe that these represent Eastern European and ex URSS countries. The Asian countries are shown in purple on the top of the plot and a mixture of high growth countries of Asia and America are highlighted in green emerald on the top of the plot. Then, there is a cluster of mostly African countries highlighted light green on the bottom left. The two remaining clusters (red and green) are more convoluted. Indeed, they both share slow growth and fast growth countries in Africa and South America.

Regarding the export countries clusters, also in this case the main drivers can be attributed to geographical proximity and economic activity. We can easily identify the European cluster and a few ex URSS countries in magenta, purple, violet, blue, light blue, and light green on the top of the plot. Moreover, we can observe that there are dissimilarities in how different European groups cluster together. We can observe that Eastern European and ex URSS countries cluster together as well as ex Yugoslavia countries. There are then the Mediterranean countries clustering together and the North Eastern countries. In the European macro-cluster, it is possible to notice a cluster composed by the Francophone countries, i.e. France, Belgium, and Luxembourg. On the bottom of the plot, it is possible to observe the Asian counties in turquoise and emerald green, and the Arabic countries highlighted in cyan. At the center right of the plot, the South American countries are shown in orange. There is a small cluster highlighted in green composed by Chile, South Africa, and Zimbabwe. The

---

[5] A set of additional figures is reported in Appendix C.

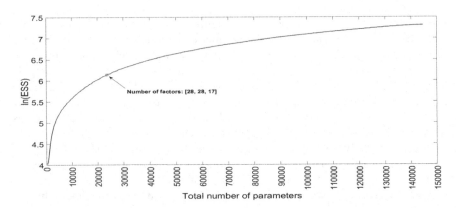

Fig. 1. Logarithm of the Explained Sum of Squares vs the number of model parameters. The circle corresponds to the specification at the elbow of the plot.

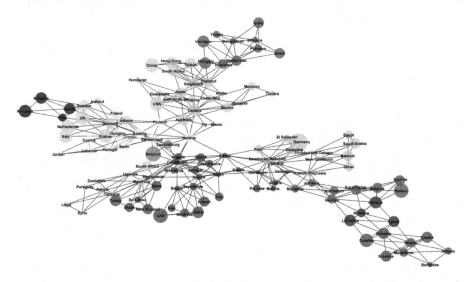

Fig. 2. HHSC clustering for imports. The size of the node is proportional to the total rate of growth of products the country imported between 1986 and 2018. (Color figure online)

first two have enhanced trading agreements while South Africa is the leading exporter and importer for Zimbabwe. The other two clusters (in red and yellow) are mixed and contain the leading importers and exporters, which do not necessarily follow geographical proximity. Indeed, it is important to mention that even though most of the clusters' countries can be explained by geographical proximity, import/export size, or growth rates, there are some of them, especially the world leading importers/exporters, which do not always follow this pattern. This is because their interrelations with other countries exhibit deeper connections, which go beyond their geographical positions.

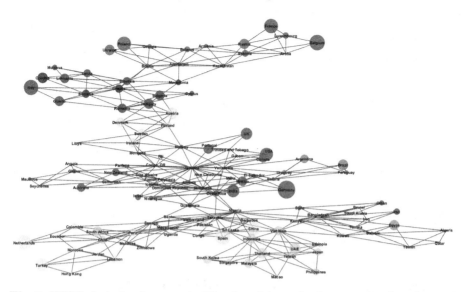

Fig. 3. HHSC clustering for exports. The size of the nodes is proportional to the total rate of growth of products the country exported between 1986 and 2018. (Color figure online)

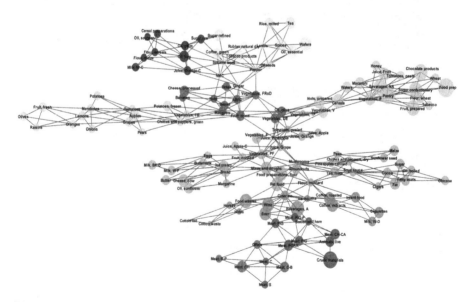

Fig. 4. HHSC clustering for products. The size of the nodes is proportional to the total rate of growth of each product all the countries imported and exported between 1986 and 2018. (Color figure online)

With respect to the products cluster, Fig. 4 shows that the DBHT has good performance also in this case. Indeed, the fruits and vegetable clusters are on the center left part of the plot highlighted in turquoise and red, while on the bottom of the plot there are two meat-related clusters highlighted in magenta and violet, with the second one more related to pig meat type of meat. We can then observe the beverages cluster in light blue and an hyper cluster, highlighted in orange, which connects the other clusters. Indeed, this hyper cluster has different products, but those products are similar to the nearest clusters. Finally, there are other two mixed clusters which correspond to high growth/high amount exported products in green and light green.

Finally, in Fig. 5 we report the modes' covariance matrices estimated by using the flip-flop Algorithm 1. As it is possible to see, the data is heteroskedastic, as the diagonal elements of the covariance matrices have different magnitudes. It is important to highlight that the static modes have diagonal covariance matrices, meaning that the model correctly factorizes the data while the time mode has a negative first-order autocovariance, in line with an autoregressive specification.

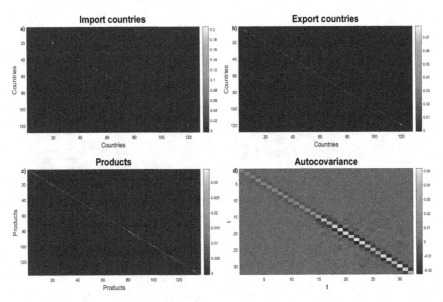

Fig. 5. Covariance of the error tensors modes estimated via the flip-flop algorithm: a) imports mode, b) exports mode, c) products mode, d) Autocovariance matrix. (Color figure online)

4   Conclusions

In this paper, we have proposed a new methodology, the Higher-order Hierarchical Spectral clustering (HHSC), to cluster multidimensional data by means of a 2-steps approach. In the first step, we decompose the multidimensional data via the Tucker decomposition, while in the second step, we use the DBHT algorithm on the factor loading matrices. We can appreciate that the clusters retrieved by this methodology can be easily explained by economic and geographical factors. Therefore, the tensor factor model in combination with hierarchical clustering is a promising tool to extract clusters and analyse the bilateral food network. Moreover, to better understand and predict the specific factors which drive the formation of clusters, an econometric model based on the Multinomial Logit can also be implemented to link the clusters to economic variables, in particular to understand cases where geographical proximity and rate of growth are not enough to explain the clustering results, e.g. Germany. Finally, the model can be extended to perform a forecasting analysis. This can be done by exploiting the dynamics of the core factor tensor by assuming a Tensor Autoregressive model [5,6]. A further extension would consist in adopting a fully dynamic setting in the spirit of data assimilation through a Kalman filter approach [18]. The HHSC algorithm proposed in this paper is general enough to be used to exploit information contained in a variety of empirical networks which evolve over time with multidimensional interactions, e.g. ecological networks, financial networks.

A   Time Complexity

The HHSC algorithm is composed by two components, i.e. Tucker decomposition and the DBHT clustering algorithm. Assuming a third order tensor $\mathcal{X} \in \mathbb{R}^{I_1 \times I_2 \times I_3}$ with $I_1 = I_2 = I_3 = I$, the time complexity of the DBHT algorithm is of the order $\mathcal{O}(I^3)$ [22] for each mode of the tensor. Regarding the Tucker decomposition with Tucker rank $\in \mathbb{R}^{R_1 \times R_2 \times R_3}$ with $R_1 = R_2 = R_3 = R$, the time complexity is in the order of $\mathcal{O}(I^3 R + I R^4 + R^6)$ [19]. Being the HHSC the combination of the two algorithms, also the time complexity follows. However, being R of much smaller dimension of I, the latter dominates the algorithm's running time.

B   Data Filtering

The complete dataset corresponds to 255 countries, 425 products, and 33 years. We first filter the countries that were inactive for more than 10 years and the products for which there were no transactions for more than 10 years. We then filter the dataset with respect to sparseness. In particular, we filter out countries and products for which the density is less than 1%. This resulted in a final dataset of 128 countries, 137 products, and 33 years.

C   Clustering with Growth Related Size of the Nodes

In this Appendix we report the same clustering plots reported in Figs. 2, 3 and 4 in which the dimension of the nodes is proportional to the total amount (in dollars) exchanged during the period analysed (Figs. 6, 7 and 8).

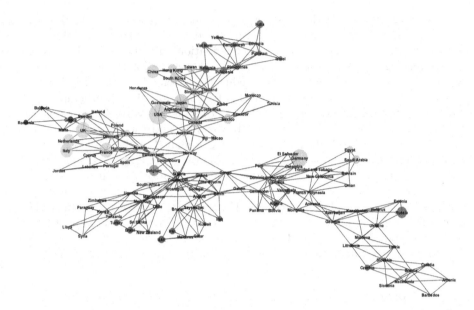

Fig. 6. HHSC clustering for the imports. The size of the nodes is proportional to the amount of products (in \$) the country import between 1986 and 2018. (Color figure online)

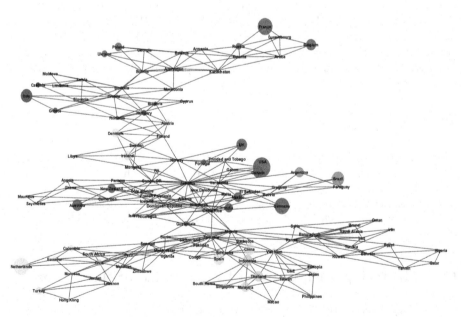

Fig. 7. HHSC clustering for exports. The size of the nodes is proportional to the amount of products (in \$) the country exported between 1986 and 2018. (Color figure online)

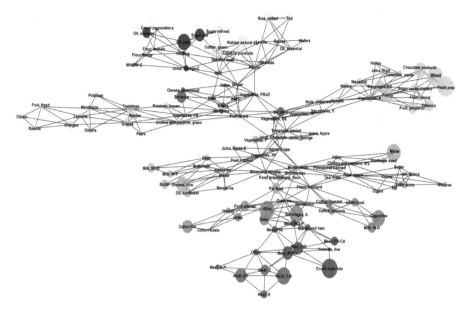

Fig. 8. HHSC clustering for products. The size of the nodes is proportional to the amount of the each product (in $) all the countries imported and exported between 1986 and 2018. (Color figure online)

References

1. Angelini, O., Di Matteo, T.: Complexity of products: the effect of data regularisation. Entropy **20**(11), 814 (2018)
2. Bader, B.W., Kolda, T.G., Harshman, R.A.: Temporal analysis of social networks using three-way DEDICOM. Technical report, Sandia National Laboratory (2006)
3. Bai, J., Ng, S.: Determining the number of factors in approximate factor models. Econometrica **70**(1), 191–221 (2002)
4. Brandi, G.: Decompose et Impera: tensor methods in high-dimensional data. Ph.D. thesis, LUISS Guido Carli (2018)
5. Brandi, G., Di Matteo, T.: A new multilayer network construction via tensor learning. In: KrzhizhanovskayaKrzhizhanovskayKrzhizhanovskayaKrzhizhanovskay Krzhizhanovskaya, V.V., et al. (eds.) ICCS 2020. LNCS, vol. 12142, pp. 148–154. Springer, Cham (2020). https://doi.org/10.1007/978-3-030-50433-5_12
6. Brandi, G., Di Matteo, T.: Predicting multidimensional data via tensor learning. J. Comput. Sci. **53**, 101372 (2021)
7. Brandi, G., Gramatica, R., Di Matteo, T.: Unveil stock correlation via a new tensor-based decomposition method. J. Comput. Sci. **46**, 101116 (2020)
8. Cattell, R.B.: The scree test for the number of factors. Multivar. Behav. Res. **1**(2), 245–276 (1966)
9. De Lathauwer, L., De Moor, B., Vandewalle, J.: A multilinear singular value decomposition. SIAM J. Matrix Anal. Appl. **21**(4), 1253–1278 (2000)
10. Ercsey-Ravasz, M., Toroczkai, Z., Lakner, Z., Baranyi, J.: Complexity of the international agro-food trade network and its impact on food safety. PLoS ONE **7**(5), e37810 (2012)

11. Escaith, H., Gaudin, H.: Clustering value-added trade: structural and policy dimensions. World Trade Organization Economic Research and Statistics Division Staff Working Paper No. ERSD-2014-08 (2014)
12. Garlaschelli, D., Di Matteo, T., Aste, T., Caldarelli, G., Loffredo, M.I.: Interplay between topology and dynamics in the world trade web. Eur. Phys. J. B **57**(2), 159–164 (2007). https://doi.org/10.1140/epjb/e2007-00131-6
13. Harshman, R.A.: Foundations of the PARAFAC procedure: models and conditions for an "explanatory" multimodal factor analysis. UCLA Working Papers in Phonetics (1970)
14. Hoff, P.D.: Separable covariance arrays via the Tucker product, with applications to multivariate relational data. Bayesian Anal. **6**(2), 179–196 (2011)
15. Kolda, T.G., Bader, B.W.: Tensor decompositions and applications. SIAM Rev. **51**(3), 455–500 (2009)
16. Kroonenberg, P.M., De Leeuw, J.: Principal component analysis of three-mode data by means of alternating least squares algorithms. Psychometrika **45**(1), 69–97 (1980). https://doi.org/10.1007/BF02293599
17. Lin, X., Dang, Q., Konar, M.: A network analysis of food flows within the United States of America. Environ. Sci. Technol. **48**(10), 5439–5447 (2014)
18. Nadler, P., Arcucci, R., Guo, Y.K.: Data assimilation for parameter estimation in economic modelling. In: 2019 15th International Conference on Signal-Image Technology & Internet-Based Systems (SITIS). IEEE, November 2019
19. Phan, A., Cichocki, A., Tichavský, P.: On fast algorithms for orthogonal tucker decomposition. In: 2014 IEEE International Conference on Acoustics, Speech and Signal Processing (ICASSP), pp. 6766–6770 (2014)
20. Puma, M.J., Bose, S., Chon, S.Y., Cook, B.I.: Assessing the evolving fragility of the global food system. Environ. Res. Lett. **10**(2), 024007 (2015)
21. Shutters, S.T., Muneepeerakul, R.: Agricultural trade networks and patterns of economic development. PLoS ONE **7**(7), e39756 (2012)
22. Song, W.M., Di Matteo, T., Aste, T.: Hierarchical information clustering by means of topologically embedded graphs. PLoS ONE **7**(3), e31929 (2012)
23. Spearman, C.: "general intelligence," objectively determined and measured. Am. J. Psychol. **15**(2), 201 (1904)
24. Stock, J.H., Watson, M.W.: Forecasting using principal components from a large number of predictors. J. Am. Stat. Assoc. **97**(460), 1167–1179 (2002)
25. Thurstone, L.L.: Multiple-Factor Analysis: A Development and Expansion of The Vectors of Mind. University of Chicago Press, Chicago (1947)
26. Torreggiani, S., Mangioni, G., Puma, M.J., Fagiolo, G.: Identifying the community structure of the food-trade international multi-network. Environ. Res. Lett. **13**(5), 054026 (2018)
27. Tucker, L.R.: Some mathematical notes on three-mode factor analysis. Psychometrika **31**(3), 279–311 (1966). https://doi.org/10.1007/BF02289464
28. Verma, A., Angelini, O., Di Matteo, T.: A new set of cluster driven composite development indicators. EPJ Data Sci. **9**(1), 8 (2020)
29. Verma, A., Vivo, P., Di Matteo, T.: A memory-based method to select the number of relevant components in principal component analysis. J. Stat. Mech: Theory Exp. **2019**(9), 093408 (2019)
30. Von Luxburg, U.: A tutorial on spectral clustering. Stat. Comput. **17**(4), 395–416 (2007). https://doi.org/10.1007/s11222-007-9033-z
31. Wu, F., Guclu, H.: Global maize trade and food security: implications from a social network model. Risk Anal. **33**(12), 2168–2178 (2013)

Towards Data-Driven Simulation Models for Building Energy Management

Juan Gómez-Romero(ID) and Miguel Molina-Solana(✉)(ID)

Department of Computer Science and AI, Universidad de Granada, Granada, Spain
{jgomez,miguelmolina}@decsai.ugr.es

Abstract. The computational simulation of physical phenomena is a highly complex and expensive process. Traditional simulation models, based on equations describing the behavior of the system, do not allow generating data in sufficient quantity and speed to predict its evolution and make decisions accordingly automatically. These features are particularly relevant in building energy simulations. In this work, we introduce the idea of deep data-driven simulation models (D3S), a novel approach in terms of the combination of models. A D3S is capable of emulating the behavior of a system in a similar way to simulators based on physical principles but requiring less effort in its construction—it is learned automatically from historical data—and less time to run—no need to solve complex equations.

Keywords: Data-driven simulation model · Deep learning · Building energy management

1 Introduction

According to a 2019 report by the consulting firm ABI Research [14], in the next five years, it is expected that more than 100,000 companies around the world will use simulation software, implying a business volume of over 2,500 million dollars annually in 2025.

However, computer simulation of physical phenomena, such as meteorology, energy transfer, or nuclear reactions, is costly. On the one hand, to create a simulation model of a system, it is necessary that the relationship between inputs and outputs is known and can be expressed in a calculable way. Thus, these models are generally created manually by coding the equations that describe the physical behavior of the system. On the other hand, running a complex simulation model can take several hours (or even days) and require large amounts of computational resources. Consequently, in most common problems, it is impossible to use these models to predict the evolution of the system in real-time and automatically make decisions from these simulations.

Taking the energy behavior of buildings (residential and non-residential) as an example, a physical simulation model characterizes the response of its components to the action of the equipment and the environmental conditions, employing differential equations that reproduce the energy transfer laws. These models

© Springer Nature Switzerland AG 2021
M. Paszynski et al. (Eds.): ICCS 2021, LNCS 12746, pp. 401–407, 2021.
https://doi.org/10.1007/978-3-030-77977-1_32

are built with specialized applications (e.g., Modelica or EnergyPlus) by assembling predefined modules that imitate the thermal response of different structures, materials, and equipment. Leveraging a simulation model, one can build an automatic control software that estimates the behavior of the building under different operating sequences (air conditioning, lights, etc.) and selects the one that involves the lowest energy consumption while maintaining comfort conditions (indoor temperature, humidity, CO_2 concentrations, etc.).

This approach is known as Model Predictive Control (MPC) [11] and can be applied in many areas beyond energy. Solutions based on MPC offer numerous advantages over traditional reactive controllers, especially in terms of optimizing the process in the medium/long term by taking into account the inertia of the systems. However, its implementation is limited for the problems mentioned above: creating the models requires much human effort, and their execution takes too long to generate operation plans within a reasonable time dynamically [9].

Although the literature has been demonstrated the possibility of significantly improving the control of a building and reducing energy consumption [7], numerous difficulties were also encountered in extending the approach to other contexts. The main bottleneck is to develop the physical simulation models that the control algorithm uses since these are created from scratch and cannot be reused from one building to another. This problem is even more acute when trying to apply Deep Reinforcement Learning techniques in the field of energy control [6]. In these cases, traditional simulation models cannot generate data in sufficient quantity and speed to train and validate the proposed algorithms.

As an alternative to physical simulation models, some approaches have been proposed in the literature to create prediction models of the behavior of systems using Machine Learning. However, these prediction models alone are not capable of addressing various needs that a simulation model must satisfy, such as the stability of the model against minor variations in inputs, the influence of the environment on the behavior of the system, the possibility of modifying its behavior through control instructions, or the use of sensor data affected by imprecision and uncertainty.

For these reasons, we aim to develop new algorithms, based on Deep Learning, to automatically learn fast, accurate, and realistic simulation models of a physical system from data. These models could be used in numerous applications and particularly in the MPC processes for energy optimization mentioned above.

The rest of this paper summarizes the background and state of the art (Sect. 2), and describes the key concepts and approach of our proposal (Sect. 3).

2   Background

Various proposals in the literature aim to perform computational simulation of physical systems using numerical models learned from available data. This approach is a generalization of the concept of system identification, a discipline close to classical control that studies the calculation of the parameters of a predefined model to adjust its outputs to those of the real system [13]. Traditionally, works on system identification have used algebraic and statistical methods [5], including time series analysis by autoregression with ARIMA or ARIMAX [19]. This is because the knowledge and availability of these tools have been traditionally broader, although in most cases, they are not the most effective, as we recently concluded in a review of works in the field of energy efficiency [16].

It was not until recently that machine learning techniques began to be applied in system identification [3]. For example, in [1] techniques based on Gaussian kernels are used to emulate the behavior of molecules at the electronic level without the need to solve differential equations. In contrast, in [10] similar techniques are applied to recognize galaxies from spectral data analysis, a process that typically requires running multiple simulations. In both cases, the proposed solutions manage to approximate the systems accurately because 1) there is a reduced number of output variables, and 2) aspects of the problem are encoded in the own model (e.g., which variables are relevant, what structure have the kernels that define the process). On the other hand, it is not easy to extend these models to other settings, even if they are only slightly different.

Deep Learning techniques have been investigated to learn [17] and calibrate [22] data series prediction models to address these limitations in recent years. They allow the automatic extraction of system characteristics and achieve more precision in the data series results than classical techniques. Among the many possible architectures, recurrent neural networks (RNNs), which allow cycles in the calculation graph, are the most effective for modeling the temporal behavior of a dynamic system [2], improving the results of the classic autoregressive techniques [18]. On the other hand, convolutional neural networks (CNNs), which are used mainly for image processing, allow ordered data sequences to be processed [8], although their adjustment is complex even for simple problems.

Simulation using Deep Learning techniques is, therefore, a new area of research in which there are hardly any works that exploit the capabilities of deep neural networks for processing multivariate data series from sensors. Furthermore, due to their own characteristics, neural networks present various additional problems regarding the connection of the prediction models with the actual phenomenon: robustness in the face of variations in the inputs, detection of errors in the data, characterization of cause-effect relationships, etc. Thus, for example, in the field of energy efficiency of buildings, there are some results about the modeling of thermal behavior [4,12], but on a very small scale, with a very short time horizon, and with little explanatory capacity.

In contrast, RNNs have been very successful in the field of natural language processing, as they are capable of learning predictive models of language. These networks are widely used in machine translation so that the network obtains the phrase in the target language that most closely matches the phrase in the source language. This type of architecture, called sequence-to-sequence [15], implements a procedure that first encodes the input phrase as a sequence of numbers (embeddings) and then decodes them to form the output phrase. Various improvements to this architecture, such as transformers [20], are close to the precision of a human translator in non-specialized domains.

3   Proposal

The fundamental concept we based our proposal on is that of a data-driven simulation model (DDS). A DDS model is capable of emulating the behavior of a system in a similar way to that of traditional simulators based on physical principles, but requiring less effort in its construction—it is automatically learned from historical data—and less time for its execution—no need to perform complex calculations. Recent advances in the area of Deep Learning suggest that it is possible to create DDS models based on deep neural networks[1], which we will call D3S (deep data-driven simulation models), improving the prediction capabilities of current time series algorithms. The D3S concept is very close to that of the digital twin [21], highlighting the essential properties of creation from data and the use of neural networks. Although neural networks are usually highly costly in terms of training, inference (simulation, in our use case) can be performed efficiently and quickly.

Formulation: A general simulation model can be seen as a computational process that transforms several inputs—corresponding to the previous states of the system, the applied control signals, and the expected conditions of the environment from the current instant—into an output that represents the sequence of n states through which the system passes from the current state to a final state (Fig. 1). Usually, the environmental conditions are not known a priori and can be estimated or even unknown. As for the control signals, they can be the result of an automatic optimization process. In the simplest case, the simulator would only consider a previous state and a following state. Likewise, it may happen that the system is not controllable or that the environmental conditions are not relevant. In more complex cases, in addition to the previous states, it would be necessary to incorporate the environmental conditions and the previous control actions. For simplicity, and without loss of generality, we will keep the general formulation of Fig. 1.

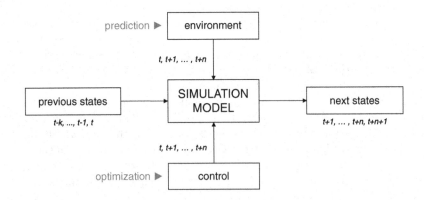

Fig. 1. Schematic view of a general simulation model.

---

[1] See, for instance, ICLR 2021's workshop "Deep Learning for Simulation (SIMDL)".

Approximation: We formalize the data-driven simulation problem as a multivariate time series prediction problem. Thus, the simulation model learns to predict a series of data representing the following states from the series of data representing the previous states, the environment, and the control, each of them including observations of several variables. As explained, classical data series analysis techniques are insufficient to solve this problem since they present difficulties in predicting more than one variable at the output. They also have limitations when it comes to capturing non-linear relationships between inputs and outputs. For these reasons, there is a need for new techniques in Deep Learning for handling data series (Fig. 2). A source of inspiration is automatic translation, which obtains the sequence of words in the target language that best represents the sequence of words in the original language, taking into account the context and the lexical-semantic relationships between them. Similarly, we propose the creation of algorithms capable of transforming the sequences of input values (environment, previous states, and control actions) into sequences of values (following states) with a relationship that is not necessarily linear considering the interrelation between states, control, and environment.

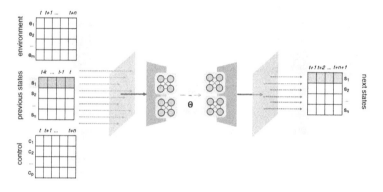

Fig. 2. Schematic view of our proposal for a Data-driven deep simulation model (D3S).

Challenges: This approach to data-driven simulation faces several challenges. In the following, we discuss several aspects that must be taken into account in building energy management:

1. Data availability: Data-driven simulations require a considerable amount and diversity of data to be performed. While data is generally available in modern buildings from SCADA systems, it is less commonly representative of exceptional or anomalous situations. Therefore, the scale of data needed by modern Deep Learning techniques, particularly transformer architectures, may exceed what can be obtained from a regular building.
2. Physics constraints: Physical systems are subject to bounding conditions, which should be incorporated into a D3S model. Neural networks can incorporate such conditions as regularizations, i.e., penalties in loss functions. However, it is not trivial to translate from one language to another, i.e., PDEs to

regularization terms. Hence, there is a growing interest in physics-informed machine learning, which investigates new neural architectures that integrate PDEs as additional optimization targets.

3. Dynamic behavior: Simulation models must evolve as their underlying process. This fact is commonplace in buildings, subject to renovations, aging, and changes in uses. Consequently, detecting model shift and activating recalibration must be automatically performed. Studies on continuous learning and data assimilation with Deep Learning suggest that this goal is achievable, but it may also impact the models' stability.

4. Explainability: Building management systems are cyber-physical systems that involve human operators' participation and sometimes the realization of critical tasks (e.g., CO_2 control). Therefore, the use of D3S for automated (or semi-automatic) decision-making requires at least some explanation of the outputs. Besides, experts' experience may be beneficial to bootstrap the training of D3S (e.g., by pre-identified relevant variables, periodicity, or spurious correlations). The growing body of knowledge on explainable AI should play a role here to make D3S more interpretable.

5. Computational cost: Learning a D3S model remains an expensive process that may require substantial computational resources, time, and energy. Accordingly, measuring the environmental impact of these models is essential to evaluate energy savings precisely. Model reuse from one building to another by transfer learning followed by fine-tuning could significantly reduce the data and energy needed to train the models.

References

1. Brockherde, F., Vogt, L., Li, L., Tuckerman, M.E., Burke, K., Müller, K.R.: Bypassing the kohn-sham equations with machine learning. Nat. Commun. **8**, 1–10 (2017). https://doi.org/10.1038/s41467-017-00839-3

2. Che, Z., Purushotham, S., Cho, K., Sontag, D., Liu, Y.: Recurrent neural networks for multivariate time series with missing values. Sci. Rep. **8**, 1–12 (2018). https://doi.org/10.1038/s41598-018-24271-9

3. Chiuso, A., Pillonetto, G.: System identification: a machine learning perspective. Ann. Rev. Control Robot. Auton. Syst. **2**, 281–304 (2019). https://doi.org/10.1146/annurev-control-053018-023744

4. Ferracuti, F., et al.: Data-driven models for short-term thermal behaviour prediction in real buildings. Appl. Energy **204**, 1375–1387 (2017). https://doi.org/10.1016/j.apenergy.2017.05.015

5. Gevers, M.: A personal view of the development of system identification: a 30-year journey through an exciting field. IEEE Control. Syst. **26**, 93–105 (2006). https://doi.org/10.1109/MCS.2006.252834

6. Gómez, J., Molina-Solana, M.: Towards self-adaptive building energy control in smart grids. In: NeurIPS 2019 Workshop Tackling Climate Change with Machine Learning. Vancouver, Canada, December 2019. https://www.climatechange.ai/papers/neurips2019/49

7. Gómez-Romero, J., et al.: A probabilistic algorithm for predictive control with full-complexity models in non-residential buildings. IEEE Access **7**, 38748–38765 (2019). https://doi.org/10.1109/ACCESS.2019.2906311

8. Kasim, M.F., et al.: Building high accuracy emulators for scientific simulations with deep neural architecture search (2020)
9. Killian, M., Kozek, M.: Ten questions concerning model predictive control for energy efficient buildings. Build. Environ. **105**, 403–412 (2016). https://doi.org/10.1016/j.buildenv.2016.05.034
10. Kwan, J., et al.: Cosmic emulation: fast predictions for the galaxy power spectrum. The Astrophysical Journal **810** (2015). https://doi.org/10.1088/0004-637X/810/1/35
11. Lee, J.H.: Model predictive control: review of the three decades of development. Int. J. Control Autom. Syst. **9**, (2011). https://doi.org/10.1007/s12555-011-0300-6
12. Liu, Y., Dinh, N., Sato, Y., Niceno, B.: Data-driven modeling for boiling heat transfer: using deep neural networks and high-fidelity simulation results. Appl. Thermal Eng. **144**, 305–320 (2018). https://doi.org/10.1016/j.applthermaleng.2018.08.041
13. Ljung, L.: System Identification: Theory for the User. Prentice Hall, Hoboken (1999)
14. Loten, A.: More manufacturers bet on simulation software. Wall Street J. (2020). https://www.wsj.com/articles/more-manufacturers-bet-on-simulation-software-11582240105
15. Luong, M.T., Le, Q.V., Sutskever, I., Vinyals, O., Kaiser, L.: Multi-task sequence to sequence learning. In: International Conference on Learning Representations (ICLR 2016) (2016). http://arxiv.org/abs/1511.06114
16. Molina-Solana, M., Ros, M., Ruiz, M.D., Gómez-Romero, J., Martin-Bautista, M.J.: Data science for building energy management: a review. Renew. Sustain. Energy Rev. **70**, 598–609 (2017). https://doi.org/10.1016/j.rser.2016.11.132
17. Raissi, M., Perdikaris, P., Karniadakis, G.E.: Physics-informed neural networks: a deep learning framework for solving forward and inverse problems involving nonlinear partial differential equations. J. Comput. Phys. **378**, 686–707 (2019). https://doi.org/10.1016/j.jcp.2018.10.045
18. Shi, H., Xu, M., Li, R.: Deep learning for household load forecasting–a novel pooling deep RNN. IEEE Trans. Smart Grid **9**, 5271–5280 (2018). https://doi.org/10.1109/TSG.2017.2686012
19. Shumway, R.H.: Time Series Analysis and its Applications. Springer, Cham (2017). https://doi.org/10.1007/978-3-319-52452-8
20. Vaswani, A., et al.: Attention is all you need. In: Proceedings of the 31st Conference on Neural Information Processing Systems. Long Beach, CA, USA, December 2017
21. Wagg, D.J., Worden, K., Barthorpe, R.J., Gardner, P.: Digital twins: state-of-the-art and future directions for modeling and simulation in engineering dynamics applications. ASME J. Risk Uncertainty Part B **6**(3) (2020). https://doi.org/10.1115/1.4046739
22. Zhu, J., Hu, S., Arcucci, R., Xu, C., Zhu, J., ke Guo, Y.: Model error correction in data assimilation by integrating neural networks. Big Data Mining and Analytics **2**, 83–91 (2019). https://doi.org/10.26599/BDMA.2018.9020033

Data Assimilation Using Heteroscedastic Bayesian Neural Network Ensembles for Reduced-Order Flame Models

Maximilian L. Croci[✉], Ushnish Sengupta, and Matthew P. Juniper

Department of Engineering, University of Cambridge, Cambridge, UK
mlc70@cam.ac.uk

Abstract. The parameters of a level-set flame model are inferred using an ensemble of heteroscedastic Bayesian neural networks (BayNNEs). The neural networks are trained on a library of 1.7 million observations of 8500 simulations of the flame edge, obtained using the model with known parameters. The ensemble produces samples from the posterior probability distribution of the parameters, conditioned on the observations, as well as estimates of the uncertainties in the parameters. The predicted parameters and uncertainties are compared to those inferred using an ensemble Kalman filter. The expected parameter values inferred with the BayNNE method, once trained, match those inferred with the Kalman filter but require less than one millionth of the time and computational cost of the Kalman filter. This method enables a physics-based model to be tuned from experimental images in real time.

Keywords: Bayesian inference · Deep learning · Thermoacoustics · Data assimilation

1 Introduction

1.1 Thermoacoustics

The prediction and control of thermoacoustic instability is a persistent challenge in jet and rocket engine design [1]. In gas turbines, the drive towards lower NO_x emissions has led to the use of lean premixed combustion, which is particularly susceptible to thermoacoustic instabilities [2]. Thermoacoustic instability is caused by the heat release rate and the pressure being in phase during combustion [3]. Heat release rate fluctuations are caused by flame surface area fluctuations, which in turn are caused by velocity perturbations and flame dynamics [4–7]. Any physics-based model must therefore contain the flame's response to velocity perturbations. This response can be calculated using detailed CFD simulations of the flame. However, these CFD simulations are expensive. In this paper we use data to tune the parameters of physics-based reduced-order models, in order to reduce the cost while retaining as much accuracy as possible.

M. L. Croci and U. Sengupta—Equal contribution.

© Springer Nature Switzerland AG 2021
M. Paszynski et al. (Eds.): ICCS 2021, LNCS 12746, pp. 408–419, 2021.
https://doi.org/10.1007/978-3-030-77977-1_33

The simplest physics-based model sets the heat release rate fluctuation to be a linear multiple of the velocity perturbation at the base of the flame some time earlier. This delay models the time taken for perturbations to travel down the flame. This is known as the $n - \tau$ model [8]. It is too simple for our purposes because it cannot simulate the flame dynamics. In this paper we model the flame as the zero contour of a continuous function that advects with the flow. This is known as the G-equation model [9]. This allows the flame dynamics to be simulated cheaply but the parameters of this model need to be assimilated from experimental data in order to render the model quantitatively accurate. The ensemble Kalman filter [10] (EnKF) has been used previously to assimilate data into the G-equation model [11,12]. The EnKF performs Bayesian inference to infer the variables (the state and parameters) of the G-equation model by statistically combining model forecasts with measurements of the variables. The EnKF in principle can be used online: Bayesian inference is performed whenever measurements become available. When used for data assimilation of our experiments of a conical Bunsen flame, however, the computational requirements of the EnKF render online Bayesian inference impossible: measurements are available every $O(10^{-3})$ seconds while forecasting between data assimilation steps takes $O(10^1)$ seconds. This study proposes an alternative method for practical online assimilation of data into the G-equation model of the Bunsen flame.

1.2 Bayesian Deep Learning

Bayesian deep learning refers to the use of deep learning algorithms, such as deep neural networks (NNs) and deep Gaussian processes (GPs), for Bayesian inference [14,15]. Bayesian NNs [16] replace the point estimates of each of the weights and biases with Gaussian probability distributions, with means and variances learned during training. The outputs can then be inferred from the inputs and the distribution of every weight and bias in the NN. Unfortunately, Bayesian NNs of practical size are too expensive to train [17]. More recently, ensembles of deep NNs have been used to perform approximate Bayesian inference [18–20], with the approximation improving with increasing width of the NN's hidden layers. These Bayesian NN ensembles (BayNNEs) learn the mean and variance of the posterior distribution of the outputs given the inputs. When multiple outputs are to be inferred, heteroscedastic BayNNEs learn the means and variances of each output, without assuming a common variance for all outputs. This study uses heteroscedastic BayNNEs to infer the parameters of the G-equation model given experimental observations.

2 Methods

2.1 Bunsen Flame Experiment

Figure 1 shows the Bunsen experiment setup: a Bunsen burner is placed inside a transparent duct and images of the flame are taken with a high-speed camera at

$f_s = 2500$ frames per second and a resolution of 1200×800 pixels. The flame is forced acoustically with a speaker from $250\,\text{Hz}$ to $450\,\text{Hz}$. The gas composition (methane, ethene and air) and flow rate are varied using mass flow controllers. By varying the forcing frequency and amplitude and gas composition and flow rate, flames with different aspect ratios, propagation speeds and degrees of cusping of the flame edge are observed. In some cases, the flame edge cusping leads to pinch-off at the flame tip. For each of the 270 different flame operating conditions, 500 images are taken.

The flame images are processed and the flame edge is extracted as a radial location x, which is a singularly-valued function of the axial co-ordinate y: $x = f(y)$. First, the pixel intensities are thresholded and the flame location x for every vertical co-ordinate y is found by weighted interpolation of the thresholded pixels, where the weights are the pixel intensities. Next, splines with 28 knots are used to smooth $x(y)$. Each flame image is therefore converted into a 90×1 vector of flame edge x locations $\mathbf{x}$. The y co-ordinates are the same for all flames, so are discarded. Observation vectors $\mathbf{z}$ are created by stacking 10 consecutive $\mathbf{x}$ vectors. These observation vectors are the inputs to the neural networks. All 500 images of each Bunsen flame are processed in this way.

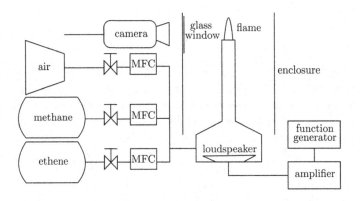

Fig. 1. Diagram of the Bunsen flame experiment setup: a Bunsen burner is placed inside a tube and a high-speed camera takes images of the flame through a glass window. The fuel composition (air, methane and ethene) and flow rate are controlled using mass flow controllers (MFCs). The flame is forced acoustically using a loudspeaker.

2.2   Flame Edge Model

In this paper the flame edge is defined to be the $G = 0$ contour (or level-set) of a scalar field $G(x, y, t)$. Regions of negative and positive G correspond to unburnt and burnt gases respectively (the magnitude of G has no physical significance). The evolution of G is governed by:

$$\frac{\partial G}{\partial t} + \mathbf{v} \cdot \nabla G = s_L |\nabla G|, \tag{1}$$

where $\mathbf{v}$ is a prescribed velocity field and s_L is the laminar flame speed: the speed at which the flame edge propagates normal to itself into the reactants. The flame speed s_L is a function of the unstretched (adiabatic) flame speed s_L^0, the flame curvature κ and the Markstein length $\mathcal{L}$, and is insensitive to pressure variations:

$$s_L = s_L^0 \left(1 - \mathcal{L}\kappa\right). \tag{2}$$

The unstretched flame speed s_L^0 depends only on the flame chemistry. The velocity field $\mathbf{v}$ comprises a parabolic base flow profile $V(x)$ and superimposed continuity-obeying velocity perturbations $u'(x, y, t)$ and $v'(x, y, t)$:

$$\mathbf{v} = (V(x) + v')\mathbf{j} + u'\mathbf{i}, \tag{3}$$

$$\frac{V(x)}{V} = 1 + \alpha \left(1 - 2\left(\frac{x}{R}\right)^2\right), \tag{4}$$

$$\frac{v'(x, y, t)}{V} = \epsilon \sin\left(\mathrm{St}\left(\frac{Ky}{R} - t\right)\right), \tag{5}$$

$$\frac{u'(x, y, t)}{V} = -\frac{\epsilon K \mathrm{St} x}{\beta R} \cos\left(\mathrm{St}\left(\frac{Ky}{R} - t\right)\right), \tag{6}$$

where α determines the shape of the base flow profile ($\alpha = 0$ is uniform flow, $\alpha = 1$ is Poiseuille flow), ϵ is the amplitude of the vertical velocity perturbation with phase speed V/K, $\mathrm{St} = 2\pi f R\beta/V$ is the Strouhal number with forcing frequency f and flame radius R, and β is the aspect ratio of the unperturbed flame. The parameters $K, \epsilon, \mathcal{L}, \alpha, \mathrm{St}$ and β are tuned to fit an observed flame shape. Figure 2 shows a diagram of the flame edge under the G-equation model. This model allows cusps to form at the flame edge and pockets of unburnt reactants to detach from the flame tip, as is observed in some experiments. It has proven to be a versatile flame edge model in several previous studies, despite having only a few parameters [21].

2.3 Forced Cycle Library

A library of flame edge locations is created with the flame edge model at known parameter values $K, \epsilon, \mathcal{L}, \alpha,$ St, β and f/f_s in the same format as the observation vectors, $\mathbf{z}$. The parameter values are sampled using quasi-Monte Carlo sampling to ensure good coverage of the parameter space. The parameters are sampled from the following ranges: $0.0 < K \le 1.5$, $0.0 < \epsilon \le 1.0$, $0.02 \le \mathcal{L} \le 0.08$, $0.0 \le \alpha \le 1.0$, $2.0 \le \beta \le 10.0$ and $0.08 \le f/f_s \le 0.20$. The values of St are calculated by additionally sampling $0.002 \le R \le 0.004$ m and $1 \le V \le 5$ m/s and calculating $\mathrm{St} = 2\pi f R\beta/V$. The parameters are sampled 8500 times, normalised to between 0 and 1 and recorded in target vectors $\{\mathbf{t} = [K, \epsilon, \mathcal{L}, \alpha, \mathrm{St}, \beta]\}$.

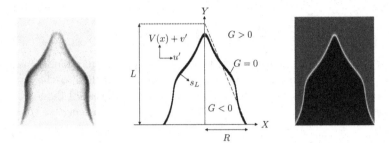

Fig. 2. *Left*: An image of a Bunsen flame. *Middle*: In the G-equation model, the flame edge is represented by the $G = 0$ contour (or level-set) of a continuous scalar field $G(x, y, t)$. Unburnt and burnt gases are regions where $G < 0$ and $G > 0$ respectively. The flame edge travels normal to itself into the unburnt gases with speed s_L. The flame edge advects under the prescribed velocity field, which comprises continuity-obeying velocity perturbations $u'(x, y, t)$ and $v'(x, y, t)$ superimposed onto a steady base flow profile $V(x)$. *Right*: A G field solution in LSGEN2D, a level-set solver. Blue and red regions are unburnt and burnt gases respectively, and the thin white band represents the flame edge.

For each of the 8500 parameter configurations, LSGEN2D [22] iterates the G field (1) until the solution is periodic and then stores 200 snapshots from one period of the forced cycle. For each snapshot the flame edge is found by interpolating G to find the contour $G = 0$. The same procedure as for the experimental images is then followed to find $x = f(y)$. Observation vectors $\mathbf{z}$ are created by stacking 10 consecutive $\mathbf{x}$ vectors. There are 200 observation vectors created from every cycle, resulting in a library of 1.7×10^6 observation-target parameter pairs $\{(\mathbf{z}, \mathbf{t})\}$. The neural networks are trained to recognise the parameter values from the observation vectors, $\mathbf{z}$.

2.4   Inference Using Heteroscedastic Bayesian Neural Network Ensembles

We assume that the posterior probability distribution of the parameters, given the observations, can be modelled by a neural network, $p_\theta(\mathbf{t}|\mathbf{z})$, with its own parameters $\boldsymbol{\theta}$. We assume that this posterior distribution has the form:

$$p_\theta(\mathbf{t}|\mathbf{z}) = \mathcal{N}\left(\boldsymbol{\mu}(\mathbf{z}), \boldsymbol{\Sigma}(\mathbf{z})\right), \tag{7}$$

where $\boldsymbol{\Sigma}(\mathbf{z})$, the posterior covariance matrix of the parameters given the data, is diagonal with $\boldsymbol{\sigma}^2(\mathbf{z})$ on its diagonal. This enforces our assumption that the parameters are mutually independent, given the observations $\mathbf{z}$. We use an ensemble of $M = 20$ neural networks. The architecture of each neural network is shown in Fig. 3. Each neural network comprises an input layer, four hidden layers with ReLU activations and two output layers: one for the mean vector $\boldsymbol{\mu}(\mathbf{z})$ and one for the variance vector $\boldsymbol{\sigma}^2(\mathbf{z})$. The output layer for the mean uses a sigmoid activation to restrict outputs to the range $(0, 1)$. The output layer for the variance uses an exponential activation to ensure positivity. Each neural network in the ensemble is initialised with unique weights $\boldsymbol{\theta}_{j,anc}$ sampled from a Gaussian prior distribution $\mathcal{N}(0, \frac{1}{N_H})$ and biases $\mathbf{b}_{j,anc}$ sampled from a uniform prior distribution in the range $[-\frac{1}{\sqrt{N_H}}, \frac{1}{\sqrt{N_H}}]$.

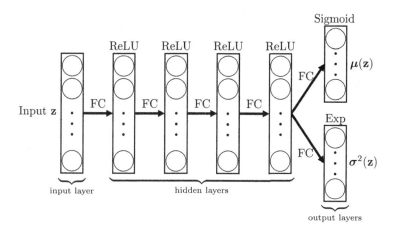

Fig. 3. Architecture of each neural network in the ensemble of 20. The input and hidden layers have 900 nodes each, while each output layer has 6 nodes each. All layers are fully connected (FC). Rectified Linear Unit (ReLU) activation functions are used for the hidden layers and sigmoid and exponential (Exp) activation functions are used for the mean and variance output layers respectively.

For a single observation $\mathbf{z}$, the j-th neural network in the ensemble produces a mean and variance estimate of the G-equation parameters:

$$\boldsymbol{\mu}_j(\mathbf{z}), \boldsymbol{\sigma}_j^2(\mathbf{z}). \tag{8}$$

This is achieved by minimising the loss function $\mathfrak{L}_j$:

$$
\begin{aligned}
\mathfrak{L}_j = (\boldsymbol{\mu}_j(\mathbf{z}) - \mathbf{t})^T \boldsymbol{\Sigma}_j(\mathbf{z})^{-1} (\boldsymbol{\mu}_j(\mathbf{z}) - \mathbf{t}) + \log(|\boldsymbol{\Sigma}_j(\mathbf{z})|) \\
+ (\boldsymbol{\theta}_j - \boldsymbol{\theta}_{anc,j})^T \boldsymbol{\Sigma}_{prior}^{-1} (\boldsymbol{\theta}_j - \boldsymbol{\theta}_{anc,j}).
\end{aligned}
\tag{9}
$$

The first two terms of the loss function are the negative logarithm of the normalised Gaussian likelihood function up to an additive constant. The third term is a regularising term that penalises deviation from prior anchor values $\boldsymbol{\theta}_{anc,j}$. The NNs produce samples from the posterior distribution. This is called randomised maximum a-posteriori (MAP) sampling [18].

For a single observation vector $\mathbf{z}$, the prediction from the ensemble of neural networks is therefore a distribution of M Gaussians, each centred at their respective means $\boldsymbol{\mu}_j(\mathbf{z})$. Following similar treatment in [23], this distribution

is then approximated by a single multivariate Gaussian posterior distribution
$p(\mathbf{t}|\mathbf{z}) \approx \mathcal{N}(\boldsymbol{\mu}(\mathbf{z}), \boldsymbol{\Sigma}(\mathbf{z}))$ with mean and covariance

$$\boldsymbol{\mu}(\mathbf{z}) = \frac{\Sigma_j \boldsymbol{\mu}_j(\mathbf{z})}{M}, \quad \boldsymbol{\Sigma}(\mathbf{z}) = \text{diag}\left(\boldsymbol{\sigma}^2(\mathbf{z})\right), \tag{10}$$

$$\boldsymbol{\sigma}^2(\mathbf{z}) = \frac{\Sigma_j \boldsymbol{\sigma}_j^2(\mathbf{z})}{M} + \frac{\Sigma_j \boldsymbol{\mu}_j^2(\mathbf{z})}{M} - \left(\frac{\Sigma_j \boldsymbol{\mu}_j(\mathbf{z})}{M}\right)^2. \tag{11}$$

This is repeated for every observation vector $\mathbf{z}$. The posterior distribution $p(\mathbf{t}|\mathbf{z}_i)$
with the smallest total variance $\sigma_{i,\text{tot}}^2 = ||\boldsymbol{\sigma}^2(\mathbf{z}_i)||_1$ is chosen as the best guess
to the true posterior. The M parameter samples from the chosen posterior are
used for re-simulation, which allows us to check the predicted flame shapes and
to calculate the normalised area variation over one cycle.

2.5 Inference Using the Ensemble Kalman Filter

The Kalman filter iteratively performs Bayesian inference to find the probability
distribution of the state of a system given noisy observations of the system and
an imperfect model of the system dynamics. In this study, the state comprises
the location of the flame edge and the parameters K and ϵ. These parameters are
assumed to be independent given the observations of the flame edge and constant
for each of the Bunsen flame experiments. The flame edge is modelled using the
G-equation (1). The ensemble Kalman filter [10] (EnKF) evolves an ensemble of
simulations forward in time. The covariance matrix of these ensembles is assumed
to approximate the covariance matrix of the state evolved over the same period
of time. The EnKF is more practical when the state contains many variables
and the evolution is nonlinear. In this study, the state contains $O(10^2)$ variables
and the governing equation (1) is nonlinear.

The parameters $\mathcal{L}, \alpha, \beta$ and St are calculated by solving the G-equation (1)
when steady and do not need to be inferred with the EnKF. This reduces the
cost of the EnKF but increases the number of steps compared with the BayNNE
method. The forcing frequency f is manually set when running the Bunsen
flame experiments. The unperturbed laminar flame speed s_L^0 is calculated using
Cantera[1] and knowledge of the methane and ethene flow rates. V is calculated
from s_L^0 and β: $V = s_L^0 \sqrt{\beta^2 - 1}$. The Strouhal number can then be calculated:
St $= 2\pi f \beta L/V$. Ref. [12] contains details about the implementation of the EnKF.

An ensemble size of 32 is used in this study. A multiplicative inflation factor
of 1% is chosen to mitigate the underestimation of the error covariances due
to the finite ensemble size [24]. Once the EnKF has converged, the parameters
K and ϵ calculated by each ensemble member are recorded. These are samples
from the posterior distribution of the parameters given all the flame x-location
vectors: $p(K, \epsilon|\mathbf{x}_1, \mathbf{x}_2, \ldots, \mathbf{x}_N)$. This differs from the BayNNE, which was given
$\mathbf{x}$-location vectors in groups of 10.

---

[1] Cantera is a suite of tools for problems involving chemical kinetics, thermodynamics,
and transport processes [13].

3   Results and Discussion

The ensemble of 20 Bayesian neural networks is trained on the forced cycle library for 5000 epochs, with a 80 : 20 train-test split and an Adam optimiser with learning rate 10^{-3}. Training takes approximately 12 h per neural network on an NVIDIA P100 GPU. The ensemble is then evaluated on the observations of every Bunsen flame and the estimate of the parameters with the lowest total uncertainty is selected for re-simulation. The evaluation takes $O(10^{-3})$ seconds on an Intel Core i7 processor on a laptop.

For each Bunsen flame test case, the ensemble Kalman filter technique requires an hour to calculate estimates of $\mathcal{L}, \alpha, St$ and β on an Intel Core i7 processor on a laptop, followed by 2 h of data assimilation on a pair of Intel Xeon Skylake 6142 processors[2] to produce estimates of K and ϵ.

Figures 4 and 5 show the results of inference on two different Bunsen flames. Both techniques produce good parameter estimates, in that the predicted flame shapes are in good agreement with the experiments. Furthermore, the BayNNE predicts normalised area variation curves at least as accurate as those predicted by the EnKF. However, the EnKF's predictions of K and ϵ are more confident than the BayNNE's predictions of all 6 parameters. The difference between the predicted uncertainties of the EnKF and BayNNE can be explained by the difference between the posterior distributions calculated by both techniques. The EnKF calculates $p(K, \epsilon | \mathbf{x}_1, \mathbf{x}_2, \dots, \mathbf{x}_{500})$: the probability distribution of the parameters K and ϵ given the location vectors $\mathbf{x}$ from all 500 images. For the BayNNE technique, the posterior $p(\mathbf{t}|\mathbf{z}_i) = p(\mathbf{t}|\mathbf{x}_i, \mathbf{x}_{i+1}, \dots, \mathbf{x}_{i+9})$ with the smallest total variance is chosen. This is the probability distribution of the parameters $\mathbf{t}$ given only 10 image location vectors, as opposed to the 500 used by the EnKF. Therefore, it is expected that the BayNNE technique produces more uncertain parameter estimates.

This raises the question as to whether it is possible to increase the certainty of the parameters by providing the BayNNE with more data. It is not possible to infer a posterior $p(\mathbf{t}|\mathbf{z}_1, \mathbf{z}_2, \dots, \mathbf{z}_N)$ from the individual posteriors $p(\mathbf{t}|\mathbf{z}_i)$ without knowledge of the dependence between any two observations, $p(\mathbf{z}_i|\mathbf{z}_j)$. Two observation vectors are not independent, because the information gained from a first observation restricts the expected subsequent observations to a likely set of forced cycle states. One solution is to increase the number of location vectors $\mathbf{x}$ in each observation vector, which increases the computational cost of training the neural networks. Another solution is a recurrent neural network, which can have variable length sequences of data as inputs. Future work will focus on the development of a Bayesian recurrent neural network solution to this problem.

---

[2] The EnKF is fully parallelised: the processors have 32 cores in total, one for each member in the ensemble.

To summarise, once the BayNNE has been trained on the forced cycle library it can be used to reliably infer all 6 parameters of the flame edge model based on 10 consecutive snapshots of the Bunsen flame experiments. If more snapshots are provided, this method finds the sequence of 10 snapshots that minimises the uncertainty in the parameters. This differs from the EnKF which requires hours to infer the parameters of each Bunsen flame. The BayNNE technique therefore provides similarly accurate parameter estimates at a fraction of the computational cost.

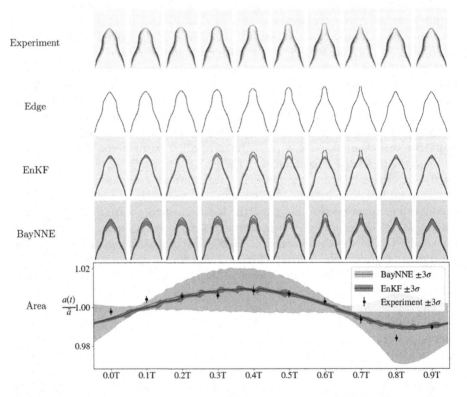

Fig. 4. Results of inference on a flame with flow rates of ethene 0.20 NL/min, methane 0.40 NL/min, air 4.50 NL/min and forcing 450 Hz. Top row: Bunsen flame whose parameters are to be estimated. Second row: the detected flame edge. Third row: re-simulated flames using EnKF estimated parameters. Fourth row: re-simulated flames using BayNNE estimated parameters. Bottom: normalised surface area variations over one period.

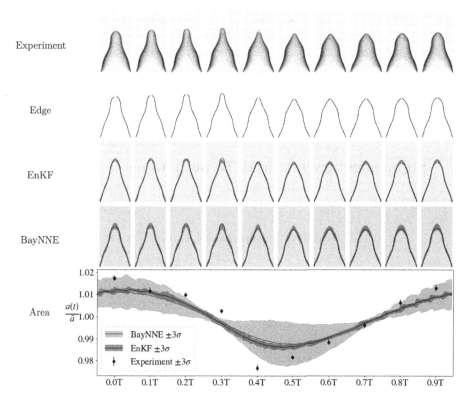

Fig. 5. Results of inference on a flame with flow rates of ethene 0.40 NL/min, methane 0.20 NL/min, air 6.00 NL/min and forcing 425 Hz. Top row: Bunsen flame whose parameters are to be estimated. Second row: the detected flame edge. Third row: re-simulated flames using EnKF estimated parameters. Fourth row: re-simulated flames using BayNNE estimated parameters. Bottom: normalised surface area variations over one period.

4 Conclusions

This study proposes a method to assimilate data from a Bunsen flame experiment into a kinematic model of a flame edge. The model parameters and their uncertainties are inferred using heteroscedastic Bayesian neural network ensembles. The neural networks are trained on a library of synthetic flame edge observations created using a level-set solver, LSGEN2D. Once trained, the Bayesian neural network ensemble accurately predicts the parameters and uncertainties from 10 consecutive images of the Bunsen flame. If more images are provided, this method selects the 10 consecutive images that give the smallest uncertainty in the parameters. This method is more than 6 orders of magnitude faster than the ensemble Kalman filter, which produces the same expected values of the parameters but lower variances. The Bayesian neural network ensemble also infers parameters of the model which cannot be inferred with the ensemble Kalman filter. Future work will focus on improving the parameter and uncertainty estimates by using more flame image data without necessarily increasing the size of the neural networks.

Disclosure of Funding. This project has received funding from the UK Engineering and Physical Sciences Research Council (EPSRC) award EP/N509620/1 and from the European Union's Horizon 2020 research and innovation program under the Marie Skłodowska-Curie grant agreement number 766264.

A   Supplementary material: Hyperparameter settings

See Table 1.

Table 1. Hyperparameter settings.

Hyperparameter	Value
Training	
Train-test split	80:20
Batch size	2048
Epochs	5000
Optimiser	Adam
Learning rate	10^{-3}
Architecture	
Input units	900
Hidden layers	4
Units per hidden layer	900
Output layers	2
Units per output layer	6
Ensemble size	20

References

1. Juniper, M.P., Sujith, R.: Sensitivity and nonlinearity of thermoacoustic oscillations. Ann. Rev. Fluid Mech. **50**, 661–689 (2018)
2. Keller, J.J.: Thermoacoustic oscillations in combustion chambers of gas turbines. AIAA 33-12 (1995)
3. Strutt, J.W.: The Theory of Sound, vol. II. Macmillan and Co., London (1878)
4. Smart, A.E., Jones, B., Jewel, N.T.: Measurements of unsteady parameters in a rig designed to study reheat combustion instabilities. AIAA 26-88 (1976)
5. Pitz, R.W., Daily, J.W.: Experimental study of combustion in a turbulent free shear layer formed at a rearward facing step. AIAA 81-106 (1981)

6. Smith, D.A., Zukoski, E.E.: Combustion instability sustained by unsteady vortex combustion. AIAA 85-1248 (1985)
7. Poinsot, T., Trouve, A., Veynante, D., Candel, S., Esposito, E.: Vortex-driven acoustically coupled combustion instabilities. J. Fluid Mech. **177–220**, 265–292 (1987)
8. Crocco, L.: Research on combustion instability in liquid propellant rockets. In: Symposium (International) Combustion, vol. 12, no. 1, pp. 85–99 (1969)
9. Williams, F.A.: Turbulent combustion. In: The Mathematics of Combustion, pp. 97–131 (1985)
10. Evensen, G.: Sequential data assimilation with a nonlinear quasi-geostrophic model using Monte Carlo methods to forecast error statistics. J. Geophys. Res. **99**, 10143–10162 (1994)
11. Yu, H., Juniper, M.P., Magri, L.: Combined state and parameter estimation in level-set methods. J. Comput. Phys. **399**, 108950 (2019)
12. Yu, H., Juniper, M.P., Magri, L.: A data-driven kinematic model of a ducted premixed flame. Proc. Combust. Inst. **399**, 6231–6239 (2020)
13. Goodwin, D.G., Speth, R.L., Moffat, H.K., Weber, B.W.: Cantera: an object-oriented software toolkit for chemical kinetics, thermodynamics, and transport processes (2018). https://www.cantera.org. Version 2.4.0
14. Gal, Y.: Uncertainty in deep learning. Ph.D. thesis (2016)
15. Damianou, A.C., Lawrence, N.D.: Deep Gaussian processes. In: Proceedings of the 16th International Conference on Artificial Intelligence and Statistics (AISTATS) (2013)
16. MacKay, D.J.C.: Information Theory, Inference and Learning Algorithms. Cambridge University Press, Cambridge (2003)
17. Gal, Y., Ghahramani, Z.: Dropout as a Bayesian approximation: representing model uncertainty in deep learning. In: Proceedings of the 33rd International Conference on Machine Learning. NY, USA, New York (2016)
18. Pearce, T., Zaki, M., Brintrup, A., Anastassacos, N., Neely, A.: Uncertainty in neural networks: Bayesian ensembling. In: International Conference on Artificial Intelligence and Statistics (AISTATS) (2020)
19. Sengupta, U., Croci, M.L., Juniper, M.P.: Real-time parameter inference in reduced-order flame models with heteroscedastic Bayesian neural network ensembles. In: Machine Learning and the Physical Sciences Workshop at the 34th Conference on Neural Information Processing Systems (NeurIPS) (2020)
20. Sengupta, U., Amos, M., Hosking, J.S., Rasmussen, C.E., Juniper, M.P, Young, P.J.: Ensembling geophysical models with Bayesian neural networks. In: Advances in Neural Information Processing Systems (NeurIPS) (2020)
21. Kashinath, K., Li, L.K.B., Juniper, M.P.: Forced synchronization of periodic and aperiodic thermoacoustic oscillations: lock-in, bifurcations and open-loop control. J. Fluid Mech. **838**, 690–714 (2018)
22. Hemchandra, S.: Dynamics of turbulent premixed flames in acoustic fields. Ph.D. thesis (2009)
23. Lakshminarayanan, B., Pritzel, A., Blundell, C.: Simple and scalable predictive uncertainty estimation using deep ensembles. In: Advances in Neural Information Processing Systems (NeurIPS), pp. 6402–6413 (2017)
24. Luo, X., Hoteit, I.: Robust ensemble filtering and its relation to covariance inflation in the ensemble Kalman filter. Monthly Weather Rev. **139**(12), 3938–3953 (2011)

A GPU Algorithm for Outliers Detection in TESS Light Curves

Stefano Fiscale[1]([✉])(iD), Pasquale De Luca[1,2](iD), Laura Inno[1,3](iD),
Livia Marcellino[1](iD), Ardelio Galletti[1](iD), Alessandra Rotundi[1](iD),
Angelo Ciaramella[1](iD), Giovanni Covone[4](iD),
and Elisa Quintana[5](iD)

[1] Science and Technology Department, Parthenope University of Naples,
Naples, Italy
stefano.fiscale001@studenti.uniparthenope.it,
{laura.inno,livia.marcellino,ardelio.galletti,rotundi,
angelo.ciaramella}@uniparthenope.it
[2] Department of Computer Science, University of Salerno, Fisciano, Italy
deluca@ieee.org
[3] INAF-Osservatorio Astronomico di Capodimonte, Salita Moraliello, Napoli, Italy
[4] Department of Physics "Ettore Pancini", University of Naples Federico II,
Naples, Italy
giovanni.covone@unina.it
[5] NASA Goddard Space Flight Center, Greenbelt, MD, USA
elisa.quintana@nasa.gov

Abstract. In recent years, Machine Learning (ML) algorithms have proved to be very helpful in several research fields, such as engineering, health-science, physics etc. Among these fields, Astrophysics also started to develop a stronger need of ML techniques for the management of big-data collected by ongoing and future all-sky surveys (e.g. Gaia, LAMOST, LSST etc.). NASA's Transiting Exoplanet Survey Satellite (TESS) is a space-based all-sky time-domain survey searching for planets outside of the solar system, by means of transit method. During its first two years of operations, TESS collected hundreds of terabytes of photometric observations at a two minutes cadence. ML approaches allow to perform a fast planet candidates recognition into TESS light curves, but they require assimilated data. Therefore, different pre-processing operations need to be performed on the light curves. In particular, cleaning the data from inconsistent values is a critical initial step, but because of the large amount of TESS light curves, this process requires a long execution time. In this context, High-Performance computing techniques allow to significantly accelerate the procedure, thus dramatically improving the efficiency of the outliers rejection. Here, we demonstrate that the GPU-parallel algorithm that we developed improves the efficiency, accuracy and reliability of the outliers rejection in TESS light curves.

Keywords: ML · Light curves · GP-GPU · Parallel algorithm · HPC

© Springer Nature Switzerland AG 2021
M. Paszynski et al. (Eds.): ICCS 2021, LNCS 12746, pp. 420–432, 2021.
https://doi.org/10.1007/978-3-030-77977-1_34

1   Introduction

Machine Learning (ML) can be defined as computational methods exploiting experience to improve performance or to make accurate predictions. Experience refers to the past ground-truth available to the learner. Typically, ML approaches are based on two main steps: the *pre-processing* phase - retrieve and assimilate specific information by starting from an initial raw dataset; *training* phase - where an ad-hoc model is trained by using the above standardized data. In order to compute an adequate forecast, several data assimilation techniques are performed. In order to avoid any inconsistency in the results, anomalous values in the raw data must be detected and removed.

This is achieved by using specific *data-cleaning* techniques, such as removal of inconsistent observations, filtering-out outliers, missing data handling. In exoplanets surveys, pre-processing procedures are very helpful due to random nature of retrieved data [1]. In particular, the NASA's Transiting Exoplanet Survey Satellite, TESS, collected a large dataset of high-precision photometric observations for more than 200,000 stars [2,3].

Given the low signal to noise ratio (SNR) that characterizes transit signals, dataset are often corrupted by different instrumental *systematic* error, which needs to be removed, in order to perform a correct analysis. In this work, we focus on the rejection of outliers in TESS light curves by adopting the $Z-$Score method, one of the most effectively and widely adopted to this purpose.

Most pre-processing pipelines including outliers rejection are currently implemented by following a sequential approach that, due to the huge data dimension, requires several waiting hours prior to the training phase.

It is well-known that High-Performance Computing (HPC) offers a powerful tool to overcome this issue, thanks to its advanced parallel architectures. In particular, the high computational power of Graphics Processing Units (GPUs) allows to analyse a huge data volume by following the Single Instruction Multiple Thread (SIMT) paradigm. Moreover, thanks to novel GPUs architectures, the numerical stability of each computation is more guaranteed with respect to any CPU computations [4].

In this work, we propose a GPU-parallel algorithm based on the $Z-$Score method for outliers detection in TESS light curves. The parallel implementation exploits the Compute Unified Device Architecture (CUDA) framework [5] for achieving an appreciable gain of performance. Therefore, a consistent workload distribution has been performed by an ad-hoc Domain Decomposition (DD) approach. Moreover, in order to accelerate the reading-writing memory operations, a suitable memory strategy has been designed.

The paper is organized as follows. In Sect. 2 related works are reviewed. Section 3 recalls some preliminaries about TESS light curves and the $Z-$Score method. In Sect. 4, the underlying domain decomposition strategy and the GPU-CUDA parallel algorithm are provided. The experiments discussed in Sect. 5 confirm the efficiency of the proposed implementation in terms of performance. Finally, our conclusions are presented in Sect. 6.

2   Detection of Transits in TESS Light Curves

This section briefly summarises the purpose of TESS space mission and the different analysis of its light curves presented in the literature. After the first planets were discovered outside the solar system in 1992 [19], and the subsequent successful observations of additional ones, astronomers started an intense and multi-approached search for exoplanets. With the advance of newest technologies, the search for exoplanets has become one of the most dynamic research field in astronomy.

TESS is the most recent transit survey to join the exoplanets-hunting field. It is a space-borne NASA mission launched in 2018, whose principal objective is to discover transiting Earths and Super-Earths orbiting nearby, bright dwarf stars [6]. In order to achieve its goal, TESS employs a "stare and step" observation strategy: its four on-board cameras observe a 24-by-96-degree sector of the sky for 27 days, then the spacecraft is reoriented to observe the next sector, which has a marginal overlap with the previous one. Over the past two years, TESS used this strategy to tile the entire sky with a total of 26 observation sectors.

The time-series photometric observations collected by TESS have been analysed in several studies with different computational approaches. Here, we focus on those based on ML techniques.

In particular, in [7,8] ML approaches based on Deep Neural Networks (DNNs) are applied to predict and classify stellar properties from TESS noisy and sparse time series data. Among the family of DNNs, the authors exploit a powerful model (a Convolutional Neural Network) for classifying TESS light curves. Nevertheless, before training on the data, several pre-processing steps are required. The pre-processing pipelines embedded into the ML algorithms mentioned above rely on a sequential approach to the problem, thus requiring long execution times, somehow affecting the performance of the whole process.

Many efforts have been done in the last years for improving performances. In particular, in [9] the framework BATMAN is presented, together with a state of the art study. BATMAN is a Python package for modeling exoplanet transit and eclipse light curves. It supports calculation of light curves for any radially symmetric stellar limb darkening law, using a new integration algorithm for models that cannot be quickly calculated analytically. The code uses C extension modules to speed up model calculation and it is developed for multi-core environment, by means of the OpenMP library.

3   Mathematical Background

TESS performs a photometric collections of a large number of stars with a regular time sampling, i.e. a two minutes cadence over a baseline of 27 days. These images are read-out and made available to the community as target pixel files (TPFs) and light curves [10], i.e. tables of the flux emitted by the source over time (Fig. 1).

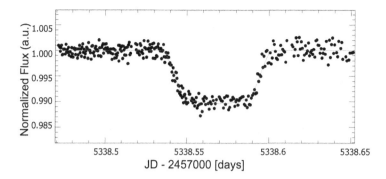

Fig. 1. The plot shows the the normalized flux of a star measured by TESS over the course of approximately one hour. The decrease in the flux at the center is due to an exoplanet tranist.

When an exoplanet transits in front of its host star (as seen from Earth), a portion of the star light is blocked out and a decrease in the observed flux is measured [11]. This phenomenon is described by a complex model, including quadratic or nonlinear limb darkening effects [20,21]. However, we can simplify it here for the sake of clarity by considering the ratio of the observed variation in flux, ΔF, to the stellar flux, F, proportional to:

$$\frac{\Delta F}{F} \propto \frac{R_p^2}{R_*^2} \tag{1}$$

where R_p and R_* are the planetary and stellar radii respectively, with the radius of the star related to its luminosity L_* and temperature T_* by the Stefan-Boltzmann law:

$$L_* = 4\pi R_*^2 \sigma_{Boltz} T_*^4.$$

This relation shows that depending on the star luminosity and the planet size, the detection of the transit can be difficult because of the low contrast, making the removal of possible *outliers* critical to the analysis.

Outliers can be defined as extreme entries hovering around the fringes of a normal distribution that are unlikely part of the population of interest, defined as the one within ± 3.0 standard deviations σ from the mean μ [12].

Let us define:

$$\mathcal{D} = \{X_0, X_1, \ldots, X_{d-1}\} \tag{2}$$

where d is the cardinality of $\mathcal{D}$, i.e. the number of light curves we want to pre-process. For each light curve $X_i, i = 0, \ldots, d-1$ of size N, it is: $X_i \sim U[a,b] \quad (a,b \in \mathbb{R})$, where U is a uniform distribution within the boundary values $[a,b]$. Thus, we can set the class:

$$\mathcal{C} = \begin{cases} 1, & \text{outlier} \\ 0, & \text{not-outlier} \end{cases} \tag{3}$$

In our case, $X_i = \{x_0, x_1, \ldots, x_{N-1}\}$ refers to the measured flux F and the range $[a, b]$ represents the boundary values of TESS light curves.

The most widely adopted technique for attributing the class 0 or 1 to any data point $x_j \in X_i$ is the Z−Score method, defined as follows:

$$Z_j = \frac{x_j - \mu}{\sigma}. \tag{4}$$

Assuming as threshold the value $\pm 3.0\ \sigma$, we assign the value $\lambda_j = C$ at each data point x_j according to the following rule:

$$\lambda_j = \begin{cases} 1, & |Z_j| > 3 \\ 0, & |Z_j| \leq 3. \end{cases} \tag{5}$$

The above discussion allows us to introduce the following scheme, Algorithm 1, to solve the numerical problem.

Algorithm 1. Sequential algorithm

    Input: $\mathcal{D}, T, N$

1: `matD[i]` $= X_i$
2: **for** each i in `matD` **do**
3:    compute: μ_i, σ_i
4:    `x[j]` $= X_i^{(j)}$ % Z-Score
5:    **for** each j=1 to N **do**
      % compute λ_j % as in (4)
6:      $\lambda_{X^{(i)}} \leftarrow \lambda_j$
7:    **end for**
8:    $\lambda_{\mathcal{D}} \leftarrow \lambda_{X^{(i)}}$
9: **end for**
    Output: $\lambda_{\mathcal{D}}[\ ]$

The above algorithm is designed by the following steps:

STEP 1 Loading each light curve X_i as row into matrix `matD`. For each row, both μ and σ are computed.

STEP 2 Starting from previous computed information, a new *loop-for* starts for computing Z−Score value for each $x_j \in X_i$. By using (5), a value of C is assigned to x_j. Hence, it is added to $\lambda_{X^{(i)}}$, which contains the overall classification results.

STEP 3 In order to obtain the ensemble $\lambda_{\mathcal{D}}$, a collection operation is performed on each $\lambda_{X^{(i)}}$, where $\lambda_{\mathcal{D}}$ contains the classification results of every $\lambda_{X^{(i)}}$.

Starting from good results achieved, according to aims of Z−Score method as in [22], we observed that executing the algorithm sequentially even on the latest-generation CPU requires very long execution times, due to polynomial computational complexity. In particular its upper bound is $\mathcal{O}(dN^2)$.

4 GPU-Parallel Algorithm

In order to improve the performance of the Algorithm 1, we decided to take advantage of the computational power offered by the most modern HPC architectures [14–17]. We developed a parallel implementation through GPU architecture, based on a suitable Domain Decomposition (DD) strategy. Moreover, in order to distribute the overall work to a GPU grid of threads and then to design a good DD, the shared-memory is exploited. In fact, with a proper management of this kind of memory, available from the NVIDIA-CUDA environment for GPUs that we used, a considerable improvement, in terms of execution times, has been achieved.

We started by the following consideration: light curves of the TESS dataset $\mathcal{D}$ are related to different stars, which means that there is no correlation between data of two different light curves. Thanks to this, data distribution can be made by assigning at each CUDA-block the data structure corresponding to a single light curve, using d CUDA-block to process all data in a parallel-embarrassingly way. Each block will exploit its own shared memory to store and computing data of each single light curve.

More specifically, in each block th threads work, following the SIMT paradigm, on N elements of each array $B_i, i = 0, ..., d - 1$, corresponding to a light curve, indicated with X_i in the previous section. The workload distribution is organized by assigning to each thread a sub-set of data of size L, where:

$$L = \begin{cases} \frac{N}{th} & \text{if } \mathrm{mod}(N,th) = 0 \\ \frac{N}{th} + 1 & \text{if } \mathrm{mod}(N,th) \neq 0 \end{cases} \tag{6}$$

The described DD schema is illustrated in Fig. 2.

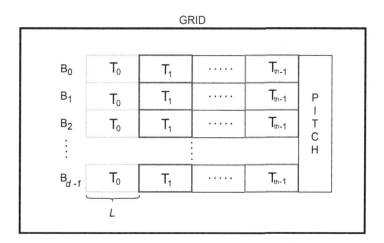

Fig. 2. Domain decomposition strategy with pitch allocation method.

The d light curves are distributed respectively on d blocks, in each of them th threads are activated to process each light curve. In particular, for each threads a chunk of L is processed. Moreover, for each row, the columns number is computed by using the pitch method for an efficient memory allocation and storage. In fact, the pitch method provides to compute the length each row should have so that the number of memory accesses to manage (fetch/insert) data is minimized, [18] and this can be done, exploiting the CUDA routine `cudaMallocPitch`. After the data distribution, in each block the values μ and σ have to be computed, in parallel by all threads activated, and stored efficiently using the fast read/write access of the shared memory. Therefore, the global values of μ and σ will be used to compute the Z-score of each light curve element, in each CUDA block. The parallel procedure, which includes the implemented CUDA kernel, is reported in Algorithm 2.

Algorithm 2. GPU algorithm

Input: $\mathcal{D}, th$

STEP 1: % *Domain Decomposition.*

1: L = (6) % *chunk size*

2: i = `threadIdx.x+(blockDim.x × blockIdx.x)` % *thread id*

3: $X_k \leftarrow \mathcal{D}[k]$% *load light curve k*

STEP 2:% *Mean value computation*

4: **for** each thread i **do**

5:    x_i %*local sums*

6:    $count = count + 1$

7: **end for**

master thread 0:

8: $\mu_k \leftarrow \frac{1}{count_i} \sum_{i=0}^{th} x_i$

STEP 3: *Threads synchronization and Z-score computation*

9: **for** each thread i **do**

10:    $sq_i = \sum_{j \in L}(x_j - \mu)^2$

11: **end for**

master thread 0:

12: $\sigma_k \leftarrow \sqrt{\frac{\sum_{i=0}^{th}(sq_i)}{count_i}}$

13: **for** each thread i **do**

14:    **for** $j_i \in L_i$ **do**

    %*compute* $Z-Score$ *as in Eq.* (4)

15:       $\lambda_{k(i)} \leftarrow \lambda_j$

16:    **end for**

17: **end for**

STEP 4:% *Data collection*

18: **for** each thread i **do**

19:    $\lambda_{\mathcal{D}} \leftarrow \lambda_{k(i)}$

20: **end for**

    Output: $\lambda_D[\,]$

Main operations of Algorithm 2 are summarized in the following steps:

STEP 1 - Data distribution: the Domain Decomposition approach describe previously is here used by computing for each thread its id-number and its portion of the data set $\mathcal{D}$. Therefore, each block $k = 0, ..., d - 1$ of threads loads locally a row of $\mathcal{D}$ by extracting the related light curve stored into X_k, which is the data-structure representing B_k, the $k - th$ light curve.

STEP 2 - Mean value computation: each thread computes a local sum related to its chunk. Hence, starting from each chunk's sum, the master thread computes the global sum and the mean value, which is just stored into the shared memory.

STEP 3 - Threads synchronization and Z-score computation: after the parallel computation and in order to preserve the memory consistent, all threads of each block have to be synchronized by using the CUDA routine __syncthreads(). Therefore, in a similar way the standard deviation is computed. Finally, the $Z-$Score value is computed for each element within the chunk (line 14), leveraging on shared-memory data as in Eq. (4).

STEP 4 - Data collection: the final operation is to collect the local λ values, computed by each thread. In particular, in order to perform a correct memory access and avoid any memory contention, each thread i copies into the global $\lambda_{\mathcal{D}}$, in the position L_i, the local results.

5   Results

In this section we discuss different experimental tests in order to highlighting the gain of performance achieved by the GPU-parallel algorithm with respect to the sequential CPU-based one. Two different metrics are used: we first show a comparison between the two implementations in terms of execution times; then we present a comparison in terms of Gflops.

The parallel code runs on a machine with the following technical specifications:

- 1 x CPU Intel i7 860 with 4 cores, 2 threads per core, 2.80 GHz, 8 GB of RAM
- 1 x GPU NVIDIA Quadro K5000 with 1536 CUDA Cores, 706 MHz Core GPU Clock, and 4 GB 256-bit, GDDR5 memory configuration.

In order to exploit the overall parallel powerful nature of GPUs, the CUDA framework has been adopted. The high programming flexibility of this framework allow us to model specific memory strategies for improving the performance. In fact, different experimental tests with different ad-hoc memory-based strategies will be considered. The main aim of these tests is to find the best strategy so that we can confirm the reliability of our proposed algorithm. Two different GPU memory strategy will be discussed, the first one with basic global memory data allocation, the second one making use of the *pitch* technique.

The following tests have been performed by repeating the executions 10 times.

Test 1: CPU vs GPU Execution Times Comparison

The first experimental test shows the gain in execution time of the GPU version with respect to CPU-version.

Table 1. Execution times: CPU vs GPU for different input sizes. $N = 16741$, block $\times$ thread $= d \times 1024$. Execution times: CPU vs GPU for different input sizes. $N = 16741$, block $\times$ thread $= d \times 1024$.

d	Sequential time (s)	GPU time (s)
5 000	1.591	0.142
10 000	2.970	0.284
15 000	4.392	0.426
20 000	6.523	0.521
25 000	8.114	0.724

Table 1 highlights the appreciable gain of performance of the GPU-based version with respect to the CPU version. As d increases, the execution time of the GPU-based algorithm decreases of 90% with respect to that of the sequential algorithm on the CPU Here, the memory strategy is based on global memory storing of $\mathcal{D}$ without pitch. In particular the reliability of parallel algorithm is confirmed by varying the input data problem. The CUDA configuration block $\times$ threads is fixed to $d \times 1024$, according to Domain Decomposition strategy adopted. Next text deals with find the best CUDA configuration.

Test 2: GPU Execution Times Comparison with Different Memory Strategies.

Despite the appreciable performance obtained in previous experimental test, we observed that the execution times can slow down by choosing a different memory strategy. In particular the next test adopt the global memory allocation with *pitch* technique.

Table 2. Execution times: GPU with different memory strategies by varying d, block $\times$ thread $= d \times 1024$.

d	GPU no pitch (s)	GPU pitch (s)
5 000	0.142	0.125
10 000	0.284	0.260
15 000	0.426	0.384
20 000	0.521	0.503
25 000	0.724	0.672

Table 2 shows the execution times for two different memory strategies. In particular, the first column includes the results obtained with a GPU version with global memory storing strategy without the pitch method, while in the second column the ones with the pitch method. Indeed, an improving lower than 10% has been achieved. More in details, the pitch method is able to determine the length that each row should have. Therefore, the number of memory accesses to manage (fetch/insert) data is minimized.

Test 3: GPU Times with Shared-Memory Strategy. Here, we show a comparison with another memory strategy applied to our parallel algorithm.

Table 3. Execution times: GPU with shared-memory strategies by varying d, `block` $\times$ `thread` $= d \times 1024$.

d	GPU GB_t (s)	GPU SM_t (s)
5 000	0.125	0.121
10 000	0.260	0.255
15 000	0.384	0.375
20 000	0.503	0.495
25 000	0.672	0.617

where GB_t and SM_t are, respectively, the GPU times with the global memory strategy and the shared-memory strategy, both with the pitch method. Table 3 confirms that our domain decomposition approach combined with ad-hoc memory-based strategy is more performing with respect to previous memory configurations. In particular, the shared-memory is involved in our algorithm due to its hardware position close to the threads. In other words, the access time is clearly reduced. In fact, according to the CUDA architecture, the shared-memory is placed close to each blocks threads of grids. Moreover, by considering the advantages achieved in the previous test, the pitch method was adopted in this memory storing strategy as well.

Test 4: GPU Times with Different CUDA Configurations. Here we show the execution times achieved by varying the CUDA configurations and input size problem. In particular, thanks to the appreciable results of last memory strategy just shown, the next test is based on this configuration in order to retrieve the best parallelism available. According to pitch-based memory strategy applied, the parameter N is set to 16768.

Table 4 is used to find the ideal CUDA configuration in order to confirm the best exploit of overall GPU parallelism offered. We observe that the best configuration relies on 1024 threads with the shared-memory strategy. In particular, according to coalescing rules of CUDA, a large threads number slow down the performance due to hardware resource limits. In other words, the number of Streaming Multiprocessor of our GPU is saturated, and an implicit overhead is

Table 4. GPU execution times comparison by varying input size and CUDA threads configuration, `block` × `threads` = $d \times t$.

	# Threads				
d	$t = 128$	$t = 256$	$t = 512$	$t = 1024$	$t = 2048$
5 000	0.193	0.180	0.146	0.121	0.302
10 000	0.405	0.378	0.306	0.255	0.785
15 000	0.598	0.558	0.452	0.375	0.815
20 000	0.785	0.732	0.592	0.495	0.845
25 000	0.984	0.918	0.744	0.617	0.875

introduced. In fact, the remaining number of blocks will be scheduled in pipeline mode when the run-time blocks complete the related computation.

Test 5: FLOPS Comparison. In this test, in order to confirm the gain of performance with respect to the sequential implementation, we analyse an addiction theoretical metric, i.e. the performance analysis in Giga floating point operations per second (Gflops). The results obtained are referred to the previous tests and to the best observed CUDA configuration.

Table 5. Performance in terms of Gflops.

Input size	Gflops CPU	Gflops GPU
5 000	2.634	146.446
10 000	5.645	257.568
15 000	8.590	276.080
20 000	10.823	264.989
25 000	12.915	328.541

Table 5 shows the gain in terms of Gflops, comparing the operation number per seconds (CPU vs GPU) in several execution by varying the input size. We observe an appreciable enhancement of performance obtained by exploiting the GPU architecture. Indeed, the table shows an increasing of performance of about 37×, in terms of Gflops, for all the executions.

The results presented above confirm the reliability and growing of the performance related to the parallel implementation.

6   Conclusion

In this work, we presented a GPU-parallel algorithm based on $Z-$Score for detecting outliers in TESS light curve. The method we presented here allows to significantly improve the efficiency of the outliers detection and cleaning on TESS data. The same approach can be employed to all the remaining pre-processing steps (such as e.g. noise reduction, flattening and folding of the light curves) required to produce assimilated data, which are necessary for ML applications. Therefore, we anticipate that a complete pre-processing pipeline based on GPU-parallel computation will significantly contribute to accelerate the following processes required during Machine Learning flow operations.

References

1. Shallue, C.J., Vanderburg, A.: Identifying exoplanets with deep learning: a five-planet resonant chain around kepler-80 and an eighth planet around kepler-90. Astronomical J. **155**(2), 94 (2018)
2. Ricker, G.R., et al.: Transiting exoplanet survey satellite (TESS). American Astronomical Society Meeting Abstracts, vol. 215 (2010)
3. Wright, J.T., et al.: The exoplanet orbit database. Publications of the Astronomical Society of the Pacific, vol. 123, no. 902, p. 412 (2011)
4. Krüger, J., Westermann, R.: Linear algebra operators for GPU implementation of numerical algorithms. In: ACM SIGGRAPH 2005 Courses, pp. 234-es (2005). ISO 690
5. https://developer.nvidia.com/cuda-zone
6. Parker, J.J.K., NASA Goddard Space: Transiting Exoplanet Survey Satellite (TESS) Flight Dynamics Commissioning Results and Experiences (2018)
7. Osborn, H.P., et al.: Rapid classification of TESS planet candidates with convolutional neural networks. Astronomy Astrophys. **633**, A53 (2020)
8. Yu, L., et al.: Identifying exoplanets with deep learning. III. Automated triage and vetting of TESS candidates. Astronomical J. **158**(1), 25 (2019)
9. Kreidberg, L.: batman: BAsic transit model cAlculatioN in Python. Publi. Astronomical Soc. Pacific **127**(957), 1161 (2015)
10. https://tasoc.dk/docs/EXP-TESS-ARC-ICD-TM-0014-Rev-F.pdf
11. Abukhaled, M., Guessoum, N., Alsaeed, N.: Mathematical modeling of light curves of RHESSI and AGILE terrestrial gamma-ray flashes. Astrophys. Space Sci. **364**(8), 1–16 (2019). https://doi.org/10.1007/s10509-019-3611-3
12. Osborne, J.W.: Best Practices in Data Cleaning: A Complete Guide to Everything you Need to do Before and After Collecting your Data. Sage, Thousand Oaks
13. Bansal, D., Cody, D., Herrera, C., Russell, D., Campbell, M.R.: Light curve analysis for transit of Exoplanet Qatar-1b. Baylor University Department of Astrophysics (2015)
14. De Luca, P., Galletti, A., Ghehsareh, H.R., Marcellino, L., Raei, M.: A GPU-CUDA framework for solving a two-dimensional inverse anomalous diffusion problem. In: Foster, I., Joubert, G.R., Kučera, L., Nagel, W.E., Peters, F. (eds.) Parallel Computing: Technology Trends, Advances in Parallel Computing. vol. 36. pp. 311–320. IOS Press (2020)

15. Amich, M., De Luca, P., Fiscale, S.: Accelerated implementation of FQSqueezer novel genomic compression method. In: 2020 19th International Symposium on Parallel and Distributed Computing, ISPDC, pp. 158–163. IEEE, July 2020

16. Marcellino, L., et al.: Using GPGPU accelerated interpolation algorithms for marine bathymetry processing with on-premises and cloud based computational resources. In: Wyrzykowski, R., Dongarra, J., Deelman, E., Karczewski, K. (eds.) PPAM 2017. LNCS, vol. 10778, pp. 14–24. Springer, Cham (2018). https://doi.org/10.1007/978-3-319-78054-2_2

17. Montella, R., Di Luccio, D., Troiano, P., Riccio, A., Brizius, A., Foster, I.: WaComM: a parallel water quality community model for pollutant transport and dispersion operational predictions. In: 2016 12th International Conference on Signal-Image Technology & Internet-Based SystemsSITIS, pp. 717–724. IEEE, November 2016

18. https://docs.nvidia.com/cuda/cuda-runtime-api/group_CUDART_MEMORY.html

19. Wolszczan, A., Frail, D.A.: A planetary system around the millisecond pulsar PSR1257 + 12. Nature **355**, 145 (1992). https://doi.org/10.1038/355145a0

20. Torres, G., Winn, J.N., Holman, M.J.: Improved parameters for extrasolar transiting planets. Astrophys. J. **677**, 1324 (2008). https://doi.org/10.1086/529429

21. Mandel, K., Agol, E.: Analytic light curves for planetary transit searches. Astrophys. J. Lett. **580**, L171 (2002). https://doi.org/10.1086/345520

22. Zucker, S., Giryes, R.: Shallow transits-deep learning. i. feasibility study of deep learning to detect periodic transits of exoplanets. Astronom. J. **155**(4), 147 (2018)

Data-Driven Deep Learning Emulators for Geophysical Forecasting

Varuni Katti Sastry[1], Romit Maulik[2], Vishwas Rao[2(✉)], Bethany Lusch[2], S. Ashwin Renganathan[2], and Rao Kotamarthi[2,3]

[1] Naperville, USA
[2] Argonne National Laboratory, Lemont, IL, USA
{rmaulik,vhebbur,blusch,srenganathan,vkotamarthi}@anl.gov
[3] University of Chicago, Chicago, IL, USA

Abstract. We perform a comparative study of different supervised machine learning time-series methods for short-term and long-term temperature forecasts on a real world dataset for the daily maximum temperature over North America given by DayMET. DayMET showcases a stochastic and high-dimensional spatio-temporal structure and is available at exceptionally fine resolution (a 1 km grid). We apply projection-based reduced order modeling to compress this high dimensional data, while preserving its spatio-temporal structure. We use variants of time-series specific neural network models on this reduced representation to perform multi-step weather predictions. We also use a Gaussian-process based error correction model to improve the forecasts from the neural network models. From our study, we learn that the recurrent neural network based techniques can accurately perform both short-term as well as long-term forecasts, with minimal computational cost as compared to the convolution based techniques. We see that the simple kernel based Gaussian-processes can also predict the neural network model errors, which can then be used to improve the long term forecasts.

Keywords: Data-driven forecasting · Geophysical systems · Deep learning · LSTM · Gaussian process regression

1 Introduction

Forecasting the maximum temperature is a crucial capability for several applications relevant to agriculture, energy, industry, tourism and the environment. Improved accuracy in short and long-term forecasting of the air temperature has significant implications for cost-effective energy policy, infrastructure development, and downstream economic consequences [1,2]. Existing state of the

This material was based upon work supported by the U.S. Department of Energy, Office of Science, Office of Advanced Scientific Computing Research (ASCR) under Contract DE-AC02-06CH11347.

M. Paszynski et al. (Eds.): ICCS 2021, LNCS 12746, pp. 433–446, 2021.
https://doi.org/10.1007/978-3-030-77977-1_35

art temperature forecasting methods rely on solving large scale partial differential equations (PDE), which generally requires the utilization of large computing resources and are therefore limited by access and considerations of energy-efficiency.

Machine learning methods promise to provide forecasts whose accuracy is comparable to the traditional methods at a much lower computational cost. This also allows for the possibility of using large ensemble-based forecasts that provide confidence intervals, which was hitherto considered expensive and impractical due to the large computational costs associated with the PDE-based methods. For these reasons, there has been a great degree of interest in building machine learning 'emulators' or 'surrogate models' for various geophysical data sets. There have been several studies on the use of machine learning for accelerating geophysical forecasts in recent times. Several rely on using machine learning methods to devise parameterizations for processes that contribute a significant cost to the numerical simulation of the weather and climate [3–7]. Other studies have looked at complete system emulators (i.e., forecasting from data alone) with a view to forecast without any use of and consequent limitations of equation based methods [8–13]. Other studies have looked at utilizing historical information for forecasting specific processes using data from the process alone [14–16]. Opportunities and perspectives for the use of data-driven methods for the geosciences may be found in [17,18]. In this paper, we introduce a purely data-driven method for forecasting the maximum air temperature over the North American continent. In addition, we also provide an interpretable extension for improving the accuracy of these models using probabilistic machine learning. We achieve this by obtaining a low-dimensional affine subspace of the temperature on which a reduced system is evolved. Both dimensionality reduction and system evolution are performed using data-driven techniques alone with the former requiring a proper-orthogonal decomposition (POD) and the latter using a variety of time-series emulation methods. Among the methods investigated in this study, we include the long short-term memory (LSTM) neural network [19] which has previously been used for surrogate modeling of various geophysical and engineering applications [15,20,21]. In addition, we assess the viability of competing time-series forecast methods such as the bidirectional LSTM, previously deployed for a shallow-water equation surrogate [22]. In contrast to previous literature, we assess, for the first time, novel methods such as the sequence-to-sequence encoder-decoder model and the temporal convolutional network for geophysical emulation. These approaches forecast the evolution of POD-coefficients in the truncated POD space. Previously, extensive studies of this method with the LSTM have revealed issues related to stability when using observation data sets due to a relatively low signal to noise ratio. Therefore, we also introduce a error-correction module based on Gaussian process regression (GPR) that extends the viability of these compressed emulation methods and thereby obtain a more accurate forecast for a longer time into the future. Our experiments ascertain that the use of a probabilistic model to learn the bias in the neural network further enhances fidelity of forecasts in comparison to the standard, deterministic, neural network based approach.

2 Proper Orthogonal Decomposition

Projection-based reduced order models (ROMs) can effectively compress a high dimensional model while still preserving its spatio-temporal structure. The dimensionality reduction is performed through the projection step where the high dimensional model is projected onto a set of optimally chosen bases [23,24]. The mechanics of a POD-ROM (POD-based ROMs) can be illustrated for a state variable $\mathbf{t} \in \mathbb{R}^N$, where N represents the size of the computational grid. The POD-ROM then approximates $\mathbf{t}$ as the linear expansion on a finite set of k orthonormal basis vectors $\{\boldsymbol{\phi}_i \in \mathbb{R}^N, \ i = 1, \ldots, k\}$, alternatively called as POD bases. That is,

$$\mathbf{t} \approx \sum_{i=1}^{k} \tilde{t}_i \boldsymbol{\phi}_i, \tag{1}$$

where $\tilde{t}_i \in \mathbb{R}$ is the ith component of $\tilde{\mathbf{t}} \in \mathbb{R}^k$, which are the coefficients of the basis expansion. The $\{\boldsymbol{\phi}_i\}$ are the POD *modes*. POD modes in Eq. 1 can be shown to be the left singular vectors of the snapshot matrix (obtained by stacking M snapshots of $\mathbf{t}$), $\mathbf{T} = [\mathbf{t}_1, \ldots, \mathbf{t}_M]$, extracted by performing a singular value decomposition (SVD) on $\mathbf{T}$ [25,26]. That is,

$$\mathbf{T} \underset{\text{svd}}{=} \mathbf{U}\boldsymbol{\Sigma}\mathbf{V}^\top, \tag{2}$$

where $\mathbf{U} \in \mathbb{R}^{N \times M}$ and $\boldsymbol{\Phi}_k$ represent the first k columns of $\mathbf{U}$ after truncating the last $M - k$ columns based on the relative magnitudes of the cumulative sum of their singular values. The total L_2 error in approximating the snapshots via the truncated POD basis is then given as

$$\sum_{j=1}^{M} \left\| \mathbf{t}_j - (\boldsymbol{\Phi}_k \boldsymbol{\Phi}_k^\top) \mathbf{t}_j \right\|_2^2 = \sum_{i=k+1}^{M} \sigma_i^2, \tag{3}$$

where σ_i is the singular value corresponding to the ith column of $\mathbf{U}$ and is also the ith diagonal element of $\boldsymbol{\Sigma}$. It is well known that the POD bases are L_2-optimal and present a good choice for an efficient compression of high-dimensional data.

An important point to note is the effect of premature truncation when representing dynamics in the POD space. For real-world problems that are advection-dominated, an impractically high number of POD-bases need to be retained to faithfully represent the flow-field. However, this limits the gains of data-driven or equation-based surrogate modeling from the perspective of computational efficiency as well as accuracy. Therefore, it is necessary to devise evolution strategies that can preserve the effects of the truncated scales on the resolved ones. In reduced-order modeling parlance, this is often referred to as a model that is equipped with closure. In this study, we leverage ideas from the application of time-delay embedded machine learning techniques with analogs to the Mori-Zwanzig formalism to account for closure implicitly [27,28]. We remind the reader that the Mori-Zwanzig formalism proposes the use of memory to account for errors due to coarse-graining of systems. In the following sections, we shall introduce time-delay embedded forecasting techniques to handle forecasting in the reduced-space spanned by truncated POD basis vectors.

3   Models

Figure 1 represents a block diagram of the two-stage model, with the first stage comprising of a neural network based forecasting model, followed by a second stage error correction model based on GPR. The neural network model is trained on the POD coefficients of the training set to obtain an m–step forecasting. The predicted data is compared against the true coefficients to get the error e. The true POD coefficients of the training set and the corresponding error e from the neural network model are used in the error correction model to develop a function mapping between the two entities. This function mapping is used to estimate the "predicted error" e' which is then added to the predicted coefficients from the neural network model during the deployment phase. This results in an improved m-step forecasting of the POD coefficients. First, we briefly describe different machine learning approaches that we use within our two-stage model.

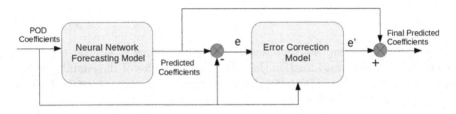

Fig. 1. Block diagram of the two-stage forecasting model.

3.1   Forecasting Neural Network Models

In this paper, we consider a set of neural network models and compare the forecasting abilities of those models against each other. We consider two major families of neural network models. The first is Recurrent Neural Network (RNN) based algorithms such as vanilla LSTM, Bi-directional LSTM, and a sequence-to-sequence (Seq2Seq) Encoder-Decoder model. We also consider convolution neural network (CNN) based algorithms such as one-dimensional convolutional neural network (1D-CNN) and its variations such as vanilla-temporal convolutional neural network (TCN) and stacked-TCN. We briefly discuss these models below.

Recurrent Neural Network Based Models: RNNs have been the most obvious choice in recent times to model sequential data given their ability to characterise the temporal dependencies [29,30]. LSTMs are by far the most well known variant of RNNs [31], where the gated structure helps in learning long-term dependencies, by allowing the flow of information across several time steps. For this reason, they have been extensively used in time series predictions especially applications such as weather forecasting [32–34]. Bi-directional LSTM is a variant of LSTM, where one set of LSTMs is used across the forward direction of the sequential data, while another set is applied to the same sequence in the reverse direction. Bi-LSTM is known to outperform vanilla LSTM [32,35,36], due to the fact that they tend to capture the temporal structure better than the latter by considering correlations in both forward and backward directions. For a prediction task, which is framed as generating a window of outputs in the future based on a window of inputs from the past, the bi-directional element of the modeling strategy ensures all components of the input contribute to the forecast of all components of the output. Though Seq2Seq encoder-decoder models traditionally have been associated with natural language processing (NLP) applications [37], they have been shown to be well suited for time series based forecasting problems too [38].

Convolutional Network Based Models: Though traditionally convolutional neural networks have been applied to image data [39,40], there are several variants of CNNs adapted to learn the temporal information from time series data [41]. Among them, the 1d-CNN is the simplest of the CNN models where the output is the convolution of input and the convolutional kernel over a single temporal dimension. We also use the more advanced TCNs that employ dialated causal convolution [42,43]. Such a network is seen to have information propogated over a larger number of time steps as compared to the recurrent models [44]. We further use a stacked-TCN architecture by stacking multiple TCN residual blocks.

3.2   Error Correction Model

The predictions from the neural network model can be improved upon with a error correction model. This is especially helpful in long term forecasting using neural network models with feedback, as discussed in detail in the following sections [45]. We use GPR as the primary error correction tool to improve the forecasts from the neural network models. The Gaussian process (GP) model is a well known non-parametric supervised learning method especially for regression problems [46]. Instead of point estimates, GP takes a Bayesian approach to providing a probability distribution for a set of possible functions that fit the training data. It specifies a prior distribution over the functions that characterizes the assumptions we make about the underlying functionality of the data yet to be seen. Upon observing the training data, the prior distribution is updated through their likelihood function. This resultant distribution, called the posterior distribution can be used to compute the predictive posterior distribution on the new unobserved (test) data. The prior distribution that we assume is completely defined by the choice of the kernel. We use a simple squared exponential kernel which can effectively forecast the errors, while being computationally efficient. We use the `sklearn`, a machine learning toolbox in Python, to implement GPR for our purposes.

4   Dataset and Numerical Results

We use the DayMET Version 3 model data which provides estimates of daily weather parameters for the North America region [47]. The data has 1x1-km spatial resolution and a temporal resolution of 1 calendar day. For most methods, we use the 2014–2015 data for training and 2016 data for validation purposes. The test forecasts are performed for the year 2017. For experimental study, we choose the maximum air temperature abbreviated as `tmax` to be the quantity of interest. We note that this procedure can be seamlessly extended to other physical quantities too. The raw data is preprocessed to remove all the oceanic points, leaving us with a snapshot of 201928 grid points mapping to the North American continental land mass. The POD basis set is computed on the snapshot data through the truncated SVD procedure with length k. The goal of POD is to construct the orthogonal basis such that the resulting linear subspace can effectively represent the spatial and the temporal dynamics of `tmax`. A truncation of length $k = 5$ results in 92% of the energy being conserved, implying the need for some time-delay based prediction model to account for unresolved scales. The coefficient matrix generated from the POD procedure is of dimension $T \times k$, where T represents the number of temporal observations in the time series and k is the number of modes. Since we consider approximately two years of training data, $T_{train} = 730$. $T_{validation} = T_{test} = 365$, mapping to one year of temporal data for validation and test procedure respectively. The data is preprocessed using a traditional minmax scaler to scale each of the modes in the training data individually. Then the same scaling parameters are used to project the validation and test data. Figure 3 presents the histogram of the POD coefficients across the primary mode (mode 0) and the successive secondary modes (modes 1–3) for the DayMet data. As seen from the figure, the primary mode has a multimodal character, while all the secondary modes appear to be Gaussian. Figure 4 shows the autocorrelation plot for each of the 0th-3rd modes, which suggests that the primary mode observes a higher degree of autocorrelation compared to the secondary modes. This coefficient matrix, which holds the temporal information across different modes, is the input to the neural network models. We transform this time series dataset into a supervised learning problem by using the standard sliding window walk-forward method and perform a multivariate multi-step forecasting. We structure each sample of the training and validation data to have n time steps of input data (known as the window length) and to have the next m timesteps ($n + 1$ to $n + m$) be the labelled output data. We consider two scenarios for building the training data required for the supervised learning models.

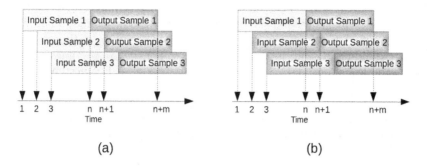

Fig. 2. Input-Output samples for a) no feedback b) with feedback

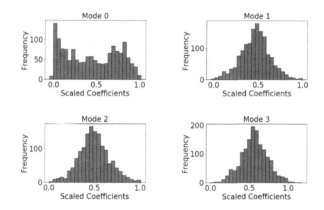

Fig. 3. Histogram of POD coefficients for primary mode (mode 0) and three secondary modes (mode 1–3).

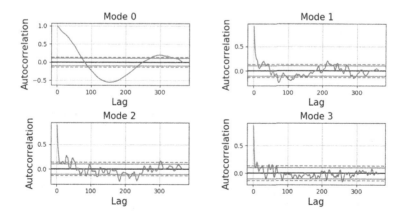

Fig. 4. Autocorrelation plots of POD coefficients for primary mode and secondary modes for test data.

Without Feedback: Here we scan through the entire training and validation datasets and build the samples by the sliding window method. For example, the input of the first sample consists of data from $1 : n$ timesteps, and the corresponding output is the data from $n + 1 : n + m$ timesteps. The second sample's input data is the data from $2 : n + 1$ timesteps, and its corresponding output is from $n + 2 : n + m + 1$, and so on. This input-output structure is used to train different neural network models. For the prediction phase, the input is structured similarly to the one in the training phase. The model forecasts are evaluated and compared against the true values. This method can be effectively used to perform short-term forecasting, as the model deployment is made completely on the historical true values (Fig. 2).

With Feedback: This method has a training phase similar to the one above. It differs from the above method during the deployment phase, where the model forecast of the previous window is fed as the input to the next window. For example, to begin with the first n timesteps are input to the model, to forecast the next m timesteps. These m timesteps of predicted data are evaluated against the true values and input to the model to produce the next set of predictions and so on. Such a method helps in long-term forecasting where the entire forecasting window is broken down

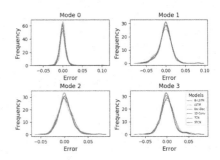

Fig. 5. Comparision of error density of test data without feedback over different models across the primary and secondary modes.

into smaller forecasting periods. For the sake of simplicity, we consider a sliding interval of 1 while building the input samples for the model, and the number of input and output timesteps, that is, the window length, is $n = m = 7$ so that we are conducting weekly forecasts. We use a batch size of 32 and a learning rate of 0.0001 with the Adam optimizer, and these parameters are kept constant across the different neural network models. At first, we compare the forecast errors across the different models, without feedback. Figure 5 shows the error distributions for the test data for the different models, namely - LSTM, Bi-LSTM, Encoder-Decoder LSTM, 1D-Convolution, TCN, and STCN. The error distribution of each mode is shown separately. We see that, while a simple Encoder-Decoder LSTM architecture is able to forecast quite accurately over the short term, the difference in the relative error for each of the modes is marginal over the different models. Therefore, we take a look at the training and the validation loss to evaluate the performance of the different models. Figures 6 and 7 present the training and validation loss (we use mean squared error as the loss function) respectively during the training phase. It can be seen from the figures and Table 1 that the Bi-LSTM, Encoder-Decoder, LSTM and 1D-Convolution are faster

Table 1. Training time statistics for different models with early stopping enabled

Model	Mean epoch time(s)	Total epochs	Total time to train(s)
LSTM	0.60	923	553.8
Bi-LSTM	1.146	655	750.6
Enc-Dec	0.81	624	505.4
1D-Conv	0.21	547	114.8
TCN	1.70	1033	1756
STCN	3.28	1797	5894

Table 2. Mean absolute error statistics for training, validation and test data over different modes for i) without feedback, ii) with feedback and before error correction, iii) with feedback and error correction

	w/o feedback			w/ feedback, before EC			w/ feedback, before EC		
	Train.	Val.	Test	Train.	Val.	Test	Train.	Val.	Test
Mode 0	0.05	0.04	0.05	0.15	0.15	0.16	0.10	0.10	0.09
Mode 1	0.11	0.10	0.10	0.13	0.10	0.12	0.03	0.02	0.03
Mode 2	0.11	0.09	0.10	0.11	0.12	0.12	0.03	0.03	0.03
Mode 3	0.12	0.11	0.12	0.13	0.10	0.14	0.04	0.04	0.04

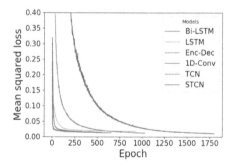

Fig. 6. Mean squared error training loss for the various models

Fig. 7. Mean squared error validation loss for the various models

and more efficient to train than the other temporal convolution models that take longer to train with a very noisy training loss.

Next, we consider the prediction with feedback models that provide long term forecasts. Figure 8 provides a comparision of the distributions of the relative errors across the different neural network models. It is natural to see an increase in the relative errors compared to the models without feedback (an increase of almost an order of magnitude). Across the different models, notice that the TCN based models exibit higher standard deviations in their error distributions. The relative errors for each of the secondary modes exhibit a normal distribution, and

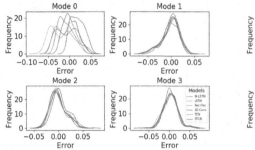 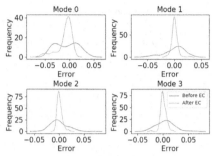

Fig. 8. Comparision of error density of test data with feedback over different models across the primary and secondary modes.

Fig. 9. Comparision of error density of test data with feedback for Bi-LSTM model across the primary and secondary modes, before and after error correction model is applied.

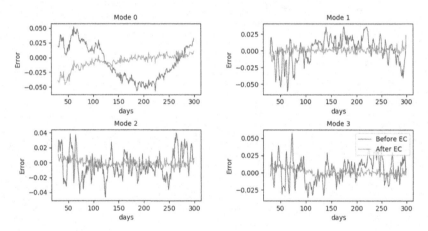

Fig. 10. Error of test data for the first 300 d of Bi-LSTM model with feedback, across the primary and secondary modes before and after the error correction is applied.

we observe a bi-modal distribution for the primary mode, which encourages us to use an error correction model over the errors generated from the neural network model. Figures 9 and 10 compare the error distribution and mean absolute error for the Bi-LSTM model with feedback, before and after the error correction is applied. As seen in Figs. 3 and 9, for mode 0, both the coefficients and the errors from the neural network forecasts exhibit a bi-modal distribution, due to which the GPR does not fit the error data well for the first few days. But for long-term forecasting, we see that the GPR fits the error data well and the overall error distribution for each of the modes is more concentrated around the mean (0) after the error correction model is deployed, as compared to the error distribution from the neural network model forecasts that have a higher

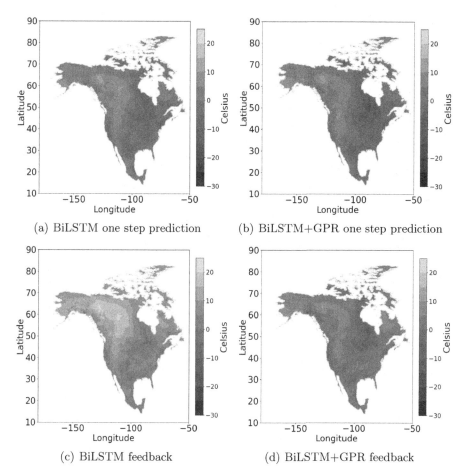

(a) BiLSTM one step prediction   (b) BiLSTM+GPR one step prediction

(c) BiLSTM feedback   (d) BiLSTM+GPR feedback

Fig. 11. Error contours displaying representative emulation performance on testing data set on the 30th of May, 2017. The first row contours indicate results from one step predictions using a BiLSTM without (a) and with GPR corrections (b). The second row displays error contours from a feedback application of the machine learning frameworks where there is no correction (c) and where a GPR is used to adjust the outputs of the BiLSTM. The use of the GP model is seen to improve predictions significantly for this case.

variance. Notice from Table 2, that the mean absolute error for the no-feedback case is relatively small compared to the with-feedback scenario, which can be reduced by using an error correction model.

Figure 11 shows the viablity of the forecast frameworks in physical space where the GPR-based error correction is seen to enhance accuracy particularly when the emulator is applied recursively with feedback.

5   Conclusion

In this work, we present multiple data-driven approaches for forecasting maximum air temperature in the North-American region. We combine dimensionality reduction and system evolution using different flavors of deep learning models and deploy these techniques on real data. We mainly consider two scenarios in training and validating the supervised learning models. First, we consider a scenario where the training data is continuously enriched with new measurements *(without feedback)*, which is useful for producing short-term forecasts. The second scenario is developed for a longer forecast horizon *(with feedback)*.

In this scenario, the output of the supervised learning model from one cycle is used as the input in the next cycle and, as expected in such cases, the errors tend to accumulate and the method suffers from higher inaccuracies and instabilities than in the 'without feedback' case. We improve the 'with feedback' case by deploying a error correction term using GPR, and this vastly improves the forecasts. The forecast error in both primary and secondary modes is much smaller and hence allows for accurate forecasts for a much longer horizon. We have demonstrated these methods by carrying out extensive tests on DayMet data. In the future we will study the evolution of the discrete Lyapunov exponents to understand the stability of these methods and improve them. Additionally, we will also deploy state-of-the-art machine learning models to study multiple correlated physical quantities. Finally, we will also explore the use of Gaussin-mixtures as an error correction model to improve the error forecasts.

References

1. Ciscar, J.-C., et al.: Physical and economic consequences of climate change in Europe. Proc. Natl. Acad. Sci. **108**(7), 2678–2683 (2011)
2. Gosling, S.N., Lowe, J.A., McGregor, G.R., Pelling, M., Malamud, B.D.: Associations between elevated atmospheric temperature and human mortality: a critical review of the literature. Clim. Change **92**(3), 299–341 (2009)
3. Gentine, P., Pritchard, M., Rasp, S., Reinaudi, G., Yacalis, G.: Could machine learning break the convection parameterization deadlock? Geophys. Res. Lett. **45**(11), 5742–5751 (2018)
4. Brenowitz, N.D., Bretherton, C.S.: Prognostic validation of a neural network unified physics parameterization. Geophys. Rese. Lett. **45**(12), 6289–6298 (2018)
5. Rasp, S., Pritchard, M.S., Gentine, P.: Deep learning to represent subgrid processes in climate models. Proc. Natl. Acad. Sci. **115**(39), 9684–9689 (2018)
6. Nooteboom, P.D., Feng, Q.Y., López, C., Hernández-García, E., Dijkstra, H.A.: Using network theory and machine learning to predict el niño. Earth Syst. Dyn. **9**(3), 969–983 (2018)
7. Moosavi, A., Rao, V., Sandu, A.: Machine learning based algorithms for uncertainty quantification in numerical weather prediction models. J. Comput. Sci. 101295 (2021)
8. Liu, Y., et al.: Application of deep convolutional neural networks for detecting extreme weather in climate datasets. arXiv:1605.01156 (2016)

9. Scher, S.: Toward data-driven weather and climate forecasting: Approximating a simple general circulation model with deep learning. Geophys. Res. Lett. **45**(22), 12–616 (2018)
10. Weyn, J.A., Durran, D.R., Caruana, R.: Improving data-driven global weather prediction using deep convolutional neural networks on a cubed sphere. J. Adv. Model. Earth Syst. **12**(9), e2020MS002109 (2020)
11. Rasp, S., Thuerey, N.: Purely data-driven medium-range weather forecasting achieves comparable skill to physical models at similar resolution. arXiv:2008.08626 (2020)
12. Chattopadhyay, A., Nabizadeh, E., Hassanzadeh, P.: Analog forecasting of extreme-causing weather patterns using deep learning. J. Adv. Model. Earth Syst. **12**(2), e2019MS001958 (2020)
13. Rodrigues, E.R., Oliveira, I., Cunha, R., Netto, M.: Deepdownscale: a deep learning strategy for high-resolution weather forecast. In: 2018 IEEE 14th International Conference on e-Science (e-Science), pp. 415–422. IEEE (2018)
14. Shi, X., Chen, Z., Wang, H., Yeung, D.-Y. Wong, W.-K., Woo, W.-C.: Convolutional lstm network: a machine learning approach for precipitation nowcasting. arXiv:1506.04214 (2015)
15. Maulik, R., Egele, R., Lusch, B., Balaprakash, P.: Recurrent Neural Network Architecture Search for Geophysical Emulation. IEEE Press (2020)
16. Skinner, D.J., Maulik, R.: Meta-modeling strategy for data-driven forecasting. arXiv:2012.00678 (2020)
17. Karpatne, A., Ebert-Uphoff, I., Ravela, S., Babaie, H.A., Kumar, V.: Machine learning for the geosciences: challenges and opportunities. IEEE Trans. Knowl. Data Eng. **31**(8), 1544–1554 (2018)
18. Dueben, P.D., Bauer, P.: Challenges and design choices for global weather and climate models based on machine learning. Geosci. Model Dev. **11**(10), 3999–4009 (2018)
19. Hochreiter, S., Schmidhuber, J.: Long short-term memory. Neural Comput. **9**(8), 1735–1780 (1997)
20. Rahman, S.M., Pawar, S., San, O., Rasheed, A., Iliescu, T.: Nonintrusive reduced order modeling framework for quasigeostrophic turbulence. Phys. Rev. E **100**(5), 053306 (2019)
21. Mohan, A.T., Gaitonde, D.V.: A deep learning based approach to reduced order modeling for turbulent flow control using lSTM neural networks. arXiv:1804.09269 (2018)
22. Maulik, R., Lusch, B., Balaprakash, P.: Non-autoregressive time-series methods for stable parametric reduced-order models. Phys. Fluids **32**(8), 087115 (2020)
23. Taira, K.: Modal analysis of fluid flows: applications and outlook. AIAA J. 1–25 (2019)
24. Benner, P., Gugercin, S., Willcox, K.: A survey of projection-based model reduction methods for parametric dynamical systems. SIAM Rev. **57**(4), 483–531 (2015)
25. Holmes, P., Lumley, J.L., Berkooz, G., Rowley, C.W.: Turbulence, Coherent Structures, Dynamical Systems and Symmetry, vol. 36. Princeton University, New Jersey (1998)
26. Chatterjee, A.: An introduction to the proper orthogonal decomposition. Curr. Sci. **78**, 808–817 (2000)
27. Maulik, R., Mohan, A., Lusch, B., Madireddy, S., Balaprakash, P., Livescu, D.: Time-series learning of latent-space dynamics for reduced-order model closure. Phys. D: Nonlinear Phenom. **405**, 132368 (2020)

28. Ma, C., Wang, J., et al.: Model reduction with memory and the machine learning of dynamical systems. arXiv:1808.04258 (2018)
29. Yu, Y., Si, X., Hu, C., Zhang, J.: A review of recurrent neural networks: LSTM cells and network architectures. Neural Comput. **31**(7), 1235–1270 (2019)
30. Petneházi, G.: Recurrent Neural Networks for Time Series Forecasting. arXiv:1901.00069 (2018)
31. Hochreiter, S., Schmidhuber, J.: Long short-term memory. Neural Comput. **9**(8), 1735–1780 (1997)
32. Karevan, Z., Suykens, J.A.K.: Transductive lstm for time-series prediction: an application to weather forecasting. Neural Netw. **125**, 1–9 (2020)
33. Zhang, Q., Wang, H., Dong, J., Zhong, G., Sun, X.: Prediction of sea surface temperature using long short-term memory. IEEE Geosci. Remote Sens. Lett. **05** (2017)
34. Broni-Bediako, C., Katsriku, F., Unemi, T., Shinomiya, N., Abdulai, J.-D., Atsumi, M.: El niño-southern oscillation forecasting using complex networks analysis of lstm neural networks. vol. 01 (2018)
35. Schuster, M., Paliwal, K.K.: Bidirectional recurrent neural networks. IEEE Trans. Signal Process. **45**(11), 2673–2681 (1997)
36. Namini, S.S., Tavakoli, N., Namin, A.S.: The performance of LSTM and biLSTM in forecasting time series. vol. 12, pp. 3285–3292 (2019)
37. Socher, R., Lin, C., Ng, A., Manning, C.: Parsing natural scenes and natural language with recursive neural networks, vol. 01, pp. 129–136 (2011)
38. Shen, G., Kurths, J., Yuan, Y.: Sequence-to-sequence prediction of spatiotemporal systems. Chaos: Interdiscip. J. Nonlinear Sci. **30**(2), 023102 (2020)
39. Ciresan, D., Meier, U., Masci, J., Gambardella, L.M., Schmidhuber, J.: Flexible, high performance convolutional neural networks for image classification, vol. 07, pp. 1237–1242 (2011)
40. Krizhevsky, A., Sutskever, I., Hinton, G.E.: Imagenet classification with deep convolutional neural networks. Commun. ACM **60**(6), 84–90 (2017)
41. Bai, S., Kolter, J., Koltun, V.: An empirical evaluation of generic convolutional and recurrent networks for sequence modeling, vol. 03 (2018)
42. Yan, J., Mu, L., Wang, L., Ranjan, R., Zomaya, A.Y.: Temporal convolutional networks for the advance prediction of ENSO. Sci. Rep. **10**(1), 1–15 (2020)
43. Remy, P.: Temporal convolutional networks for keras. https://github.com/philipperemy/keras-tcn (2020)
44. Bai, S., Kolter, J.Z., Koltun, V.: An empirical evaluation of generic convolutional and recurrent networks for sequence modeling. arXiv:1803.01271 (2018)
45. Sapsis, T.P., Majda, A.J.: Blending modified gaussian closure and non-gaussian reduced subspace methods for turbulent dynamical systems. J. Nonlinear Sci. **23**(6), 1039–1071 (2013)
46. Rasmussen, C.E.: Gaussian processes in machine learning. In Summer School on Machine Learning, pp. 63–71. Springer, Berlin (2003)
47. Thornton, M.M., Thornton, P.E., Wei, Y., Mayer, B.W., Cook, R.B., Vose, R.S.: Daymet: Annual climate summaries on a 1-km grid for north America, version 3. Ornl Daac, Oak Ridge, Tennessee, USA (2016)

NVIDIA SimNet™: An AI-Accelerated Multi-Physics Simulation Framework

Oliver Hennigh, Susheela Narasimhan, Mohammad Amin Nabian$^{(\boxtimes)}$, Akshay Subramaniam, Kaustubh Tangsali, Zhiwei Fang, Max Rietmann, Wonmin Byeon, and Sanjay Choudhry

Nvidia, Santa Clara, CA, USA
{ohennigh,susheelan,mnabian,asubramaniam,ktangsali,zhiweif, mrietmann,wbyeon,schoudhry}@nvidia.com

Abstract. We present SimNet, an AI-driven multi-physics simulation framework, to accelerate simulations across a wide range of disciplines in science and engineering. Compared to traditional numerical solvers, SimNet addresses a wide range of use cases - coupled forward simulations without any training data, inverse and data assimilation problems. SimNet offers fast turnaround time by enabling parameterized system representation that solves for multiple configurations simultaneously, as opposed to the traditional solvers that solve for one configuration at a time. SimNet is integrated with parameterized constructive solid geometry as well as STL modules to generate point clouds. Furthermore, it is customizable with APIs that enable user extensions to geometry, physics and network architecture. It has advanced network architectures that are optimized for high-performance GPU computing, and offers scalable performance for multi-GPU and multi-Node implementation with accelerated linear algebra as well as FP32, FP64 and TF32 computations. In this paper we review the neural network solver methodology, the SimNet architecture, and the various features that are needed for effective solution of the PDEs. We present real-world use cases that range from challenging forward multi-physics simulations with turbulence and complex 3D geometries, to industrial design optimization and inverse problems that are not addressed efficiently by the traditional solvers. Extensive comparisons of SimNet results with open source and commercial solvers show good correlation. The SimNet source code is available at https://developer.nvidia.com/simnet.

1 Introduction

Simulations are pervasive in every domain of science and engineering. However, they become computationally expensive as more geometry details are included and as model size, the complexity of physics or the number of design evaluations increases. Although deep learning offers a path to overcome this constraint, supervised learning techniques are used most often in the form of traditional data driven neural networks (e.g., [1,2]). However, generating data can be an

© Springer Nature Switzerland AG 2021
M. Paszynski et al. (Eds.): ICCS 2021, LNCS 12746, pp. 447–461, 2021.
https://doi.org/10.1007/978-3-030-77977-1_36

expensive and time consuming process. Furthermore, these models may not obey the governing physics of the problem, involve extrapolation and generalization errors, and provide unreliable results.

In comparison with the traditional solvers, neural network solvers [3–5] can not only do parameterized simulations in a single run, but also address problems not solvable using traditional solvers, such as inverse or data assimilation problems and real time simulation. They can also be embedded in the traditional solvers to improve the predictive capability of the solvers. Training of neural network forward solvers can be supervised only based on the governing laws of physics, and thus, unlike the data-driven deep learning models, neural network solvers do not require any training data. However, for data assimilation or inverse problems, data constraints are introduced in the loss function.

Rapid evolution of GPU architecture suited for AI and HPC, as well as introduction of open source frameworks like Tensorflow have motivated researchers to develop novel algorithms for solving PDEs (e.g., [3,5–8]). Recently, a number of neural network solver libraries are being developed (e.g., TensorFlow-based DeepXDE [9], Keras-based SciANN [10], and Julia-based NeuralPDE.jl [11]), aiming at making these solvers more accessible. Although the existing research studies and libraries played a crucial role in advancing the neural network solvers, the attempted examples are mostly limited to simple 1D or 2D domains with straightforward governing physics, and the neural network solvers in their current form still struggle to solve real-world applications that involve complex 3D geometries and multi-physics systems. In this paper we present SimNet, that aims to address the current computational challenges with neural network solvers. As an example, SimNet enables design optimization of a FPGA heat sink (see Fig. 1) through a single network training without any training data. In contrast, the traditional solvers are not capable of simulating geometries with several design parameters in a single run.

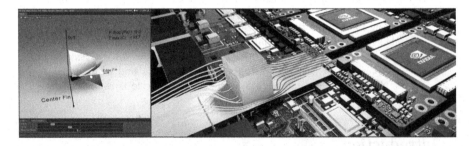

Fig. 1. Design optimization of an FPGA heat sink using SimNet. The center and side fin heights are the two design variables.

Our Contributions: Several research studies have recently been published demonstrating solution of PDEs using neural networks. However, our experience has shown that they do not converge well when used as forward solvers for industrial problems due to the gradients, singularities and discontinuities introduced by complex geometries or physics. Our main contributions in this paper are to offer several novel features to address these challenges - Signed Distance Functions (SDFs) for loss weighting, integral continuity planes for flow simulation, advanced neural network architectures, point cloud generation for real world geometries using constructive geometry module as well as STL module and finally parameterization of both geometry and physics. Additionally, for the first time to our knowledge, we solve high Reynolds number flows (by adopting the RANS equations and the zero-equation turbulence model [12]) in industrial applications without using any data.

2 Neural Network Solvers

A neural network solver approximates the solution to a given PDE and a set of boundary and initial constraints using a feed-forward fully-connected neural network. The model is trained by constructing a loss function for how well the neural network is satisfying the PDE and constraints. A schematic of the structure of a neural network solver is shown in Fig. 2.

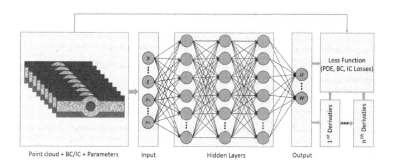

Fig. 2. A schematic of the structure of a neural network solver.

Let us consider the following general form of a PDE:

$$\mathcal{N}_i[u](\mathbf{x}) = f_i(\mathbf{x}), \quad \forall i \in \{1, \cdots, N_\mathcal{N}\}, \mathbf{x} \in \mathcal{D},$$
$$\mathcal{C}_j[u](\mathbf{x}) = g_j(\mathbf{x}), \quad \forall j \in \{1, \cdots, N_\mathcal{C}\}, \mathbf{x} \in \partial\mathcal{D}, \tag{1}$$

where $\mathcal{N}_i$'s are general differential operators, $\mathbf{x}$ is the set of independent variables defined over a bounded continuous domain $\mathcal{D} \subseteq \mathbb{R}^D, D \in \{1, 2, 3, \cdots\}$, and $u(\mathbf{x})$ is the solution to the PDE. $\mathcal{C}_j$'s denote the constraint operators and usually cover the boundary and initial conditions. $\partial\mathcal{D}$ also denotes a subset of the domain boundary that is required for defining the constraints. We seek to approximate the solution $u(\mathbf{x})$ by a neural network $u_{net}(\mathbf{x})$ that, in it's most simple form, takes the following form:

$$u_{net}(\mathbf{x};\theta) = \mathbf{W}_n\{\phi_{n-1} \circ \phi_{n-2} \circ \cdots \circ \phi_1 \circ \phi_E\}(\mathbf{x}) + \mathbf{b}_n, \quad \phi_i(\mathbf{x}_i) = \sigma\left(\mathbf{W}_i\mathbf{x}_i + \mathbf{b}_i\right), \quad (2)$$

where $\mathbf{x} \in \mathbb{R}^{d_0}$ is the input to network, $\phi_i \in \mathbb{R}^{d_i}$ is the ith layer of the network, $\mathbf{W}_i \in \mathbb{R}^{d_i \times d_{i-1}}, \mathbf{b}_i \in \mathbb{R}^{d_i}$ are the weight and bias of the ith layer, θ denotes the set of network's trainable parameters, i.e., $\theta = \{\mathbf{W}_1, \mathbf{b}_1, \cdots, \mathbf{b}_n, \mathbf{W}_n\}$, n is the number of layers, and σ is the activation function. We suppose that this neural network is infinitely differentiable, i.e. $u_{net} \in C^{\infty}$. ϕ_E is an input encoding layer. More advanced architectures will be introduced in Sect. 3.3.

In order to train this neural network, we construct a loss function that penalizes over the divergence of the approximate solution $u_{net}(\theta)$ from the PDE in Eq. 1, and such that the constraints are encoded as penalty terms. To this end, we define the following residuals:

$$\begin{aligned} r_{\mathcal{N}}^{(i)}\left(\mathbf{x}; u_{net}(\theta)\right) &= \mathcal{N}_i[u_{net}(\theta)]\left(\mathbf{x}\right) - f_i\left(\mathbf{x}\right), \\ r_{\mathcal{C}}^{(j)}\left(\mathbf{x}; u_{net}(\theta)\right) &= \mathcal{C}_j[u_{net}(\theta)]\left(\mathbf{x}\right) - g_j\left(\mathbf{x}\right), \end{aligned} \quad (3)$$

where $r_{\mathcal{N}}^{(i)}$ and $r_{\mathcal{C}}^{(j)}$ are the PDE and constraint residuals, respectively. The loss function then takes the following form:

$$\mathcal{L}_{res}(\theta) = \sum_{i=1}^{N_{\mathcal{N}}} \int_{\mathcal{D}} \lambda_{\mathcal{N}}^{(i)}(\mathbf{x})\left\|r_{\mathcal{N}}^{(i)}\left(\mathbf{x}; u_{net}(\theta)\right)\right\|_p d\mathbf{x} + \sum_{j=1}^{N_{\mathcal{C}}} \int_{\partial\mathcal{D}} \lambda_{\mathcal{C}}^{(j)}(\mathbf{x})\left\|r_{\mathcal{C}}^{(j)}\left(\mathbf{x}; u_{net}(\theta)\right)\right\|_p d\mathbf{x}, \quad (4)$$

where $\|\cdot\|_p$ denotes the p-norm, and $\lambda_{\mathcal{N}}^{(i)}, \lambda_{\mathcal{C}}^{(j)}$ are weight functions that control the loss interplay between within and across different terms. The network parameters θ are optimized iteratively using variants of the stochastic gradient descent method. At each iteration, the integral terms in the loss function are approximated using a regular or Quasi-Monte Carlo method, and using a batch of samples from the independent variables $\mathbf{x}$. Automatic differentiation is commonly used to compute the required gradients in $\nabla\mathcal{L}_{res}(\theta)$.

3   SimNet Overview

SimNet is a Tensorflow based neural network solver and offers various APIs that enable the user to leverage the existing functionality to build their own applications on the existing modules. An overview of SimNet architecture is presented in Fig. 3. The geometry modules, PDE module, and data are used to fully specify the physical system. The user also specifies the network architecture, optimizer and learning rate schedule. SimNet then constructs the neural network solver, forms the loss function, and unrolls the graph efficiently to compute the gradients. The SimNet solver then starts the training or inference procedure using TensorFlow's built-in functions on a single or cluster of GPUs. The outputs are saved in the form of CSV or VTK files and can be visualized using TensorBoard and ParaView.

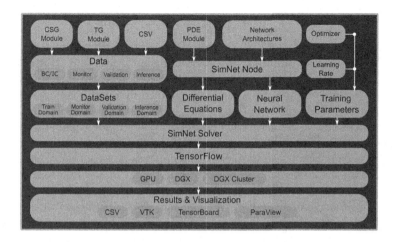

Fig. 3. SimNet structure.

3.1 Geometry Modules

SimNet contains Constructive Solid Geometry (CSG) and Tessellated Geometry (TG) modules. With SimNet's CSG module, constructive geometry primitives can be defined and Boolean operations performed. This allows the creation and parameterization of a wide range of geometries. The TG module uses tesselated geometries in the form of STL or OBJ files to work with complex geometries. One area of considerable interest is how to weight the loss terms in the overall loss function. SimNet offers spatial loss weighting, where each weight parameter can be a function of the spatial inputs. In many cases we use the Signed Distance Function (SDF) for this weighting. Assuming $\mathcal{D}_x$ is the spatial subset of the input domain $\mathcal{D}$ with boundaries $\partial \mathcal{D}_x$, the SDF-based weight function is defined as

$$\lambda(\mathbf{x_s}) = \begin{cases} d(\mathbf{x_s}, \partial \mathcal{D}_x) & \mathbf{x_s} \in \mathcal{D}_x, \\ -d(\mathbf{x_s}, \partial \mathcal{D}_x) & \mathbf{x_s} \in \mathcal{D}_x^c. \end{cases} \tag{5}$$

Here, $\mathbf{x}_s$ is the spatial inputs, and $d(\mathbf{x_s}, \partial \mathcal{D}_x)$ represents the Euclidean distance between $\mathbf{x}_s$ and it's nearest neighbor on $\mathcal{D}_x$. If the geometry has sharp corners this often results in sharp gradients in the solution of the PDE. Weighting by the SDF tends to mitigate the deleterious effects of sharp local gradients, and often results in an improvement in convergence speed and accuracy. Both of the SimNet geometry modules allow for the SDF and its spatial derivatives to be computed. CSG uses SDF functions to implicitly define the geometry. To accelerate the computation of the SDF on tessellated meshes of complex geometries, we developed a custom library that leverages NVIDIA's OptiX for both inside/outside (sign) testing and distance computation. The sign test uses ray intersection and triangle normal alignment (via dot product). The distance testing is done by using the bounded volume hierarchy (BVH) interface provided by OptiX, which yields excellent performance and accuracy for distance computations.

3.2   PDE Module

The PDE module in SimNet consists of a variety of differential equations including the Navier-Stokes, diffusion, advection-diffusion, wave, and elasticity equations. To make this module extensible for the user to easily define their own PDEs, SimNet uses symbolic mathematics enabled by SymPy [13]. A novel contribution of SimNet is the adoption of the zero-equation turbulence model [12], and this is the first time a neural network solver is made capable of simulating flows with high Reynolds numbers, as shown in the next section. Moreover, for fluid flow simulation, we propose the use of integral continuity planes. For some problems involving channel flow, we found that, in addition to solving the Navier-Stokes equations in differential form, specifying the mass flow (for compressible flows) or volumetric flow rate (for incompressible flows) through some of the planes in the domain helps in satisfying the continuity equation better and faster and improving the accuracy further. Assuming there is no leakage of flow, we can guarantee that the flow exiting the system must be equal to the flow entering the system, and also equal to the flow passing from any plane parallel to the inlet plane throughout the channel.

3.3   Network Architectures

In addition to the standard fully connected networks, SimNet offers more advanced architectures, including the Fourier feature and Modified Fourier feature networks, and Sinusoidal Representation Networks (SiReNs) [14] to alleviate the spectral bias [15] in neural networks and improve convergence. The Fourier feature network in SimNet is a variation of the one proposed in [16] with trainable encoding, and takes the form in Eq. 7 with the following encoding

$$\phi_E = \left[\sin\left(2\pi \mathbf{f} \times \mathbf{x}\right); \cos\left(2\pi \mathbf{f} \times \mathbf{x}\right) \right]^T, \tag{6}$$

where $\mathbf{f} \in \mathbb{R}^{n_f \times d_0}$ is the trainable frequency matrix and n_f is the number of frequency sets. The modified Fourier feature network is SimNet's novel architecture, where two transformation layers are introduced to project the Fourier features to another learned feature space, and are then used to update the hidden layers through element-wise multiplications, similar to its standard fully connected counterpart in [8]. It is shown in the next section that this multiplicative interaction between the Fourier features and hidden layers can improve the training convergence and accuracy. The hidden layers in this architecture take the following form

$$\phi_i(\mathbf{x}_i) = \left(1 - \sigma\left(\mathbf{W}_i \mathbf{x}_i + \mathbf{b}_i\right)\right) \odot \sigma\left(\mathbf{W}_{T_1} \phi_E + \mathbf{b}_{T_1}\right) + \sigma\left(\mathbf{W}_i \mathbf{x}_i + \mathbf{b}_i\right) \odot \sigma\left(\mathbf{W}_{T_2} \phi_E + \mathbf{b}_{T_2}\right), \tag{7}$$

where $i > 1$ and $\{\mathbf{W}_{T_1}, \mathbf{b}_{T_1}\}, \{\mathbf{W}_{T_2}, \mathbf{b}_{T_2}\}$ are the parameters for the two transformation layers.

4 Use Cases

In this section, we present four use cases for SimNet to illustrate its capabilities. Although SimNet is capable of simulating transient flows using the continuous-time sampling approach [3], the first three use cases are time-independent. A more efficient and accurate approach based on the convolutional LSTMs for transient simulations as well as integration of two-equation turbulence models for turbulent simulations are under development. For the entire networks in this section, the architectures consist of 6 layers, each with 512 units. Swish [17] nonlinearities are used in the fully connected, Fourier feature, and modified Fourier feature networks (except for the Fourier layers). For the simulations presented in Sects. 4.2 to 4.4, the standard fully connected architecture is used. Adam optimizer with an initial learning rate of 10^{-4} and an exponential decay is used. We use Monte Carlo integration for computing the loss function in Eq. 4. Moreover, we use integral continuity planes for channel flows. For the simulations in use cases 4.1 to 4.3, we use the SDF for weighting the PDE residuals. It must be noted that use cases 4.1 to 4.3 are solved in the forward manner without using any training data. Please refer to the SimNet user guide for details of the problem setup.

4.1 Turbulent and Multi-physics Simulations

Using an FPGA heat sink example, we demonstrate the SimNet's capability in accurately solving multi-physics problems involving high Reynolds number flows. The heat sink geometry placed inside a channel is depicted in Figs. 4a, 4b. This particular geometry is challenging to simulate due to thin fin spacing that causes sharp gradients that are difficult to learn for a neural network solver. Using the zero-equation turbulence model, we solve a conjugate heat transfer problem with a flow at $Re = 13{,}239$. Generally, simulation of high-Re flows are particularly difficult due to the chaotic fluctuations of the flow field properties that are caused by instabilities in the shear layer. Due to the one-way coupling between the heat and incompressible flow equations, two separate neural networks are trained for flow (trained first) and the temperature (trained next) fields. This approach is useful for one-way coupled multi-physics problems to achieve significant speed-up.

We simulate this conjugate heat transfer problem with different architectures and also with symmetry boundary conditions. Loss curves are shown in Fig. 5. This figure also includes the flow convergence results for a Fourier feature model without SDF loss weighting and a standard fully connected model, showing that they fail to provide a reasonable convergence and highlighting the importance of SDF loss weighting and advanced architectures. The streamlines and temperature profile obtained from the modified Fourier feature model are shown in Fig. 4c. A comparison between the SimNet and OpenFoam results for flow and temperature fields is also presented in Fig. 6. Results for the pressure drop and peak temperature are presented in Table 1. The OpenFoam simulation was performed using a conjugate heat solver based on the SIMPLE algorithm and the differences between the commercial solver and OpenFoam peak temperatures are likely due to the differences in the solvers and the schemes used in these two simulations.

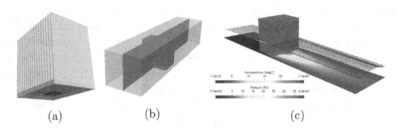

(a)   (b)   (c)

Fig. 4. FPGA heat sink example. (a) heat sink geometry; (b) Simulation domain (with symmetry plane); (c) SimNet results for streamlines and temperature.

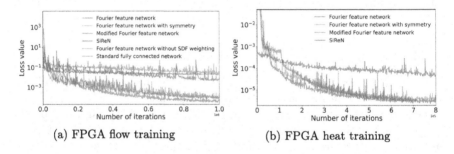

(a) FPGA flow training   (b) FPGA heat training

Fig. 5. Loss curves for FPGA training using different architectures.

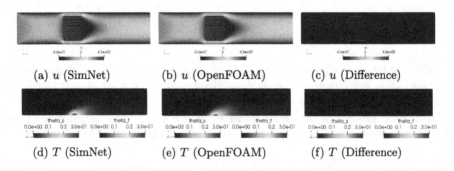

(a) u (SimNet)   (b) u (OpenFOAM)   (c) u (Difference)

(d) T (SimNet)   (e) T (OpenFOAM)   (f) T (Difference)

Fig. 6. A comparison between the SimNet (with modified Fourier feature network) and OpenFoam results for FPGA on a 2D slice of the domain.

Table 1. FPGA pressure drop and peak temperature from various models.

Case description	P_{drop} (Pa)	T_{peak} $(°C)$
SimNet: Fourier network (axis spectrum)	25.47	73.01
SimNet: Fourier network (partial spectrum) with symmetry	29.03	72.36
SimNet: Modified Fourier network	29.17	72.52
SimNet: SiReN	29.70	72.00
OpenFOAM solver	27.82	56.54
Commercial solver	24.04	72.44

4.2   Blood Flow in an Intracranial Aneurysm

We demonstrate the ability of SimNet to work with STL geometries from a CAD system. Using the SimNet's TG module, we simulate the flow inside a patient specific geometry of an aneurysm depicted in Fig. 7a. The SimNet results for the distribution of velocity magnitude and pressure developed inside the aneurysm are shown in Figs. 7c and 7d, respectively. Using the same geometry, the authors in [18] solve this as an inverse problem using concentration data from the spectral/hp-element solver Nektar. We solve this problem as a forward problem without any data. When solving the forward CFD problem with non-trivial geometries, one of the key challenges is getting the flow to develop correctly, especially inside the aneurysm sac. The streamline plot in Fig. 7b shows that SimNet successfully captures the flow field very accurately.

4.3   Design Optimization for Multi-physics Industrial Systems

SimNet can solve several, simultaneous design configurations in a multi-physics, design space exploration problem much more efficiently than traditional solvers. This is possible because unlike a traditional solver, a neural network trains with multiple design parameters in a single training run. Once the training is complete, several geometry or physical parameter combinations can be evaluated using inference as a post-processing step, without solving the forward problem again. Such throughput enables more efficient design optimization and design space exploration tasks for complex systems in science and engineering. Here, we train a conjugate heat transfer problem over the Nvidia's NVSwitch heat sink whose fin geometry is variable, as shown in Fig. 8 (nine geometry variables in total). Details on the problem setup and training can be found in SimNet user guide. Forward solution of parameterized, complex geometry with turbulent fluid flow between thinly spaced fins and no training data makes this problem extremely challenging for the neural networks. Following the training, we perform a design optimization to find out the most optimal fin configuration that minimizes the peak temperature while satisfying a maximum pressure drop constraint. The fluid and heat neural networks in this example consist of 12 variables,

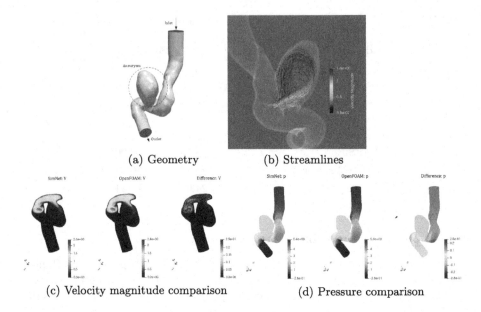

(a) Geometry   (b) Streamlines

(c) Velocity magnitude comparison   (d) Pressure comparison

Fig. 7. SimNet results for the aneurysm problem, and a comparison between the Sim-Net and OpenFOAM results for the velocity magnitude and pressure.

i.e. three spatial variables and nine geometry parameter variables. Using Sim-Net, we train these two parameterized neural networks, and then use the trained models to compute the pressure drops and peak temperatures corresponding to 4 million random geometry realizations. Figure 9 shows the streamlines and temperature profile for the optimal NVSwitch geometry.

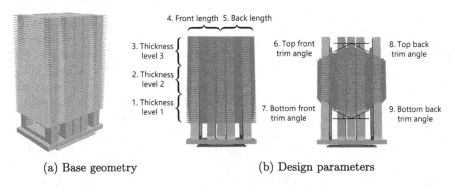

(a) Base geometry   (b) Design parameters

Fig. 8. NVSwitch base geometry and design parameters.

By parameterizing the geometry, SimNet accelerates this design optimization task by several orders of magnitude when compared to traditional solvers, which are limited to single geometry simulations. This also suggests that SimNet can

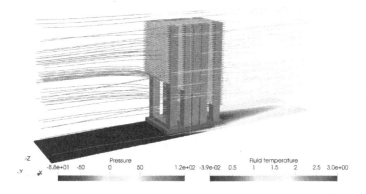

Fig. 9. SimNet results for the optimal NVSwitch geometry.

provide significant time savings when other design optimization methods, such as gradient-based design optimization, are used. The total compute time required by OpenFOAM, a commercial solver, and SimNet (including the training time) for this design optimization task is reported in Table 2. The OpenFOAM and commercial solver runs are run on 22 CPU processors, and the SimNet runs are on 8 V100 GPUs. To confirm the accuracy of the SimNet parameterized model, we take the NVSwitch base geometry and compare the SimNet results (obtained from the parameterized model) for pressure drop and peak temperature with the OpenFOAM and commercial solver results, reported in Table 3.

Table 2. Total compute time for the NVSwitch heat sink design optimization.

Solver	OpenFOAM	Commercial solver	SimNet
Compute time (x 1000 h)	405935	137494	3

Table 3. A comparison for the solver and SimNet results for NVSwitch pressure drop and peak temperature.

Property	OpenFOAM single run	Commercial solver single run	SimNet parameterized run
Pressure drop (Pa)	133.96	128.30	109.53
Peak temperature $(°C)$	41.55	43.57	39.33

4.4   Inverse Problems

Many applications in science and engineering involve inferring unknown system characteristics given measured data from sensors or imaging for certain dependent variables describing the behavior of the system. Such problems usually involve solving for the latent physics using the PDEs as well as the data. This is done in SimNet by combining the data with PDEs to decipher the underlying physics.

Here, we demonstrate the ability of SimNet to solve data assimilation and inverse problems on a transient flow past a 2D cylinder example. This example is adopted from [19]. Given the data consisting of the scattered concentration of a passive scalar in the flow domain at different times, the task is to infer the flow velocity and pressure fields as well as the entire concentration field of the passive scalar. In reality, the data is collected using measurements but for the purpose of this example, synthetic data generated by OpenFOAM is used. We construct a model with a hybrid data and physics-driven loss function. Specifically, we require the neural network prediction for the passive scalar concentration to fit to the measurements, and also satisfy the governing laws of the system that includes the transient Navier-Stokes and advection-diffusion equations. Here, the quantities of interest are also modeled as trainable variables, and are inferred by minimizing the hybrid loss function. A comparison between the SimNet results and the ground truth for a snapshot of the flow velocity, pressure, and passive scalar concentration fields is presented in Fig. 10.

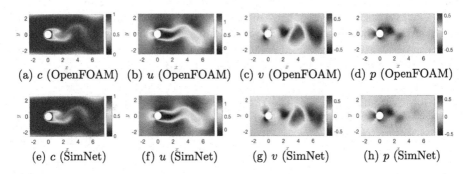

Fig. 10. A comparison between the SimNet and OpenFOAM results for a snapshot of the flow velocity, pressure, and passive scalar concentration fields. This example is adopted from [19].

5   Performance Upgrades and Multi-GPU Training

SimNet supports multi-GPU/multi-node scaling to enable larger batch sizes while time per iteration remains nearly constant, as shown in Fig. 11a. Therefore, the total time to convergence can be reduced by scaling the learning rate linearly with the number of GPUs, as suggested in [20]. Doing so without a warmup would cause the model to diverge since the initial learning rate can be very large. Figure 11b shows the Limerock results for large batch training acceleration on A100 using learning rate scaling. SimNet also supports TensorFloat-32 (TF32), a new math mode available on NVIDIA A100 GPUs. Based on our experiments on the FPGA problem, using TF32 provides up to 1.6x and 3.1x speed-up over FP32 on A100 and V100 GPUs, respectively. Moreover, SimNet supports kernel fusion using XLA that, based on our experiments, can accelerate a single training iteration in SimNet by up to 3.3x.

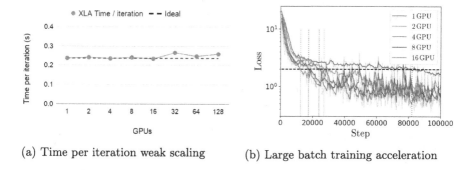

(a) Time per iteration weak scaling   (b) Large batch training acceleration

Fig. 11. SimNet's scaling performance results.

6 Conclusion

SimNet is an end-to-end AI-driven simulation framework with unique, state-of-art architectures that enables rapid training of forward, inverse, and data assimilation problems for real world geometries and multiple physics types with or without any training data. SDF is used for loss weighting, which is shown to significantly improve the convergence in cases where the geometry has sharp corners and results in sharp solution gradients. SimNet's TG module enables the import tessellated geometries from CAD programs. For channel flow problems, continuity is imposed globally and locally to further improve the convergence and accuracy. SimNet enables the neural network solvers to simulate high Reynolds number flows for industrial applications. To the authors knowledge, this is the first such application of neural network solvers for RANS simulation of turbulent flows.

SimNet is designed to be flexible so that users can leverage the functionality in the existing toolkit and focus on solving their problem well rather than re-creating the tools. There are various APIs that enable the user to implement their own equations to simulate the physics, their own geometry primitives or importing complex tessellated geometries, or a variety of domains/boundary conditions. The geometry parameterization in the CSG module allows the neural network to address the entire range of all given parameters in a single training, as opposed to the traditional simulations that run one at a time. The inference for any design configuration can then be completed in real time. This accelerates the simulation with neural network solvers by orders of magnitude.

Acknowledgments. We would like to thank Doris Pan, Anshuman Bhat, Rekha Mukund, Pat Brooks, Gunter Roth, Ingo Wald, Maziar Raissi, Jose del Aguila Ferrandis, and Sukirt Thakur for their assistance and feedback in SimNet development. We also acknowledge Peter Messemer, Mathias Hummel, Tim Biedert and Kees Van Kooten for integration with Omniverse.

References

1. Guo, X., Li, W., Iorio, F.: Convolutional neural networks for steady flow approximation. In: Proceedings of the 22nd ACM SIGKDD International Conference on Knowledge Discovery and Data Mining, pp. 481–490 (2016)
2. Hennigh, O.: Lat-Net: compressing lattice Boltzmann flow simulations using deep neural networks. arXiv preprint arXiv:1705.09036 (2017)
3. Raissi, M., Perdikaris, P., Karniadakis, G.E.: Physics-informed neural networks: a deep learning framework for solving forward and inverse problems involving nonlinear partial differential equations. J. Comput. Phys. **378**, 686–707 (2019)
4. Lagaris, I.E., Likas, A., Fotiadis, D.I.: Artificial neural networks for solving ordinary and partial differential equations. IEEE Trans. Neural Netw. **9**(5), 987–1000 (1998)
5. Sirignano, Justin, Spiliopoulos, Konstantinos: Dgm: a deep learning algorithm for solving partial differential equations. J. Comput. Phys. **375**, 1339–1364 (2018)
6. Kharazmi, E., Zhang, Z., Karniadakis, G.E.: hp-vpinns: variational physics-informed neural networks with domain decomposition. arXiv preprint arXiv:2003.05385 (2020)
7. Zhu, Y., Zabaras, N., Koutsourelakis, P.-S., Perdikaris, P.: Physics-constrained deep learning for high-dimensional surrogate modeling and uncertainty quantification without labeled data. J. Comput. Phys. **394**, 56–81 (2019)
8. Wang, S., Teng, Y., Perdikaris, P.: Understanding and mitigating gradient pathologies in physics-informed neural networks. arXiv preprint arXiv:2001.04536 (2020)
9. Lu, L., Meng, X., Mao, Z., Karniadakis, G.E.: DeepXDE: a deep learning library for solving differential equations. arXiv preprint arXiv:1907.04502 (2019)
10. Haghighat, E., Juanes, R.: Sciann: a keras wrapper for scientific computations and physics-informed deep learning using artificial neural networks. arXiv preprint arXiv:2005.08803 (2020)
11. Rackauckas, C., Nie, Q.: Differentialequations.jl — a performant and feature-rich ecosystem for solving differential equations in julia. J. Open Res. Softw. **5**(1) (2017). https://app.dimensions.ai
12. Wilcox, D.C., et al.: Turbulence Modeling for CFD. volume 2. DCW industries La Canada, CA (1998)
13. Meurer, A., et al. Sympy: symbolic computing in python. PeerJ Comput. Sci. **3**, e103 (2017)
14. Sitzmann, V., Martel, J., Bergman, A., Lindell, D., Wetzstein, G.: Implicit neural representations with periodic activation functions. arXiv preprint arXiv:2006.09661 (2020)
15. Rahaman, N., et al.: On the spectral bias of neural networks. In: International Conference on Machine Learning, pp. 5301–5310 (2019)
16. Tancik, M., et al.: Fourier features let networks learn high frequency functions in low dimensional domains. arXiv preprint arXiv:2006.10739 (2020)
17. Ramachandran, P., Zoph, B., Le, Q.V.: Searching for activation functions. arXiv preprint arXiv:1710.05941 (2017)

18. Raissi, M., Yazdani, A., Karniadakis, G.E.: Hidden fluid mechanics: learning velocity and pressure fields from flow visualizations. Science **367**(6481), 1026–1030 (2020)
19. Raissi, M., Wang, Z., Triantafyllou, M.S., Karniadakis, G.E.: Deep learning of vortex-induced vibrations. J. Fluid Mech. **861**, 119–137 (2019)
20. Goyal, P., et al.: Accurate, large minibatch SGD: training imagenet in 1 hour. arXiv preprint arXiv:1706.02677 (2017)

MeshFree Methods and Radial Basis Functions in Computational Sciences

Analysis of Vortex Induced Vibration of a Thermowell by High Fidelity FSI Numerical Analysis Based on RBF Structural Modes Embedding

Alessandro Felici, Antonio Martínez-Pascual, Corrado Groth, Leonardo Geronzi, Stefano Porziani, Ubaldo Cella, Carlo Brutti, and Marco Evangelos Biancolini[✉] [iD]

Department of Enterprise Engineering, University of Rome Tor Vergata, Rome, Italy
biancolini@ing.uniroma2.it

Abstract. The present paper addresses the numerical fluid-structure interaction (FSI) analysis of a thermowell immersed in a water flow. The study was carried out implementing a modal superposition approach into a computational fluid dynamics (CFD) solver. The core of the procedure consists in embedding the structural natural modes, computed by a finite element analysis (FEA), by means of a mesh morphing tool based on radial basis functions (RBF). In order to minimize the distortion during the morphing action and to obtain a high quality of the mesh, a set of corrective solutions, that allowed the achievement of a sliding morphing on the duct surface, was introduced. The obtained numerical results were compared with experimental data, providing a satisfying agreement and demonstrating that the modal approach, with an adequate mesh morphing setup, is able to tackle unsteady FSI problems with the accuracy needed for industrial applications.

Keywords: Morphing · Radial basis functions · Multi-physics · RBF · FSI · Fluid-structure interaction · Vortex-induced vibration · Modal superposition · Sliding morphing · Thermowell

1 Introduction

Today the need of developing multi-physics approaches to address complex design challenges is rising. The numerical methods adopted must involve coupled-field analyses that allow evaluating the combined effects of the multiple physical phenomena acting on a given system. One of the most interesting and with a wide range of application multi-physics phenomenon is the interaction between a fluid and a structure. This interaction can occur for several reasons: it can be the working principle of the system; it can be aimed to the lightweight design of the structure or it can be exploited to finely tune the design.

The fluid structure interaction (FSI) mechanism plays a fundamental role in a wide range of engineering fields, such as automotive, aerospace, marine, civil and biomedical. The numerical approaches adopted to couple the computational fluid dynamics (CFD)

© Springer Nature Switzerland AG 2021
M. Paszynski et al. (Eds.): ICCS 2021, LNCS 12746, pp. 465–478, 2021.
https://doi.org/10.1007/978-3-030-77977-1_37

and computational structural mechanics (CSM) codes can be classified according to two main categories: the monolithic approach and the partitioned approach. In the former the fluid dynamics and the structural dynamics models are solved simultaneously within a unified solver; in the latter they are solved separately, with two distinct solvers. Whatever type of approach is chosen, the deformation of the CFD mesh is needed in order to accommodate the shape changes of the structure. In the present work a mesh morphing algorithm based on radial basis functions (RBF) will be used to update the CFD mesh according to the deformed shape of the structure. The FSI approach here proposed allows to adapt the mesh to the shape of the deformable structure by a superposition of its natural modes during the progress of the CFD computation. This method proved in the past its efficiency and reliability in many studies for both steady [1, 2] and unsteady flows [3, 4]. Its main limit is that it cannot be directly employed in problems involving non-linearities of any kind, contact or pre-stressed components.

The underlying idea of the proposed workflow is that at each time-step the fluid forces over the structure surface, together with inertial loads, are computed as modal forces to determine the amplitude of each modal shape. Superimposing the modal shapes, the overall deformation of the structure is deducible at each instant and can be imposed by mesh morphing. The method is implemented to investigate an industrial problem: the vortex induced vibration of a thermowell immersed in a fluid flow.

Thermowells are cylindrical fittings used to protect temperature sensors (as for example thermometers or thermocouples) installed in industrial processes. In such setup the fluid transfers heat to the thermowell wall which, in turn, transfers heat to the sensor. The usage of a thermowell, other than protecting the sensor from the pressure and chemical effects of the process fluid, allows to easily replace the sensor without draining the vessel or the piping. Thermowells, however, are subjected to the risk of flow-induced vibrations generated by vortex shedding which might lead to bending fatigue failure. Hence, in modern applications involving high strength piping and elevated fluid velocity, the dynamics of the system have to be carefully evaluated to foresee ad-hoc countermeasure to limit this phenomenon, such as for example twisted square thermowells. A numerical method able to reliably reproduce the fluid and structural coupling is therefore needed to quickly evaluate different designs and reduce the time to market of new products.

In the present work, authors will first give an introduction of the proposed FSI approach, then its application to an industrial problem will be illustrated and finally the study results will be detailed and discussed.

2   Theoretical Background

2.1   Unsteady FSI Using Modal Superposition

The FSI approach used in this work is based on the modal theory [3, 5]: the structural deformation can be thought as a linear superimposition of the modal shapes of the body itself, so that by importing the modal shapes in the CFD solver with a mesh morphing tool, the fluid dynamic numerical configuration can be made implicitly aeroelastic.

For a generic n-degrees-of-freedom system (for example n masses virtually positioned on the nodes of the mesh for a FEM structural analysis), the second order differential system of equations of motion can be written in matrix form as (1) [6]:

$$[M]\ddot{y} + [C]\dot{y} + [K]y = Q \tag{1}$$

Where: $[M]$ is the mass matrix, $[C]$ is the damping matrix, $[K]$ is the stiffness matrix, Q is the external forces vector (that may vary in time), y is the generalized coordinates vector. Being the modal shapes linearly independent, a new reference system in which the equations can be uncoupled is introduced so that each contribution of a mode to the total structure deformation is isolated:

$$q = [v]^{-1}y \tag{2}$$

Where: q is the vector of modal coordinates and $[v]$ is the modal matrix, whose columns are the mass-normalized natural modes. By substituting this definition in Eq. (1), pre-multiplying by $[v]^T$ both terms and retaining the mass normalization of the modal shapes, the following equation is obtained:

$$[I]\ddot{q} + [v]^T[C][v]\dot{q} + [v]^T[K][v]q = [v]^T Q \tag{3}$$

The term on the right-hand side of the equation is referred to as modal force vector, computed integrating the projection of the nodal forces on the surfaces onto the relevant mode. The solution to this equation, describing the temporal evolution of each modal coordinate, can be deduced by recurring to the Duhamel's integral:

$$
\begin{aligned}
q(t) = {} & e^{-\varsigma\omega_n t}\left[q_0\cos(\omega_d t) + \tfrac{\dot{q}_0 + \varsigma\omega_n q_0}{\omega_d}\sin(\omega_d t)\right] \\
& + e^{-\varsigma\omega_n t}\left\{\tfrac{1}{m\omega_d}\int_0^t e^{-\tfrac{b(t-\tau)}{2m}}f(\tau)\sin[\omega_d(t-\tau)]d\tau\right\}
\end{aligned}
\tag{4}
$$

Where: ω_n is the natural circular frequency of the considered mode, ς is the damping factor, ω_d is the damped circular frequency, q_0 and $\dot{q}_0$ are the boundary conditions (respectively initial modal coordinates and initial modal velocities), m is the modal mass (unitary if the mass normalization is implemented).

2.2 RBF Mesh Morphing

RBF are mathematical functions able to interpolate, on a distance basis, the scalar information known at defined source points of a domain in which the functions are not zero valued. They can be defined in an n-dimensions space and are function of the Euclidean norm of the distance between two points in the space. The interpolation function is composed of a radial function ϕ and a polynomial term h, whose degree depends on the chosen basis. This polynomial term is added to assure uniqueness of the problem and polynomial precision, allowing to prescribe rigid body translations. If m is the number of source points, the interpolation function can be written as follows:

$$s(x) = \sum_{i=1}^{m}\gamma_i\phi\left(x - x_{k_i}\right) + h(x) \tag{5}$$

Where x_{k_i} is the position vector of the i-th source point.

A radial basis interpolation exists if the coefficients γ and the weights of the polynomial term allow to guarantee the exact function values at the source points and the polynomial term satisfies the orthogonality conditions. Mathematically:

$$s\left(x_{k_i}\right) = g_i, 1 \leq i \leq m \tag{6}$$

$$\sum_{i=1}^{m} \gamma_i P\left(x_{k_i}\right) = 0 \tag{7}$$

Where: g_i is the known value of the function at the i-th source point; (7) has to be written for all polynomials P with a degree less than or equal to that of polynomial h [7]. The minimal degree of polynomial h depends on the chosen RBF.

A unique interpolation solution exists if the basis function is a conditionally positive definite function [8]. If the basis functions are conditionally positive definite of order less than or equal to two [9], a linear polynomial h can be used:

$$h(x) = \beta_1 + \beta_2 x_1 + \beta_3 x_2 + \ldots + \beta_{n+1} x_n \tag{8}$$

The values for the coefficients of the RBF and the weights of the linear polynomial h can be obtained by solving the system:

$$\begin{bmatrix} A & P \\ P^T & 0 \end{bmatrix} \begin{pmatrix} \gamma \\ \beta \end{pmatrix} = \begin{pmatrix} g \\ 0 \end{pmatrix} \tag{9}$$

Where $[A]$ is the interpolation matrix obtained by calculating all the radial interactions among source points, with the radial distances between them:

$$A_{ij} = \phi\left(x_{k_i} - x_{k_j}\right), 1 \leq i \leq m, 1 \leq j \leq m \tag{10}$$

$[P]$ is the constraint matrix that results from balancing the polynomial contribution. If the space in which the RBF is defined is the physical one, it contains the coordinates of the source points in the space:

$$[P] = \begin{bmatrix} 1 & x_{k_1} & y_{k_1} & z_{k_1} \\ 1 & x_{k_2} & y_{k_2} & z_{k_2} \\ \vdots & \vdots & \vdots & \vdots \\ 1 & x_{k_m} & y_{k_m} & z_{k_m} \end{bmatrix} \tag{11}$$

It is clear that the RBF interpolation works for scalar fields. For the smoothing problem (in which, formally, a vector field is prescribed) each component of the displacement filed prescribed at the source points is interpolated, once the weights and the coefficients of the system have been obtained solving the system (9) for each component (i.e. the three directions in space x, y and z), as follows:

$$\begin{cases} s_x(x) = \sum_{i=1}^{m} \gamma_i^x \phi\left(||x - x_{k_i}||\right) + \beta_1^x + \beta_2^x x + \beta_3^x y + \beta_4^x z \\ s_y(x) = \sum_{i=1}^{m} \gamma_i^y \phi\left(||x - x_{k_i}||\right) + \beta_1^y + \beta_2^y x + \beta_3^y y + \beta_4^y z \\ s_z(x) = \sum_{i=1}^{m} \gamma_i^z \phi\left(||x - x_{k_i}||\right) + \beta_1^z + \beta_2^z x + \beta_3^z y + \beta_4^z z \end{cases} \tag{12}$$

The meshless nature of the morphing method appears clear, because the final configuration of the controlled nodes only depends on their original position. Therefore, grid points are moved regardless of element type or connection. This allows the implementation of the RBF to prescribe deformations to the surface mesh and for volume mesh smoothing. In literature RBF were successfully employed in a broad spectrum of applications, from optimizations based on zero order [10] or higher order methods [11–13], evolutionary optimization mimicking nature [14], load transfer [15], but also for ice accretion [16], fracture mechanics studies [17–19] and structural results post-processing [20–22].

The morphing approach allows to apply shape modifications directly to the numerical domain avoiding a remeshing procedure, saving time and preserving computational consistency. When thinking about the number of shape modifications that occur during an unsteady FSI analysis, the advantages of the mesh morphing approach are evident.

2.3 Modal FSI Implementation

Starting from the undeformed configuration, the flexible components of the system are modelled and studied by means of a structural modal analysis in order to extract a selected number of eigenvectors. The obtained modes are used to generate an RBF solution for each shape. In this step the far field conditions and the rigid surfaces need to be constrained, whereas the FEM results need to be mapped on the deformable surfaces of the CFD domain. The RBF solutions obtained constitute the modal base that, opportunely amplified, allows to represent the structural deformation under load generating an intrinsically aeroelastic domain. This process is known as "RBF structural modes embedding". To speed up the mesh morphing step, the deformations associated with each modal shape are stored in memory allowing a morphing action cost that is in the order of a single CFD iteration.

The proposed FSI modelling technique falls into the class of weak approaches because, for an unsteady analysis, loads are considered frozen during each time-step. The modal forces are computed on the prescribed surfaces (i.e. the deformable ones) by projecting the nodal forces (pressure and shear stresses) onto the modal shapes. The mesh is updated during the progress of the CFD computation every prescribed number of iterations, according to the computed modal coordinates. The mesh morphing tool used is RBF Morph^{TM} [23]. CFD and FEM solvers adopted are Ansys Fluent and Ansys Mechanical 2020 R1.

3 Experimental Investigation

The investigated industrial problem concerns the vortex induced vibration of a thermowell immersed in a fluid flow. The case study was experimentally measured and recorded by Emerson Electric Co. [24], the multinational corporation that owns the manufacturer of the studied thermowell. The aim of the experiment was to evaluate the flow induced vibrations of the traditional cylindrical thermowell design. The sensor, 470.219 mm in length, was equipped with an accelerometer in the tip and immersed in a water flow inside a 152.4 mm diameter pipe. The water velocity ranged from 0 m/s to 8.5 m/s. The accelerometer allowed the reconstruction of the evolution of the tip displacement. The gathered results are summarized in Fig. 1, in terms of the root mean square of the tip displacement as a function of the fluid velocity.

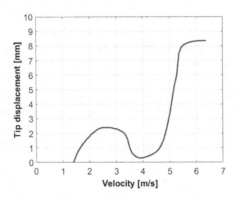

Fig. 1. Experimental results, RMS tip displacement vs fluid velocity

The presence of two lock-in regions is observed: an in-line vibration lock-in and a transverse vibration lock-in region. In the in-line vibration lock-in region the maximum root mean square tip displacement in the streamwise direction is 2.33 mm, registered with a 2.44 m/s fluid velocity. In the transverse vibration lock-in region the maximum root mean square tip displacement in the cross-flow direction is 8.3 mm, registered with a 6.4 m/s fluid velocity. The vibration is induced by organized vortices that shed in sheets along the axial length of the stem and involve the generation of alternating forces. If the shedding frequency approaches a natural frequency of the thermowell or its half (generating the transverse or the in-line vibrations respectively) a failure of the sensor might occur. The failure conditions were reached for the cylindrical thermowell at velocity larger than 6.4 m/s.

The aim of this work was to numerically capture the transverse vibration lock-in region of the cylindrical thermowell.

4   Numerical Analysis

4.1   Modal Analysis

Figure 2 reports the configuration of the numerical domain. It consists of a pipe with a diameter of 152.4 mm having a 50.8 mm aperture which houses a 470.219 mm long cylindrical thermowell whose diameter is 16.764 mm. The exposed length of the thermowell is about 143 mm.

The thermowell is made of a 304/304L dual rated steel with a density of 7750 kg/m^3, a Young's modulus of 200 GPa and a Poisson's ratio of 0.3 [25]. It was discretized by a uniform mesh made of 34456 20-noded hexahedrons for a total of 148675 nodes. The sensor was modelled as a cantilever beam. The first six modes were extracted from the FEM modal analysis and adopted to populate the modal base adopted for the FSI analysis. The shapes of the six modes are reported in Fig. 3.

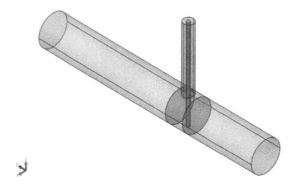

Fig. 2. CAD model of the analyzed system

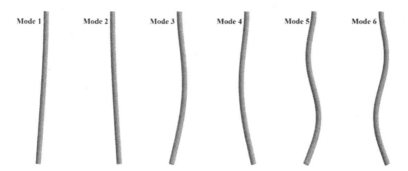

Fig. 3. First six modal shapes

4.2 RBF Solutions Setup

The shapes of the modes were extracted in terms of displacements of the mesh nodes belonging to the surface of the sensor normalized with respect to the mass. To generate the RBF solution for each natural mode a so-called *two-step* technique was employed [26]. This procedure provides a smoother solution and better quality of the morphed mesh. In the first step an RBF solution is generated applying the nodal displacement corresponding to the selected modal shape to the surface mesh of the sensor only, so that the surfaces are exactly morphed in the desired modified configuration, i.e. the one imposed by modal deformation. In the second step the first RBF solution is imposed as a motion law to the thermowell surface together with additional morphing set-up and the surrounding volume domain is then morphed accordingly.

The proximity of the tip of the thermowell to the boundary wall of the pipe caused a challenging problem. In fact the large displacements that the thermowell is expected to experience due to the vortex induced vibrations, combined to the requirement of maintaining a cylindrical shape of the pipe wall, would involve a significant distortion of the mesh if the duct wall nodes were imposed as fixed. The effect of such setup is evident comparing the starting (undeformed) mesh (Fig. 4 on the left) with the mesh resulting from a lateral displacement of the sensor in the region of the thermowell tip (Fig. 4 on the right).

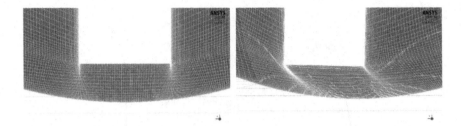

Fig. 4. Starting (left) and deformed (right) mesh around the tip

To avoid the high mesh distortion, the surface nodes contained in a *shadow area* (the portion of the duct surface defined by the projection of the tip of the thermowell) [27] had to follow the tip of the sensor during the morphing action. This task was accomplished assigning to the shadow area an appropriate rotation around the pipe axis and a translation in the direction of the axis itself (in two additional RBF solutions), in order to always keep it under the tip. The problem related to such corrective solutions is the absence of source points on the duct surface outside the shadow area that caused the loss of direct control on the morphing of the surface itself. This lack of control causes the distortion of the cylindrical duct surface visible in Fig. 5 (obtained with a 15° rotation of the shadow area, corresponding to the maximum expected shadow rotation during the unsteady FSI calculation).

Fig. 5. Cylinder distortion

To solve this issue a third RBF corrective solution derived from an STL-target solution was implemented. An STL-target motion type allowing to project the selected nodes onto a target surface (and therefore recovering the cylindricity of the duct) was assigned to the nodes belonging to the distorted portion of the cylinder (after the 15° rotation of the shadow). By tracking the position of the affected nodes in the three available meshes (the undeformed one, the distorted one after the maximum shadow rotation and the recovered one after the STL-target), it was possible to build a new RBF solution in which the source points were the nodes extracted from the starting duct surface mesh and their displacement was calculated as the difference between their corresponding position in the recovered mesh and in the distorted mesh.

By doing so it was possible to setup a recovery solution defined starting from the undeformed mesh and with an associated displacement able to recover the portion of the displacement imposed by the rotation of the shadow area that caused the distortion of the cylindrical surface. Therefore, if the rotation of the shadow area is applied with an amplification factor lower than the maximum one (i.e. 15°), the cylindricity of the duct can be recovered over-imposing this STL-target-derived solution with a proportional amplification factor.

Figure 6 reports the volume mesh around the thermowell tip obtained after morphing applying the described correction procedure. Figure 6 also shows the morphed surface mesh in the shadow region. The improvement in morphing quality, the correct positioning of the shadow area and the preservation of the cylindricity of the duct can be noticed.

Fig. 6. Deformed mesh around the tip (left) and deformed surface mesh (right), with the corrective solutions

All the exposed RBF solutions were built using the bi-harmonic function, that allowed to achieve a high-quality morphing able to guarantee a low mesh distortion and the preservation of the boundary layers [28]:

$$\phi(r) = r \tag{13}$$

4.3 CFD and FSI Setups

The fluid dynamic domain was discretized by a structured and multiblock mesh composed of 3.16 M hexahedrons. In order to solve the wall boundary layer up to the wall the thickness of the first layer of cells has been set to obtain a nondimensional wall distance y^+ lower than one. The adopted turbulence model was the SST k-ω. The velocity-inlet boundary condition was set to the inlet imposing a flow velocity equal to 6.4 m/s. A pressure-outlet condition was set at the outlet. Unsteady incompressible RANS calculation was run with a time-step set to 10^{-4} s. The structural damping ratio was set to 0.041, using the guidance found in literature [29]. The mesh is updated every time step computing the modal coordinates and the amplification factors of the corrective solutions.

4.4 Damping Ratio

The value of the damping ratio of the system was not available for the experimental reference. As it is expected in the 0.01–0.07 range [29], a parametric study of its sensitivity with respect to the computed amplitude and the computed frequency has been carried out.

Table 1 reports the results of the sensitivity study. It can be clearly stated that a 0.041 damping ratio, inside the expected range, is able to reproduce the experimental results with sufficient accuracy. For this reason, such value was chosen.

Table 1. Results of the parametric study

Damping ratio	Maximum RMS transverse tip displacement at dynamic steady state [mm]	Relative error [%]
0.01	Not reached	–
0.02	Not reached	–
0.05	6.45	22.3
0.04	8.48	−2.17
0.041	8.304	−0.048

4.5 FSI Analysis Results

The simulation ran on a HPC dense node equipped with 256 GB of RAM and four Intel® Xeon® Gold 6152 CPUs, each of them featuring 22 cores @ 2.1 GHz. Out of the overall 88 cores, 30 were used to run the simulation until a dynamic constant periodic state was reached, after about 5 days of computation. Extrapolating the results obtained in previous studies [30, 31], the time needed to face the simulation with the same time-step size and simulated flow time by means of a full two-way coupling approach would have been approximately 60 days. The time needed to face the simulation of the rigid case with the same time-step size and simulated flow time is about 5 days.

In Figs. 7 and 8 the contours of the velocity magnitude on a plane perpendicular to the thermowell axis are displayed at two different flow-times corresponding to the maximum transverse displacements, both in the positive and in the negative direction.

In Fig. 9 the temporal evolution of the side force on the thermowell is displayed; Fig. 9 also depicts the temporal evolution of the transverse tip displacement. It can be observed that the maximum RMS transverse tip displacement is 8.304 mm, in good agreement with the available experimental data. The distributions of power spectral density of the two signals (the temporal evolution of the side force and of the transverse tip displacement) as a function of the frequency are shown in Fig. 10. For both signals a dominant frequency of 48.8 Hz was observed, confirming the correct capture of the lock-in condition.

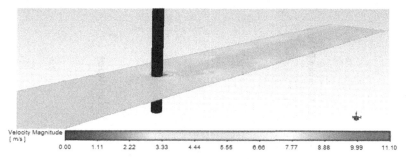

Fig. 7. Velocity magnitude contours at $t = 0.8425$ s

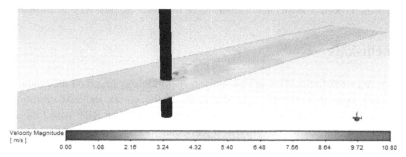

Fig. 8. Velocity magnitude contours at $t = 0.8525$ s

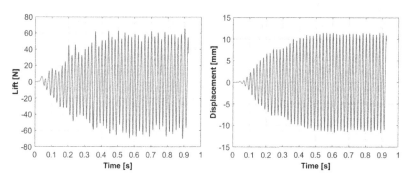

Fig. 9. Temporal evolution of the lift on the thermowell (left) and of the transverse tip displacement (right)

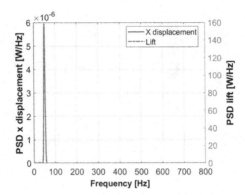

Fig. 10. Power spectral density distributions of the lift and of the transverse tip displacement

5   Conclusions

The presented work focused on an FSI analysis methodology based on the modal super-position approach. It was applied to the study of the vortex induced vibration of a thermowell. The problem of mesh adaptation was faced by an RBF based mesh morphing technique that allowed to provide a particular fast and robust configuration. The method consists in the extraction of a set of modal shapes, by a structural modal analysis, to be used for the implementation of a parametric mesh to be amplified according to modal coordinates computed during the progress of the CFD computation. The numerical configuration obtained is then an intrinsically aeroelastic fluid dynamic domain.

The studied configuration represents a particular challenging problem for the mesh morphing tool. The proximity of fixed and moving boundaries, in fact, involves strong distortion of the mesh which significantly limit the tolerable displacement. The adopted morphing software allowed to setup a particular efficient set of corrective solutions that made possible to manage very large displacement if compared to the dimension involved. The results of the implemented unsteady FSI analysis were compared to experimental data providing good agreement with measurements.

Further developments of this work might include gathering more comprehensive experimental data concerning the frequency of the thermowell vibration and the damping ratio of the system. The analysis could also be extended to other fluid velocity to capture the lock-in region of the in-line vibration and the lock-off regions. Other mesh setups, turbulence models and transition criteria could be investigated and compared to the results obtained with the current configuration.

References

1. Biancolini, M.E., Cella, U., Groth, C., Genta, M.: Static aeroelastic analysis of an aircraft wind-tunnel model by means of modal RBF mesh updating. J. Aerosp. Eng. **29**(6) (2016). https://doi.org/10.1061/(ASCE)AS.1943-5525.0000627

2. Biancolini, M.E., Cella, U., Groth, C., Chiappa, A., Giorgetti, F., Nicolosi, F.: Progresses in fluid-structure interaction and structural optimization numerical tools within the EU CS RIBES project. Comput. Meth. Appl. Sci. **49**, 529–544 (2019). https://doi.org/10.1007/978-3-319-89890-2_34

3. Di Domenico, N., Groth, C., Wade, A., Berg, T., Biancolini, M.E.: Fluid structure interaction analysis: vortex shedding induced vibrations. Procedia Struct. Integr. **8**, 422–432 (2018). https://doi.org/10.1016/j.prostr.2017.12.042

4. Groth, C., Cella, U., Costa, E., Biancolini, M.E.: Fast high fidelity CFD/CSM fluid structure interaction using RBF mesh morphing and modal superposition method. Airc. Eng. Aerosp. Technol. **91**(6), 893–904 (2019). https://doi.org/10.1108/AEAT-09-2018-0246

5. RBF Morph: RBF Morph User's Guide (2020)

6. Meirovitch, L.: Fundamentals of Vibrations. McGraw-Hill, McGraw-Hill Higher Education (2001)

7. Beckert, A., Wendland, H.: Multivariate interpolation for fluid-structure-interaction problems using radial basis functions. Aerosp. Sci. Technol. **5**(2), 125–134 (2001)

8. van Zuijlen, A.H., de Boer, A., Bijl, H.: Higher-order time integration through smooth mesh deformation for 3D fluid–structure interaction simulations. J. Comput. Phys. **224**(1), 414–430 (2007)

9. Jin, R., Chen, W., Simpson, T.W.: Comparative studies of metamodelling techniques under multiple modelling criteria. Struct. Multi. Optim. **23**(1), 1–13 (2001)

10. Biancolini, M.E., Costa, E., Cella, U., Groth, C., Veble, G., Andrejašič, M.: Glider fuselage-wing junction optimization using CFD and RBF mesh morphing. Aircr. Eng. Aerosp. Technol. **88**(6), 740–752 (2016). https://doi.org/10.1108/AEAT-12-2014-0211

11. Papoutsis-Kiachagias, E.M. et al.: Combining an RBF-based morpher with continuous adjoint for low-speed aeronautical optimization applications. In: ECCOMAS Congress 2016 - Proceedings of the 7th European Congress on Computational Methods in Applied Sciences and Engineering, vol. 3 (2016)

12. Groth, C., Chiappa, A., Biancolini, M.E.: Shape optimization using structural adjoint and RBF mesh morphing. Procedia Struct. Integr. **8**, 379–389 (2018). https://doi.org/10.1016/j.prostr.2017.12.038

13. Papoutsis-Kiachagias, E.M., Porziani, S., Groth, C., Biancolini, M.E., Costa, E., Giannakoglou, K.C.: aerodynamic optimization of car shapes using the continuous adjoint method and an RBF morpher. Comput. Meth. Appl. Sci. **48**, 173–187 (2019). https://doi.org/10.1007/978-3-319-89988-6_11

14. Porziani, S., Groth, C., Waldman, W., Biancolini, M.E.: Automatic shape optimisation of structural parts driven by BGM and RBF mesh morphing. Int. J. Mech. Sci. **189**, (2021)

15. Biancolini, M.E., Chiappa, A., Giorgetti, F., Groth, C., Cella, U., Salvini, P.: A balanced load mapping method based on radial basis functions and fuzzy sets. Int. J. Numer. Meth. Eng. **115**(12), 1411–1429 (2018). https://doi.org/10.1002/nme.5850

16. Groth, C., Costa, E., Biancolini, M.E.: RBF-based mesh morphing approach to perform icing simulations in the aviation sector. Aircr. Eng. Aerosp. Technol. **91**(4), 620–633 (2019). https://doi.org/10.1108/AEAT-07-2018-0178

17. Giorgetti, F., et al.: Crack propagation analysis of near-surface defects with radial basis functions mesh morphing. Procedia Struct. Integr. **12**, 471–478 (2018). https://doi.org/10.1016/j.prostr.2018.11.071

18. Pompa, E., et al.: Crack propagation analysis of ITER Vacuum Vessel port stub with Radial Basis Functions mesh morphing. Fusion Eng. Des. **157**, (2020)

19. Groth, C., et al.: High fidelity numerical fracture mechanics assisted by RBF mesh morphing. Procedia Struct. Integr. **25**, 136–148 (2020)

20. Chiappa, A., Groth, C., Biancolini, M.E.: Improvement of 2D finite element analysis stress results by radial basis functions and balance equations. Int. J. Mech. **13**, 90–99 (2019)

21. Chiappa, A., Groth, C., Brutti, C., Salvini, P., Biancolini, M.E.: Post-processing of 2D FEM Q1 models for fracture mechanics by radial basis functions and balance equations. Int. J. Mech. **13**, 104–113 (2019)
22. Chiappa, A., Groth, C., Reali, A., Biancolini, M.E.: A stress recovery procedure for laminated composite plates based on strong-form equilibrium enforced via the RBF Kansa method. Compos. Struct. 112292 (2020)
23. Biancolini, M.E.: Mesh morphing and smoothing by means of Radial Basis Functions (RBF): a practical example using fluent and RBF morph. Handbook of Research on Computational Science and Engineering: Theory and Practice, IGI Global, Hershey, PA (2012)
24. Emerson (2017). https://www.emerson.com/en-us/asset-detail/rosemount-twisted-square-a-new-twist-on-thermowell-design-1800740
25. Boyer, H.E., Gall, T.L. (eds.): Metals Handbook. American Society for Metals, Materials Park, OH (1985)
26. RBF Morph: RBF Morph – Modelling Guidelines and Best Practices Guide (2020)
27. Costa, E., Biancolini, M. E., Groth, C., Caridi, D., Lavedrine, J., Dupain, G.: Unsteady FSI analysis of a square array of tubes in water crossflow. Flexible Engineering Toward Green Aircraft, Springer, New York (2020) https://doi.org/10.1007/978-3-030-36514-1_8
28. Biancolini, M.E., Chiappa, A., Cella, U., Costa, E., Groth, C., Porziani, S.: Radial basis functions mesh morphing - a comparison between the Bi-harmonic Spline and the Wendland C2 radial function. In: International Conference on Computational Science, pp. 294–308 (2020). https://doi.org/10.1007/978-3-030-50433-5_23
29. Adams, V., Askenazi, A.: Building Better Products with Finite Element Analysis. OnWord Press, Santa Fe, N. M. (1999)
30. Geronzi, L., et al.: Advanced Radial Basis Functions mesh morphing for high fidelity Fluid-Structure Interaction with known movement of the walls: simulation of an aortic valve. In: International Conference on Computational Science, pp. 280–293 (2020). https://doi.org/10.1007/978-3-030-50433-5_22
31. Geronzi, L., et al.: High fidelity fluid-structure interaction by radial basis functions mesh adaption of moving walls: a workflow applied to an aortic valve. J. Comput. Sci. **51**, (2021). https://doi.org/10.1016/j.jocs.2021.101327

Automatic Optimization Method Based on Mesh Morphing Surface Sculpting Driven by Biological Growth Method: An Application to the Coiled Spring Section Shape

Stefano Porziani(✉), Francesco De Crescenzo, Emanuele Lombardi,
Christian Iandiorio, Pietro Salvini, and Marco Evangelos Biancolini

University of Rome "Tor Vergata", Via del Politecnico 1, 00133 Rome, Italy
porziani@ing.uniroma2.it

Abstract. The increasing importance of optimization in manufacturing processes led to the improvement of well established optimization techniques and to the development of new and innovative approaches. Among these, an approach that exploits surface stresses distribution to obtain an optimized configuration is the Biological Growth Method (BGM). Coupling this method with surface sculpting based on Radial Basis Functions (RBF) mesh morphing had proven to be efficient and effective in optimizing specific mechanical components. In this work, the automatic, meshless and constrained parameter-less optimization approach is applied to a classical mechanical component and then compared with a parameter-based shape optimisation result.

Keywords: Parameter-less optimization · Mesh morphing ·
Automatic surface sculpting · Coiled spring

1 Introduction

Optimization of manufactured mechanical components is an important phase of every production process. The optimal configuration achievement is a process which requires a lot of design efforts and that can be lowered adopting numerical techniques. Computer Aided Design (CAD), can support designers in every phase of the product manufacturing. Finite Element Method (FEM) gave an important speed up to design tasks, allowing designer to numerically test performances of different configurations before realising a test prototype. These activities, however, can be very time consuming: mesh morphing [3, 8, 22] had been introduced in the design process to obtain shape variation without the need to generate a modified geometry. Thanks to the high reliability of the Radial Basis Functions (RBF) based mesh morphing, this mesh-less technique

© Springer Nature Switzerland AG 2021
M. Paszynski et al. (Eds.): ICCS 2021, LNCS 12746, pp. 479–491, 2021.
https://doi.org/10.1007/978-3-030-77977-1_38

had been successfully adopted into various engineering workflows. Among the various engineering applications that had benefits in adopting mesh morphing it is possible to report Fluid Structure Interaction (FSI) [5,12,16], or crack front propagation prediction [9], as reported in [10] and [6].

Mesh morphing acts directly on the numerical model, by modifying the calculation grid nodes position without considering the underlying geometry used to generate the grid. This shape modification can also be applied using numerical results obtained with the same numerical model to be optimized, exploiting, for example, adjoint data [11,18] or using the Biological Growth Method (BGM) [20]. BGM mimics the way natural tissues, such as tree trunks and bones, evolve to mitigate stress peaks on their surfaces. The BGM has been successfully employed in mechanical component shape optimization [20], and had proven its reliability in optimization also compared with parameter-based optimization methods [19] and with other parameter-less methods [20].

In the present work, the optimization procedure [21] based on BGM and using RBF mesh morphing to automatically sculpt a mechanical component surfaces is first presented and then used to optimize a cross section of the coiled spring, comparing results with the ones achieved adopting the classical circular cross section and the ones computed with a parameter-based optimisation.

The tools adopted for the generation, optimization and analysis of the numerical model are included in the ANSYS Workbench Finite Element Analysis (FEA) framework [1]. The RBF Morph ACT extension [2] is used to apply RBF based mesh morphing driven by the BGM algorithm.

2   Recall on the Theoretical Background

In this section an overview on parameter-less optimization methodology is given, describing first the RBF based mesh morphing procedure (Sect. 2.1), then the BGM used to drive the morphing action (Sect. 2.2), and then concluding with the description of the coupling of these two techniques to obtain a surface sculpting optimization procedure (Sect. 2.3).

2.1   RBF Based Mesh Morphing

RBF was at first employed as a mathematical tool to interpolate multidimensional data [7]: this set of scalar function allow to interpolate data in every point of the definition space using known values at specific points, also called source points. A generic interpolation function can be written as:

$$s(\boldsymbol{x}) = \sum_{i=1}^{N} \gamma_i \varphi \left(\| \boldsymbol{x} - \boldsymbol{x}_{k_i} \| \right) + h(\boldsymbol{x}) \tag{1}$$

In Eq. (1) $\boldsymbol{x}_{k_i}$ are the source points defined in the space $\mathbb{R}^n$ and $\boldsymbol{x}$ are the points at which the function is evaluated, called also target points. φ is the radial function, which is a scalar function of the Euclidean distance between each source point and the target point considered; most used radial functions are reported in Table 1 , in which $r = (\|\boldsymbol{x} - \boldsymbol{x}_{k_i}\|)$. γ_i are the weights of the radial basis which are to be evaluated solving a linear system of equations, whose order is equal to the number of source points introduced. The polynomial part h is added to guarantee the existence and the uniqueness of the solution. In mesh morphing applications, a linear polynomial can be used:

$$h(\boldsymbol{x}) = \beta_1 + \beta_2 x + \beta_3 y + \beta_4 z \qquad (2)$$

in which β_i coefficients are evaluated together with γ_i weights solving the RBF system (see for reference [21]). Once solved, the RBF coefficients and polynomial weights are used to interpolate each imposed displacement component as an independent scalar field:

Table 1. Most common radial functions.

RBF type	Equation
Spline type (Rn)	$r^n, \ n \ odd$
Thin plate spline	$r^n log(r), \ n \ even$
Multiquadric (MQ)	$\sqrt{1 + r^2}$
Inverse multiquadric (IMQ)	$\frac{1}{\sqrt{1+r^2}}$
Inverse quadric (IQ)	$\frac{1}{1+r^2}$
Gaussian (GS)	e^{-r^2}

$$\begin{cases} s_x(\boldsymbol{x}) = \displaystyle\sum_{i=0}^{N} \gamma_i^x \varphi\left(\|\boldsymbol{x} - \boldsymbol{x}_i\|\right) + \beta_1^x + \beta_2^x x + \beta_3^x y + \beta_4^x z \\[2em] s_y(\boldsymbol{x}) = \displaystyle\sum_{i=0}^{N} \gamma_i^y \varphi(\|\boldsymbol{x} - \boldsymbol{x}_i\|) + \beta_1^y + \beta_2^y x + \beta_3^y y + \beta_4^y z \qquad (3) \\[2em] s_z(\boldsymbol{x}) = \displaystyle\sum_{i=0}^{N} \gamma_i^z \varphi(\|\boldsymbol{x} - \boldsymbol{x}_i\|) + \beta_1^z + \beta_2^z x + \beta_3^z y + \beta_4^z z \end{cases}$$

In mesh morphing, source points are the mesh nodes on which the displacement is imposed, whilst the target nodes are the whole set of nodes that have to be morphed in order to obtain the new numerical model shape.

2.2   Biological Growth Method

BGM allows to perform optimization using as driving quantity surface stress of the considered component. It moves from biological tissues observation: they evolve at surfaces by adding layers when an activation stress is reached.

In [13] and in [17] an extension of this natural mechanism is introduced: as material is added to surfaces where high stresses are present, so material can be removed if acting stresses are low. [13] demonstrated that using photo-elastic techniques and BGM, stresses level can be modified so that uniform stress acts on the boundary of a stress raiser. In [17] a bi-dimensional study that can reproduce the natural evolution of biological structures is illustrated and suggested to be used in optimization workflows: the authors computed the volumetric growth ($\dot{\varepsilon}$) according to the von Mises stress (σ_{Mises}) and a threshold stress (σ_{ref}). The latter one was chosen according to the allowable stress for the specific design:

$$\dot{\varepsilon} = k \left(\sigma_{Mises} - \sigma_{ref} \right) \tag{4}$$

Waldman and Heller [24] proposed a more complex model for layer growth, which has been successfully employed in shape optimization of holes in airframe structures. The proposed equation to evaluate the nodal displacement is reported in Eq. (5):

$$d_i^j = \left(\frac{\sigma_i^j - \sigma_i^{th}}{\sigma_i^{th}} \right) \cdot s \cdot c \, , \, \sigma_i^{th} = max(\sigma_i^j) \text{ if } \sigma_i^j > 0 \text{ or } \sigma_i^{th} = min(\sigma_i^j) \text{ if } \sigma_i^j < 0 \tag{5}$$

The model by Waldman and Heller moves the i-th boundary node of the j-th region by a distance d_i^j, computed using (5), where σ_i^j is the normal stress in the tangential direction, σ_i^{th} is the stress threshold; c is and arbitrary characteristic length and s is a scaling factor.

In the present work another formulation for BGM is used and implemented in the framework of ANSYS Mechanical, whose functionalities were enhanced by the RBF Morph ACT Extension. RBF Morph BGM approach implementation has been presented in [4]. To each target node (i.e. the set of nodes to be moved in order to perform optimization) a displacement (S_{node}) is imposed along the surface normal direction (inward or outward); the displacement value is calculated using Eq. (6), in which σ_{node} is the stress value for each node, σ_{th} is a threshold value for stress chosen by user, σ_{max} and σ_{min} are the maximum and minimum value for stress evaluated in the current set of source nodes. d is the maximum offset between the nodes on which the maximum and the minimum stress are evaluated. This parameter is defined by the user to control the nodes displacement whilst limiting the possible distortion of the mesh:

$$S_{node} = \frac{\sigma_{node} - \sigma_{th}}{\sigma_{max} - \sigma_{min}} \cdot d \tag{6}$$

Equation (6) allows to impose a displacement for nodes on the surface to be optimized that can be either inward, in case the stress for the current target node is lower than the stress chosen by user as threshold, or outward, in case stress on target node is higher than threshold one.

2.3   Parameter-Less Based Optimisation

The RBF based mesh morphing technique described in Sect. 2.1 and the BGM described in Sect. 2.2 can be coupled so that an automatic optimization approach can be defined, according to the following steps:

1. from the baseline geometry CAD description a finite element model is generated, by discretizing the geometry and setting up the numerical model by adding load and constraints; the FEM solution is then evaluated;
2. BGM routines retrieve nodal stress for target nodes on the model surface to be sculpted, user set both σ_{th} and d parameters of Eq. (6) and S_{node} displacement value for each target node is evaluated;
3. the RBF problem is set up by using the BGM evaluated displacements as values to be interpolated (i.e. surface nodes are used as source nodes). User can complete or improve RBF problem set up by adding more source points (e.g. imposing zero displacement value for those nodes of the model to be maintained fixed);
4. mesh morphing is applied to the FEM model and numerical solution is re-evaluated;
5. new evaluated stress on target nodes are analyzed: if further optimization can be performed the procedure can be repeated starting from point 2; if no additional optimzation steps can be performed, an optimized configuration is reached.

In the proposed workflow, user has to define only two parameters: the stress threshold σ_{th} and the maximum displacement d. The first can be described as the stress value on which the algorithm will try to uniform stress levels on the target nodes; the second parameter is the amount of offset between source nodes in each optimization step: the smaller the value the higher will be the number of steps needed to optimize stress levels and the lower will be the risk of generating distorted mesh which cannot be analyzed by FEM solver.

2.4   Parameter-Based Shape Optimisation

The parameter-based optimization can be performed with RBF mesh morphing by prescribing actions to specific groups of nodes (scaling, translation, surface offset, ...) so that the shape of surfaces and of the volume mesh is updated accordingly. The entity of such actions (scaling factors, component of translation, amount of offset ...) are then combined and controlled so that a certain number of new shapes is generated by Design of Experiment (DoE) and optimal performances are then computed on the response surface [19].

3   Coiled Springs Background

Helical springs are key components of many mechanical systems and have been long studied for decades. Stress distribution on helical springs is not uniform and both academics and industry researchers are focused on the optimization of wire cross-section by means of stress equalization along cross-section boundary. Thus, many shapes have been proposed to reduce mass or to extend safety and fatigue life of the component. As a matter of fact, most of the engine valve springs have non circular, "ovate" sections. Such optimal shape is obtained as a result of *ad-hoc* numerical optimization algorithms where the stresses are computed using numerical methods like finite or boundary elements. Examples of shape optimisation are given in [14, 15].

Optimum spring design would require to meet specifications in terms of stiffness, maximum load, design stress and some geometrical constraints, like solid height and outer and inner radius. As a first step, it is investigated the possibility to use BGM to equalize stresses at cross-section boundaries of a baseline coil, with two different constraints:

1. outer radius is fixed, inside surface is sculpted
2. inner radius is fixed, outside surface is sculpted

The optimized geometries are then compared to equivalent circular cross-section of same stiffness and outer/inner radius. A further comparison is made with a circular cross-section with same stiffness and swept volume, i.e. with same amount of material (and thus weight).

3.1 Equivalent Circular Section

Since the coil is flat and the spring index is moderately large (>6), it is possible to apply basic spring design formulas, as found in [23]. The stiffness of a single coil is:

$$K = \frac{Gd^4}{8D^3} \tag{7}$$

where d is the cross-section diameter, D is mean coil diameter and G is the shear modulus. Due to curvature effect and direct shear, the maximum tangential stress occurs at the inner radius:

$$\tau_{in} = \frac{8PD}{\pi d^3} \left(\frac{4c - 1}{4c - 4} + \frac{0.615}{c} \right) \tag{8}$$

on the contrary, the minimum stress along the boundary occurs at the outer radius:

$$\tau_{out} = \frac{8PD}{\pi d^3} \left(\frac{4c + 1}{4c + 4} - \frac{0.615}{c} \right) \tag{9}$$

The design must satisfy the prescribed stiffness and, depending on the case, outer or inner radius.

Constraint on Outer Diameter. For this scenario the spring design must satisfy the following conditions:

$$D + d = D_e \quad \text{and} \quad \frac{Gd^4}{8D^3} = K \tag{10}$$

Combining the constraint equations together it is found that the cross-section diameter must satisfy:

$$Gd^4 - 8(D_e - d)^3 K^* = 0 \tag{11}$$

Constraint on Internal Diameter. When the internal diameter is kept constant the geometric constraint writes:

$$D - d = D_i \tag{12}$$

leading to:

$$Gd^4 - 8(D_i + d)^3 K^* = 0 \tag{13}$$

Solving equations gives the diameter of the equivalent wire for prescribed stiffness and geometric constraint.

Constraint on Spring Volume. For a given volume and stiffness coil and wire diameters are given as:

$$D = \left(\frac{2GV^2}{\pi^4 K}\right)^{\frac{1}{5}} \tag{14}$$

$$d = \sqrt{\frac{4V}{\pi^2 D}} \tag{15}$$

3.2   Numerical Model of the Coiled Spring

The numerical model represented half coiled spring: it has been modelled in ANSYS Workbench Framework, modelling a 4 mm diameter wire with 20 mm coil radius. At both ends of the coil, cross section area has been increased and connected to the central part of the coil with fillets (see Fig. 1a), in order to mitigate constraints and load application induced stress concentration, ensuring that at the internal section of the modeled wire those effects are not influencing evaluated stress levels. The model was discretized into 74200 parabolic solid

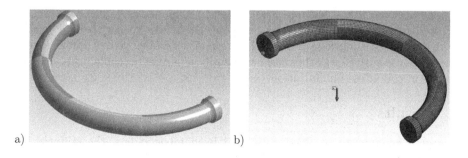

a)   b)

Fig. 1. a) geometry of the modeled half coil, b) load and constraints applied.

elements, resulting in 306569 mesh nodes. A $5N$ external load was applied aligned with the coil axis and connected using ANSYS Mechanical 'Remote Force' load option to mesh nodes on one coil end, whilst the nodes on the other coil end were constrained fixing all degrees of freedom (Fig. 1b).

The BGM based optimization on the half coil geometry was performed according the two optimization constraints described in Sect. 3. In the first one, nodes on the outer coil surface have been fixed and nodes on the inner coil surface have been sculpted (Fig. 2a); in the second one, nodes on the outer coil surface have been sculpted and nodes on the inner coil surface have been constrained (Fig. 2b). For both optimization, the parameters d and S_{node} (Eq. (6)) have been set as 1.2% of the wire diameter and the 80% of the maximum initial stress for Equivalent von Mises Stress acting on coil surfaces. For each optimization, 20 BGM driven mesh morphing iterations were performed.

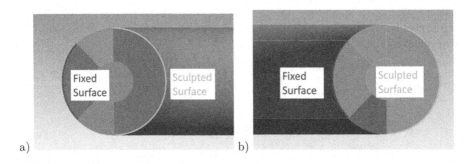

Fig. 2. RBF set up for sculpted coil surfaces (section view): a) inner ones, b) outer ones.

A parameter-based optimization was performed only modifying the inner coil surface, by selecting two groups of nodes (Fig. 3): the first one at the inner point of the coil surface (point 1 in Fig. 3) and the second one placed at 45° in the coil cross section with respect to the inner point location (points 2 in Fig. 3). Both nodes groups were imposed to move along the coil surface normal. The optimization was performed exploiting the ANSYS Desing Explorer optimization tool. The design space was defined setting the displacement range for both nodes groups between 0 and 0.2 mm; the Design of Experiment (DoE) was created according to the Latin Hypercube Sampling Design approach; the DoE results were then used to generate a Kriging response surface with variable kernel. The response surface was then used to identify a parameter configuration which results in coil stress minimization.

4   Results

For both configurations, fixed outer and fixed inner radius, the stiffness has been derived from parameter-less resulting models model and equivalent cross-sections have been calculated using Eqs. (11) and (13); same weight coils have

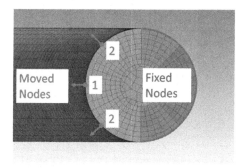

Fig. 3. RBF set up for the parameter-based optimization (section view).

been computed using Eqs. (14) and (15); for the case of fixed outer radius a parameter-based solution was computed as a reference. Optimised shapes are "quasi-elliptical" and can be described in terms of mean diameter D and vertical and horizontal axis of, d_z and d_r respectively.

4.1 Sculpting Inner Surface

When sculpting at the inner radius, the BGM surface sculpting approach moves nodes outward the coil surface, adding material. Optimized section has 3.73% lower maximum stress and 0.6% larger volume with respect to the equivalent circular cross-section. Initial, optimised and equivalent cross-sections are shown in Fig. 4a, stress levels distribution is depicted in Fig. 5a and relevant parameters are listed in Table 2. It can be seen that optimised shape has higher efficiency (41%) than the equivalent circular cross-section (38%), since BGM optimization procedure is adding material where it is more needed. Spring efficiency ϵ is defined as the ratio between the elastic energy stored over the energy the spring would store if all the volume was at maximum stress. Straight torsion bar has 0.5 efficiency, in this case the circular section efficiency is lower because of curvature and direct shear effect. Maximum stress on same weight spring is 4% higher than that on the optimized shape, showing that the optimization is not only improving efficiency but it is also reducing maximum stress.

Table 2. Results for inner surface sculpting - $K = 4.27\ 10^4\ [\frac{N}{m}]$

	d_r	d_z	D	τ_{in}/P	τ_{out}/P	ϵ	A	V
	[mm]	[mm]	[mm]	[MPa/N]	[MPa/N]	[−]	mm²	mm³
Optimized	4.26	3.85	39.74	1.635	1.273	41%	13.13	1643
Equivalent	4.10	4.10	39.90	1.696	1.280	38%	13.19	1653
Same weight	4.09	4.09	39.80	1.702	1.515	38%	13.14	1643

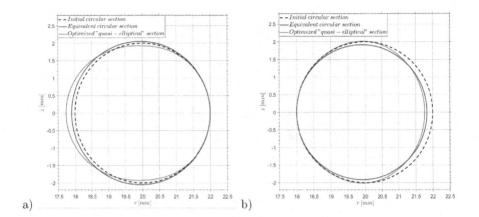

Fig. 4. Cross sections comparison for: a) inner surface sculpting optimization, b) outer surface sculpting optimization.

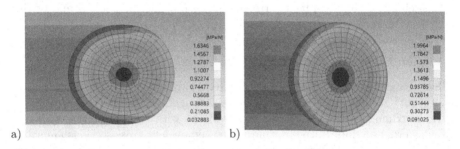

Fig. 5. Stress distribution in section optimized: a) sculpting inner surface, b) sculpting outer surface.

4.2   Sculpting Outer Surface

When sculpting at the outer radius, the BGM moves nodes inward the coils surface, removing material, and reducing maximum stress by 3% and increasing cross-section swept volume by 2.3%, with respect to the equivalent circular cross-section. Baseline, optimised and equivalent circular cross-sections are shown in Fig. 4b, stress levels distribution is depicted in Fig. 5b and relevant parameters are listed in Table 3. It can be seen that in this case the optimisation is not improving the efficiency and is only slightly reducing the maximum stress when compared to an equivalent circular section. More important, the optimized spring performs worse than the same weight spring. In this case the optimization is neither improving the efficiency, nor reducing the maximum stress.

4.3   Optimization Method Comparison

In order to complete the proposed method presentation, a final comparison with the parameter-based response surface optimization method is given. The comparison has been performed with the optimized shape obtained sculpting inner

Table 3. Results for outer surface sculpting - $K = 3.44\,10^4\,[\frac{N}{m}]$

	d_r	d_z	D	τ_{in}/P	τ_{out}/P	ϵ	A	V
	[mm]	[mm]	[mm]	[MPa/N]	[MPa/N]	[−]	[mm²]	[mm³]
Optimized	3.76	4.02	39.76	1.996	1.598	38%	11.78	1476
Equivalent	3.83	3.83	39.83	2.055	1.579	38%	11.53	1443
Same weight	3.88	3.88	39.81	1.986	1.777	38%	11.80	1476

surface (see Sect. 4.1 and Fig. 4a). Results for both optimization methods are comparable in terms of maximum surface stress (see Fig. 6a and Fig. 6b). On the other hand, the cross-section area in this configuration is 1.75% higher than the cross-section obtained with the BGM method. The final spring also has a higher stiffness (+3.76%) with respect to the section obtained with the parameters-less optimization and 38% efficiency value, demonstrating how the full reshaping freedom of the parameter-less approach allows to gain slight better results if compared with a 2 parameters optimisation result.

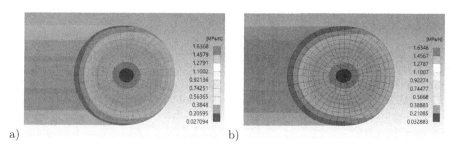

a)   b)

Fig. 6. Stress distribution in section optimized: a) sculpting inner surface with parameter-based method, b) sculpting inner surface with parameter-less method (BGM).

5 Conclusions

In the present work an innovative optimization procedure has been presented. This procedure has been realised by combining an RBF based mesh morphing procedure and a stress level optimization approach based on the observation of natural structures stress mitigation strategies. The reliability of the approach has proven in both implementing in an automatic surface sculpting optimization strategy and in obtaining optimized component shapes.

The focus of the present work was pointed to the optimization of a widely investigated mechanical component, in order to compare the methodology results with the well known and established analytical description of the component itself.

Two optimization constraint were imposed in two optimization tasks: in the first the BGM driven surface sculpting was applied to the inner surface of the coil constraining the outer surface, in the second optimization task the sculpting was applied to the outer surface and the constraint was applied to the inner surface. The first optimization set up gave a final optimized shape that performs better than the circular cross-section, since the BGM driven surface sculpting acts only in adding material on inner surface, where stresses are higher. On the other hand, the optimization through the sculpting of the outer surface failed in improving the design of the component, since BGM is removing material not subject to lowest stress (material should be removed at the core of the cross-section), neither adding it where stress is maximum (i.e. at the inner surface).

Finally, the best results obtained, that are produced by sculpting the inner surface, are compared with the ones computed by a parameter-based optimization of the same region. The obtained shape has comparable performances with respect to the one optimized with the BGM method, but the section efficiency is lower. Furthermore, the parameter-based optimization required more user efforts in setting up of the required DoE and Response Surface model needed to minimize the coil surface stress.

Considering these results, it is possible to say that the proposed automatic surface sculpting optimization approach gave successful results in optimizing the surface coil stress levels.

Acknowledgements. The work here presented is developed within the research project "SMART MAINTENANCE OF INDUSTRIAL PLANTS AND CIVIL STRUCTURES BY 4.0 MONITORING TECHNOLOGIES AND PROGNOSTIC APPROACHES - MAC4PRO", sponsored by the call BRIC-2018 of the National Institute for Insurance against Accidents at Work - INAIL.

References

1. ANSYS, Inc. http://www.ansys.com/products/structures. Accessed 29 July 2020
2. RBF Morph srl. http://www.rbf-morph.com/act-module/. Accessed 29 July 2020
3. Biancolini, M.E.: Mesh morphing and smoothing by means of radial basis functions (RBF): a practical example using Fluent and RBF Morph. In: Handbook of Research on Computational Science and Engineering: Theory and Practice, pp. 347–380. IGI Global (2011)
4. Biancolini, M.E.: Fast Radial Basis Functions for Engineering Applications. Springer, Cham (2017). https://doi.org/10.1007/978-3-319-75011-8
5. Biancolini, M.E., Cella, U.: An advanced RBF morph application: coupled CFD CSM aeroelastic analysis of a full aircraft model and comparison to experimental data. In: MIRA International Vehicle Aerodynamics Conference, Grove, pp. 243–258 (2010)
6. Biancolini, M.E., Chiappa, A., Giorgetti, F., Porziani, S., Rochette, M.: Radial basis functions mesh morphing for the analysis of cracks propagation. Procedia Struct. Integrity **8**, 433–443 (2018)
7. Davis, P.J.: Interpolation and Approximation. Blaisdell Publishing Company, New York (1963)

8. De Boer, A., Van der Schoot, M., Bijl, H.: Mesh deformation based on radial basis function interpolation. Comput. Struct. **85**(11–14), 784–795 (2007)
9. Galland, F., Gravouil, A., Malvesin, E., Rochette, M.: A global model reduction approach for 3D fatigue crack growth with confined plasticity. Comput. Methods Appl. Mech. Eng. **200**(5–8), 699–716 (2011)
10. Giorgetti, F., et al.: Crack propagation analysis of near-surface defects with radial basis functions mesh morphing. Procedia Struct. Integrity **12**, 471–478 (2018)
11. Groth, C., Chiappa, A., Biancolini, M.E.: Shape optimization using structural adjoint and RBF mesh morphing. Procedia Struct. Integrity **8**, 379–389 (2018)
12. Groth, C., Cella, U., Costa, E., Biancolini, M.E.: Fast high fidelity CFD/CSM fluid structure interaction using RBF mesh morphing and modal superposition method. In: Aircraft Engineering and Aerospace Technology (2019). https://doi.org/10.1108/AEAT-09-2018-0246
13. Heywood, R.B.: Photoelasticity for Designers. Pergamon Press, Oxford (1969)
14. Imaizumi, T., Ohkouchi, T., Ichikawa, S.: Shape optimization of the wire cross section of helical springs. SAE Tech. Paper **920775**, 775 (1990)
15. Kamiya, N., Kita, E.: Boundary element method for quasi-harmonic differential equation with application to stress analysis and shape optimization of helical spring. Comput. Struct. **37**(1), 81–86 (1990). https://www.sciencedirect.com/science/article/pii/004579499090199C
16. Lombardi, M., Parolini, N., Quarteroni, A.: Radial basis functions for inter-grid interpolation and mesh motion in FSI problems. Comput. Methods Appl. Mech. Eng. **256**, 117 (2013)
17. Mattheck, C., Burkhardt, S.: A new method of structural shape optimization based on biological growth. Int. J. Fatigue **12**(3), 185–190 (1990)
18. Papoutsis-Kiachagias, E.M., Porziani, S., Groth, C., Biancolini, M.E., Costa, E., Giannakoglou, K.C.: Aerodynamic optimization of car shapes using the continuous adjoint method and an RBF morpher. In: Minisci, E., Vasile, M., Periaux, J., Gauger, N.R., Giannakoglou, K.C., Quagliarella, D. (eds.) Advances in Evolutionary and Deterministic Methods for Design, Optimization and Control in Engineering and Sciences. CMAS, vol. 48, pp. 173–187. Springer, Cham (2019). https://doi.org/10.1007/978-3-319-89988-6_11
19. Porziani, S., Groth, C., Biancolini, M.E.: Automatic shape optimization of structural components with manufacturing constraints. Procedia Struct. Integrity **12**, 416–428 (2018)
20. Porziani, S., Groth, C., Mancini, L., Cenni, R., Cova, M., Biancolini, M.E.: Optimisation of industrial parts by mesh morphing enabled automatic shape sculpting. Procedia Struct. Integrity **24**, 724–737 (2019)
21. Porziani, S., Groth, C., Waldman, W., Biancolini, M.E.: Automatic shape optimisation of structural parts driven by BGM and RBF mesh morphing. Int. J. Mech. Sci. 105976 (2020). http://www.sciencedirect.com/science/article/pii/S0020740320306184
22. Staten, M.L., Owen, S.J., Shontz, S.M., Salinger, A.G., Coffey, T.S.: A comparison of mesh morphing methods for 3D shape optimization. In: Proceedings of the 20th International Meshing Roundtable, pp. 293–311. Springer, Cham (2011). https://doi.org/10.1007/978-3-642-24734-7_16
23. Wahl, A.M.: Mechanical Springs. McGraw-Hill Book Company Inc., New York (1963)
24. Waldman, W., Heller, M.: Shape optimisation of holes in loaded plates by minimisation of multiple stress peaks. Technical Report DSTO-RR-0412, Aerospace Division, Defence Science And Technology Organisation, Melbourne (2015)

Multiscale Modelling and Simulation

Verification, Validation and Uncertainty Quantification of Large-Scale Applications with QCG-PilotJob

Bartosz Bosak[1(✉)], Tomasz Piontek[1(✉)], Paul Karlshoefer[2], Erwan Raffin[2], Jalal Lakhlili[3], and Piotr Kopta[1(✉)]

[1] Poznań Supercomputing and Networking Center, Poznań, Poland
{bbosak,piontek,pkopta}@man.poznan.pl
[2] CEPP – Center for Excellence in Performance Programming, Atos, France
[3] Max-Planck Institute for Plasma Physics - Garching, Munich, Germany

Abstract. Efficient execution of large-scale and extremely demanding computational scenarios is a challenge for both the infrastructure providers and end-users, usually scientists, that need to develop highly scalable computational codes. Nevertheless, at this time, on the eve of exa-scale supercomputers, the particular role has to be given also to the intermediate software that can help in the preparation of applications so they can be efficiently executed on the emerging HPC systems. The efficiency and scalability of such software can be seen as priorities, however, these are not the only elements that should be addressed. Equally important is to offer software that is elastic, portable between platforms of different sizes, and easy to use. Trying to fulfill all the above needs we present QCG-PilotJob, a tool designed to enable flexible execution of numerous potentially dynamic and interdependent computing tasks in a single allocation on a computing cluster. QCG-PilotJob is built on many years of collaboration with computational scientists representing various domains and it responses to the practical requirements of real scientific use-cases. In this paper, we focus on the recent integration of QCG-PilotJob with the EasyVVUQ library and its successful use for Uncertainty Quantification workflows of several complex multiscale applications being developed within the VECMA project. However, we believe that with a well-thought-out design that allows for fully user-space execution and straightforward installation, QCG-PilotJob may be easily exploited in many other application scenarios, even by inexperienced users.

1 Introduction

The success of scientific research can be evaluated based on its applicability for solving real-world problems. Not surprisingly, before computational simulation codes are used in production, their robustness needs to be strictly proven. This applies to both, the quality of the code itself and, even more importantly, the

M. Paszynski et al. (Eds.): ICCS 2021, LNCS 12746, pp. 495–501, 2021.
https://doi.org/10.1007/978-3-030-77977-1_39

reliability of the generated results. To this end, scientists employ VVUQ procedures to verify, validate, and precisely quantify the uncertainty of calculations.

The inherent characteristic of the majority of available techniques is multiple evaluations of models using different input parameters selected from the space of possible values. This is a computationally demanding scenario even for evaluations of traditional single-scale applications, but in the case of multiscale simulations that consist of many coupled single-scale models, the problem becomes a real challenge. There is a need to support simultaneous execution of single-scale models that may have extremely large and different resource requirements, some single-scale models may need to be attached dynamically, the number of evaluations of single-scale models may not be known in advance, to name just a few difficulties. Within the VECMA project[1] we are trying to resolve all these issues with efficient and flexible tools. One of key elements in VECMA toolkit[2] [4] is EasyVVUQ[3] [8], the user-facing library that brings VVUQ methods to many different use-cases. However, EasyVVUQ itself abstracts from the aspects of execution of model evaluations on computational resources and outsources this topic to the external tools. One of them is QCG-PilotJob[4] (QCG-PJ) developed by Poznan Supercomputing and Networking Center as part of QCG (Quality in Cloud and Grid) middleware [7]. Its functionality, based on the idea of so called *Pilot Job*, which manages a number of subordinate tasks, is essential for the flexible and efficient execution of VVUQ scenarios that inherently consists of many tasks, from which some may be relatively small.

The rest of this paper is structured as follows. In Sect. 2 we present a brief overview of related work. In Sect. 3 we describe basic objectives and functionality of QCG-PJ. Next, in Sect. 4, we introduce a few schemes of integration between EasyVVUQ and QCG-PJ that have been developed in the VECMA project and then we present a range of application use-cases that already use QCG-PJ. The results of performance tests conducted so far are outlined in Sect. 5. Finally, in Sect. 6, we conclude and share main plans for the future.

2   Related Work

The problem of efficient and automated execution of a large-number of tasks on computing clusters managed by queuing systems is known from decades. One of the most recognized systems that deal particularly with the pilot job style of execution on computing resources is RADICAL-Pilot [6], being developed as part of the RADICAL-Cybertools suite. In contrast to QCG-PJ, which can be easily installed in a user's home directory, RADICAL-Pilot is not a self-contained component and needs to be integrated with external services. There are also several solutions having some commonalities with the QCG-PJ idea, but they are focused primarily on the workflow orchestration rather than on the efficiency

---

[1] https://www.vecma.eu.

[2] https://www.vecma-toolkit.eu.

[3] https://github.com/UCL-CCS/EasyVVUQ.

[4] https://github.com/vecma-project/QCG-PilotJob.

and flexibility of computing on HPC machines. An example of a mature system is here Kepler [2], which addresses the need for the HTC execution of parts of the workflows on clusters. Further examples are Swift-T [9] and Parsl [1], which share a common goal to effectively support data-oriented workflows, or Dask[5], which aims to enable parallel processing for analytical workflows.

3 Objectives

The access to HPC systems is regulated by the policies of resource providers and restricted by local resource management system configurations as well as their implementations. For example, the policy at SuperMUC-NG cluster installed at the Leibniz Supercomputing Centre[6], in order to promote large-scale computing, allows users to submit and run only a small number of jobs at the same time. Smaller HPC installations may be less restrictive, but in general, the large tasks have a priority over small tasks, and the rule remains the same: there is no way to flexibly schedule many jobs with basic mechanisms. If users want to efficiently run a huge number of conceptually different tasks they need to employ solutions that can mitigate the regulations on a level of single allocation. One of the possible approaches is to define a processing scheme in a scripting language, but this is neither generic nor flexible and possibly prone to many bugs and inefficiency. The other, recommended approach is the utilisation of specialised software, like QCG-PJ.

Functionality

The basic idea of QCG-PJ is to bring an additional tasks management level within the already created allocation. As it is presented in Fig. 1, from the queuing system's perspective, QCG-PJ is only a single regular task, but for a user, it is a second-level lightweight resource management system that can be administered and used on an exclusive basis.

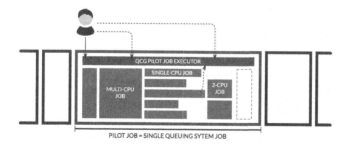

Fig. 1. The general computation scheme in QCG-PilotJob

---

[5] https://dask.org.

[6] https://doku.lrz.de/display/PUBLIC/Job+Processing+with+SLURM+on+Super MUC-NG.

As SLURM manages resources of a cluster, QCG-PJ manages resources of an allocation and ensures that tasks are scheduled efficiently. That being said, users or alternatively client software components can interact with it in a similar way as with any queuing system. Through a dedicated lightweight service called QCG-PJ Scheduler, they can submit new tasks with specific requirements, list submitted tasks or cancel them.

In order to enhance usability, QCG-PJ offers two ways of submission of tasks: firstly, it is possible to provide task definitions in a form of JSON file and submit this file from a command line at startup of the QCG-PJ, and secondly, it is possible to use the provided python API for dynamic creation and management of jobs directly from a running python program.

Moreover, the tool provides a few supplementary built-in features that can be recognised as particularly useful for selected scenarios. Among others, it allows defining dependencies between tasks as well as it offers resume mechanism to support fault tolerance at a workflow level.

Architecture Towards Exascale

One of the biggest challenges at the very beginning of QCG-PJ development was to ensure its ability to meet requirements defined by extremely demanding multiscale applications. In order to reach the performance of hundreds of petaflops or even higher, these applications ultimately call for, it was particularly important to design an appropriate architecture. First of all, such architecture should be scalable: the system should be easy to use, even on a laptop, but also easily extendable, portable and efficient once the use-cases grow up to require HPC resources. Consequently, the natural choice was to propose a hierarchical structure of components, where top-level services are released from the high-intensive processing that can be performed in a distributed way by low-level services. In consequence, the QCG-PJ architecture includes a concept of partition, which reflects a subset of resources that are managed separately and can be dynamically attached to the optional top-level QCG-PJ Scheduling Queue service. This is presented in Fig. 2.

In the presented full-scale deployment scenario, QCG-PJ Scheduling Queue is an entry point to the system and keeps global information about all tasks that should be processed. One or multiple QCG-PJ Scheduler services, associated with the elementary partitions, can request Scheduling Queue for a portion of tasks to execute. Once the tasks are completed, the schedulers report this information back to the central service. Consequently, resources coming from a single or many allocations, also from a single or many HPC clusters, can be robustly integrated and offered as a single logical concept, while the communication overhead is minimised.

Having a closer look at the logic present within a single partition, two elements should be noted. The first of them is the possibility to reserve a core for QCG-PJ Scheduler. This option is useful when processing done by the Scheduler service significantly influences the actual computations. The second element is the Node Launcher service. This component is designed to improve the startup efficiency of single-core tasks.

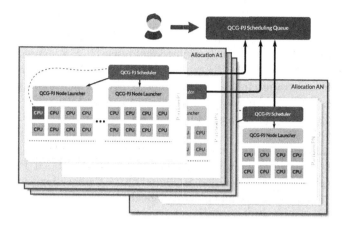

Fig. 2. QCG-PilotJob architecture overview

4   Use Cases

Integration with EasyVVUQ

EasyVVUQ is a tool for domain experts who work on concrete VVUQ scenarios related to their applications. We argue that these experts shouldn't spend their valuable time to set-up the logic of execution of EasyVVUQ workflows on computing resources. Rather, they should focus on purely scientific or engineering aspects. To this end, VECMA toolkit provides a few approaches for the integration of EasyVVUQ with QCG-PJ so it can be efficient and natural for the domain scientist. Currently, there are the following possibilities:

Direct integration with EasyVVUQ: It is the most straightforward type of integration, where QCG-PJ is transparently employed in EasyVVUQ as one of its internal execution engines. Although at the moment of writing, this type of integration is not yet completed and doesn't benefit from more advanced features offered by QCG-PJ, e.g. iterative tasks, it is expected to be the preferred one at some point in the future.

Integration through the EQI library: EQI[7], which stands for EasyVVUQ-QCGPilotJob Integrator, is a lightweight library designed to bring optimised processing schemes to selected types of highly-demanding EasyVVUQ workflows. It makes use of advanced functionalities of QCG-PJ, like resume mechanism and iterative jobs.

Integration through FabSim3: QCG-PJ has been integrated with the FabSim3 automation toolkit[8] in order to support demanding application campaigns. Since FabSim3 internally uses EasyVVUQ, the combined execution of EasyVVUQ and QCG-PJ is also possible.

---

[7] https://github.com/vecma-project/EasyVVUQ-QCGPJ.
[8] https://github.com/djgroen/FabSim3.

Applications

At the moment of writing, there are already several application teams from VECMA that use QCG-PJ for their professional research. For instance, scientists from Max-Planck Institute for Plasma Physics use EasyVVUQ to quantify the propagation of uncertainty in a fusion turbulence model which is computationally expensive. It is a 3D parallel code and needs 512 to 16384 MPI cores [5]. Running a UQ campaign on such models require a very large number of jobs. Thanks to EQI and QCG-PJ, it was possible to execute the required simulations in a single batch allocation. In a similar way, the QCG-PJ tool has been employed for UQ of the UrbanAir application developed by PSNC [10]. It is also worth mentioning recent studies, where FabSim3 and QCG-PJ have been employed for UQ performed on the CovidSim epidemiological code [3].

5   Performance Evaluation

Ultimately, the performance of QCG-PJ will be the most determining factor for its usability. Since the early days of its development, scalability and accessibility are evaluated repeatedly on large European supercomputers such as SuperMUC-NG at LRZ and Eagle/Altair at PSNC.

Thus far, test runs involving 100 dual-socket nodes, which equates to around 5000 CPU cores, showed very promising results, with more than 99% of time spent in the user-defined pilot jobs. More specifically, 20.000 pilot jobs with a runtime of five minutes each kept 99.2% of the available resources occupied.

With these promising results, we plan on conducting experiments with actual scientific applications which involve much larger node counts, alongside an exhaustive scalability study. Additionally, these tests were conducted by users which are not directly involved in the development of QCG-PJ, which in turn further contributed to the accessibility of the API.

6   Summary and Future Work

In this paper, we shortly introduced the concepts and features of QCG-PilotJob system and depicted how it is used by VECMA project to support demanding VVUQ scenarios. The progress made to several large-scale applications, when they successfully employed QCG-PJ, allows us to rank the current usability of QCG-PJ relatively high. Nevertheless, there are still ongoing works aimed to enhance the quality and functionality of the software. In regards to the former, since individual SLURM configurations can pose challenges for QCG-PJ and require its adaptation and customization, we are in the process of extensive tests of the tool on high-end European clusters. For instance, PSNC is in the process of deploying QCG-PJ to SURF (Amsterdam) and ARCHER2 (Edinburgh). Ultimately, we want to ensure that QCG-PJ is easily deployable and works efficiently on a large variety of machines and configurations. In terms of new functionality, our aim is to complete the implementation of the global Scheduling Queue service as well as to provide a dedicated monitoring solution.

Acknowledgments. This work received funding from the VECMA project realised under grant agreement 800925 of the European Union's Horizon 2020 research and innovation programme. We are thankful to the Poznan Supercomputing and Networking Center for providing its computational infrastructure. We are also grateful to the VECMA partners for the invaluable motivation.

References

1. Babuji, Y., et al.: Parsl: pervasive parallel programming in python. In: Proceedings of the 28th International Symposium on High-Performance Parallel and Distributed Computing, pp. 25–36. HPDC 2019, New York, NY, USA (2019). https://doi.org/10.1145/3307681.3325400

2. Cabellos, L., Campos, I., del Castillo, E.F., Owsiak, M., Palak, B., Płóciennik, M.: Scientific workflow orchestration interoperating HTC and HPC resources. Comput. Phys. Commun. **182**(4), 890–897 (2011). https://doi.org/10.1016/j.cpc.2010.12.020

3. Edeling, W., et al.: The impact of uncertainty on predictions of the CovidSim epidemiological code. Nat. Comput. Sci. **1**(2), 128–135 (2021). https://doi.org/10.1038/s43588-021-00028-9

4. Groen, D., et al.: VECMAtk: a scalable verification, validation and uncertainty quantification toolkit for scientific simulations. Philos. Trans. R. Soc. A: Mathe. Phys. Eng. Sci. **379**(2197), 20200221 (2021). https://doi.org/10.1098/rsta.2020.0221

5. Luk, O., Hoenen, O., Bottino, A., Scott, B., Coster, D.: Compat framework for multiscale simulations applied to fusion plasmas. Comput. Phys. Commun. (2019). https://doi.org/10.1016/j.cpc.2018.12.021

6. Merzky, A., Turilli, M., Titov, M., Al-Saadi, A., Jha, S.: Design and performance characterization of RADICAL-PILOT on leadership-class platforms (2021)

7. Piontek, T., et al.: Development of science gateways using QCG – lessons learned from the deployment on large scale distributed and HPC infrastructures. J. Grid Comput. **14**, 559–573 (2016)

8. Richardson, R., Wright, D., Edeling, W., Jancauskas, V., Lakhlili, J., Coveney, P.: EasyVVUQ: a library for verification, validation and uncertainty quantification in high performance computing. J. Open Res. Softw. **8**, 1–8 (2020)

9. Wozniak, J.M., Armstrong, T.G., Wilde, M., Katz, D.S., Lusk, E., Foster, I.T.: Swift/t: large-scale application composition via distributed-memory dataflow processing. In: 2013 13th IEEE/ACM International Symposium on Cluster, Cloud, and Grid Computing, pp. 95–102 (2013). https://doi.org/10.1109/CCGrid.2013.99

10. Wright, D.W., et al.: Building confidence in simulation: applications of EasyVVUQ. Adv. Theory Simul. **3**(8), 1900246 (2020)

Towards a Coupled Migration and Weather Simulation: South Sudan Conflict

Alireza Jahani[1(✉)], Hamid Arabnejad[1], Diana Suleimanova[1],
Milana Vuckovic[2], Imran Mahmood[1], and Derek Groen[1]

[1] Department of Computer Science, Brunel University London,
Uxbridge UB8 3PH, UK
alireza.jahani@brunel.ac.uk
[2] Forecast Department, European Centre for Medium-Range Weather
Forecasts (ECMWF), Reading, UK

Abstract. Multiscale simulations present a new approach to increase the level of accuracy in terms of forced displacement forecasting, which can help humanitarian aid organizations to better plan resource allocations for refugee camps. People's decisions to move may depend on perceived levels of safety, accessibility or weather conditions; simulating this combination realistically requires a coupled approach. In this paper, we implement a multiscale simulation for the South Sudan conflict in 2016–2017 by defining a macroscale model covering most of South Sudan and a microscale model covering the region around the White Nile, which is in turn coupled to weather data from the Copernicus project. We couple these models cyclically in two different ways: using file I/O and using the MUSCLE3 coupling environment. For the microscale model, we incorporated weather factors including precipitation and river discharge datasets. To investigate the effects of the multiscale simulation and its coupling with weather data on refugees' decisions to move and their speed, we compare the results with single-scale approaches in terms of the total validation error, total execution time and coupling overhead.

Keywords: Agent-based modelling · Multiscale simulation · Refugee movements · Data coupling

1 Introduction

Internal conflicts, environmental disasters, or severe economic circumstances force people to displace from their homes [1]. For instance, people still struggling with the continuation of violence and instability which led to escalating food insecurity and drastic economic decline in South Sudan after the civil crisis in 2013. All these had resulted in the displacement of people who became forced migrants to find safety in camps located in neighbouring countries [2]. By mid-December 2016, more than 3 million South Sudanese had been forced to

© Springer Nature Switzerland AG 2021
M. Paszynski et al. (Eds.): ICCS 2021, LNCS 12746, pp. 502–515, 2021.
https://doi.org/10.1007/978-3-030-77977-1_40

flee their homes. Hence, one in four people in South Sudan had been uprooted, their lives disrupted, their homes destroyed, and their livelihoods decimated [3]. United Nations Office for the Coordination of Humanitarian Affairs (OCHA) identified 7.5 million people out of a population of 12 million in need of humanitarian assistance [4]. Computational models can forecast refugees' arrival time and counts to the camps. It helps humanitarian aid organizations to allocate enough resources for refugees [4]. Among the computational models, agent-based modelling (ABM) can provide such insights and information [1].

ABM hybrid techniques are the best way to understand the complex decision-making processes of social systems like agent behaviours and their relationships [5–7]. A lot of efforts have been made in ABM to simulate forced displacement [8–11]. More recent research in ABM's hybrid techniques is integration and data coupling in simulation, particularly their varied number of approaches. Groen et al. [12] identified four popular approaches to couple and integrate simulation models including, multi-scale integration, multi-paradigm integration, multi-platform/multi-architecture integration and multi-processing integration. Among these approaches, multiscale simulations are more inherent for scientific problems like forced displacement forecasting to create more accurate models. However, multiscale simulations face several challenges. In general, formulating generic frameworks for multiscale modelling and simulation is a big challenge [13]. To study the effects of policy decisions on ABM, Gilbert et al. [14] and Suleimenova et al. [15] examined ABM in complex systems, such as human movement, to provide insights for governments, stakeholders and policymakers. Searle et al. [16] proposed a generic framework by designing an ABM to simulate conflict instances and decisions behind the movement of refugees fleeing conflict-affected areas.

In more detail, Alowayyed et al. [17] investigated computational challenges regarding coupling between a range of scales. Incorporating external factors affecting conflict events is another challenge that needs to be tackled [12]. Furthermore, due to incomplete or small size datasets, forecasting forced displacement does still suffer major challenges like outdated statistical methods and poor refugees arrival estimations [18,19].. Besides, despite the necessity of investigating the effects of climate, weather conditions and seasonal factors on refugees movement, there is very limited research in the literature to identify how they influence movements, particularly adverse conditions that might restrict possible migration paths. A study showed the correlation coefficient between arrivals in different countries and different weather-related variables [20]. Abel et al. [21] stated that climate change will increase the number of refugees fleeing from conflicts and also low precipitation level will increase conflicts which in turn cause rising the number of refugees. Black et al. [22] studied the drivers of refugees movements through different climatic related problems, such as sea level, fluctuations and intensity of storms and rainfall patterns, temperature rise and changes in weather conditions.

In [23], we presented the FLEE agent-based simulation approach where a complex system is modelled as a set of autonomous decision-making agents that behave accordingly with their environment based on a set of rules. Each agent in FLEE acts as a forcibly displaced person and tries to move between locations, attempting to reach the safety zone (i.e., camps). In this paper, we focus on implementing coupling ability which allows us to connect simulations of multiple scales of movement in different regions regarding their different circumstances like new rules, policy decisions or weather conditions. To test the proposed coupling model and investigate the effects of the aforementioned decisions and changes to our FLEE ABM assumptions, we select the South Sudan conflict as our test scenario and construct a multiscale (macro-scale and micro-scale) model to explore the establishment of data coupling between such models.

The rest of the paper is set out as follows. In Sect. 2, we explain a multiscale Flee model. Section 3 discusses the effort on the coupling approaches for multiscale simulations. We explain the South Sudan multiscale simulation and coupling with weather datasets in Sect. 4. Section 5 presents the experimental preliminary results with discussion. Finally, Sect. 6 concludes and briefly outlines future work.

2   Flee: A Multiscale Approach

The developed multiscale simulation prototype in this work is based on the Flee code[1][23]. The Flee code is an ABM kernel, written in Python 3, which predicts the distribution of fleeing refugees across target camps. Flee is optimised for simplicity and flexibility, and support simulations with 100,000s of agents on a single desktop.

Our proposed prototype divides the whole model into two sub-models, namely, macroscale and microscale models. Each sub-model is executed independently and agents pass between them during the simulation. In this model, each location in the location graph, where agents pass through the coupling interface, should be registered as coupled locations. In addition to coupled locations, all microscale model's conflict locations should be added to the macroscale model as ghost locations. It means that although they are added to the macroscale model, they don't have any link to other macroscale locations and this is why they are named ghosts locations. They are a special type of coupling locations where (a) the macroscale model inserts agents into these locations according to the normal FLEE agent insertion algorithm and (b) at each time step, the coupling interface transfers all agents from each ghost location to the microscale model. Figure 1 illustrates the schematic scale Separation Map for data coupling between macroscale and microscale models.

3   Coupling Approaches

In this section, we highlight the use of two different cyclic (two-way) coupling approaches to interconnect macroscale and microscale models: coupling through file I/O and coupling using MUSCLE3. We also describe the acyclic (one-way) coupling with the ECMWF Climate Data Store, which we have used to incorporate weather data into the microscale model.

---

[1] http://www.github.com/djgroen/flee.

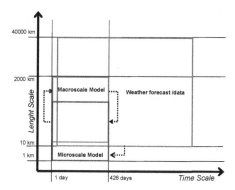

Fig. 1. Scale Separation Map of our model. The macro- and micro-scale model have identical time scales and overlapping spatial scales, and are coupled cyclically. In addition, the micro-scale model receives data from the weather forecast data source (or from ECMWF Climate Data Store in the case of historical data).

3.1   File Coupling

File I/O is a coupling approach to exchange data between two sub-models. By establishing this approach, exchanged data, such as the number of new agents added to each location, can be passed between sub-models models using a local shared file system. In this work, the number of all agents passing between sub-models through coupled and ghost locations are stored in the format of CSV files. Both sub-models fill their coupled CSV files, in a parallel fashion, to make sure that both sub-models are synchronized in terms of simulation time steps when all necessary coupled inputs files are checked at the start of each iteration. Compared to other approaches, this method is straightforward to implement and debug and easy to maintain, but lacks flexibility and can lead to high I/O overhead for large problems.

3.2   MUSCLE3 Coupling

MUSCLE3 [24], the Multiscale Coupling Library and Environment, aims to simplify the scale-separated coupled simulation. It contains two main components: the MUSCLE library, i.e., `libmuscle`, and the MUSCLE Manager. The `libmuscle` handles the data exchange over the network between each sub-model in a peer-to-peer fashion. The MUSCLE Manager sets up sub-model instances configuration and coordinates the connections between them. A model can be described to the Manager using yMMSL, a YAML-based serialisation of the Multiscale Modelling and Simulation Language (MMSL). Figure 2(a) shows the architecture of MUSCLE3 and Fig. 2(b) represents the used yMMSL in our design.

To implement our coupling strategy with MUSCLE3, we defined two main compute elements, `macro` and `micro`, which represent the macro and micro models, and two manager elements, `micro_manager` and `macro_manager`, which handle the inputs from multiple instances of each sub-models. By starting the simulation, each lunched sub-model will be registered into the coupling system by MUSCLE3 manager. In this example, 10 concurrent macro and micro sub-model will be executed. Each sub-model instance will simulate the agent's movement between locations on each day. To Exchange the data, since we have multiple instances, we designed a manager sub-model to (a) gather data from each instance of the sub-models, (b) combine the founded newAgents per location by each instance into one, and (c) pass to the other model, e.g., macro_manager will collect and combine data from all macro instances, and pass to all micro instances.

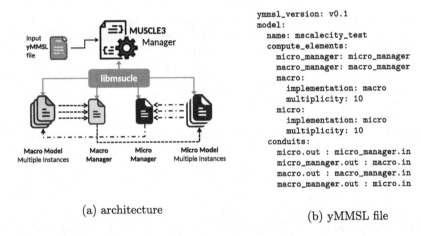

```
ymmsl_version: v0.1
model:
  name: mscalecity_test
  compute_elements:
    micro_manager: micro_manager
    macro_manager: macro_manager
    macro:
      implementation: macro
      multiplicity: 10
    micro:
      implementation: micro
      multiplicity: 10
  conduits:
    micro.out : micro_manager.in
    micro_manager.out : macro.in
    macro.out : macro_manager.in
    macro_manager.out : micro.in
```

(a) architecture

(b) yMMSL file

Fig. 2. Implemented Macro-Micro coupling approach by MUSCLE3

In particular, MUSCLE 3 provides valuable features: coupling different submodel instances, spatial and temporal scale separation and overlap, settings management, and combining features. At the time of writing, we have established these features and we are scrutinizing the simulation to ensure coupling rules are scientifically robust. We plan to perform a performance test of the different coupling approaches once this scrutiny exercise has concluded.

4   South Sudan Multiscale Simulation

For the South Sudan multiscale simulation, we use a cyclic (two-way) coupling between a more approximate model that captures most of South Sudan as a macroscale model (see Fig. 3a) comprising 8 regions of South Sudan and 14 camps in 4 neighbouring countries, including Uganda, Kenya, Sudan and Democratic Republic of Congo (DRC), and a more detailed model that captures the region around the White Nile as a micro model (see Fig. 3b). In the microscale model, we aim to capture key walking routes, roads, and river crossings in the

mountainous areas in eastern South Sudan. We also increase the level of detail
in terms of locations and incorporate a broader range of relevant phenomena,
such as weather conditions. The microscale model focuses on forced migrant
movements from Upper Nile and Jonglei regions towards Ethiopian camps in
Gambela. We create both models for the same conflict period between 1 June
2016 and 31 July 2017. More detailed maps of macroscale and microscale models'
locations are depicted in Figs. 3a and 3b wherein each, red points represent conflict locations, yellow points are towns and green points show camps. Besides, to
couple macroscale and microscale models, four coupled locations co-exist in both
models for passing agents between both models. Moreover, as described before,
the microscale model has additional algorithm assumptions which include three
types of routes: drive, walk and river that affect the agents' movement speed.

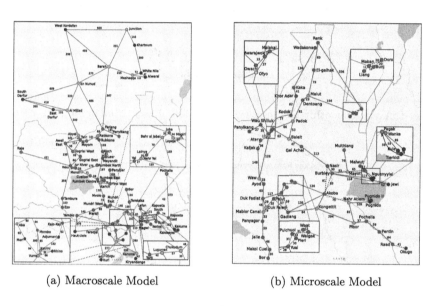

(a) Macroscale Model   (b) Microscale Model

Fig. 3. South Sudan location graphs. (Color figure online)

4.1   Weather Coupling in South Sudan Microscale Model

South Sudan experiences a tropical climate, characterized by a rainy season
which differs by location but generally falls between April and November, followed by a drier season. Annual rainfall ranges from 700–1,300 mm in the north
of the country, to 1,200–2,200 mm in the southern upland areas. Most of this
rainfall occurs in the wet season, therefore monthly rainfall averages less than
10 mm in the dry season and above 200 mm in the rainy season in the Bahr el
Ghazal and Eastern Equatoria. The temperature averages are normally high,
above 25 °C, and exceeding 35 °C in March which is the hottest month which
is depicted in Fig. 4. Since freezing temperature is non-existent in a tropical

climate, and storms and strong winds are rare, floods and droughts represent South Sudan most frequent natural disasters in the past decades. They are also the most damaging natural disasters in South Sudan in terms of the number of affected people, as seen in Fig. 5. Flooding mainly occurs between July and September, when heavy rains fall in most parts of the country, leading to the flooding of the Nile River tributaries [25]. Therefore, we consider using precipitation and river discharge most influential on refugee movement in our simulation.

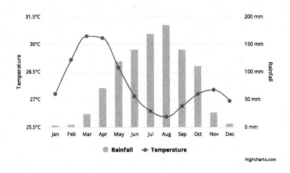

Fig. 4. Average monthly temperature and rainfall of South Sudan for 1991–2016

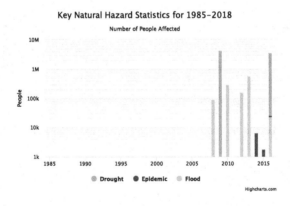

Fig. 5. Key natural hazard statistics of South Sudan for 1985–2018

The purpose of coupling with ECMWF's weather forecasts is to improve the simulation model forecasts through the inclusion of such data. Therefore, to couple the microscale model with weather datasets like river discharge and precipitation levels and study their effects on agents' movement, we have to answer how they can affect refugees' decision to move from a location and their speed?. More importantly, all these assumptions need to be reflected in a rule set for coupling microscale model with weather datasets. For this aim, we examine our prototype in the real case of South Sudan's conflict with real data provided by UNHCR, ACLED, and of course weather data provided by ECMWF.

To couple the microscale model with the weather datasets, the overall strategy is the static file coupling. We have analysed 40 years of precipitation data for South Sudan and surrounding areas, to identify the precipitation range for each location. This range is used to set the thresholds which trigger the agents' movement speed changes accordingly. The data is retrieved from the Climate Data Store (CDS) by using CDSAPI for the years 2016 and 2017. Daily aggregations and conversion from m/day to mm/day were calculated for the period of simulation 01/06/2016 to 31/07/2017 using xarray Python library. The data is prepared for three smaller regions of interest: Upper Nile, Jonglei and Gambelais which is saved as CSV files - one file for each day to be compatible with other input files for the microscale model (see Fig. 6).

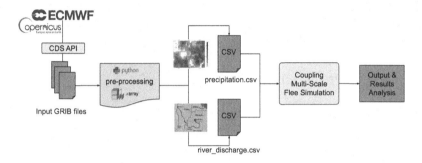

Fig. 6. Coupled weather and migration simulation

The datasets used for these calculations include:

- Daily average precipitation data per month, calculated from 40 years ERA5 climate reanalysis, for the South Sudan simulation locations. It consists of two parameters:
  - N - number of rainy days within that month
  - tp - Total precipitation
- Daily precipitation data (ERA5 climate reanalysis [26]) for the microscale model for the South Sudan area:
  - Latitude range = 11.75–6.0, Longitude range = 31.0–35.25
  - Time range: 01-06-2016 till 31-07-2017

Table 1. Sample structure for precipitation input file

Day	Bor_Akobo	Juba_Bor	Renk_Alwaral	...
0	0.1	0.465	3.6	...
1	...	...	...	...
...	...	...	...	...

The steps of the implemented weather coupling are as follows:

1. Using Precipitation data for 40 years to identify total precipitation range the given date and location's latitude and longitude.
2. Calculating midpoint location for all routes in input data.
3. Creating precipitation.csv in the format as shown in Table 1, including the total precipitation for each microscale model routes midpoints.
4. Going through the precipitation.csv to find total precipitation for each link at the given time step.
5. Calculating the movement speed by using two thresholds. The first threshold is 5 mm which means if total precipitation is below this amount, the speed will not change. For the second threshold, if the total precipitation is bigger than 75% of the average level of that location and bigger than 15 mm, the link will be considered closed. Any other precipitation levels which will be in the middle of these thresholds means the route distance will be doubled. Table 2 summarizes these assumptions.

Table 2. Level of precipitation vs Move_speed

Move_speed	No change in speed	Double the distance	Close the link
$tp < X1$	$X1 < tp < X2$	$X2 < tp$
Low level	High level	Vey high level

Furthermore, we use daily river discharge data from Global flood forecasting system (GloFAS [27]) to explore the threshold for closing the route considering values of river discharge for return periods of 2, 5 and 20 years. Currently, because of having only one crossing route, we only use a simple rule with one threshold to define the river distance. If the river discharge at the midpoint of the given route is more than the average of that point in history, $8000\,\mathrm{m^3/s}$, the link will be closed (see Fig. 7). The datasets used in this part are:

- River discharge data (GloFAS historical reanalysis dataset) for the same temporal and horizontal range as total precipitation
- River discharge data for 2, 5 and 20 years return period filtered to White Nile area, also calculated from GloFAS historical reanalysis dataset

In summary, to couple the South Sudan multiscale model with weather datasets, we take the following steps. We construct macroscale and microscale models individually for the South Sudan conflict. We incorporate more location types in the microscale model to increase the level of detail (e.g. forwarding hubs and marker locations). We also add new route types, such as key walking routes, driving roads, and crossing rivers. Then, we interlink the two models, using two coupling approaches, File Coupling and Model coupling with MUSCLE3. Then, we integrate the weather datasets into the microscale model, including precipitation level and river discharge provided by Copernicus Climate Change Service (C3S) and ECMWF.

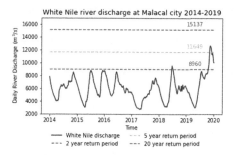

Fig. 7. River discharge values at Malacal city crossing for the period between 2014 and 2019

5   Results and Discussion

To demonstrate a highly detailed model of the South Sudan conflict and forecast forced displacement, we take five approaches to investigate the coupling within multiscale simulations and also coupling with external datasets like weather datasets and their effects on refugees' speed and their decisions to move or stay. These approaches are:

1. Singlescale Simulation (Uncoupled Serial Mode)
2. Multiscale Agent-based Simulation (File Coupling)
3. Multiscale Agent-based Simulation (MUSCLE3 Coupling)
4. Multiscale Agent-based Simulation + Weather Coupling (File Coupling)
5. Multiscale Agent-based Simulation + Weather Coupling (MUSCLE3 Coupling).

In the first approach, we only simulate the whole South Sudan location graph using FLEE rule set 2.0 in serial mode or singlescale only for comparison with multiscale approaches. It means that the serial mode doesn't follow the multiscale ruleset which needs dividing the whole model into sub-models. In approaches 2 and 3, we investigate multiscale simulations without weather coupling by using File I/O and MUSCLE3 coupling between sub-models. In approaches 4 and 5, we incorporate weather data into our multiscale simulations and again we use File I/O and MUSCLE3 coupling to pass agents between sub-models. The simulation is performed on a node with 32 cores, and a total of 12 GB of memory. For the comparison, Table 3 illustrates the results of these simulations on the Eagle supercomputer, including the total execution time, total validation error and coupling overhead for the aforementioned approaches.

Also, we deliberately avoid minimizing the validation error by calibrating existing model parameters against data. Because, it might lead to over-fitting which not only reduces the reusability of our simulations in new contexts, but also makes it highly sensitive to the (often incomplete) validation data sources we use. Therefore, we mostly incorporate data sources as model input, and combine them with our general knowledge and qualitative data about human behaviour.

Table 3. Comparison of the simulation approaches

Approaches		Total validation	Total execution	Coupling
		Error	Time (hh:mm)	Overhead
Whole South Sudan (Serial Mode)		0.431	02:21	1
Multiscale simulation	File coupling	0.507	01:25	0.60
	MUSCLE3 cloupling	0.507	01:25	0.56
Multiscale simulation	File coupling	0.510	09:01	3.83
+ Weather coupling	MUSCLE3 coupling	0.509	09:17	3.95

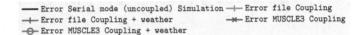

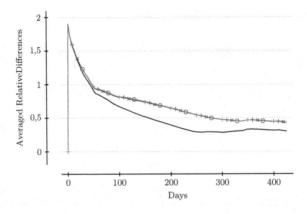

Fig. 8. Overview of the averaged relative differences for the taken approaches (Color figure online)

As results show, the total validation error of the serial mode approach, which is 0.431, is lower than multiscale approaches (0.507). However, the total execution time of multiscale approaches, which is 01:25 for both second and third approaches, is lower than serial mode (02:21) and they are pretty much lower than the approaches coupled with weather datasets (09:01 and 09:17). Hence, we can claim that despite having slightly higher validation error, multiscale approaches are much faster than the serial mode approach. Furthermore, to explain the reasons for high execution time for the weather coupled approaches, we have to point out that they ran longer in the early versions, and after a lot of optimizations we got to this time. To justify this, it can be said that perhaps due to the coupling and integration with static datasets and repeated reading of such data at each time step, the execution time of these approaches increases. The third column in Table 3 demonstrates these as the Coupling Overhead which assumes the serial mode as a benchmark to calculate the deviations of other approaches. Therefore, the coupling overhead for serial mode is 1. The values

0.60 and 0.56 complying faster pace of multiscale approaches without weather coupling, and values 3.83 and 3.95 are for weather coupled approaches which comply with their lower pace comparing to the serial mode.

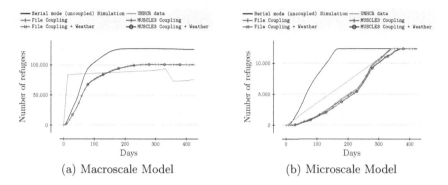

(a) Macroscale Model   (b) Microscale Model

Fig. 9. Sample camps simulation results for 426 days from 1 Jun 2016 till 31 July 2017

Nevertheless, the overall validation error for the South Sudan simulation is relatively high, because many people do not use the main routes and their choice of destination is strongly affected by the weather conditions. We are in the process of incorporating these phenomena in these simulations, and the validation errors presented for the multiscale simulations represent preliminary, not final, results.

The overview plots for the averaged relative differences for Serial mode (uncoupled) simulation (black line), file coupling (red line), file coupling (green line) file coupling + weather (red line), MUSCLE3 coupling (violet line) and MUSCLE3 coupling + weather (blue line) for the aggregated macroscale and microscale models are illustrated in Fig. 8. However, because both file and MUSCLE3 coupling approaches do the same data exchange and the weather coupled models are very limited in this case, all the coupled results are overlapping in this figure. Besides, the simulation results for sample camps at both macro (Kakuma) and micro level (Okugo) regions are depicted to show the difference between taken approaches at the camp arrival forecasting level (see Fig. 9a and 9b).

6 Conclusion

In this paper, we presented a multiscale simulation approach for modelling forced migration in South Sudan, and described how different coupling approaches have an effect on the total execution time, validation error and coupling overhead. We investigated file I/O based and MUSCLE3 based coupling approaches. Also, we integrated a weather data source with our microscale model, to determine realistic agent movements e.g. the changes in road accessibility due to flooding. Our multiscale models result in a higher validation error than our single-scale models, which points towards the need to add further details in our coupling implementation and do a larger-scale validation exercise. Besides, the location graph needs to be revised and the coupled locations should be reconsidered to properly simulate refugees' movement in multiscale models. In terms of runtime, we find out that the macro-micro model coupling outperforms the weather coupling, that needs to be improved considerably.

In general, modelling the South Sudan conflict while taking all these aspects into account is a highly demanding endeavour, mainly because the collection of input and validation data is often very challenging. We, therefore, see this work as a first major step in a sequence of iterations towards a highly detailed population displacement model of this conflict.

Acknowledgements. This work has been supported by the HiDALGO project and has been partly funded by the European Commission's ICT activity of the H2020 Programme under grant agreement number: 824115. This paper expresses the opinions of the authors and not necessarily of the European Commission. The European Commission is not liable for any use that may be made of the information contained in this paper.

References

1. Groen, D.: Development of a multiscale simulation approach for forced migration. In: Shi, Y., et al. (eds.) ICCS 2018. LNCS, vol. 10861, pp. 869–875. Springer, Cham (2018). https://doi.org/10.1007/978-3-319-93701-4_69
2. IOM: South Sudan 2017 Humanitarian Response Plan (January - December 2017), Technical report, UN Office for the Coordination of Humanitarian Affairs (2017)
3. IOM: 2017 consolidated appeal - IOM South Sudan, Technical report, International Organization for Migration (2017)
4. OCHA: South Sudan: $1.6 billion needed to provide life-saving assistance and protection to 5.8 million people across the country (2017)
5. Anderson, J., Chaturvedi, A., Cibulskis, M.: Simulation tools for developing policies for complex systems: Modeling the health and safety of refugee communities. Health Care Manag. Sci. **10**(4), 331–339 (2007)
6. Johnson, R.T., Lampe, T.A., Seichter, S.: Calibration of an agent-based simulation model depicting a refugee camp scenario. In: Proceedings of the 2009 Winter Simulation Conference (WSC), pp. 1778–1786. IEEE (2009)
7. Bishai, D., Paina, L., Li, Q., Peters, D.H., Hyder, A.A.: Advancing the application of systems thinking in health: why cure crowds out prevention. Health Res. Policy Syst. **12**(1), 1–12 (2014)
8. Lemos, C., Coelho, H., Lopes, R.J., et al.: Agent-based modeling of social conflict, civil violence and revolution: state-of-the-art-review and further prospects. In: EUMAS, pp. 124–138, Toulouse (2013)
9. Crooks, A.T., Wise, S.: GIS and agent-based models for humanitarian assistance. Comput. Environ. Urban Syst. **41**, 100–111 (2013)
10. Simon, M., Schwartz, C., Hudson, D., Johnson, S.D.: A data-driven computational model on the effects of immigration policies. Proc. Natl. Acad. Sci. **115**(34), E7914–E7923 (2018)

11. Klabunde, A., Willekens, F.: Decision-making in agent-based models of migration: state of the art and challenges. Eur. J. Popul. **32**(1), 73–97 (2016)
12. Groen, D., Bell, D., Arabnejad, H., Suleimenova, D., Taylor, S.J.E., Anagnostou, A.: Towards modelling the effect of evolving violence on forced migration. In: 2019 Winter Simulation Conference (WSC), pp. 297–307. IEEE (2019)
13. Chopard, B., Falcone, J.-L., Kunzli, P., Veen, L., Hoekstra, A.: Multiscale modeling: recent progress and open questions. Multiscale Multidisc. Model. Exp. Des. **1**(1), 57–68 (2018)
14. Gilbert, N., Ahrweiler, P., Barbrook-Johnson, P., Narasimhan, K.P., Wilkinson, H.: Computational modelling of public policy: reflections on practice. J. Artif. Soc. Soc. Simul. **21**(1), 14 (2018)
15. Suleimenova, D., Groen, D.: How policy decisions affect refugee journeys in South Sudan: a study using automated ensemble simulations. J. Artif. Soc. Soc. Simul. (2020)
16. Searle, C., van Vuuren, J.: Modelling forced migration: a framework for conflict-induced forced migration modelling according to an agent-based approach. Comput. Env. Urban Syst. **85**, 101568 (2021)
17. Alowayyed, S., Groen, D., Coveney, P.V., Hoekstra, A.G.: Multiscale computing in the exascale era. J. Comput. Sci. **22**, 15–25 (2017)
18. Disney, G., Wiśniowski, A., Forster, J.J., Smith, P.W.F., Bijak, J.: Evaluation of existing migration forecasting methods and models, Report for the Migration Advisory Committee: Commissioned research. University of Southampton, ESRC Centre for Population Change (2015)
19. Edwards, S.: Computational tools in predicting and assessing forced migration. J. Refugee Stud. **21**, 347–359 (2008)
20. Ahmed, M.N., et al.: A Multi-scale approach to data-driven mass migration analysis. In: SoGood@ ECML-PKDD, p. 17 (2016)
21. Abel, G.J., Brottrager, M., Cuaresma, J.C., Muttarak, R.: Climate, conflict and forced migration. Glob. Environ. Chang. **54**, 239–249 (2019)
22. Black, R., Adger, W.N., Arnell, N.W., Dercon, S., Geddes, A., Thomas, D.: The effect of environmental change on human migration. Glob. Environ. Chang. **21**, S3–S11 (2011)
23. Suleimenova, D., Bell, D., Groen, D.: A generalized simulation development approach for predicting refugee destinations. Sci. Rep. **7**(1), 1–13 (2017)
24. Veen, L.E., Hoekstra, A.G.: MUSCLE3 Readthedocs (2020)
25. WorldBank: South sudan historical climate data (2016). https://climateknowledge portal.worldbank.org/country/south-sudan/climate-data-historical
26. Hersbach, H., et al.: Era5 hourly data on single levels from 1979 to present (2018). https://doi.org/10.24381/cds.adbb2d47l. Accessed 11 Feb 2021
27. Harrigan, S., Zsoter, E., Barnard, C.W.F., Salamon, P., Prudhomme, C.: River discharge and related historical data from the global flood awareness system, v2.1 (2019). https://doi.org/10.24381/cds.a4fdd6b9. Accessed 11 Feb 2021

Evaluating WRF-BEP/BEM Performance: On the Way to Analyze Urban Air Quality at High Resolution Using WRF-Chem+BEP/BEM

Veronica Vidal[1,2](✉) ⓘ, Ana Cortés[1]ⓘ, Alba Badia[2]ⓘ, and Gara Villalba[2]ⓘ

[1] Departament d'Arquitectura de Computadors i Sistemes Operatius,
Universitat Autònoma de Barcelona, Bellaterra, Spain
{veronica.vidal,ana.cortes}@uab.cat
[2] Institut de Ciència i Tecnologia Ambientals, Universitat Autònoma de Barcelona,
Bellaterra, Spain
{alba.badia,gara.villalba}@uab.cat

Abstract. Air pollution exposure is a major environmental risk to health and has been estimated to be responsible for 7 million premature deaths worldwide every year. This is of special concern in cities, where there are high levels of pollution and high population densities. Not only is there an urgent need for cities to monitor, analyze, predict and inform residents about the air quality, but also to develop tools to help evaluate mitigation strategies to prevent contamination. In this respect, the Weather Research and Forecasting model coupled with chemistry (WRF-Chem) is useful in providing simulations of meteorological conditions but also of the concentrations of polluting species. When combined with the multi-layer urban scheme Building Effect Parameterization (BEP) coupled with the Building Energy Model (BEM), we are furthermore able to include urban morphology and urban canopy effects into the atmosphere that affect the chemistry and transport of the gases. However, using WRF-Chem+BEP/BEM is computationally very expensive especially at very high urban resolutions below 5 km. It is thus indispensable to properly analyze the performance of these models in terms of execution time and quality to be useful for both operational and reanalysis purposes. This work represents the first step towards this overall objective which is to determine the performance (in terms of computational time and quality of results) and the scalability of WRF-BEP/BEM. To do so, we use the case study of Metropolitan Area of Barcelona and analyze a 24-h period (March 2015) under two with different Urban schemes (Bulk and BEP/BEM). We analyze the execution time by running the two experiments in its serial configuration and in their parallel configurations using 2, 4, 8, 16, 32 and 64 cores. And the quality of the results by comparing to observed data from four meteorological stations in Barcelona.

Keywords: WRF-BEP/BEM · WRF-Chem · Scalability · Air quality · Urban scale

© Springer Nature Switzerland AG 2021
M. Paszynski et al. (Eds.): ICCS 2021, LNCS 12746, pp. 516–527, 2021.
https://doi.org/10.1007/978-3-030-77977-1_41

1   Introduction

Understanding urban atmosphere behavior is fundamental to know how air pollution exposure affects citizens. Air pollution exposure is a significant environmental risk to health in highly populated areas and is responsible for an estimated 7 million people deaths every year worldwide [6,17], which is 3.5 times more deaths than for COVID-19 in 2020 [3]. Countries can reduce serious diseases by reducing air pollution levels in their major cities and making them more sustainable. Simulation and analysis of the urban atmosphere and air pollution in our cities can help the public to be more informed and urban planners to make better decisions. However, urban air quality models are complex, mainly due to the diversity of spatio-temporal scales on which the phenomena occur. In particular, two essential scales are involved [8]:

1. An 'urban' scale of a few tens of kilometers (city size), where the primary pollutants are emitted;
2. A 'meso' scale of a few hundreds of kilometers, where the secondary pollutants are formed and dispersed.

The dispersion of pollutants depends on the structure of the urban boundary layer and its interactions with the rural boundary layer and the synoptic flow. Being a nonlinear system, it is common to use numerical models to study air pollution problems. In order to compute the mean and turbulent transport and the chemical transformations of pollutants, several meteorological variables are needed (wind, turbulent coefficients, temperature, pressure, humidity), which can be interpolated from measurements or computed with mesoscale circulation models. These models must, indeed, ideally be able to represent the two main scales (the 'urban' and the 'meso') involved. Since the horizontal dimensions of the domain are on the order of the mesoscale (100 km), to keep the number of grid points compatible with the CPU time cost, the horizontal grid resolution of such (mesoscale) models ranges, in general, between several hundreds of metres and a few kilometres. This means that it is not possible to resolve the city structure in detail (buildings or blocks), but that the effects of the urban surfaces must be parameterised. Another obstacle to a complete resolution of the city structure is given by the difficulty to provide the necessary input data [8].

For that purpose, as a first step, we use the Weather Research and Forecasting (WRF) model with the incorporation of the urban canopy model Building Energy Parameterization and Building Energy Model (BEP/BEM) to take into account the urban morphology [11,12]. After that, we will use WRF coupled with chemistry (WRF-Chem; [4]), which adds calculations with chemical substances to the weather simulations, including additional emission files and chemical traces. However, WRF-Chem coupled with BEP/BEM is computationally very expensive, either for reanalysis or operational purposes and especially at very high urban resolutions, so that it makes indispensable understand WRF-Chem + BEM/BEM computational behaviour helping its users and developers to better exploit computational resources using computational strategies in

High Performance Computing (HPC) platforms. This work represents the first step towards this global objective. Concretely, this paper presents the WRF-BEP/BEM scalability. The quality of the results provided has been analyzed for a study case of March 11th, 2015, in Barcelona's urban area (Spain).

This paper is organised as follows. Section 2 describes the Weather Research and Forecasting (WRF) model and the Building Effect Parameterization coupled to the Building Energy Model (BEP/BEM) and describes the experimental study case. Section 3 presents the model scalability and it also includes an analysis of the quality results by comparing the model outputs to the observations. Finally, Sect. 4 summarizes the main conclusions and future work.

2   Data and Methods

2.1   Modelling System

Weather Research and Forecasting Model (WRF):   The Weather Research and Forecasting (WRFV4) model is an atmospheric modeling system designed for both research and numerical weather prediction [14]. The Advanced Research WRF (ARW) dynamics solver integrates the compressible, non-hydrostatic Euler equations with several physics schemes and dynamics options designed to phenomena at regional scale [14]. The WRF-ARW model was developed by NCAR among other organisations and is designed to be an efficient massively-parallel code to be able to take advantage of advanced high-performance computing systems [13].

Building Effect Parameterization Coupled to the Building Energy Model (BEP/BEM):   As more than half of the world's population lives in urban areas and this proportion is expected to grow up to 64–69% by 2050 [18], there is a growing interest in simulating the urban atmosphere and its complex dynamics. For that reason, with the aim of reproducing the effects of the urban canopy on the urban boundary layer (UBL) dynamics [12], new features has been included in the main WRF parameterization [1,9]. The level of detail and the degree of complexity of these urban parameterizations depends on the number and nature of the processes described and their integration with the primitive equations of the mesoscale model.

The bulk scheme is the simplest approach, included in the Noah Land Surface Model, which modifies several parameters such as roughness length, albedo, volumetric heat capacity and soil thermal conductivity to better represent the reduced wind speed and increased heat storage capacity of urban areas. This scheme estimates heat and momentum fluxes based on the Monin-Obukhov Similarity Theory (MOST) and does not take into account the heterogeneity within the city, i.e., these values apply to all urban areas, regardless of their urban structure [9,10]. Urban canopy schemes were developed and included in WRF as physics options. These more advanced urban schemes were explicitly designed to represent city morphology (e.g., building and street canyon geometry) and surface characteristics (e.g., albedo, heat capacity, emissivity, urban/vegetation fraction). The currently available urban canopy schemes are:

- Single Layer Urban Canopy Model (SLUCM - [7]);
- Building Effect Parameterization (BEP, [8]): a multi-layer scheme;
- BEP coupled to the Building Energy Model (BEP/BEM, [8,12]: the second generation of BEP considers energy consumption in buildings (heating/cooling) for a more accurate effect on urban heat budget.

BEP parameterizes a 3D urban morphology in a multi-layer model grid, being capable of estimating the heat fluxes from roofs, ground, and walls, individually [8] and computing the impact of buildings on the airflow and turbulence (term included in the conservation equation for the Turbulent Kinetic Energy – TKE), as well as the source/sinks of heat by solving the energy budget for each surface [8,9]. Unlike BEP that keeps the indoor temperature constant, BEP/BEM calculates the anthropogenic heat generated by air conditioning systems and the heat exchanges between the building's interior and the outer atmosphere [10,12]. In this work, BULK and BEP/BEM options are evaluated to study the computational performance of such schemes (see Table 1) in the Barcelona urban region.

Table 1. Parameterizations used for each experiment (same as in [10])

WRF schemes	Bulk	BEP/BEM
Urban scheme	Included in the Noah (LSM) Land Surface Model	BEP/BEM
Land Surface Model	Noah LSM	
PBL scheme	Bougeault-Lacarrère PBL (BouLac), designed to use with urban schemes	
Microphysics	WRF Single Moment 6-class scheme	
Long- and short-wave radiation	Rapid Radiative Transfer Model for General circulation models (RRTMG) scheme	

Atmospheric Chemistry Using WRF-Chem: WRF-Chem is a full online atmospheric chemistry model with many options and some interactions with physics aerosols affecting radiation and microphysics. Typically it requires emission source maps as additional inputs [4,14]. Although the complete study will include this feature, in this work, as it was previously mentioned, we do not consider the Chem module.

2.2   Experimental Study Case

The study case selected was the day March 11, 2015 which corresponds to a day within a period of high temperatures in Catalonia (North-East of Spain). From the 1st to the 13th of March 2015 the meteorological situation in Catalonia was marked by a strong anticyclone that stabilised the atmosphere and caused a significant increase in temperature. It is worth mentioning that in certain stations with more than 20 years of data, recorded the maximum temperature value for all recorded series for the first half of March [15]. Due to the anticyclone and the high pressure situation, urban air pollution also increased in Barcelona making this a good period to study both the urban heat (WRF-BEP/BEM) as well as urban pollution and atmospheric chemistry (WRF-Chem). This work focuses on urban heat and scaling simulations of WRF and WRF-BEP/BEM. To simulate the study region, two-way nested domains were defined. The parent domain (Iberian Peninsula) with a horizontal resolution of 9 km × 9 km (WE: 1350 km, NS: 1305 km), followed by a finer domain comprehending the Catalonia region at 3 km × 3 km horizontal resolution (WE: 354 km; NS: 354 km) (Fig. 1). Vertically, both domains are described by 57 layers (model top pressure: 0.1 hPa).

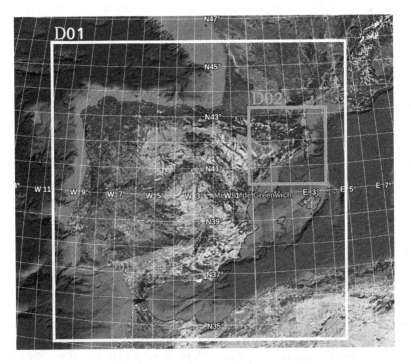

Fig. 1. Simulation domain configuration. D01: Iberian Peninsula (IP), D02: Catalonia (CAT) with 9 km × 9 km and 3 × 3 km horizontal resolution, respectively.

As initial and boundary meteorological conditions to the parent domain we use ERA5 reanalysis [5,16], from the European Center for Medium-Range

Weather Forecasts (ECMWF), which provides hourly data and a good resolution of 0.25° (31 km).

3 Results

3.1 Quality Results

The observational data used to evaluate the WRF output was provided by the Meteorological Service of Catalonia (SMC). We compared data from four stations located in the Metropolitan Area of Barcelona (see Table 2).

Table 2. Meteorological stations

Station acronym	Measurement	Altitude above ground (m)	Latitude	Longitude
Badalona	T, RH	50	41.452	2.248
Raval	T, RH, W	40	41.384	2.168
Zuni	T, RH, W	10	41.379	2.105
ObsFabra	T, RH, W	411	41.418	2.124

Temperature (T): In Fig. 2, we can observe the temperature daily profile comparison from the stations of Badalona, Raval, Zona Universitaria (ZUni), and Observatori Fabra (ObsFabra). Figure 2(a) shows the observed temperature against the obtained temperature from bulk and BEP/BEM simulations. From 00 to 03h the simulations estimate an increase in the temperature while we observe a reduction of the observational value. During this period time, both implementations underestimate the temperature. From 6 o'clock, we can see that both models estimate an increase of the temperature until reaching a peak at noon. The simulations and observations follow a similar temperature variation trend from this hour on. Figure 2(b) shows that BEP/BEM tends to overestimate the temperature value more than BULK except between 9 and 18 o'clock. Neither of the two models accurately describes the observations since they both underestimate the T by about two degrees at midday. Figure 2(c) shows that the two schemes give higher temperature values during the whole day. The trend of the simulated temperature variation is close to the observed one. In Fig. 2(d), the observed temperature is approximately constant from 00 to 09h, while the two schemes predict an increase in temperature from 00 till to 12 h. The two simulations describe correctly the evolution of the temperature. From 18h, the simulations underestimate again, estimating a decrease in temperature while the observations indicate that the temperature remained constant.

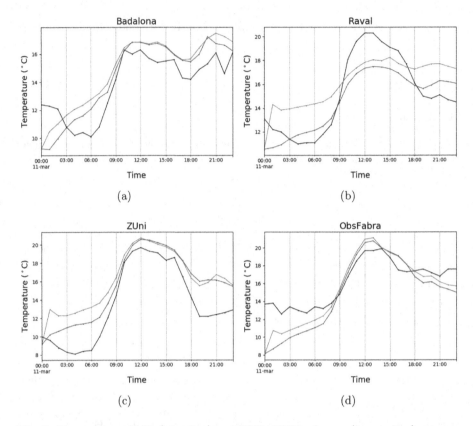

Fig. 2. Temperature. Bulk (blue line) and BEP+BEM schemes (orange line) comparison with meteorological observation (black line) (Color figure online)

Relative Humidity (RH): In Fig. 3(a), we can see both simulations tend to underestimate the RH except from 00 to 02 h that underestimate it. During these hours of the day RH is high in the four stations, after compare with ERA5 global model it seems that this high RH's come from initial conditions. Figure 3(b) shows that the RH is underestimated from 03 to 09 h and from 17 to 00 h, while in central hours, between 9 and 17 h, the simulation estimates are very close to reality. Moreover, BULK is maintained with a higher temperature all day long. In Fig. 3(c) we can see similar case to Raval, but in this case, BEP/BEM estimates higher relative humidity than bulk, being closer to the value of the observations. Figure 3(d) presents that the model estimates the observed relative humidity better at low RH, while in night hours, the simulation estimates and the observations diverge.

Wind Speed (WS): In the Figs. 4(a) and 4(b), both simulations describe the wind speed trend, but BEPBEM fits better to reality in this simulation. In Fig. 4(b) BEP/BEM fits again very well and better than bulk, which

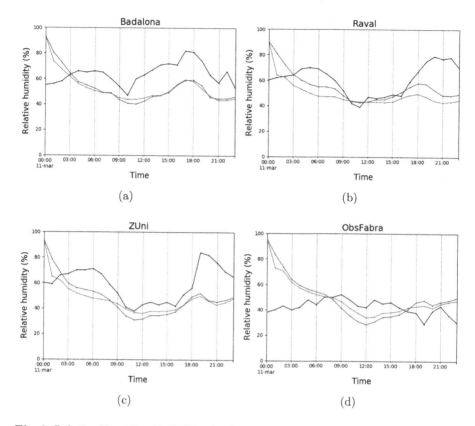

Fig. 3. Relative Humidity. Bulk (blue line) and BEP/BEM schemes (orange line) comparison with meteorological observation (black line) (Color figure online)

overestimates in 1.5 m/s the wind speed value. Figures 4(c) and 4(d) show that BEP/BEM underestimates the wind speed. Instead, Bulk at (Fig. 4(c)) overestimates it from 00 to 9 o'clock, in central hours of the day it follows a similar tendency and from 20 o'clock it overestimates the wind speed. In Fig. 4(d) the tendency at the end of the day, from 18 h, tends slightly to observation values.

For more results in this region using WRF-BEP/BEM see the study carry out by Ribeiro et al. (2021) [10].

3.2   Scalability Results

The computing platform used for these experiments using distributed memory is a multi-core system composed of 2 sockets with AMD EPYC 7551 32-Core Processors each and a total of 128 CPUs (64 cores and 128 threads). In order to analyse the parallelization and the scalability of WRF-BEP/BEM, we have executed the two experiments in its serial configuration and in their parallel configurations using 2, 4, 8, 16, 32 and 64 cores.

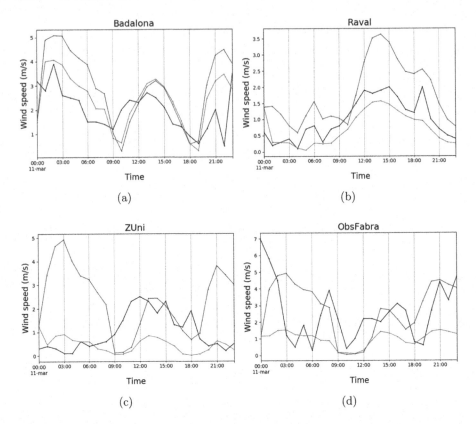

Fig. 4. Wind speed. Bulk (blue line) and BEP/BEM schemes (orange line) comparison with meteorological observation (black line) (Color figure online)

WRF-BEP/BEM has been parallelized using MPI. The MPI parallelization is done through patches; that is, the domain is divided into a fixed number of parts according to the number of cores available for a given simulation. These p ince each patch needs information from its neighbors' patches to run the model, each patch includes extra points (called halo) to incorporate those points from the four borders that are required to execute on one iteration of the model. After finishing each iteration, the patches must exchange the results from the points in the halo. Therefore, a synchronization barrier is required. This scheme implies that all MPI processes proceed in a synchronized fashion what can imply non-depreciable communication time if the patch size is not well evaluated [2].

Figure 5 shows execution times for each experiment. As we can see, BEP/BEM scheme needs higher execution times, especially for his serial configuration, in the parallel configuration there are less difference in time execution between both schemes but is still higher in BEP/BEM. At 32 cores and above, we can see that allocating more resources does not result in a reduction of execution time proportional to the resources invested. From 32 cores to 64 cores

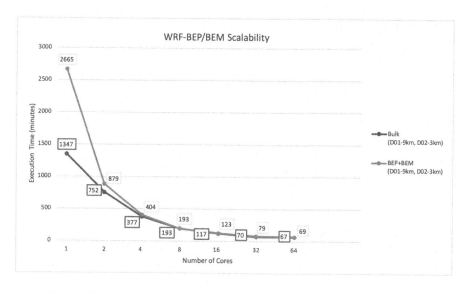

Fig. 5. WRF-BEP/BEM execution time when running a 24 h simulation

only have a time execution improvement of 10 min with BEP/BEM and 3 min with Bulk.

Figure 6 also shows that Bulk scales well up to 32 cores where the curve starts to flatten, and we do not have the performance improvement we would expect at 64 cores, getting a speedup of ×20. In WRF-BEP/BEM, although the curve also begins to flatten, we are getting a speedup of ×39 with 64 cores.

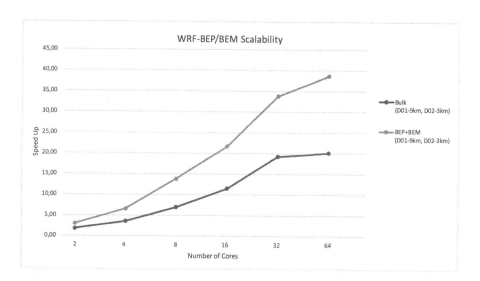

Fig. 6. WRF and WRF-BEP/BEM speed up when executing a 24 h simulation.

4   Conclusions and Future Steps

Urban air quality models are complex, mainly due to the diversity of spatio-temporal scales on which these phenomena occur. To evaluate the air quality in urban zones, the combination of both WRF-Chem model including BEP/BEM. However, these coupled models are computationally very expensive, especially at very high urban resolutions and for these reason we start evaluating the WRF-BEP/BEM performance.

This work represents the first step towards this global objective. Concretely, this paper has presented the WRF-BEP/BEM scalability. The quality of the results provided has been analyzed for a study case of March 11th, 2015, in Barcelona's urban area (Spain).

The analysis performed of both implementations, Bulk and BEP/BEM, describes the evolution of temperature and humidity similarly. They have some differences in wind speed, more accurate estimation of BEP/BEM in this simulation. However, these are preliminary results, and we need to simulate more days and analyze more stations to reach better quality results.

Regarding the scalability analysis, the main problem we face is that BEP/BEM is computationally more expensive than Bulk. Moreover, in this work, we only considered two domains with a 3 km resolution, and BEP/BEM substantially increases the resource requirements by adding the third domain with a resolution of 1 km. For this reason, the execution time spent by BEP/BEM slightly differs from the time invested in Bulk. BEP/BEM configuration scales better than Bulk, which reaches its maximum performance when using 32 cores. At the same time, BEP/BEM could still reduce the execution time by increasing the number of cores.

Future steps will be oriented to complete the Bulk and BEP/BEM's scalability study introducing also a third domain of 1 km-resolution. After that, we will continue our study introducing WRF-Chem and WRF-Chem+BEP/BEM configuration and studying their scalabilities and computational performance, always taking into account the simulations' quality results. Once we have the complete WRF-Chem+BEP/BEM scalability study, we will analyze and evaluate WRF-Chem+BEP/BEM bottlenecks and apply computational strategies to run WRF-Chem+BEP/BEM at urban resolutions while using moderated execution times.

Acknowledgments. This research has been supported by MINECO-Spain under contract TIN2017-84553-C2-1-R, and by the Spanish government under grant PRE2018-085425.

References

1. Chen, F., et al.: The integrated WRF/urban modelling system: development, evaluation, and applications to urban environmental problems. Int. J. Climatol. **31**(2), 273–288 (2011). https://doi.org/10.1002/joc.2158

2. Farguell, A., Cortés, A., Margalef, T., Miró, J.R., Mercader, J.: Scalability of a multi-physics system for forest fire spread prediction in multi-core platforms. J. Supercomput. **75**(3), 1163–1174 (2019). https://doi.org/10.1007/s11227-018-2330-9
3. GOOGLE: Coronavirus (COVID-19). https://news.google.com/covid19/map
4. Grell, G.A., et al.: Fully coupled "online" chemistry within the WRF model. Atmos. Environ. **39**(37), 6957–6975 (2005). https://doi.org/10.1016/j.atmosenv.2005.04. 027
5. Hersbach, H., et al.: The ERA5 global reanalysis. Q. J. Roy. Meteorol. Soc. **146**(730), 1999–2049 (2020). https://doi.org/10.1002/qj.3803
6. Kuehn, B.: Who: more than 7 million air pollution deaths each year. JAMA, J. Am. Med. Assoc. **311**, 1486 (2014). https://doi.org/10.1001/jama.2014.4031
7. Kusaka, H., Kimura, F.: A simple single-layer urban canopy model for atmospheric models: comparison with multi-layer and SLAB models. Bound.-Layer Meteorol. **101**(ii), 329–358 (2001). https://doi.org/10.1023/A:1019207923078
8. Martilli, A., Clappier, A., Rotach, M.W.: An urban surface exchange parameterisation for mesoscale models. Bound.-Layer Meteorol. **104**(2), 261–304 (2002). https://doi.org/10.1023/A:1016099921195
9. de la Paz, D., Borge, R., Martilli, A.: Assessment of a high resolution annual WRF-BEP/CMAQ simulation for the urban area of Madrid (Spain). Atmos. Environ. **144**, 282–296 (2016). https://doi.org/10.1016/j.atmosenv.2016.08.082
10. Ribeiro, I., Martilli, A., Falls, M., Zonato, A., Villalba, G.: Highly resolved WRF-BEP/BEM simulations over Barcelona urban area with LCZ. Atmos. Res. **248**, 105220 (2021). https://doi.org/10.1016/j.atmosres.2020.105220
11. Salamanca, F., Krpo, A., Martilli, A., Clappier, A.: A new building energy model coupled with an urban canopy parameterization for urban climate simulations-part I. Formulation, verification, and sensitivity analysis of the model. Theor. Appl. Climatol. **99**(3–4), 331–344 (2010). https://doi.org/10.1007/s00704-009-0142-9
12. Salamanca, F., Martilli, A.: A new building energy model coupled with an urban canopy parameterization for urban climate simulations-part II. Validation with one dimension off-line simulations. Theor. Appl. Climatol. **99**(3–4), 345–356 (2010). https://doi.org/10.1007/s00704-009-0143-8
13. Shainer, G., et al.: Weather research and forecast (WRF) model: performance analysis on advanced multi-core HPC clusters. In: The 10th LCI International Conference on High Performance Clustered Computing, pp. 1–14 (2009)
14. Skamarock, W., Al., E.: A description of the advanced research WRF model version 4. NCAR (2019)
15. SMC: Climatic report. Servei Meteorològic de Catalunya. Departament de Territori i Sostenibilitat, March 2015. https://www.meteo.cat/wpweb/climatologia/el-clima-ara/butlleti-mensual/
16. Tetzner, D., Thomas, E., Allen, C.: A validation of ERA5 reanalysis data in the southern Antarctic peninsula-Ellsworth land region, and its implications for ice core studies. Geosciences **9**(7), 289 (2019). https://doi.org/10.3390/geosciences9070289
17. WHO: Air pollution deaths per year. https://www.who.int/news-room/air-pollution
18. Wilmoth, J.: Global demographic projections: future trajectories and associated uncertainty. https://www.un.org/en/development/desa/population/commission/ pdf/48/sideEvents/14April2015_GlobalPopulationProjections_Presentation.pdf

Pathology Dynamics in Healthy-Toxic Protein Interaction and the Multiscale Analysis of Neurodegenerative Diseases

Swadesh Pal[1] and Roderick Melnik[1,2](✉)

[1] MS2Discovery Interdisciplinary Research Institute, Wilfrid Laurier University, Waterloo, Canada
rmelnik@wlu.ca
[2] BCAM-Basque Center for Applied Mathematics, Bilbao, Spain

Abstract. Neurodegenerative diseases are frequently associated with aggregation and propagation of toxic proteins. In particular, it is well known that along with amyloid-beta, the tau protein is also driving Alzheimer's disease. Multiscale reaction-diffusion models can assist in our better understanding of the evolution of the disease. Based on a coarse-graining procedure of the continuous model and taking advantage of the brain data connectome, a computationally challenging network mathematical model has been described where the edges of the network are the axonal bundles in white-matter tracts. Further, we have modified the heterodimer model in such a way that it can now capture some of the critical characteristics of this evolution such as the conversion time from healthy to toxic proteins. Finally, we have analyzed the modified model theoretically and validated the theoretical findings with numerical simulations.

Keywords: Alzheimer's disease · Coupled multiscale models · Amyloid-beta and tau proteins · Neurodegenerative disorders · Holling type-II

1 Introduction

Alzheimer's disease (AD) is an example of a neurodegenerative disease, associated with aggregation and propagation of toxic proteins [1]. Initially, the "amyloid cascade hypothesis" has dominated for the treatments [2,3]. However, due to the failures of large clinical trials, researchers started focussing on some other mechanisms. It is now well accepted that the tau-protein (τP) is a viable alternative to the "amyloid cascade hypothesis".

Authors are grateful to the NSERC and the CRC Program for their support. RM is also acknowledging support of the BERC 2018–2021 program and Spanish Ministry of Science, Innovation and Universities through the Agencia Estatal de Investigacion (AEI) BCAM Severo Ochoa excellence accreditation SEV-2017-0718 and the Basque Government fund AI in BCAM EXP. 2019/00432.

© Springer Nature Switzerland AG 2021
M. Paszynski et al. (Eds.): ICCS 2021, LNCS 12746, pp. 528–540, 2021.
https://doi.org/10.1007/978-3-030-77977-1_42

The τP plays a prominent role as a secondary agent in the disease development. For example, (i) frontotemporal lobar degeneration is mostly dominated by τP spreading [4], (ii) neurofibrillary tangles (NFT) are correlated in brain atrophy in AD [5,6], (iii) lower τP concentration prevents neuronal loss [7], (iv) τP also reduces neural activity [8], etc. This helps to explain the relative lack of clinical improvements. There is an open debate in the literature on the roles of $A\beta$ proteopathy and τP tauopathy in AD but it is clear by now that "the amyloid $-\beta-\tau$ nexus is central to disease-specific diagnosis, prevention and treatment" [9]. In recent years, many researchers have focussed on $A\beta$ and τP interaction. Moreover, in neurodegenerative diseases, the protein-protein interactions become a key for understanding both spreading and toxicity of the proteins [10–13]. There are some crucial observations specific to AD [10,14]: (i) the seeding of new toxic τP is enhanced by the presence of $A\beta$, (ii) the toxicity of $A\beta$ depends on the presence of τP, and (iii) $A\beta$ and τP amplify each other's toxic effects.

Mathematical models are widely used for the interpretation of biological processes, and this field is not an exception. Building on earlier advances [15–20], we analyze AD by a deterministic mathematical modelling approach and predict the dynamics of the disease based on several novel features of our model. Recall that the heterodimer model describes the interaction between the healthy and toxic proteins [21–25]. In the heterodimer model, with an increase in the healthy protein density, the toxic protein conversion increases. We have modified that linear conversion term to include nonlinear effects via Holling type-II functional response. In this case, the toxic protein's conversion rate remains constant with an increase in the healthy protein density. This modification incorporates the reaction time (conversion from healthy protein to toxic protein). AD is itself a complex and multiscale disease. The introduction of the reaction time, along with the conventional wave propagation times, reflects a multiscale character of the modified model. We have considered two modified coupled heterodimer systems for healthy-toxic interactions for both proteins, $A\beta$ and τP, along with a single balance interaction term. This study identifies two types of disease propagation modes depending on the parameters: primary tauopathy and secondary tauopathy.

Finally, we note that a network mathematical model can be constructed from the brain data by a coarse-graining procedure of the continuous model (e.g., [14]). In this case, we need to define the network nodes in the region or domain where we are interested to see the dynamics. The edges of the network are the axonal bundles in white-matter tracts. Here, the network model in the brain connectome becomes an undirected weighted graph, and the weights of the graph are used to construct the adjacency matrix and hence the Laplacian of the graph. Studying AD in the whole brain connectome is computationally very challenging. One of the efficient and logical ways to proceed is to investigate AD in the brain connectome by fixing some crucial nodes and edges, which we are currently undertaking. In the current manuscript, we provide a brief description of such a network model, but our main focus here is to establish the speed of wave propagation of toxic fronts of the two modes of primary and secondary tauopathies for the modified reaction-diffusion model.

This manuscript is organized as follows. In Sect. 2, we briefly discuss the heterodimer model and its modification. Temporal behaviour of the modified model is analyzed in Sect. 3, focussing on possible stationary points and linear stability. In Sect. 4, we provide results on the wave propagation described by a simplified model, specific to stationary states. Section 6 is devoted to a detailed analysis of the AD propagation in terms of primary and secondary tauopathies. Concluding remarks are given in Sect. 7.

2   Mathematical Model

We first consider the usual heterodimer model for the healthy and toxic variants of the proteins $A\beta$ and τP. Let Ω be a spatial domain in $\mathbb{R}^3$. For $\mathbf{x} \in \Omega$ and time $t \in \mathbb{R}^+$, let $u = u(\mathbf{x}, t)$ and $\widetilde{u} = \widetilde{u}(\mathbf{x}, t)$ be healthy and toxic concentrations of the protein $A\beta$, respectively. Similarly, we denote $v = v(\mathbf{x}, t)$ and $\widetilde{v} = \widetilde{v}(\mathbf{x}, t)$ the healthy and toxic concentrations of τP, respectively. The concentration evolution is then given by the following system of coupled partial differential equations [14, 22]:

$$\frac{\partial u}{\partial t} = \nabla \cdot (\mathbf{D_1} \nabla u) + a_0 - a_1 u - a_2 u \widetilde{u}, \tag{1a}$$

$$\frac{\partial \widetilde{u}}{\partial t} = \nabla \cdot (\widetilde{\mathbf{D}}_1 \nabla \widetilde{u}) - \widetilde{a}_1 \widetilde{u} + a_2 u \widetilde{u}, \tag{1b}$$

$$\frac{\partial v}{\partial t} = \nabla \cdot (\mathbf{D_2} \nabla v) + b_0 - b_1 v - b_2 v \widetilde{v} - b_3 \widetilde{u} v \widetilde{v}, \tag{1c}$$

$$\frac{\partial \widetilde{v}}{\partial t} = \nabla \cdot (\widetilde{\mathbf{D}}_2 \nabla \widetilde{v}) - \widetilde{b}_1 \widetilde{v} + b_2 v \widetilde{v} + b_3 \widetilde{u} v \widetilde{v}. \tag{1d}$$

In system (1), a_0 and b_0 are the mean production rates of healthy proteins, $a_1, b_1, \widetilde{a}_1$ and $\widetilde{b}_1$ are the mean clearance rates of healthy and toxic proteins, and a_2 and b_2 represent the mean conversion rates of healthy proteins to toxic proteins. The parameter b_3 is the coupling constant between the two proteins $A\beta$ and τP. Further, $\mathbf{D}_1, \widetilde{\mathbf{D}}_1, \mathbf{D}_2$ and $\widetilde{\mathbf{D}}_2$ are the diffusion tensors which characterize the spreading of each proteins. For the isometric diffusion, the diffusion tensor is $\nabla \cdot (\mathbf{D_1} \nabla u) = D_1 \Delta u$, the usual Laplacian operator (similarly for $\widetilde{u}$, v and $\widetilde{v}$). We assume that all variables and initial conditions are non-negative and also all the parameters to be strictly positive.

Here, the healthy protein is approached by the toxic protein, and after transitions, a healthy protein is converted into a toxic state. In the current formulation, we have assumed that the probability of a given toxic protein encountering healthy protein in a fixed time interval T_t, within a fixed spatial region, depends linearly on the healthy protein density. In this case, the total density of the healthy proteins u converted by the toxic proteins $\widetilde{u}$ can be expressed as $\widetilde{u} = aT_s u$, following the Holling functional response idea [26]. The parameter T_s is the time to getting contact with each other and a is a proportionality constant.

If there is no reaction time, then $T_s = T_t$ and hence we get a linear conversion rate $\tilde{u} = aT_t u$. Now, if each toxic protein requires a reaction time h for healthy proteins that are converted, then the time available to getting contact becomes $T_s = T_t - h\tilde{u}$. Therefore, $\tilde{u} = a(T_t - h\tilde{u})u$, hence $\tilde{u} = aT_t u/(1 + ahu)$, which is a nonlinear conversion rate. So, we modify the above model (1) as follows:

$$\frac{\partial u}{\partial t} = \nabla \cdot (\mathbf{D_1} \nabla u) + a_0 - a_1 u - \frac{a_2 u}{1 + e_1 u}\tilde{u}, \tag{2a}$$

$$\frac{\partial \tilde{u}}{\partial t} = \nabla \cdot (\widetilde{\mathbf{D}}_1 \nabla \tilde{u}) - \tilde{a}_1 \tilde{u} + \frac{a_2 u}{1 + e_1 u}\tilde{u}, \tag{2b}$$

$$\frac{\partial v}{\partial t} = \nabla \cdot (\mathbf{D_2} \nabla v) + b_0 - b_1 v - \frac{b_2 v}{1 + e_2 v}\tilde{v} - b_3 \tilde{u} v \tilde{v}, \tag{2c}$$

$$\frac{\partial \tilde{v}}{\partial t} = \nabla \cdot (\widetilde{\mathbf{D}}_2 \nabla \tilde{v}) - \tilde{b}_1 \tilde{v} + \frac{b_2 v}{1 + e_2 v}\tilde{v} + b_3 \tilde{u} v \tilde{v}, \tag{2d}$$

where $e_1 (= a_\beta h_\beta)$ and $e_2 (= a_\tau h_\tau)$ are dimensionless parameters. We use no-flux boundary conditions and non-negative initial conditions. Here, in model (2), the rate of conversion of the healthy protein by the toxic protein increases as the healthy protein density increases, but eventually it saturates at the level where the rate of conversion remains constant regardless of increases in healthy protein density. On the other hand, in model (1), the rate of conversion of the healthy protein by the toxic protein rises constantly with an increase in the healthy protein density.

3   Temporal Dynamics

For studying the wave propagation based on the reaction-diffusion model (2), we will first find homogeneous steady-states of the system. The homogeneous steady-states of the system (2) can be determined by finding the equilibrium points of the following system

$$\frac{du}{dt} = a_0 - a_1 u - \frac{a_2 u}{1 + e_1 u}\tilde{u}, \tag{3a}$$

$$\frac{d\tilde{u}}{dt} = -\tilde{a}_1 \tilde{u} + \frac{a_2 u}{1 + e_1 u}\tilde{u}, \tag{3b}$$

$$\frac{dv}{dt} = b_0 - b_1 v - \frac{b_2 v}{1 + e_2 v}\tilde{v} - b_3 \tilde{u} v \tilde{v}, \tag{3c}$$

$$\frac{d\tilde{v}}{dt} = -\tilde{b}_1 \tilde{v} + \frac{b_2 v}{1 + e_2 v}\tilde{v} + b_3 \tilde{u} v \tilde{v}, \tag{3d}$$

with non-negative initial conditions.

3.1   Stationary Points

The system (3) always has a disease-free state called a healthy stationary state. Depending on the parameter values, the system may possess more stationary points. We summarise each possible stationary state in the following:

1. Healthy $A\beta$ - healthy τP: It is the trivial stationary state and is given by

$$(u_1, \tilde{u}_1, v_1, \tilde{v}_1) = \left(\frac{a_0}{a_1}, 0, \frac{b_0}{b_1}, 0 \right). \tag{4}$$

This stationary state is the same for both systems, (1) and (2), due to zero toxic loads.

2. Healthy $A\beta$ - toxic τP: The stationary state of "healthy $A\beta$ - toxic τP" is given by

$$(u_2, \tilde{u}_2, v_2, \tilde{v}_2) = \left(\frac{a_0}{a_1}, 0, \frac{\tilde{b}_1}{b_2 - e_2\tilde{b}_1}, \frac{b_0(b_2 - e_2\tilde{b}_1) - b_1\tilde{b}_1}{\tilde{b}_1(b_2 - e_2\tilde{b}_1)} \right). \tag{5}$$

For the non-negativity of the stationary point (5), we must have $b_2 > e_2\tilde{b}_1$ and $b_0/b_1 \geq \tilde{b}_1/(b_2 - e_2\tilde{b}_1)$.

3. Toxic $A\beta$ - healthy τP: The stationary state of "toxic $A\beta$ - healthy τP" is given by

$$(u_3, \tilde{u}_3, v_3, \tilde{v}_3) = \left(\frac{\tilde{a}_1}{a_2 - e_1\tilde{a}_1}, \frac{a_0(a_2 - e_1\tilde{a}_1) - a_1\tilde{a}_1}{\tilde{a}_1(a_2 - e_1\tilde{a}_1)}, \frac{b_0}{b_1}, 0 \right). \tag{6}$$

For the non-negativity of the stationary point (6), we must have $a_2 > e_1\tilde{a}_1$ and $a_0/a_1 \geq \tilde{a}_1/(a_2 - e_1\tilde{a}_1)$.

4. Toxic $A\beta$ - toxic τP: Suppose $(u_4, \tilde{u}_4, v_4, \tilde{v}_4)$ is a stationary state of the "toxic $A\beta$ - toxic τP" type. In this case, we obtain $u_4 = u_3$, $\tilde{u}_4 = \tilde{u}_3$, $\tilde{v}_4 = (b_0 - b_1 v_4)/\tilde{b}_1$ and v_4 satisfy the quadratic equation

$$b_3 e_2 \tilde{u}_4 v_4^2 + (b_3\tilde{u}_4 - e_2\tilde{b}_1 + b_2)v_4 - \tilde{b}_1 = 0. \tag{7}$$

The equation (7) always has a real positive solution. For the uniqueness of v_4, we must have $b_3\tilde{u}_4 - e_2\tilde{b}_1 + b_2 \geq 0$. Also, for the positivity of $\tilde{v}_4$, we need $v_4 < b_0/b_1$.

Note that under small perturbations of any one of these stationary points, the trajectories may or may not come to that stationary point. Next, we examine this situation in more detail by the linear stability analysis.

3.2   Linear Stability Analysis

For the stability analysis, we linearize the system (3) about any of the stationary points $(u_*, \tilde{u}_*, v_*, \tilde{v}_*)$. The coefficient matrix $\mathbf{M}$ of the resulting system is the Jacobian matrix of the system (3) and is given by

$$\begin{bmatrix} -a_1 - \frac{a_2\tilde{u}_*}{(1+e_1 u_*)^2} & -\frac{a_2 u_*}{1+e_1 u_*} & 0 & 0 \\ \frac{a_2\tilde{u}_*}{(1+e_1 u_*)^2} & \frac{a_2 u_*}{1+e_1 u_*} - \tilde{a}_1 & 0 & 0 \\ 0 & -b_3 v_*\tilde{v}_* & -b_1 - \frac{b_2\tilde{v}_*}{(1+e_2 v_*)^2} - b_3\tilde{u}_*\tilde{v}_* & -\frac{b_2 v_*}{1+e_2 v_*} - b_3\tilde{u}_* v_* \\ 0 & b_3 v_*\tilde{v}_* & \frac{b_2\tilde{v}_*}{(1+e_2 v_*)^2} + b_3\tilde{u}_*\tilde{v}_* & \frac{b_2 v_*}{1+e_2 v_*} + b_3\tilde{u}_* v_* - \tilde{b}_1 \end{bmatrix}. \tag{8}$$

Now, the eigenvalues of the Jacobian matrix $\mathbf{M}$ are given by

$$\lambda_1 = -\frac{1}{2}(B + \sqrt{B^2 - 4C}), \lambda_2 = -\frac{1}{2}(B - \sqrt{B^2 - 4C}),$$

$$\lambda_3 = -\frac{1}{2}(\widehat{B} + \sqrt{\widehat{B}^2 - 4\widehat{C}}), \lambda_4 = -\frac{1}{2}(\widehat{B} - \sqrt{\widehat{B}^2 - 4\widehat{C}}),$$

where $B = a_1 + \tilde{a}_1 + a_2\tilde{u}_*/(1 + e_1u_*)^2 - a_2u_*/(1 + e_1u_*)$, $C = a_1\tilde{a}_1 + \tilde{a}_1a_2\tilde{u}_*/(1 + e_1u_*)^2 - a_1a_2u_*/(1 + e_1u_*)$, $\widehat{B} = b_1 + \tilde{b}_1 + b_3\tilde{u}_*(\tilde{v}_* - v_*) + b_2\tilde{v}_*/(1 + e_2v_*)^2 - b_2v_*/(1 + e_2v_*)$ and $\widehat{C} = b_1\tilde{b}_1 + b_3\tilde{u}_*(\tilde{b}_1\tilde{v}_* - b_1v_*) + \tilde{b}_1b_2\tilde{v}_*/(1 + e_2v_*)^2 - b_1b_2v_*/(1 + e_2v_*)$.

For each of the stationary points, we find the Jacobian matrix $\mathbf{M}$ and all its eigenvalues $\lambda_i, i = 1, 2, 3, 4$. Hence, the conclusion can be drawn easily, because for a given stationary point, if all the eigenvalues have negative real parts, this stationary point is stable, otherwise it is unstable.

4   Wave Propagation

We analyze travelling wave solutions of the spatio-temporal model (2) in one dimension ($\Omega = \mathbb{R}$) connecting any two stationary states $(u_i, \tilde{u}_i, v_i, \tilde{v}_i), i = 1, 2, 3, 4$ [21]. First, we consider the travelling wave emanating from healthy stationary state $(u_1, \tilde{u}_1, v_1, \tilde{v}_1)$ and connecting to $(u_2, \tilde{u}_2, v_2, \tilde{v}_2)$. For analysing the travelling wave fronts, we linearize the spatio-temporal model (2) around the healthy stationary state which leads to the following uncoupled system

$$\frac{\partial \tilde{u}}{\partial t} = \tilde{d}_1 \frac{\partial^2 \tilde{u}}{\partial x^2} + \frac{a_2u_1}{1 + e_1u_1} - \tilde{a}_1, \tag{9a}$$

$$\frac{\partial \tilde{v}}{\partial t} = \tilde{d}_2 \frac{\partial^2 \tilde{v}}{\partial x^2} + \frac{b_2v_1}{1 + e_2v_1} - \tilde{b}_1. \tag{9b}$$

Firstly, for the travelling wave solution, we substitute $\tilde{u}(x, t) = \tilde{u}(x - ct) \equiv \tilde{u}(z)$, $\tilde{v}(x, t) = \tilde{v}(x - ct) \equiv \tilde{v}(z)$ in (9) and will look for linear solutions of the form $\tilde{u} = C_1\exp(\lambda z)$, $\tilde{v} = C_2\exp(\lambda z)$. Then, the minimum wave speeds $c_{\min}$ are given by

$$c_\beta^{(12)} = 0 \text{ and } c_\tau^{(12)} = 2\sqrt{\tilde{d}_2\left(\frac{b_2v_1}{1 + e_2v_1} - \tilde{b}_1\right)}. \tag{10}$$

Here, $c_\beta^{(ij)}$ and $c_\tau^{(ij)}$ denote the speeds of the front from state i to the state j for the $A\beta$ fields $(u, \tilde{u})$ and τP fields $(v, \tilde{v})$, respectively.

Similarly, the minimum wave speeds for the travelling wave fronts emanating from healthy stationary state $(u_1, \tilde{u}_1, v_1, \tilde{v}_1)$ and connecting to $(u_3, \tilde{u}_3, v_3, \tilde{v}_3)$ are given by

$$c_\beta^{(13)} = 2\sqrt{\tilde{d}_1\left(\frac{a_2u_1}{1 + e_1u_1} - \tilde{a}_1\right)} \text{ and } c_\tau^{(13)} = 0. \tag{11}$$

Also, we have

$$c_\beta^{(14)} = c_\beta^{(13)} \text{ and } c_\tau^{(14)} = c_\tau^{(12)}. \tag{12}$$

Secondly, we consider the travelling wave emanating from the stationary state $(u_3, \widetilde{u}_3, v_3, \widetilde{v}_3)$ and connecting to $(u_4, \widetilde{u}_4, v_4, \widetilde{v}_4)$. We linearize the spatio-temporal model (2) around $(u_3, \widetilde{u}_3, v_3, \widetilde{v}_3)$ and repeat the same techniques to deduce that

$$c_\beta^{(34)} = 0 \text{ and } c_\tau^{(34)} = 2\sqrt{\widetilde{d}_2\left(\frac{b_2 v_3}{1 + e_2 v_3} + b_3 \widetilde{u}_3 v_3 - \widetilde{b}_1\right)}. \tag{13}$$

5   Network Mathematical Model

Based on a coarse-graining procedure of the continuous model and taking advantage of the brain data, a network mathematical model can be constructed where the edges of the network are the axonal bundles in white-matter tracts (e.g., [14]). The choice of the network nodes is carried out in the region of interest and in what follows we describe the network mathematical model corresponding to the modified continuous model (2) for the brain data connectome [27]. The latter can be modelled by the coarse-grain model of the continuous system. Specifically, it is a weighted graph $\mathcal{G}$ with V nodes and E edges defined in a domain Ω. The weights of the graph $\mathcal{G}$ are represented by the adjacency matrix $\mathbf{W}$ which provides a way to construct the graph of the Laplacian. For $i, j = 1, 2, 3, \ldots, V$, the elements of $\mathbf{W}$ are

$$W_{ij} = \frac{n_{ij}}{l_{ij}^2},$$

where n_{ij} is the mean fiber number and l_{ij}^2 is the mean length squared between the nodes i and j. We define the graph of the Laplacian $\mathbf{L}$ as

$$L_{ij} = \rho(D_{ii} - W_{ij}), \quad i, j = 1, 2, 3, \ldots, V,$$

where ρ is the diffusion coefficient and $D_{ii} = \sum_{j=1}^{V} W_{ij}$ is the elements of the diagonal weighted degree matrix. Now, we are ready to build a network mathematical model in the graph $\mathcal{G}$.

At the node j, let $(u_j, \widetilde{u}_j)$ be the concentrations of healthy and toxic $A\beta$ proteins, respectively, whereas $(v_j, \widetilde{v}_j)$ be the concentrations of healthy and toxic τP proteins, respectively. Then, for all the nodes $j = 1, 2, 3, \ldots, V$, the network equations corresponding to the continuous model (2) is a system of first order differential equations and it is given by

$$\frac{du_j}{dt} = -\sum_{k=1}^{V} L_{jk} u_k + a_0 - a_1 u_j - \frac{a_2 u_j}{1 + e_1 u_j}\widetilde{u}_j, \tag{14a}$$

$$\frac{d\widetilde{u}_j}{dt} = -\sum_{k=1}^{V} L_{jk} \widetilde{u}_k - \widetilde{a}_1 \widetilde{u}_j + \frac{a_2 u_j}{1 + e_1 u_j}\widetilde{u}_j, \tag{14b}$$

$$\frac{dv_j}{dt} = -\sum_{k=1}^{V} L_{jk} v_k + b_0 - b_1 v_j - \frac{b_2 v_j}{1 + e_2 v_j}\widetilde{v}_j - b_3 \widetilde{u}_j v_j \widetilde{v}_j, \tag{14c}$$

$$\frac{d\widetilde{v}_j}{dt} = -\sum_{k=1}^{V} L_{jk} \widetilde{v}_k - \widetilde{b}_1 \widetilde{v}_j + \frac{b_2 v_j}{1 + e_2 v_j}\widetilde{v}_j + b_3 \widetilde{u}_j v_j \widetilde{v}_j, \tag{14d}$$

with non-negative initial conditions.

6   Results and Discussion

A "healthy $A\beta$ - healthy τP" stationary state satisfies $\tilde{u} = \tilde{v} = 0$. The non-existence of a physically relevant healthy state occurs due to a failure of healthy clearance, which is either with an $A\beta$ clearance or with a τP clearance. To start with, we consider the following balance of clearance inequalities:

$$\frac{a_0}{a_1} < \frac{\tilde{a}_1}{a_2 - e_1 \tilde{a}_1}, \quad \frac{b_0}{b_1} < \frac{\tilde{b}_1}{b_2 - e_2 \tilde{b}_1}. \tag{15}$$

Now, if (15) holds for the stationary point $(u_1, \tilde{u}_1, v_1, \tilde{v}_1)$ in (4), then all the eigenvalues corresponding to the Jacobian matrix $\mathbf{M}$ have negative real parts. So, given the small amounts of the production of toxic $A\beta$ or toxic τP, or excess amounts of the production of healthy $A\beta$ or healthy τP, the system would be approaching towards the "healthy $A\beta$- healthy τP" stationary state.

Due to the failure of the clearance inequality (15), a transcritical bifurcation occurs for the homogeneous system (3). Hence, all the other stationary states $(u_i, \tilde{u}_i, v_i, \tilde{v}_i), i = 2, 3, 4$ are physically meaningful and a pathological development becomes possible. Motivated by [14], we fix the parameter values as $a_0 = a_1 = a_2 = b_0 = b_1 = b_2 = 1$ and $e_1 = e_2 = 0.1$. Now, we fix $\tilde{a}_1 = 3/4, \tilde{b}_1 = 4/3$ and we take b_3 as the bifurcation parameter. For $b_3 < 1.575$, the system has only two stationary points $(u_1, \tilde{u}_1, v_1, \tilde{v}_1)$ and $(u_3, \tilde{u}_3, v_3, \tilde{v}_3)$. The equilibrium point $(u_1, \tilde{u}_1, v_1, \tilde{v}_1)$ is saddle and $(u_3, \tilde{u}_3, v_3, \tilde{v}_3)$ is stable. A nontrivial stationary point $(u_4, \tilde{u}_4, v_4, \tilde{v}_4)$ is generated through a transcritical bifurcation at $b_3 = 1.575$. Then $(u_3, \tilde{u}_3, v_3, \tilde{v}_3)$ changes its stability to $(u_4, \tilde{u}_4, v_4, \tilde{v}_4)$ and becomes saddle (see Fig. 1).

These results could lead to a number of important observations. For example, due to the instability of the healthy stationary state of the system (3), a proteopathic brain patient would be progressing toward a disease state. The actual state would depend on the parameter values. If $b_0/b_1 \geq \tilde{b}_1/(b_2 - e_2 \tilde{b}_1)$ holds, then $(u_2, \tilde{u}_2, v_2, \tilde{v}_2)$ exists and if $a_0/a_1 \geq \tilde{a}_1/(a_2 - e_1 \tilde{a}_1)$ holds, then $(u_3, \tilde{u}_3, v_3, \tilde{v}_3)$ exists. Sometimes both the relations hold simultaneously. Also, the proteopathic state $(u_4, \tilde{u}_4, v_4, \tilde{v}_4)$ exists if $b_0/b_1 > v_4$ holds. Since, $\tilde{u}_4 = \tilde{u}_3$, we can choose b_3 in such a way that $b_3 \tilde{u}_4 - e_2 \tilde{b}_1 + b_2 \geq 0$. Therefore, to produce tau proteopathy, the stationary state $(u_2, \tilde{u}_2, v_2, \tilde{v}_2)$ is not needed. So, we study only two types of patient proteopathies: (i) primary tauopathy and (ii) secondary tauopathy.

For the primary tauopathy, which is usually related to neurodegenerative diseases such as AD, all the four stationary states exist i.e., both the conditions $b_0/b_1 \geq \tilde{b}_1/(b_2 - e_2 \tilde{b}_1)$ and $a_0/a_1 \geq \tilde{a}_1/(a_2 - e_1 \tilde{a}_1)$ hold. In this case, we have plotted the dynamics of the system (3) in Fig. 2(a). Also, for the secondary tauopathy, only three stationary states exist. Here, the inequality $a_0/a_1 \geq \tilde{a}_1/(a_2 - e_1 \tilde{a}_1)$ is true and the other inequality fails. An example of secondary tauopathy is shown in Fig. 2(b). Comparing the homogeneous systems corresponding to (1) and (2), the modified system requires less toxic load.

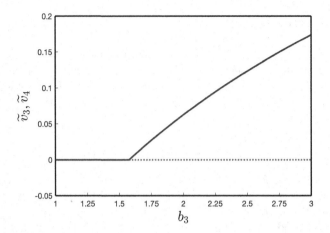

Fig. 1. Transcritical bifurcation diagram of the stationary points for the system (3). (Parameter values: $a_0 = a_1 = a_2 = b_0 = b_1 = b_2 = 1, \tilde{a}_1 = 3/4, \tilde{b}_1 = 4/3, e_1 = e_2 = 0.1$.)

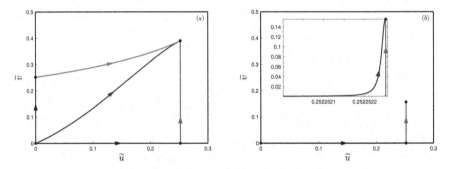

Fig. 2. Phase plane $(\tilde{u}, \tilde{v})$ with four and three stationary points for the system (3): (a) $\tilde{b}_1 = 3/4, b_3 = 0.5$ and (b) $\tilde{b}_1 = 4/3, b_3 = 3$. (Parameter values: $a_0 = a_1 = a_2 = b_0 = b_1 = b_2 = 1, \tilde{a}_1 = 3/4, e_1 = e_2 = 0.1$.)

For the wave propagation, we consider the spatial domain in one dimension $[-100, 100]$ as an example. However, the results are robust for a wide range of intervals. We take the initial condition $(u(x, 0), \tilde{u}(x, 0), v(x, 0), \tilde{v}(x, 0))$ for the primary tauopathy as $(u_3, \tilde{u}_3, v_3, \tilde{v}_3)$ for $-100 \le x \le -95$, $(u_1, \tilde{u}_1, v_1, \tilde{v}_1)$ for $-95 < x < 95$ and $(u_2, \tilde{u}_2, v_2, \tilde{v}_2)$ for $95 \le x \le 100$. On the other hand, the initial condition $(u(x, 0), \tilde{u}(x, 0), v(x, 0), \tilde{v}(x, 0))$ for the secondary tauopathy has been taken as $(u_3, \tilde{u}_3, v_3, \tilde{v}_3)$ for $-100 \le x \le -95$, $(u_1, \tilde{u}_1, v_1, \tilde{v}_1)$ for $-95 < x < 95$ and $(u_2, \tilde{u}_2, v_2, 10^{-6})$ for $95 \le x \le 100$.

We have shown the wave propagation for the primary tauopathy in Fig. 3 at different time steps $t = 50, 150, 180$ and 220. Motivated by Thompson et al., we have chosen the parameter values as $a_0 = a_1 = a_2 = b_0 = b_1 = b_2 = 1, \tilde{a}_1 = \tilde{b}_1 = 3/4, b_3 = 0.5, e_1 = e_2 = 0.1, d_1 = \tilde{d}_1 = d_2 = \tilde{d}_2 = 1$ and no-flux

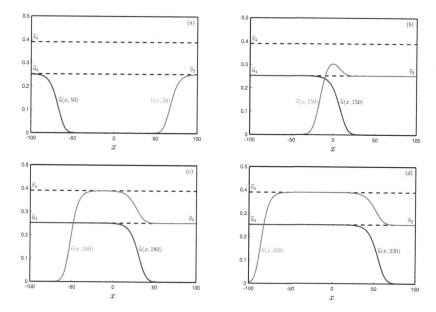

Fig. 3. Front propagations of $\widetilde{u}$ and $\widetilde{v}$ for the system (2) at different time steps: (a) $t = 50$, (b) $t = 150$, (c) $t = 180$ and (d) $t = 220$.

boundary conditions for all the variables. For these parametric values, we obtain $c_\beta^{(14)} = c_\tau^{(12)} = c_\tau^{(14)} = 0.798$ and $c_\tau^{(34)} = 1.068$. In the simulation, we have considered the toxic $A\beta$ front on the left side of the domain and toxic τP on the right. Initially, the toxic $A\beta$ front propagates to the right with speed $c_\beta^{(14)}$ and toxic τP propagates to the left with speed $c_\tau^{(12)}$. After overlapping both the fronts, τP increases its concentration and connects to $\widetilde{v}_4$. Then, the left front of the wave of τP boosts its speed to $c_\tau^{(34)}$ and moves to the left. On the other hand, the right front of the wave of τP moves with speed $c_\tau^{(14)}$, it eventually fills the domain and the entire system converges to the stable equilibrium solution $(u_4, \widetilde{u}_4, v_4, \widetilde{v}_4)$.

In Fig. 4, we plot wave propagation for the secondary tauopathy at different time steps $t = 60, 250, 400$ and 425. We have chosen the parameter values as $a_0 = a_1 = a_2 = b_0 = b_1 = b_2 = 1, \widetilde{a}_1 = 3/4, \widetilde{b}_1 = 4/3, b_3 = 3, e_1 = e_2 = 0.1, d_1 = \widetilde{d}_1 = d_2 = \widetilde{d}_2 = 1$ and no-flux boundary conditions for all the variables. For these parametric values, we obtain $c_\beta^{(14)} = 0.798$ and $c_\tau^{(34)} = 1.153$. Here, the toxic $A\beta$ front propagates to the right with speed $c_\beta^{(14)}$ and fills the domain $\widetilde{u}_4$ with negligible toxic τP (see Fig. 2(b)). However, we note that after filling toxic $A\beta$ in the entire domain, toxic τP starts to increase its concentration and connects to $\widetilde{v}_4$. It moves with the speed $c_\tau^{(34)}$ and fills the domain. Finally, the entire system converge to the stable equilibrium solution $(u_4, \widetilde{u}_4, v_4, \widetilde{v}_4)$.

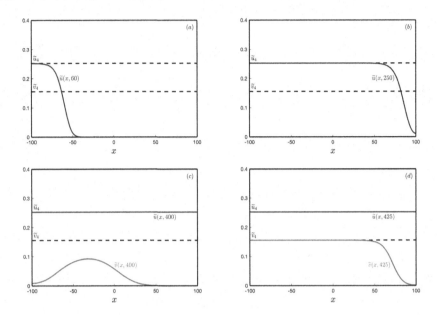

Fig. 4. Front propagations of $\tilde{u}$ and $\tilde{v}$ for the system (2) at different time steps: (a) $t = 60$, (b) $t = 250$, (c) $t = 400$ and (d) $t = 425$.

For the network model, brain connectome data is available with different resolutions in, e.g., [14]: the lowest resolution consists of 83 nodes, and the highest resolution consists of 1015 nodes. However, there is some difference in the staging area of $A\beta$ and τP in the brain connectome. A more general approach to the analysis of brain hubs in human connectomes has recently been proposed in [27]. In the context of our research on the pathology dynamics, the network model (14) can be solved numerically for the given number of nodes with non-negative initial conditions. Furthermore, we can extend our analysis on primary and secondary tauopathies for the network model as well. Finally, we note that in the analysis currently being undertaken, not only we can choose uniform parameter values for all the nodes but also different parameter values in different regions in the brain connectome, as required by a more detailed study.

7   Conclusion

We have studied a modification of the heterodimer model, which captures the conversion time from healthy to toxic proteins. For the temporal dynamics, we have carried out the linear stability analysis of all the stationary points. We have also investigated the wave speeds of the travelling wavefronts for the spatio-temporal model. Further, a computationally challenging network mathematical model has been described based on a coarse-graining procedure of the continuous model and taking advantage of the brain data connectome. In this latter model the edges of the network are the axonal bundles in white-matter tracts.

We have highlighted an efficient way to analyze such models in the context of neurodegenerative diseases such as AD.

We have obtained two clinically interesting patient proteopathies for further detailed analysis: primary and secondary tauopathies. For the case of primary tauopathy, a possible invasion of τP exists independent of the invasion of $A\beta$. On the other hand, for the secondary tauopathy, the sustained presence of toxic τP requires the company of toxic $A\beta$. These conclusions are similar for both the models (heterodimer and the modified version). However, for the same parametric values, the introduction of Holling type-II functional response decreases the concentrations of toxic τP and toxic $A\beta$ compared to the original model. Finally, a detailed analysis of different tauopathies with non-uniform parameters has been recently carried out in [28] with further developments of the network model reported here.

References

1. Alzheimer, A.: Uber eine eigenartige Erkrankung der Hirnrinde. Zentralbl Nervenh Psych. **18**, 177–179 (1907)
2. Hardy, J.A., Higgins, G.A.: Alzheimer's disease: the amyloid cascade hypothesis. Science **256**, 184–186 (1992)
3. Hardy, J., Allsop, D.: Amyloid deposition as the central event in the aetiology of Alzheimer's disease. Trends Pharmacol. Sci. **12**, 383–388 (1991)
4. Gotz, J., Halliday, G., Nisbet, R.M.: Molecular pathogenesis of the tauopathies. Annu. Rev. Pathol. **14**, 239–261 (2019)
5. Cho, H., et al.: In vivo cortical spreading pattern of tau and amyloid in the Alzheimer disease spectrum. Ann. Neurol. **80**, 247–258 (2016)
6. Jack, C.R., et al.: NIA-AA research framework: toward a biological definition of Alzheimer's disease. Alzheimer's Dementia **14**, 535–562 (2018)
7. DeVos, S.L., et al.: Tau reduction in the presence of amyloid $-\beta$ prevents tau pathology and neuronal death in vivo. Brain **141**, 2194–2212 (2018)
8. Busche, M.A., et al.: Tau impairs neural circuits, dominating amyloid $-\beta$ effects, in Alzheimer models in vivo. Threshold **30**, 50 (2019)
9. Walker, L.C., Lynn, D.G., Chernoff, Y.O.: A standard model of Alzheimer's disease? Prion **12**, 261–265 (2018)
10. Ittner, L.M., Gotz, J.: Amyloid-β and tau-a toxic pas de deux in Alzheimer's disease. Nature Rev. Neurosci. **12**, 67 (2011)
11. Jack, Jr. C.R., et al.: Tracking pathophysiological processes in Alzheimer's disease: an updated hypothetical model of dynamic biomarkers. Lancet Neurol. **12**, 207–216 (2013)
12. Kara, E., Marks, J.D., Aguzzi, A.: Toxic protein spread in neurodegeneration: reality versus fantasy. Trends in Molecular Medicine (2018)
13. Vosoughi, A., et al.: Mathematical models to shed light on amyloid-beta and tau protein dependent pathologies in Alzheimer's disease. Neuroscience **424**, 45–57 (2020)
14. Thompson, T.B., Chaggar, P., Kuhl, E., Goriely, A.: Protein-protein interactions in neurodegenerative diseases: a conspiracy theory. PLoS Comput. Biol. **16**, e1008267 (2020)

15. Fornari, S., Schäfer, A., Goriely, A., Kuhl, E.: Prion-like spreading of Alzheimer's disease within the brain's connectome. Interface R Society **16**, 20190356 (2019)
16. Fornari, S., Schäfer, A., Kuhl, E., Goriely, A.: Spatially-extended nucleation-aggregation-fragmentation models for the dynamics of prion-like neurodegenerative protein-spreading in the brain and its connectome. J. Theor. Biol. **486**, 110102 (2020)
17. Jucker, M., Walker, L.C.: Pathogenic protein seeding in Alzheimer disease and other neurodegenerative disorders. Ann. Neurol. **70**, 532–540 (2011)
18. Zheng, Y.Q., et al.: Local vulnerability and global connectivity jointly shape neurodegenerative disease propagation. PLoS Biol. **17**(11), e3000495 (2019)
19. Insel, P., Mormino, E., Aisen, P., Thompson, W., Donahue, M.: Neuroanatomical spread of amyloid β and tau in Alzheimer's disease: implications for primary prevention. Brain Commun. **2**, 1–11 (2020)
20. Moreno-Jimenez, E., Flor-Garcia, M.: Adult hippocampal neurogenesis is abundant in neurologically healthy subjects and drops sharply in patients with Alzheimer's disease. Nat. Med. **25**, 554–560 (2019)
21. Bressloff, P.C.: Waves in Neural Media. Lecture Notes on Mathematical Modelling in the Life Sciences. Springer, New York (2014). https://doi.org/10.1007/978-1-4614-8866-8
22. Matthäus, F.: Comparison of modeling approaches for the spread of prion diseases in the brain. In: Mitkowski, W., Kacprzyk, J. (eds.) Modelling Dynamics in Processes and Systems, vol. 180, pp. 109–117. Springer, Heidelberg (2009). https://doi.org/10.1007/978-3-540-92203-2_8
23. Bertsch, M., Franchi, B., Marcello, N., Tesi, M.C., Tosin, A.: Alzheimer's disease: a mathematical model for onset and progression. Math. Med. Biol. **34**, 193–214 (2016)
24. Weickenmeier, J., Kuhl, E., Goriely, A.: The multiphysics of prion-like diseases: progression and atrophy. Phys. Rev. Lett. **121**, 264–281 (2018)
25. Weickenmeier, J., Jucker, M., Goriely, A., Kuhl, E.: A physics-based model explains the prion-like features of neurodegeneration in Alzheimer's disease, Parkinson's disease, and amyotrophic lateral sclerosis. J. Mech. Phys. Solids **124**, 264–281 (2019)
26. Dawes, J.H.P., Souza, M.O.: A derivation of Holling's type I, II and III functional responses in predator-prey systems. J. Theo. Bio. **327**, 11–22 (2013)
27. Tadic, B., Melnik, R., Andjelkovic, M.: The topology of higher-order complexes associated with brain hubs in human connectomes. Scientifc Rep. **10**, 17320 (2020)
28. Pal, S., Melnik, R.: Nonlocal multiscale interactions in brain neurodegenerative protein dynamics and coupled proteopathic processes. In: Onate, E., Papadrakakis, M., Schreflfler, B. (eds.) Proceedings of the IX International Conference on Computational Methods for Coupled Problems in Science and Engineering, CIMNE, Barcelona, Coupled problems 2021 (2021). 12 p.

A Semi-implicit Backward Differentiation ADI Method for Solving Monodomain Model

Maryam Alqasemi and Youssef Belhamadia$^{(\boxtimes)}$ (iD)

Department of Mathematics, American University of Sharjah,
Sharjah, United Arab Emirates
ybelhamadia@aus.edu

Abstract. In this paper, we present an efficient numerical method for solving the electrical activity of the heart. We propose a second order alternating direction implicit finite difference method (ADI) for both space and time. The derivation of the proposed ADI scheme is based on the semi-implicit backward differentiation formula (SBDF). Numerical simulation showing the computational advantages of the proposed algorithm in terms of the computational time and memory consumption are presented.

Keywords: Monodomain model · SBDF methods · ADI methods · Mitchell–Schaeffer model · Spiral wave

1 Introduction

Mathematical modeling of biological activities has been proven to be of high importance in the modern computer age. It is an alternative tool to live experiments and can provide solutions to several biomedical problems. In electrocardiology, the most used mathematical models are the bidomain and monodomain models. The monodomain is considered as a simplified version of the bidomain model and although it lacks physiological foundation, it is widely used in the computational electrophysiology community.

Mathematically, the monodomain model consists of a single parabolic partial differential equation coupled with a system of nonlinear ordinary differential equations modeling cell ionic activity. Solving the monodomain model requires fine meshes and small time-steps as the cardiac electrical wave has a stiff wave front, which makes the numerical simulation challenging. In the literature, there are many methods to reduce the computational challenges of both the bidomain and monodomain models. The spatial and temporal discretization effects in cardiac electrophysiology have been numerically investigated in [1] and [2].

The authors acknowledge the financial support of the American University of Sharjah for the Enhanced Faculty Research Grant.

© Springer Nature Switzerland AG 2021
M. Paszynski et al. (Eds.): ICCS 2021, LNCS 12746, pp. 541–548, 2021.
https://doi.org/10.1007/978-3-030-77977-1_43

Anisotropic mesh adaptation techniques are presented in [3–5]. Operator splitting and high-order methods have been studied in [6–9] and parallel algorithms for cardiac models have been investigated in [10–12].

Recently, we demonstrated the efficiency of alternating direction implicit (ADI) finite difference methods for solving the monodomain model [13]. The computational advantages of ADI methods have been demonstrated in comparison with the standard finite difference method. The derivation of the ADI methods in [13] was based on the semi-implicit Crank–Nicolson/Adams–Bashforth (CNAB) scheme. The main goal of this paper is to derive an ADI scheme based on the semi-implicit backward differentiation formula (SBDF). The proposed method will be referred to as SBDF-ADI scheme. The semi-implicit methods, especially SBDF type methods, are considered among the best schemes for solving the bidomain and monodomain models (see [2,14]). The SBDF schemes enable higher order time stepping that is needed for accurate numerical simulations of the electrical waves of the heart.

In this paper, a second order SBDF-ADI method for the monodomain model is derived. In Sect. 2, both SBDF and SBDF-ADI schemes are illustrated in the two-dimensional case. In Sect. 3, numerical results are presented to demonstrate the order of convergence and the computational advantages of the proposed scheme in terms of the computational time and memory consumption. In all our simulations, a comparison with the standard SBDF finite difference method is presented.

2   Derivation of SBDF-ADI Method

2.1   Two-Dimensional SBDF Method

The main governing equations of the monodomain model are given by the following system

$$
\begin{cases}
\dfrac{\partial V_m}{\partial t} - \nabla \cdot (\boldsymbol{D}\nabla V_m) = I_{ion}(V_m, \boldsymbol{W}), \\[2mm]
\dfrac{\partial \boldsymbol{W}}{\partial t} = g(V_m, \boldsymbol{W}).
\end{cases}
\tag{1}
$$

Where V_m is the trans-membrane potential, $\boldsymbol{W}$ is the cellular states, and $\boldsymbol{D} = \mathrm{diag}(D_x, D_y)$ is the conductivity tensor. The functions $I_{ion}(V, \boldsymbol{W})$, and $g(V, \boldsymbol{W})$ represents the single cell model. Before we derive the SBDF-ADI method, we first present the second-order SBDF scheme by re-expressing the system (1) in the following vector form:

$$
\frac{\partial \boldsymbol{U}}{\partial t} = A\boldsymbol{U} + F(\boldsymbol{U}),
\tag{2}
$$

where

$$
\boldsymbol{U} = \begin{pmatrix} V \\ \boldsymbol{W} \end{pmatrix}, \quad
A = \begin{pmatrix} D_x \dfrac{\partial^2}{\partial x^2} + D_y \dfrac{\partial^2}{\partial y^2} & 0 \\ 0 & 0 \end{pmatrix} \text{ and }
F(\boldsymbol{U}) = \begin{pmatrix} I_{ion}(V, \boldsymbol{W}) \\ g(V, \boldsymbol{W}) \end{pmatrix}.
$$

The second-order SBDF requires to start from U^{n-1} at time t^{n-1} and U^n at time t^n as follows:

$$\frac{3U^{n+1} - 4U^n + U^{n-1}}{2\Delta t} = AU^{n+1} + 2F(U^n) - F(U^{n-1}) \tag{3}$$

Using finite difference method, the continuous space domain must be discretized into a mesh with a finite number of grid points. To ensure second order in space, we use the second-order central difference in all our numerical results. For the two-dimensional case, the scheme (3) requires solving a linear system of size $((M+1)^2, (M+1)^2)$, where M is the number of spatial steps. The main idea about ADI-type method is to reduce this system to series of a linear system of size $((M+1), (M+1))$ as is presented in the next subsection.

2.2   Two-Dimensional SBDF-ADI Method

To derive the two-dimensional SBDF-ADI algorithm proposed in this study, we must first re-express system (2) as follows

$$\frac{\partial U}{\partial t} = A_1 U + A_2 U + F(U), \tag{4}$$

where

$$A_1 = \begin{pmatrix} D_x \dfrac{\partial^2}{\partial x^2} & 0 \\ 0 & 0 \end{pmatrix} \text{ and } A_2 = \begin{pmatrix} D_y \dfrac{\partial^2}{\partial y^2} & 0 \\ 0 & 0 \end{pmatrix}.$$

The SBDF system (3) is therefore written as

$$\frac{3U^{n+1} - 4U^n + U^{n-1}}{2\Delta t} = A_1 U^{n+1} + A_2 U^{n+1} + 2F(U^n) - F(U^{n-1}) \tag{5}$$

Rearanging (5) by taking U^{n+1} terms in one side and the rest terms in the other:

$$\left(I - \frac{2\Delta t}{3} A_1 - \frac{2\Delta t}{3} A_2\right) U^{n+1} = \frac{4}{3} U^n - \frac{1}{3} U^{n-1} + \frac{2\Delta t}{3} \left(2F(U^n) - F(U^{n-1})\right). \tag{6}$$

The main idea for ADI is to use a perturbation of this equation. In our case, the perturbed form used is

$$\left(I - \frac{2\Delta t}{3} A_1\right) \left(I - \frac{2\Delta t}{3} A_2\right) U^{n+1} = \frac{4}{3} U^n - \frac{1}{3} U^{n-1} + \frac{2\Delta t}{3} \left(2F(U^n) - F(U^{n-1})\right). \tag{7}$$

Both equations, (3) and (7), are equivalent and preserve the same time order of accuracy. Now, based on the Douglas–Gunn time splitting scheme, our proposed SBDF-ADI scheme consists of the following system of equations:

$$\begin{cases} \left(I - \dfrac{2\Delta t}{3} A_1\right) U^* = \dfrac{\Delta t}{3} A_2 U^n + \dfrac{4}{3} U^n - \dfrac{1}{3} U^{n-1} + \dfrac{2\Delta t}{3} \left(2F(U^n) - F(U^{n-1})\right), \\ \left(I - \dfrac{2\Delta t}{3} A_2\right) U^{n+1} = U^* - \dfrac{2\Delta t}{3} A_2 U^n. \end{cases} \tag{8}$$

The second-order central difference is used for spatial discretization. Each linear system corresponding to each spatial direction in this scheme is of size $((M+1), (M+1))$, which allows a great gain in the computational time as is presented in the next section.

All the simulations were performed using MATLAB. All the linear systems obtained in this paper were solved by using decomposition MATLAB built-in function. This function returns the same results as mldivide (backslash operator) but in a much faster way for the presented iterative algorithms. All our MATLAB functions are optimized compared to our previous work presented in [13].

3   Numerical Results

In this section, we discuss the order of convergence in time of the proposed SBDF-ADI (8) and the standard SBDF (3) schemes. Then we discuss the performance of both methods in term of computational time and memory consumption. In all our numerical results, we will use the Mitchell–Schaeffer ionic model, given by:

$$I_{ion}(V_m, W) = \frac{1}{\tau_{in}} W V^2 (1 - V) - \frac{1}{\tau_{out}} V,$$

$$G(V_m, W) = \begin{cases} \dfrac{1 - W}{\tau_{open}} & \text{for } V < v_{gate}, \\ -\dfrac{W}{\tau_{close}} & \text{for } V \geq v_{gate}. \end{cases}$$

The convergence in time is demonstrated using similar technique presented in [13] where a regular electrical cardiac wave is considered. The parameter values used provide a fast action potential upstroke and are given in Table 1.

Table 1. Parameters used in Mitchell–Schaeffer model

Constant	Value	Constant	Value
τ_{in}	0.05	τ_{out}	1
τ_{open}	95	τ_{close}	162
v_{gate}	0.13(mV)	$D_x = D_y$	0.001

We use the following discrete norms:

$$e_{L^\infty} = \|V_h - V_r\|_{L^\infty} \text{ and } e_{L^2} = \|V_h - V_r\|_{L^2},$$

where V_r is a reference solution for the transmembrane potential obtained with a spatial discretization of 401 points ($M = 400$) in each direction and small time step ($N = 2 \times 10^6$). V_h is the numerical solution obtained with the same spatial mesh. In this example, the final time is $T = 330$. The result of this convergence test is presented in Fig. 1, where it is clearly demonstrated the second order convergence for both methods. These numerical results can be supported by analytical work similar to the work presented in [13] to show that indeed both systems are of second order in time.

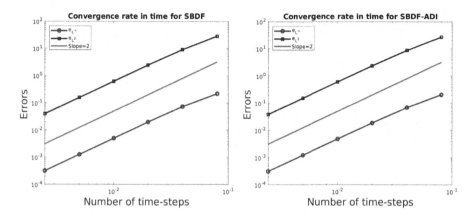

Fig. 1. Convergence order for SBDF and SBDF-ADI using Mitchell–Schaeffer model.

Now to demonstrate the performance of the proposed scheme SBDF-ADI (8), we investigate the computational time required for the simulations using both methods SBDF (3) and SBDF-ADI (8) for different values of space resolution (M) and for a fixed time step size ($N = 1.5 \times 10^4$). We consider a spiral wave dynamic in the Mitchell–Schaeffer model for the simulations. Different parameter values have been used in this case that are presented in Table 2.

Table 2. Parameters used in Mitchell–Schaeffer model

Constant	Value	Constant	Value
τ_{in}	0.3	τ_{out}	6
τ_{open}	120	τ_{close}	150
v_{gate}	0.13(mV)	$D_x = D_y$	0.001

The spiral wave is obtained with a similar method presented in [15, 16], where the initial conditions are given as follows

$$V(x,y,0) = \begin{cases} 1 & \text{if } y \leq 4, \\ 0 & \text{otherwize,} \end{cases} \quad \text{and} \quad W(x,y,0) = \begin{cases} 0.75/2 & \text{if } x \leq 4.5, \\ 0.75/2 & \text{otherwize.} \end{cases}$$

The final time in this example is $T = 1500$. The time evolution of the transmembrane potential is presented in Fig. 2.

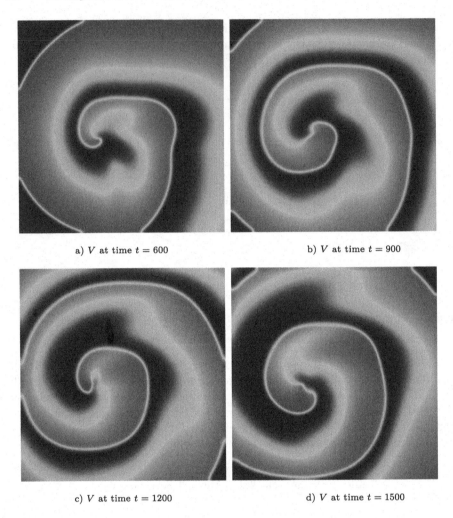

a) V at time $t = 600$ b) V at time $t = 900$

c) V at time $t = 1200$ d) V at time $t = 1500$

Fig. 2. Time evolution of spiral wave in Mitchell–Schaeffer model

It is well known that this type of spiral wave requires extremely fine mesh resolution. Therefore, the computational time required for the simulation corresponding to various space resolutions, M, is presented in Fig. 3. As can be seen, the computational time required by the proposed SBDF-ADI scheme is clearly lower than the required time for the standard SBDF method. This gain is mainly because of the size of the linear system involved in each method. In fact, the SBDF-ADI scheme involves matrices of size $(M+1, M+1)$, whereas the standard SBDF scheme requires a matrix of size $((M+1)^2, (M+1)^2)$. This also affects the memory requirement, where, for instance, in the case where $M = 600$ the memory required for the simulation using the SBDF-ADI is around 0.6 GB while the memory needed for SBDF is 1.6GB.

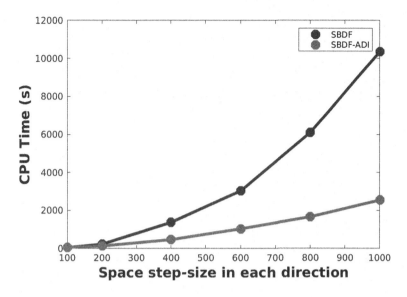

Fig. 3. Comparison of CPU time for SBDF and SBDF-ADI

4 Conclusion

In this paper, a second order alternating direction implicit finite difference method for both space and time was presented. The proposed ADI scheme was based on the semi-implicit backward differentiation formula. We showed that the proposed SBDF-ADI scheme provides the desired results while using less computational time. The advantage of the proposed SBDF-ADI scheme is that it can be extended to three-dimensional case, where the gain in computational resources will be clear as SBDF-ADI scheme involves matrices of size much smaller than the size of the matrix required for the standard SBDF scheme. The proposed SBDF-ADI scheme has also the advantage to be extended to include higher-order time and space difference schemes. This could provide accurate prediction of the cardiac electrical wave. Finite difference method has been previously used for solving the bidomain and monodomain models coupled with a realistic ionic model while using a computational geometry of a human heart. Therefore, the presented methodology can be extended to study realistic cardiac electrophysiology simulations which will be the subject of a future work.

References

1. Woodworth, L.A., Cansiz, B., Kaliske, M.: A numerical study on the effects of spatial and temporal discretization in cardiac electrophysiology. Int. J. Numer. Methods Biomed. Eng. e3443 (2021)
2. Roy, T., Bourgault, Y., Pierre, C.: Analysis of time-stepping methods for the monodomain model. Comput. Appl. Math. **39**(230), 1–32 (2020)

3. Belhamadia, Y.: A time-dependent adaptive remeshing for electrical waves of the heart. IEEE Trans. Biomed. Eng. **55**(2), 443–452 (2008)
4. Belhamadia, Y., Fortin, A., Bourgault, Y.: Towards accurate numerical method for monodomain models using a realistic heart geometry. Math. Biosci. **220**(2), 89–101 (2009)
5. Southern, J., Gorman, G.J., Piggott, M.D., Farrell, P.E., Bernabeu, M.O., Pitt-Francis, J.: Simulating cardiac electrophysiology using anisotropic mesh adaptivity. J. Comput. Sci. **1**(2), 82–88 (2010)
6. Arthurs, C.J., Bishop, M.J., Kay, D.: Efficient simulation of cardiac electrical propagation using high order finite elements. J. Comput. Phys. **231**(10), 3946–3962 (2012)
7. Vincent, K., et al.: High-order finite element methods for cardiac monodomain simulations. Front. Physiol. **6**(217), 1–9 (2015)
8. Coudière, Y., Turpault, R.: Very high order finite volume methods for cardiac electrophysiology. Comput. Math. Appl. **74**(4), 684–700 (2017)
9. Cervi, J., Spiteri, R.J.: A comparison of fourth-order operator splitting methods for cardiac simulations. Appl. Numer. Math. **145**, 227–235 (2019)
10. Pavarino, L.F., Scacchi, S.: Parallel multilevel schwarz and block preconditioners for the bidomain parabolic-parabolic and parabolic-elliptic formulations. SIAM J. Sci. Comput. **33**(4), 1897–1919 (2011)
11. Xia, Y., Wang, K., Zhang, H.: Parallel optimization of 3D cardiac electrophysiological model using GPU. Comput. Math. Method Med. **1–10**, 2015 (2015)
12. Belhamadia, Y., Briffard, T., Fortin, A.: Application of parallel anisotropic mesh adaptation for solving monodomain cardiac model. In: AIP Conference Proceedings, vol. 2343(1), p. 130013 (2021)
13. Belhamadia, Y., Rammal, Z.: Efficiency of semi-implicit alternating direction implicit methods for solving cardiac monodomain model. Comput. Biol. Med. **130**, 104187 (2021)
14. Ethier, M., Bourgault, Y.: Semi-implicit time-discretization schemes for the bidomain model. SIAM J. Numer. Analy. **46**, 2443–2468 (2008)
15. Belhamadia, Y., Fortin, A., Bourgault, Y.: On the performance of anisotropic mesh adaptation for scroll wave turbulence dynamics in reaction-diffusion systems. J. Comput. Appl. Math. **271**, 233–246 (2014)
16. Belhamadia, Y., Grenier, J.: Modeling and simulation of hypothermia effects on cardiac electrical dynamics. PloS one **14**(5), 1–23 (2019)

A Deep Learning Approach for Polycrystalline Microstructure-Statistical Property Prediction

José Pablo Quesada-Molina[1,2](✉) [ID] and Stefano Mariani[1] [ID]

[1] Department of Civil and Environmental Engineering, Politecnico di Milano,
Piazza Leonardo da Vinci 32, 20133 Milan, Italy
josepablo.quesada@polimi.it
[2] Department of Mechanical Engineering, University of Costa Rica, San Pedro Montes de Oca,
San José, Costa Rica

Abstract. Upscaling of the mechanical properties of polycrystalline aggregates might require complex and time-consuming procedures, if adopted to help in the design and reliability analysis of micro-devices. In inertial micro electro-mechanical systems (MEMS), the movable parts are often made of polycrystalline silicon films and, due to the current trend towards further miniaturization, their mechanical properties must be characterized not only in terms of average values but also in terms of their scattering. In this work, we propose two convolutional network models based on the ResNet and DenseNet architectures, to learn the features of the microstructural morphology and allow automatic upscaling of the statistical properties of the said film properties. Results are shown for film samples featuring different values of a length scale ratio, so as to assess accuracy and computational efficiency of the proposed approach.

Keywords: Multi-scale analyses · Homogenization · Scattered mechanical properties · Deep learning · Convolutional neural networks · ResNet · DenseNet

1 Introduction

Design and reliability of micro-devices like MEMS rely more and more on digital twins, that are numerical models of their mechanical parts (but also of their electronics, though this is not of concern here) allowing for all the possible epistemic uncertainties [1–4]. Even if uncertainties in MEMS readout are not targeted as a great issue, since different methods exist to compensate for them [5], it must be stated that they can overall lead to a detrimental effect on the relevant device performance indices [6, 7].

The ever-increasing need for downsizing the devices, on top of all for economic reasons, tends to enhance issues linked to the scattering in the measured response of the devices to the external actions, in most of the cases induced by a high sensitivity to micromechanical features and defects. For polysilicon microstructures, in recent works we showed how the morphology of the films and defects caused by the microfabrication process, like e.g. overetch, can be properly accounted for in stochastic analyses to cover the aforementioned scattered experimental data [8–12]. The price to attain such

© Springer Nature Switzerland AG 2021
M. Paszynski et al. (Eds.): ICCS 2021, LNCS 12746, pp. 549–561, 2021.
https://doi.org/10.1007/978-3-030-77977-1_44

an accuracy of models handled in Monte Carlo simulations, is that the statistics of the sought mechanical properties have to be computed every time being scale dependent, namely being affected by the interaction between the length scale describing the movable structure of the device and the length scale describing the film morphology (for instance, proportional to a characteristic radius of the grains).

The mentioned length scale separation is very important for computing the homogenized properties of the polysilicon films in multi-scale and, often, multi-physics analyses [13, 14]. To avoid any surrogate or smoothing procedure of the available results, upscaling through a new Monte Carlo analysis thus looks necessary whenever the considered value does not match those already investigated. Starting in [15, 16], we proposed a deep learning approach to this problem, aiming to learn the micromechanical features of the polycrystal and their role in setting the overall property of interest, see also [17–22], as anticipated not only in terms of reference values but also in terms of scattering or, in a general sense, of its probability distribution. We started by dealing with the elastic moduli of the considered textured films, and showed the importance of the *quality* of the dataset of morphology pictures used to train a convolutional network within an image recognition-like approach.

In this work, we move on by assessing the performances of different network architectures to foresee an optimization of the entire procedure, in terms of accuracy of the results, generalization capability (to catch all the length scale separation effects) and computational efficiency. The formerly used ResNet architecture [23] is here compared to the newly proposed DenseNet one [24]. Results show that both have distinctive features and can attain a remarkable accuracy, higher in the case of the DenseNet-based model.

The remainder of this paper is organized as follows. In Sect. 2, the considered scattering in the micromechanically-driven response of the polycrystalline films is discussed, and the samples used to generate the datasets are accordingly defined. Section 3 deals with the proposed methodology to specify data generation and pre-processing, the features of the adopted convolutional network-based modes and their expected effects on the outcomes for the ResNet and DenseNet architectures. Results are reported in Sect. 4 for polysilicon films characterized by different values of the length scale separation parameter in the training and test sets, to assess the already mentioned generalization capability of the approach. Some concluding remarks and proposals for future developments are finally gathered in Sect. 5.

2   Statistical Scatter in the Homogenized Polysilicon Properties at the Mesoscale

In compliance with a standard geometry of MEMS movable structures, the polysilicon microstructures to be characterized feature a rather small ratio between the size L of the polycrystalline aggregate (termed mesoscale and representing the scale over which homogenization is being carried out) and the characteristic size d of the microscopic heterogeneities (termed microscale and representing the average size of silicon grains).

Whenever a homogenization procedure is adopted within such a frame to define the overall mechanical, electromagnetic, thermal, or any other type of properties in a heterogeneous medium, the mesoscale properties themselves are termed *apparent*. *Effective* properties instead refer to samples of the heterogeneous material large enough to comply with a kind of asymptotic rule of mixture to hold. In this latter case, the material domain can be regarded as mesoscopically homogeneous and L defines the size of the corresponding Representative Volume Element (RVE), see e.g. [25–27] for polycrystalline situations related to the specific case considered in this study. Through homogenization, the overall properties of interest are supposed to be realization independent, namely independent of the micromechanical features of the polycrystal. As said, this condition is achieved only asymptotically, when $\delta = L/d \to \infty$ or, at least, when it becomes so large that grain effects lead to marginal variations in the measured response. Such a critical threshold is attained for δ on the order of $10 - 100$, depending on the property to be determined.

As reported in [27], for polysilicon a microstructure featuring $\delta \cong 60$ could be regarded as an RVE, while domains featuring smaller values do not to comply with the asymptotic values provided by standard homogenization procedures and therefore require the scattering in the results to be assessed too. Under such conditions, Statistical Volume Elements (SVEs) are instead adopted to represent the polycrystalline material samples; since each single realization can be representative of the material only in a statistical sense, Monte Carlo simulations are necessary to quantify a mean property value and also the scattering of the results around it.

3 Methodology

3.1 Input Data Generation and Pre-processing

The polycrystalline morphology here considered is an epitaxially grown one. Moving from the free surface of the substrate, each grain has a major crystal orientation almost aligned with the direction perpendicular to the surface itself. By disregarding the scattering in such a texture, in our former works we focused on a thin slice of the polycrystal and therefore simplify the geometry allowing for two-dimensional SVEs. Such a simplification can be classified as a kind of dimensionality-reduction for the problem at hand.

A set of SVEs has been digitally generated via Voronoi tessellations. Each SVE features its own grain boundary network geometry, and orientations of the crystal lattice of all the grains gathered. The Monte Carlo procedure, exploited to quantify the stochastic effects due to the grain arrangements on the probability distribution of the apparent Young's Modulus E, has been then fed by the results of a numerical homogenization carried out for all the SVEs. The obtained results represent the ground-truth data (or the labels) to be used during the training of the neural networks (NNs). Figure 1 provides an illustration of a (not-to-scale) couple of characteristic morphologies of the handled SVEs, respectively featuring a $2\,\mu\text{m} \times 2\,\mu\text{m}$ (left) or $5\,\mu\text{m} \times 5\,\mu\text{m}$ (right) size. In these single-channel images, the color encodes the in-plane lattice orientation θ of each single-crystalline domain or grain, measured relative to a global reference axis (e.g. the horizontal one in the figure).

The artificially generated images have been then pre-processed in order to maximize the available distinctive features and reduce artifacts, represented by pixels with incorrect values assigned along the grain boundaries where grains with different orientation

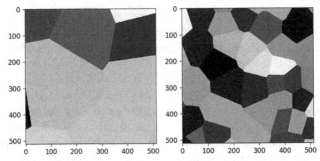

Fig. 1. Digitally generated two-dimensional polysilicon microstructures, wherein the pixel coordinates are represented along the SVE axes.

merge. A median filter with kernel size [3, 3] has been adopted, to smooth the input images and therefore reduce the mentioned artifacts. To also reduce the computational cost of training, the images have been resized to have a final resolution of 128×128 pixels, moving from the original 512×512 pixels one. Additional details regarding the generation of the input data, as well as the pre-processing steps, can be found in [16]. Figure 2 provides a sketch of the final resolution for the two microstructures gathered in Fig. 1.

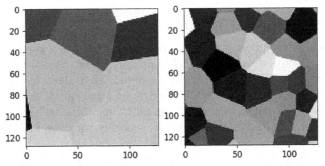

Fig. 2. Final resolution for the SVEs of Fig. 1, wherein pixel coordinates are again represented along the SVE axes.

3.2 Proposed Models and Implementation

Results discussed in [15, 16] are taken as the baseline in the current research activity. Aiming to exploit the advantages of the convolutional architecture proposed in [24], a densely connected network has been newly adopted as the specific architecture to carry on feature extraction, that is the key stage upon which the overall performance of the regression model is based. DenseNet121 is proposed to be employed for feature learning, and its performance is going to be directly compared to that of the former model based instead in the use of the residual ResNet18. Additionally, a new test set featuring SVEs with size 3 μm × 3 μm has been generated and allowed for in the analysis, in order to further assess the generalization capabilities of both models regarding intermediate length scales.

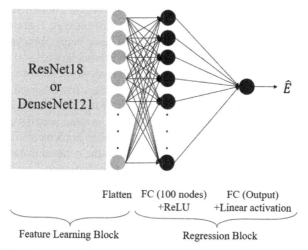

Flatten   FC (100 nodes)   FC (Output)
                    +ReLU   +Linear activation

Feature Learning Block   Regression Block

Fig. 3. Pictorial description of the adopted models: the difference between them is represented by the convolutional architecture employed as the Feature Learning Block (either the ResNet18 or the DenseNet121).

From a methodological standpoint, once each microstructural representation has been generated and preprocessed, the relevant image has been individually fed to the specific NN under consideration. With this input information, each NN model consists of an initial stage aimed at feature extraction taking place in the *feature learning block*; this activity is done thanks to the use of a convolutional network architecture, either the ResNet18 baseline or the newly proposed DenseNet121. After feature extraction, the output feature maps on the last convolutional layer undergo a flatten operation, after which the high-level features extracted are employed as input of the *regression block*, made of a standard fully connected multilayer perceptron. The aim of this block is to provide the estimation of the effective property of interest, associated to the microstructure. For comparison convenience, it is important to mention that all the design elements associated to the regression block have been kept exactly the same for both models, in order to highlight the effect of the convolutional architecture on the overall prediction and generalization capabilities of the model. Besides the regression blocks, also all the associated hyperparameters but the mini-batch size, which has been set to be the maximum number of computationally manageable samples, have been kept the same for both models. The regression block is characterized by an arrangement of a dense fully connected layer (100 nodes) + ReLU activation, followed by a fully connected output layer (1 node) + Linear activation. The total number of parameters in the ResNet18-based regression model has turned out to be 14,457,612 while the total number of parameters in the DenseNet121-based regression model has been 8,669,833. An illustration of the architecture of the two models is presented in Fig. 3: note that the output corresponds to the apparent Young's modulus prediction for each image, that is to $\hat{E}$.

Hence, the aim of the first stage of training has been to learn the individual mapping of the input data onto the effective SVE property, that is dealt with as a kind of label. After training, the statistical analysis of the estimated $\hat{E}$ values and the evaluation of the mean value and dispersion around it are readily accomplished from the set of mappings, by simply extracting the relevant statistical indicators of interest. Therefore, during the training stage a certain number of microstructures have been employed to learn the underlying mapping between the topology of the grain boundary network and the lattice orientation of each grain on one side, and the overall elastic properties of a polycrystalline aggregate on the other side. After this stage, a testing stage has followed, in which each regression model has been used to predict the value of the Young's modulus over new, unseen SVEs featuring also length scales δ different from those used during the training.

For this stage of the procedure, early stopping has been implemented as a regularization technique by monitoring the validation loss, with the patience parameter set to 50 epochs. The weights obtained at the end of the learning stage and corresponding to the minimum value of the validation loss, have been later adopted in order to assess the performance of the models on the test sets. Table 1 summarizes the main hyperparameters selected.

Results to follow have been obtained with the two models developed making use of the Keras API, based on TensorFlow. To speed up training and testing, a GeForce RTX 2080 GPU has been exploited.

Table 1. Hyperparameters used during the training of both models.

Hyperparameter	Value
Total number of epochs	500
Patience (early stopping)	50
Mini-Batch Size (B.S)	300^a, 85^b
Learning rate α	$5 \cdot 10^{-4}$
Optimizer	Adam
Loss function	MSE[c]

[a]Maximum number of computationally manageable samples for the ResNet-based model.
[b]Maximum number of computationally manageable samples for the DenseNet-based model.
[c]Mean squared error.

3.3   ResNet vs DenseNet: Addition vs Concatenation of Features

Considering the aspects of the implementation mentioned in Sect. 3.2, it is expected to note performance differences between the models due to the specific connectivity pattern displayed by the two different architectures of ResNets and DenseNets.

According to the theoretical and practical evidence presented in [24], the connectivity pattern in DenseNets is characterized by having all layers with matching feature-map sizes directly connected with each other. As explained in the original work, feature propagation and reuse is therefore by definition strengthened in this convolutional network architecture, to improve the flow of information and gradients throughout the network and substantially reduce the number of parameters to tune during training. This last aspect is linked to the relative *narrow* nature of DenseNets: each non-linear transformation at every layer adds just a few feature-maps to the network *collective knowledge*. The number of filters per layer, referred to by the authors of this architecture as growth rate k, is low if compared to the number of filters typically used in ResNet architectures: in concrete terms, it is 32 for the DenseNet121 model, while it amounts to 64, 128, 256 and 512 for the ResNet18 model.

Denoting the output of the l-th NN layer as x_l and the nonlinear transformation as $H_l(\cdot)$, with traditional convolutional networks, ResNet and DenseNet we respectively have:

$$x_l = H_l(x_{l-1}) \tag{1}$$

$$x_l = H_l(x_{l-1}) + x_{l-1} \tag{2}$$

$$x_l = H_l([x_0, x_1, \ldots, x_{l-1}]) \tag{3}$$

where $\left[x_0, x_1, \ldots, x_{l-1}\right]$ refers to the concatenation of the feature-maps produced in layers 0, 1, ...,$l - 1$. In these equations, $H_l(\cdot)$ is a composite function of consecutive operations. For example, in both ResNets and DenseNets the nonlinear transformation includes the consecutive application of three operations: a batch normalization, a rectified linear unit and a convolution.

At variance with ResNets, features in DenseNets are not combined through summation before they are passed into a layer (symbolized in Eq. 2); instead, they are combined by concatenation (symbolized in Eq. 3). This difference largely determines the characteristics of the information extraction and propagation in each model, ultimately constituting the key difference when comparing the performances of the two models here considered.

4   Results

4.1   Generalization Capability of Trained Models

The data to be handled by the two models have been split exactly in the same manner. A total of 3878 2 μm × 2 μm images ($\delta = 2$ μm/0.5 μm $= 4$) has been split into the training set and validation sets, using 75% and 25% respectively. In this way, the parameters of the model are initially fit on the training set, while simultaneously, after each epoch, the fitted model is used to predict the response over the validation set. The validation set therefore has a double purpose: it is used to tune the hyperparameters of the model and it allows for the implementation of *early stopping* as a regularization technique. Finally, two additional sets referred to as test sets have been generated using 3 μm × 3 μm images and 5 μm × 5 μm images. Clearly, although these two sets feature different δ values, specifically δ = 3 μm/0.5 μm = 6 and δ = 5 μm/0.5 μm = 10, both preserve the input size i.e. image resolution. These sets are employed for the exclusive purpose of assessing the performance (i.e. generalization) of the final model. Summarizing, the datasets are arranged as follows: 2889 2 μm × 2 μm images as training set (Mean = 150.08 GPa and SD = 5.40 GPa); 989 2 μm × 2 μm images as validation set (Mean = 150.16 GPa and SD = 5.34 GPa), and, as alternate test sets featuring a larger aggregate size, 30 3 μm × 3 μm images (Mean = 149.87 GPa and SD = 3.75 GPa) and 145 5 μm × 5 μm images (Mean = 149.34 GPa and a SD = 2.44 GPa) were generated. Within the brackets, for each set values have been reported for the relevant mean and standard deviation (SD) of E.

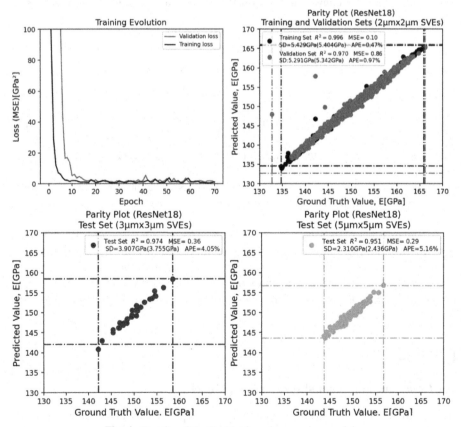

Fig. 4. Results of the ResNet-based regression model.

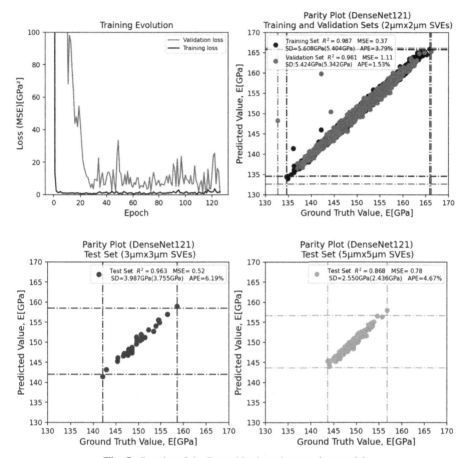

Fig. 5. Results of the DenseNet-based regression model.

Figures 4 and 5 show the obtained results with the ResNet-based and the DenseNet-based regression models, respectively. For each case, results are reported first in terms of the evolution of the training and validation loss against the epochs. To assess the performance of the models, results are reported next in terms of parity plots. In these charts, the dotted lines are used to denote the maximum and minimum label values featured by each dataset. In this way, predictions should ideally be mapped within the limits defined by these lines (i.e. inside the dotted squares). Moreover, for an optimal trained model, the data should map the identity function for every dataset, with all the dots aligned along the 45° diagonal. The performance of the regression models is assessed through the coefficient of determination, R^2 and the MSE. In the best case, e.g. predicted values exactly matching the ground-truth values, then $R^2 = 1$ and MSE $= 0$. For each dataset, the mentioned evaluation metrics are reported in the legends. The legends also include the estimation of the statistical indicator of interest (SD); these values are computed from the set of individual predictions produced by the trained model. The corresponding SD characterizing the ground-truth data appears within the brackets and the associated absolute percentage error (APE) is finally reported. Results are summarized in Table 2.

Table 2. Summary of the results

Metric	Dataset	ResNet18-based	DenseNet121-based
MSE [GPa^2]	Training set	0.1	0.37
	Validation set	0.86	1.11
	Test set 1 ($\delta = 6$)	0.36	0.52
	Test set 2 ($\delta = 10$)	0.29	0.78
R^2	Training set	0.996	0.987
	Validation set	0.970	0.961
	Test set 1 ($\delta = 6$)	0.974	0.963
	Test set 2 ($\delta = 10$)	0.951	0.868
Standard Deviation APE	Training set	0.47%	3.79%
	Validation set	0.97%	1.53%
	Test set 1 ($\delta = 6$)	4.05%	6.19%
	Test set 2 ($\delta = 10$)	5.16%	4.67%

For the case of the ResNet18-based model in Fig. 4, the learning process reached a plateau after 71 epochs, with the subsequent 50 epochs leading to small validation loss changes. This model thus shows promising generalization capabilities in setting the intended link between the microstructure and the investigated property, as confirmed by the results provided for both test sets. For the test dataset $\delta=6$, the model has effectively predicted a larger value for the statistical indicator describing the dispersion of the homogenized property, as expected for the lower value of δ associated to this test set, when compared to the one featuring $\delta = 10$.

In the case of the DenseNet121-based model in Fig. 5, the learning process reached a plateau after 126 epochs, again with the subsequent 50 epochs not leading to clear changes in the validation loss. This new model, which is linked to a lower total number of parameters to be tuned in the training process, has not displayed significant performance improvements, when compared to the ResNet18-based.

In addition, by analyzing the presented results, it can be observed that although the DenseNet121-based model requires fewer parameters it does not necessarily display a faster convergence rate, e.g. in terms of the number of epochs required to reach the reported minimum validation loss values. Moreover, the computational burden of this model, in terms of memory usage and GPU power, has been higher. Individual epochs of the DenseNet121-based model lasted about 3 times the time taken by epochs in the ResNet18-based. Furthermore, due to memory issues, the DenseNet121-based model has only allowed the use of a fraction of the mini-batch size when compared to the ResNet18-based model (as stated in Table 1), leading to a noisier convergence behavior as can be observed in the training evolution plots. This is interesting, since the lower number of trainable parameters associated to the DenseNet121-based model does not proportionally translate into larger number of computationally manageable samples in the batch size, or faster training time.

As far as the generalization capabilities are concerned, in general terms a better performance has been observed adopting the ResNet18-based architecture.

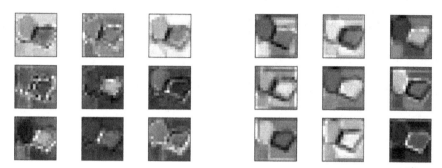

Fig. 6. Random output feature maps. ResNet18 (left) and the DenseNet121 (right).

Due to intrinsic differences between the two convolutional network architectures, DenseNet121-based model is able to extract higher level features: while for the ResNet18, the output size of the layer just before the flatten operation is [8 × 8 × 512], for the DenseNet121 this becomes [4 × 4 × 1024], featuring a larger down sampling and number of output features maps (OFMs). As an additional remark, Fig. 6 shows a comparison between the information extracted in a group of randomly selected, intermediate, same size OFMs, once both models have been trained. A qualitative comparison reveals that the DenseNet-based feature extraction better reproduces the morphology and internal homogeneity within the grains: the distribution of pixels with incorrect values along grain boundaries, observed in the previous figure in the form of pixelated fringes, appears to be thinner for the OFMs provided by the DenseNet121.

5   Conclusions

In this work, we have compared the performances of two different NN-based models to learn the micromechanical features ruling the value of the apparent stiffness of a polycrystalline aggregate. Specifically, for SVEs characterized by a mesoscale size slightly larger than the micromechanical length scale, linked e.g. to the characteristic size of the grains, the overall Young's modulus of textured polycrystalline films has been estimated in statistical terms through its mean value (though not discussed here as not significant in this work) and scattering around it.

For the feature learning stage of the models, the two different architectures adopted have been the ResNet18 and the DenseNet121. For the problem at hand, the first one has turned out to be characterized by a larger set of parameters to be tuned during the training of the NN; the second one, though featuring less parameters to tune, has resulted to be more computational expensive. Moreover, the additional computational burden associated to the second architecture did not result into higher accuracy in mapping the micromechanical features ruling the results, and so, in catching the scattering in the results attained with numerical homogenization procedures, when compared to the first one.

The proposed approach is still in its infancy and surely requires some additional work in order to optimize the proposed physics-informed NN architecture, the setting of the NN hyperparameters, and the training strategy. The goal of the project is to make this procedure available for all the devices whose response may be sensitive to microstructural features and, accordingly, to possible defects existing in their building blocks at the nano- and microscale.

Acknowledgments. Partial financial support by STMicroelectronics through the project *MaRe* (Material Reliability) is gratefully acknowledged. JPQM acknowledges the financial support by the University of Costa Rica.

References

1. Hsu, T.R.: MEMS and Microsystems: Design, Manufacture, and Nanoscale Engineering. John Wiley & Sons, Hoboken, NJ, USA (2008)
2. Brand, O., Fedder, G.K., Hierold, C., Korvink, J.G., Tabata, O., Tsuchiya, T.: Reliability of MEMS Testing of Materials and Devices. John Wiley & Sons, Hoboken, NJ, USA (2013)
3. Corigliano, A., Ardito, R., Comi, C., Frangi, A., Ghisi, A., Mariani, S.: Mechanics of Microsystems. John Wiley & Sons, Hoboken, NJ, USA (2018)
4. Weinberg, M.S., Kourepenis, A.: Error sources in in-plane silicon tuning-fork MEMS gyroscopes. J. Microelectromech. Syst. **15**, 479–491 (2006). https://doi.org/10.1109/JMEMS.2006.876779
5. De Laat, M., Pérez Garza, H., Herder, J., Ghatkesar, M.: A review on in situ stiffness adjustment methods in MEMS. J. Micromech. Microeng. **26**, 1–21 (2016). https://doi.org/10.1088/0960-1317/26/6/063001
6. Uhl, T., Martowicz, A., Codreanu, I., Klepka, A.: Analysis of uncertainties in MEMS and their influence on dynamic properties. Arch. Mech. **61**, 349–370 (2009)
7. Bagherinia, M., Mariani, S.: Stochastic effects on the dynamics of the resonant structure of a Lorentz Force MEMS magnetometer. Actuators. **8**, 36 (2019). https://doi.org/10.3390/act8020036
8. Mirzazadeh, R., Mariani, S.: Uncertainty quantification of microstructure-governed properties of polysilicon MEMS. Micromachines. **8**, 248 (2017). https://doi.org/10.3390/mi8080248
9. Mirzazadeh, R., Ghisi, A., Mariani, S.: Statistical investigation of the mechanical and geometrical properties of polysilicon films through on-chip tests. Micromachines. **9**, 53 (2018). https://doi.org/10.3390/mi9020053
10. Mariani, S., Ghisi, A., Mirzazadeh, R., Eftekhar Azam, S.: On-chip testing: a miniaturized lab to assess sub-micron uncertainties in polysilicon MEMS. Micro Nanosyst. **10**, 84–93 (2018). https://doi.org/10.2174/1876402911666181204122855
11. Mirzazadeh, R., Eftekhar Azam, S., Mariani, S.: Mechanical characterization of polysilicon MEMS: a hybrid TMCMC/POD-kriging approach. Sensors. **18**, 1243 (2018). https://doi.org/10.3390/s18041243
12. Ghisi, A., Mariani, S.: Effect of imperfections due to material heterogeneity on the offset of polysilicon MEMS structures. Sensors. **19**, 3256 (2019). https://doi.org/10.3390/s19153256

13. Mariani, S., Ghisi, A., Corigliano, A., Martini, R., Simoni, B.: Two-scale simulation of drop-induced failure of polysilicon MEMS sensors. Sensors. **11**, 4972–4989 (2011). https://doi.org/10.3390/s110504972
14. Ghisi, A., Mariani, S., Corigliano, A., Zerbini, S.: Physically-based reduced order modelling of a uni-axial polysilicon MEMS accelerometer. Sensors. **12**, 13985–14003 (2012). https://doi.org/10.3390/s121013985
15. Quesada-Molina, J.P., Rosafalco, L., Mariani, S.: Stochastic mechanical characterization of polysilicon MEMS: a deep learning approach. In: Proceedings of 6th International Electronic Conference on Sensors and Applications, vol. 42, p. 8 (2020). https://doi.org/10.3390/ecsa-6-06574
16. Quesada-Molina, J.P., Rosafalco, L., Mariani, S.: Mechanical characterization of polysilicon MEMS devices: a stochastic, deep learning-based approach. In: 2020 21st International Conference on Thermal, Mechanical and Multi-Physics Simulation and Experiments in Microelectronics and Microsystems (EuroSimE), pp. 1–8, IEEE Press. New York (2020). https://doi.org/10.1109/EuroSimE48426.2020.9152690
17. Bock, F.E., Aydin, R.C., Cyron, C.J., Huber, N., Kalidindi, S.R., Klusemann, B.: A review of the application of machine learning and data mining approaches in continuum materials mechanics. Front. Mater. **6**, 110 (2019). https://doi.org/10.3389/fmats.2019.00110
18. Cang, R., Ren, M.Y.: Deep network-based feature extraction and reconstruction of complex material microstructures. In: Proceedings of the ASME 2016 International Design Engineering Technical Conferences & Computers and Information in Engineering Conference, pp. 1–10 (2016). https://doi.org/10.1115/DETC2016-59404
19. Lubbers, N., Lookman, T., Barros, K.: Inferring low-dimensional microstructure representations using convolutional neural networks. Phys. Rev. E **96**, 1–14 (2017). https://doi.org/10.1103/PhysRevE.96.052111
20. Yang, Z., et al.: Deep learning approaches for mining structure-property linkages in high contrast composites from simulation datasets. Comput. Mater. Sci. **151**, 278–287 (2018). https://doi.org/10.1016/j.commatsci.2018.05.014
21. Cecen, A., Dai, H., Yabansu, Y.C., Kalidindi, S.R., Song, L.: Material structure-property linkages using three-dimensional convolutional neural networks. Acta Mater. **146**, 76–84 (2018). https://doi.org/10.1016/j.actamat.2017.11.053
22. Cang, R., Li, H., Yao, H., Jiao, Y., Ren, Y.: Improving direct physical properties prediction of heterogeneous materials from imaging data via convolutional neural network and a morphology-aware generative model. Comput. Mater. Sci. **150**, 212–221 (2018). https://doi.org/10.1016/j.commatsci.2018.03.074
23. He, K., Zhang, X., Ren, S., Sun, J.: Deep residual learning for image recognition. In: Proceedings of 2016 IEEE Conference on Computer Vision and Pattern Recognition (CVPR), pp. 770–778. IEEE Press, New York (2016). https://doi.org/10.1109/CVPR.2016.90
24. Huang, G., Liu, Z., Van der Maaten, L., Weinberger, K.Q.: Densely connected convolutional networks. In: 2017 IEEE Conference on Computer Vision and Pattern Recognition (CVPR), pp. 2261–2269. IEEE Press, New York (2017). https://doi.org/10.1109/CVPR.2017.243
25. Otoja-Starzewski, M.: Material spatial randomness: from statistical to representative volume element. Probab. Eng. Mech. **21**, 112–132 (2006). https://doi.org/10.1016/j.probengmech.2005.07.007
26. Kanit, T., Forest, S., Galliet, I., Mounoury, V., Jeulin, D.: Determination of the size of the representative volume element for random composites: statistical and numerical approach. Int. J. Solids Struct. **40**, 3647–3679 (2003). https://doi.org/10.1016/S0020-7683(03)00143-4
27. Mariani, S., Martini, R., Ghisi, A., Corigliano, A., Beghi, M.: Overall elastic properties of polysilicon films: a statistical investigation of the effects of polycrystal morphology. J. Multiscale Comput. Eng. **9**, 327–346 (2011). https://doi.org/10.1615/IntJMultCompEng.v9.i3.50

MsFEM Upscaling for the Coupled Thermo-Mechanical Problem

Marek Klimczak[(✉)] [iD] and Witold Cecot[(✉)] [iD]

Faculty of Civil Engineering, Cracow University of Technology, Kraków, Poland
mklimczak@L5.pk.edu.pl, plcecot@cyf-kr.edu.pl

Abstract. In this paper, we present the framework for the multiscale thermoelastic analysis of composites. Asphalt concrete (AC) was selected to demonstrate the applicability of the proposed approach. It is due to the observed high dependence of this material performance on the thermal effects. The insight into the microscale behavior is upscaled to the macroresolution by the multiscale finite element method (MsFEM) that has not been used so far for coupled problems. In the paper, we present a brief description of this approach together with its new application to coupled thermoelastic numerical modeling. The upscaled results are compared with the reference ones and the error analysis is presented. A very good agreement between these two solutions was obtained. Simultaneously, a large reduction of the degrees of freedom can be observed for the MsFEM solution. The number of degrees of freedom was reduced by 3 orders of magnitude introducing an additional approximation error of only about 6%. We also present the convergence of the method with the increasing approximation order at the macroresolution. Finally, we demonstrate the impact of the thermal effects on the displacements in the analyzed asphalt concrete sample.

Keywords: Multiscale finite element method · Upscaling · Thermoelasticity · Asphalt concrete

1 Introduction

The placement of our research, divided into two parts, is presented in this Section. Firstly, we describe the developed AC material models incorporating the thermal effects. Secondly, we present selected multiscale approaches to the modeling of asphalt concrete. Our study encompasses mainly the effects of the heterogeneous AC structure on its performance. In addition, we account for the heat transfer within this material, which affects the mechanical response.

This research was supported by the National Science Center (NCN) under grant UMO-2017/25/B/ST8/02752.

1.1 Asphalt Concrete

Asphalt concrete (AC) is a typical material used for the flexible or semi-rigid pavement structure layers. In this paper, we use this term to describe the composite made of two main ingredients: the mineral aggregate ($90 \div 95\%$ w/w, where w/w stands for the weight ratio) and the mastic ($5 \div 10\%$ w/w). The latter is understood as a mixture of the asphalt binder and a very fine mineral filler.

For the sake of simplicity, we neglect in this study the presence of the air voids, the modifiers and other possible additives since the paper focus is on the multiscale analysis rather than on the materials science. The extension of the presented in this paper framework to other asphalt mix types is possible and it does not affect its routine. The only restriction is the analysis scale, which is the continuum. We cannot directly incorporate e.g. the atomistic scale observations to the whole analysis.

In Poland, the term *asphalt concrete* precisely refers to one of the asphalt mix types, which is characterized by the uniform gradation curve. It means that all particle fractions are represented in the aggregate mix with a similar weight ratio. In this paper, we keep this Polish nomenclature. Some other asphalt mixes include SMA (stone mastic asphalt). The comparison between exemplary gradation curves for AC and SMA is shown in Fig. 1.

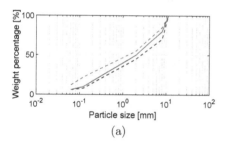
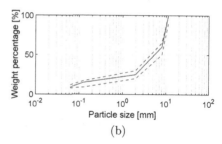

(a)   (b)

Fig. 1. Exemplary grading curves (red solid line) for (a) AC and (b) SMA - blue dotted lines denote bounding gradation curves according to Polish guidelines (Color figure online)

We are able to model any gradation curves. However, in the numerical examples we focus on AC only since its applicability is very broad comparing to other asphalt mix types designed rather for specific pavement layers. Modifying the gradation curve, the asphalt binder content and other properties, AC can be laid for all asphalt pavement layers.

The performance of AC at the continuum scale is based on the mechanisms that can be observed at lower scales. The overview of a number of complex phenomena influencing the overall AC response is presented e.g. in [11,27]. They are related either to external factors such as loading, temperature, aging, moisture, to name only a few, or the internal material structure.

The very basic description of the AC performance is provided by the linear elasticity equations. Namely, all the considered constituents are modeled as linear elastic. In some cases, when the analysis period is very short, the subject load is constant and the temperature is relatively low, such an approach may be sufficient to capture the physical response of AC. Instead of the linear elastic model, more realistic viscoelastic constitutive equations can be easily used. We have already applied MsFEM for such model [13] but in this paper, we focus on the novel MsFEM implementation, i.e. thermo-mechanical coupling.

In the number of papers, only the aggregate particles are modeled as linear elastic, whereas the asphalt binder is described using a variety of viscoelasticity [22–24, 29, 34], viscoelastoplasticity [18, 33] or even more complex models accounting for the accumulated damage [2, 12, 32]. In some studies, the adhesive (interfacial) zone between the AC constituents is also encompassed in the analysis [17, 19, 34]. Typically, a perfect bonding between the binder and the aggregate particles, as it is the case in this paper, is assumed.

It should be also remarked that due to the increasing complexity of the above models, not in all of the studies the heterogeneous AC microstructure was considered. Contrary, the effective AC material model was developed.

Temperature, which is considered as one of the major factors influencing the AC performance, is taken into account in a number of studies in several modes. Firstly, the temperature distribution in AC can be regarded as a standalone problem [9, 26, 31]. Its solution can be used in further specific analyses.

In this paper, we use the so-called *sequential coupling of thermal-stress analysis*. It is a basic approach to the solution of the thermoelasticity problem that is weakly coupled, i.e. the displacement and stress fields are temperature-dependent without the reverse relationship. This algorithm is implemented e.g. in ABAQUS. Since its common application (c.f. [30]), including this study, it is recapitulated below (Algorithm 1). When solving the final elasticity problem, the thermal strain is applied in a form of a contribution to the right-hand side of the governing differential equation.

Algorithm 1. Solve a sequentially coupled thermoelasticity problem

---

Require: define the problem (geometry, material parameters and boundary conditions)
Ensure: the FE mesh complying with the material distribution
  solve the heat transfer problem within the domain
  for $n=1$ to N_{GP} **do** {loop over Gauss points}
    compute the thermal strains resulting from the temperature distribution
    set material parameters for the computed temperature
  end for
  solve the elasticity problem with thermal strains and temperature-dependent material parameters

---

In the *fully coupled thermal-stress analysis*, the displacement/stress fields are computed simultaneously with the temperature field. It is used for the strongly coupled fields affecting each other, contrary to the weakly coupled fields analyzed using the sequential thermal-stress approach according to Algorithm 1. In order to solve such a problem, the multiscale finite element method can be also used.

1.2 Multiscale AC Analyses

As it was mentioned in the previous chapter, the complexity of the AC material models considerably limits their application to the analysis with the heterogeneous microstructure considered. Therefore, homogenized AC effective responses are needed in practice.

The multiscale modeling, that we use in this study, incorporates the information from the lower scale (or scales) to the actual analysis scale. The main goal of these approaches is to transfer this information effectively. Namely, we want to reduce the computation cost by introducing as little additional modeling error as possible.

A number of homogenization and upscaling techniques were developed for the AC modeling [8,11,12,15,20,22,23,28]. Most of them are variants of the FE^2 homogenization presented in [6]. This approach consists in a two-scale analysis using the finite element method at each level. At the macroscale Gauss points, auxiliary boundary value problems (BVP) are solved to account for the underlying microstructure. The domain used for a single BVP is called the representative volume element (RVE) and should be constructed precisely to reveal the AC microstructure. The advantage of the FE^2 homogenization is that no material model needs to be used at the macroscale. Instead, the effective quantity Q is passed from RVE using the averaging over the RVE volume V:

$$\langle Q \rangle = \frac{1}{V} \int_V Q dv \tag{1}$$

In [28], the proper orthogonal decomposition (POD) was used to reduce the problem order for the elastic AC computations. These two above-mentioned approaches were combined in [8].

The effective values of the complex modulus and phase angle of the analyzed AC specimen were obtained in [21] using the Generalized Self Consistent Scheme - an analytical model of the idealized inclusion embedded in a matrix.

Due to a number of its advantages, the FE^2 approach is the most popular. However, its limitation is the distinct separation of scales condition. In the context of the AC numerical modeling, it can be a caveat for the thinner asphalt layers. Namely, the upper pavement layers - wearing and binding course.

The upscaling approach, called the multiscale finite element method (MsFEM), was used in [14,15] by us to model the elastic and viscoelastic response of AC. A short description of MsFEM is provided hereinafter.

The upscaled/homogenized solutions for AC modeled as the thermoelastic material are not very common in the literature. General approaches, as in [3,10, 25], are rather developed.

To our best knowledge, MsFEM has never been used for such a problem. Thus, the present study aims to fill this gap, demonstrate the initial results and perspectives for further research.

2 Problem Formulation

In this study, we search for the solution of the weakly coupled thermoelasticity problem in a heterogeneous domain using the set of equations recapitulated below. For the sake of simplicity, we do not assume the time-dependency of any of the present fields. As far as the mechanical AC response is considered, the linear elasticity model is used. The temperature distribution is obtained solving the steady-state heat transfer problem. The numerical computations are performed according to Algorithm 1.

Find the displacements $u(x)$ and the temperature $\Theta(x)$ such that

$$
\begin{cases}
\mathbf{div}\boldsymbol{\sigma} + \boldsymbol{X} = \mathbf{0} & \forall\ \boldsymbol{x} \in \omega_i \subset \Omega \\
\boldsymbol{\sigma} = \boldsymbol{C}^{-1}[\boldsymbol{\varepsilon}(\boldsymbol{u}) - \boldsymbol{\varepsilon}^*] & \forall\ \boldsymbol{x} \in \omega_i \subset \Omega \\
\boldsymbol{\varepsilon} = \frac{1}{2}[\nabla\boldsymbol{u} + (\nabla\boldsymbol{u})^T] & \forall\ \boldsymbol{x} \in \omega_i \subset \Omega \\
-k\Delta\Theta = f & \forall\ \boldsymbol{x} \in \omega_i \subset \Omega \\
\boldsymbol{\varepsilon}^* = \alpha\boldsymbol{I}\Theta & \forall\ \boldsymbol{x} \in \omega_i \subset \Omega \\
+ \text{ boundary \& continuity or debonding conditions}
\end{cases}
\tag{2}
$$

where $\boldsymbol{\sigma}$ denotes the stress tensor, $\boldsymbol{X}$ are the body forces, $\boldsymbol{C}$ is the tensor of material parameters for elasticity, k is the thermal conductivity, α is the coefficient of thermal expansion and $\boldsymbol{\varepsilon}^*$ is the thermal strain tensor.

Equations 2 $_{IV,V}$, referring solely to the heat transfer problem, are solved first. Then, the thermal strains $\boldsymbol{\varepsilon}^*$ are transferred to the mechanical part contributing to the right-hand side of Eq. 2 $_I$. The weak formulation of the mechanical part, used for the finite element computations, is presented below.

Find the displacements $u(x,t) \in H_0^1(\Omega) + \hat{u}$ such that

$$
\int_\Omega \boldsymbol{\varepsilon}(\boldsymbol{v}) \colon \boldsymbol{C}^{-1}\boldsymbol{\varepsilon}(\boldsymbol{u})d\omega = \int_\Omega \boldsymbol{v} \cdot \boldsymbol{X}\,d\omega + \int_{S_\sigma} \boldsymbol{v} \cdot \hat{\boldsymbol{t}}ds + \int_\Omega \boldsymbol{\varepsilon}(\boldsymbol{v}) \colon \boldsymbol{C}^{-1}\boldsymbol{\varepsilon}^*d\omega \quad \forall \boldsymbol{v} \in H_0^1(\Omega) \tag{3}
$$

where $\boldsymbol{v}$ denotes test functions, $\hat{\boldsymbol{t}}$ are the known tractions, $\hat{\boldsymbol{u}}$ are the known displacements and H_0^1 is the Sobolev space of the functions satisfying homogeneous Dirichlet boundary conditions.

3 Multiscale Finite Element Method

MsFEM [4, 5, 7] is one of the upscaling methods. It enables to model the problem at the macroresolution incorporating the lower resolution information. The core of its idea is presented schematically in Fig. 2.

Firstly, regardless of the information on the microstructure, we discretize the domain with the coarse mesh (blue quadrilateral elements in Fig. 2). Then, we

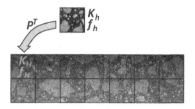

Fig. 2. MsFEM scheme illustrated for a single coarse element (Color figure online)

proceed element-wise. A fine mesh complying with the microstructure is generated within a given coarse element. Assembly within this subdomain delivers a fine mesh stiffness matrix $\boldsymbol{K}_h$ and a load vector $\boldsymbol{f}_h$. Its effective counterparts are computed using the following formulas:

$$
\begin{aligned}
\boldsymbol{K}_H &= \boldsymbol{P}^T \boldsymbol{K}_h \boldsymbol{P} \\
\boldsymbol{f}_H &= \boldsymbol{P}^T \boldsymbol{f}_h
\end{aligned}
\tag{4}
$$

Therein, matrix $\boldsymbol{P}$ stands for the prolongation operator. Its assessment requires the solution to the auxiliary boundary value problem that is used for the search of the modified coarse element shape functions. The degrees of freedom (dof) of the solution for a given shape function constitute the corresponding column of $\boldsymbol{P}$.

In the case of the thermoelasticity problem, we need to solve two sets of auxiliary BVP: for the elasticity and the heat transfer problems consequently. For the detailed description of the BVP leading to the assessment of the modified shape functions for the elasticity, we refer to our previous papers [14,15]. For the sake of brevity, only the auxiliary BVP for the heat transfer is presented below.

Given Ψ_m, *which is a coarse mesh standard scalar-valued shape function* $(m = 1, \ldots, M)$, *we look for its scalar-valued counterpart* Φ_m *that is a discrete solution of the following Dirichlet boundary value problem*

$$
\begin{cases}
\frac{\partial}{\partial x_i}\left(k\frac{\partial \Phi_m}{\partial x_i}\right) = -k^\alpha \frac{\partial^2}{\partial x_i^2}\Psi_m & \forall i = 1,2,3, \ \boldsymbol{x} \in \Omega_s^j \subset \Omega_s \subset \Omega \\
\Phi_m = \hat{\Phi}_m \quad \text{on} \quad \partial\Omega_s \\
+ \text{ interface continuity conditions}
\end{cases}
\tag{5}
$$

where k *is the thermal conductivity of the material at a given location* $\boldsymbol{x}$, k^α *is the arbitrary averaged thermal conductivity in* Ω_s, Ω_s^j *denotes the* j*-th constituent of the composite.*

Problem 5 is solved using FEM, thus the corresponding variational formulation needs to be stated. For the completeness of the problem statement, we refer to the comprehensive discussion on the definition of the Dirichlet boundary conditions for Φ_m in [14,15].

Illustration of the MsFEM routine is shown on a simple example of the 1D bar. Let us consider a prismatic bar with a length of 2. Its left half is characterized with $EA = 1$, the right one with $EA = 10$. Using 2 linear finite elements, their stiffness matrices are equal to K_L and K_R, respectively. Their assembly gives K_h. Computation of the effective matrix K_H for one linear coarse element (occupying the whole bar) requires solutions to the corresponding auxiliary problems (see [1]) to obtain prolongation operator P, consequently. The above-mentioned matrices are given below:

$$K_L = \begin{bmatrix} 1 & -1 \\ -1 & 1 \end{bmatrix}, \ K_R = \begin{bmatrix} 10 & -10 \\ -10 & 10 \end{bmatrix}, \ K_h = \begin{bmatrix} 1 & -1 & 0 \\ -1 & 11 & -10 \\ 0 & -10 & 10 \end{bmatrix}, \ P = \begin{bmatrix} 1 & 0 \\ 0.09 & 0.91 \\ 0 & 1 \end{bmatrix},$$

$$K_H = \begin{bmatrix} 0.91 & -0.91 \\ -0.91 & 0.91 \end{bmatrix}.$$

4   Numerical Results

The proposed approach is illustrated with the numerical results for the idealized AC specimen of dimensions $8\,\mathrm{cm} \times 20\,\mathrm{cm}$. Material distribution for the arbitrary gradation curve is shown in Fig. 3. As shown in Fig. 3a, the whole upper edge is subject to heating with $q = 30\,\mathrm{W/m}$, whereas temperature along the bottom edge is equal to $20\,^{\circ}\mathrm{C}$. The remaining edges are insulated. As far as the mechanical part is concerned, the bottom edge is fixed and the load of intensity $t = 100\,\mathrm{kN/m}$ is subject to the upper edge centrally for $1/5$ of its length. Remaining edges are free.

Respective material parameters are presented in Table 1. We neglect the change in the Young modulus for the aggregate particles as a function of temperature but we do account for this, crucial in the case of the mastic, phenomenon. We use the Young modulus value measured at temperature $20\,^{\circ}\mathrm{C}$ for the pure linear elastic case (see Fig. 6) and another value measured at temperature $50\,^{\circ}\mathrm{C}$. This temperature is approximately equal to the result of the heat transfer problem analyzed in Sect. 4.1. One can observe a dramatic decrease of the mastic Young modulus value at this temperature. It is two orders of magnitudes smaller than the value at the reference temperature ($20\,^{\circ}\mathrm{C}$). For the sake of simplicity, we use only one value of the Young modulus for the whole matrix in the second step of the sequentially coupled thermal-stress analysis. Practically, it is a function of the temperature distribution obtained in the first step (see Fig. 4) and it changes also spatially.

4.1   The Thermal Part

In the sequentially coupled thermal-stress analysis, the first step is to solve the heat transfer problem. The temperature and thermal strain distribution is shown in Fig. 4. Since all non-zero thermal strain tensor components are equal, we present only one of them.

As it can be observed, the discrepancies between the reference fine mesh solution and the upscaled one are extremely small. The reference solution was obtained using more than 113 000 degrees of freedom (about 220 000 finite elements) and the linear approximation was used. The upscaled solution was

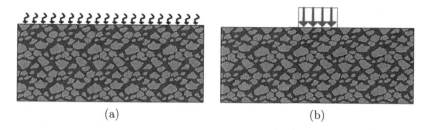

(a)   (b)

Fig. 3. Neumann boundary conditions for (a) heat transfer and (b) mechanical part

Table 1. Material parameters

Parameter	Unit	Value
Young modulus - inclusion	Pa	80e9
Young modulus - matrix (at $20\,^{\circ}$C)	Pa	10e9
Young modulus - matrix (at $50\,^{\circ}$C)	Pa	10e7
Poisson ratio - inclusion	–	0.3
Poisson ratio - matrix	–	0.3
Thermal conductivity - inclusion	W/(mK)	0.8
Thermal conductivity - matrix	W/(mK)	4.2
Coefficient of linear thermal expansion - inclusion	$10^{-6}/^{\circ}$C	8
Coefficient of linear thermal expansion - matrix	$10^{-6}/^{\circ}$C	190

obtained for a regular mesh of 2×5 coarse elements with the increasing approximation orders ($p = 1 \div 5$). The convergence of the temperature and the thermal strain errors measured in L_2 norm is shown in Fig. 5 using the logarithmic scale.

Increasing the approximation order, we observe the reduction of the L_2 error norm for this problem. It drops from about 1.4% (strain) or 1.5% (temperature) for 18 dof to about 1% (both fields) for about 300 dof (instead of about 113 000 fine mesh dof used for the reference solution). This behavior needs a short discussion. Whereas the p-convergence observed at the macroscale is a typical property of many upscaling methods, it can be observed rather for the primary field (temperature for this problem). It is not very common for other fields, derived from the primary one. Thus, this p-convergence observed for both present fields can be regarded as the MsFEM advantage. Thermal strain tensor is a linear function of temperature. However, coefficients of thermal expansion for the matrix (asphalt mastic) and inclusions (mineral aggregate) are significantly different (see Table 1). Obtaining such satisfactory results could be cumbersome for a number of upscaling methods. Such a good performance of MsFEM is possible due to the reversible relationship between macro- and microscale degrees of freedom. Thus, after obtaining macroscale ones, we can easily compute in the post-processing also the microscale degrees of freedom and proceed with this more detailed information.

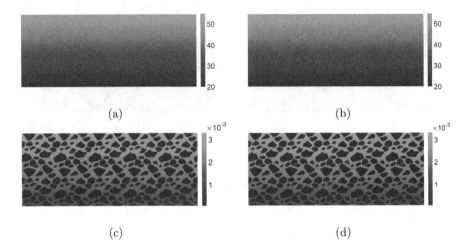

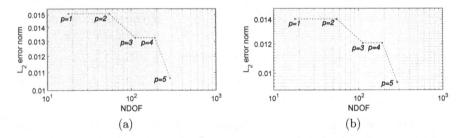

Fig. 4. Maps of temperature (a ÷ b) and thermal strain (c ÷ d) - the left column shows the upscaled solution (for $p = 5$) and the right one shows the reference fine mesh solution

Fig. 5. Error convergence for (a) temperature and (b) thermal strain

Reduction of degrees of freedom compared to the reference fine mesh solution spans from about 6300 ($p = 1$) to about 400 ($p = 5$) times with the corresponding temperature error values ranging from about 1% ($p = 5$) to about 1.5% ($p = 1$).

4.2   The Mechanical Part

The results of the elastic analysis are shown in Fig. 6. We used for this case material parameters for temperature equal to 20 °C for the mastic (matrix). A very good agreement between the reference and upscaled solutions can be observed. The p-convergence (Fig. 7) shown both for the L_2 and energy norms demonstrates the benefits of the higher order approximation used at the macroscale. Increasing the approximation order, we reduce the errors measured in L_2 and energy norms from about 20% to about 6% and from about 60% to about 20%, respectively. The corresponding reduction of the fine mesh reference solution degrees of freedom (about 220 000) spans from 18 ($p = 1$) to 572 ($p = 5$) degrees of freedom.

In Fig. 7 (c ÷ d), one can observe also the error convergence measured in the same norms with respect to the relative time, i.e. the ratio of the computational time necessary for the upscaled and reference solutions. It demonstrates the applicability of the MsFEM. The time used for the auxiliary problem solution is the price of the method. Thus, the observed speed-up is not as spectacular as the reduction of the NDOF compared to the reference fine mesh solution. The speed-up of about 40 times was obtained in this test for $p = 5$ with the corresponding error norms equal to about 6% (L_2 norm) or 20% (energy norm). For a more detailed discussion on the MsFEM computational efficiency, we refer the reader to [13,15]. Some crucial aspects of the discussion presented therein are:

- MsFEM is ready for the parallelization, the auxiliary problems can be solved independently
- formally cumbersome computations (associated with the auxiliary problems repeated for every standard shape function) reduce to the system of the algebraic equations with multiple right-hand sides
- locally assembled fine mesh stiffness matrix K_h used in Eq. 4$_I$ can be directly taken from the local auxiliary problem, no additional computational effort is necessary.

It should be remarked that the overall MsFEM computational efficiency depends mostly on the implementation, the problem being solved and the discretization at both scales. The obtained speed-up can be, thus, different at various applications.

The results of the linear elastic analysis are presented to illustrate subsequently the effect of the heat transfer within the AC sample on its overall response. The presented results are used for the forthcoming comparative study.

In the second step of the sequentially coupled thermal-stress analysis, i.e. the mechanical part, we use the temperature distribution obtained at the first step in a twofold manner. Firstly, we modify the material parameters according to the temperature distribution (in this study, we used the same parameters for the whole domain). Secondly, we compute the thermal strains according to the formula 2 $_V$ and use them to contribute to the right-hand side as presented in Eq. 3. The final results (displacements) are presented in Fig. 8. The error norms are approximately equal to those presented for the linear elastic case, which can be considered as the MsFEM advantage.

A short discussion on the accuracy of the results presented in this section is necessary. Firstly, the error convergence presented in Fig. 5 and 7 does not support strongly the idea of the higher order approximation. Particularly, in Fig. 5, the results obtained for the higher p are very similar to those for $p = 1$. It is due to a relatively smooth solution that can be efficiently captured using only the linear approximation at the macroscale. In our previous papers [1,13–15], we demonstrated a number of results in favor of the higher order approximation. Secondly, further improvement of the MsFEM accuracy can be obtained using e.g. error estimates based on the residual, which is the topic of our forthcoming paper. Using this approach, a significantly better p-convergence can be observed.

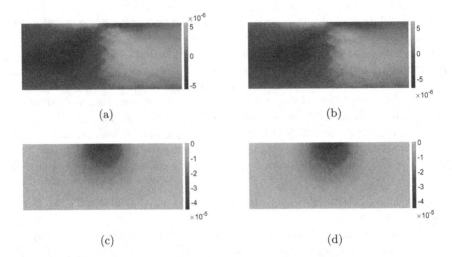

Fig. 6. Displacement maps: (a ÷ b) u_x component and (c ÷ d) u_y component - the left column shows the upscaled solution (for $p = 5$) and the right one shows the reference fine mesh solution

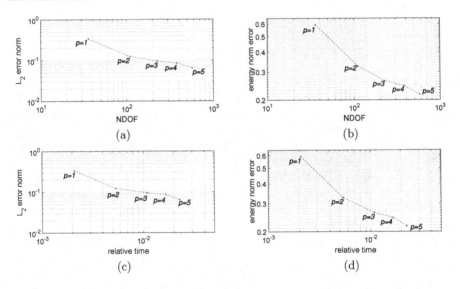

Fig. 7. Error convergence measured in (a) L_2 norm and (b) energy norm with respect to NDOF and the same convergence types (c ÷ d) with respect to the relative time (by the relative time, we mean the ratio of the computational time necessary for the upscaled and the reference fine mesh solutions)

Comparing the results presented in Figs. 6 and 8, one can observe a large modification in the AC sample response. Firstly, both of the displacement components increased substantially. Secondly, due to the applied heating, the horizontal displacements approximately equaled with the vertical ones up to the

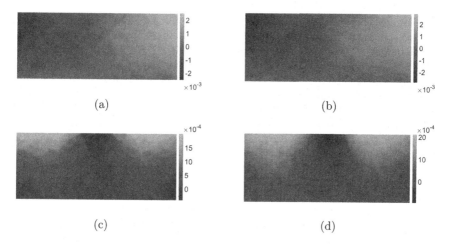

Fig. 8. Displacement maps: (a ÷ b) u_x component and (c ÷ d) u_y component - the left column shows the upscaled solution (for $p = 5$) and the right one shows the reference fine mesh solution

order of magnitude. Thirdly, the predominant vertical displacement of the upper edge reflects a very common failure of the asphalt pavement structure (or the wearing course solely). Namely, the rut is observed together with the bulged neighborhood. The upscaled solution captures this response almost as precisely as the reference one but with the large reduction of the degrees of freedom.

5   Concluding Remarks

In the paper, we presented a novel approach to upscaling for the coupled thermoelastic modeling of composites. The framework was tested on the example of asphalt concrete. However, it is general and not limited to the materials of a similar class. Our previous studies [1,13,16] refer also to other kinds of heterogeneous materials. Herein, we combined the well-known sequentially coupled thermal-stress analysis with the developed version of MsFEM. The convergence error confirmed the benefits of the higher order approximation. Reduction of about 220 00 dof to about 600 introduced modeling error of only about 6%. This error can be further reduced using the hp-adaptivity (c.f. [13]) or other approaches. To our best knowledge, MsFEM with the higher order shape functions has never been used previously for the thermoelastic analysis.

Our further research effort is to apply this framework to the unsteady heat transfer problem and thermoviscoelastic analysis of asphalt concrete. This can be very useful in the reliable modeling of flexible pavement structures.

References

1. Cecot, W., Oleksy, M.: High order FEM for multigrid homogenization. Comput. Math. Appl. **70**(7), 1391–1400 (2015)

2. Collop, A., Scarpas, A., Kasbergen, C., de Bondt, A.: Development and finite element implementation of a stress dependent elastoviscoplastic constitutive model with damage for asphalt. Transp. Res. Rec. **1832**, 96–104 (2003)

3. Eden, M., Muntean, A.: Homogenization of a fully coupled thermoelasticity problem for a highly heterogeneous medium with a priori known phase transformations. Math. Methods Appl. Sci. **40**, 3955–3972 (2017)

4. Efendiev, Y., Hou, T.: Multiscale finite element methods for porous media flows and their applications. Appl. Numer. Math. **57**, 577–596 (2007)

5. Efendiev, Y., Hou, T.: Multiscale Finite Element Methods. Springer, New York (2009). https://doi.org/10.1007/978-0-387-09496-0

6. Feyel, F., Chaboche, L.: FE^2 multiscale approach for modelling the elastoviscoplastic behaviour of long fibre SiC/Ti composite materials. Comput. Methods Appl. Mech. Eng. **183**, 309–330 (2000)

7. Hou, T., Wu, X.: A multiscale finite element method for elliptic problems in composite materials and porous media. J. Comput. Phys. **134**, 169–189 (1997)

8. Jänicke, R., Larsson, F., Runesson, K., Steeb, H.: Numerical identification of a viscoelastic substitute model for heterogeneous poroelastic media by a reduced order homogenization approach. Comput. Methods Appl. Mech. Eng. **298**, 108–120 (2016)

9. Kalluvila, J., Krishnan, A.: Structural and thermal analysis of asphalt solar collector using finite element method. J. Energy **2014**, 9 (2014). https://www.hindawi.com/journals/jen/2014/602087/. Article ID: 602087

10. Kehrer, L., Wood, J., Böhlke, T.: Mean-field homogenization of thermoelastic material properties of a long fiber-reinforced thermoset and experimental investigation. J. Compos. Mater. **54**(25), 3777–3799 (2020)

11. Kim, Y.: Modeling of Asphalt Concrete. ASCE Press, Reston (2009)

12. Kim, Y., Souza, F., Teixeira, J.: A two-way coupled multiscale model for predicting damage-associated performance of asphaltic roadways. Comput. Mech. **51**(2), 187–201 (2013). https://doi.org/10.1007/s00466-012-0716-8

13. Klimczak, M., Cecot, W.: An adaptive MsFEM for non periodic viscoelastic composites. Int. J. Numer. Meth. Eng. **114**(8), 861–881 (2018)

14. Klimczak, M., Cecot, W.: Synthetic microstructure generation and multiscale analysis of asphalt concrete. Appl. Sci. **10**(3), 765 (2020)

15. Klimczak, M., Cecot, W.: Towards asphalt concrete modeling by the multiscale finite element method. Finite Elem. Anal. Des. **171**, 103367 (2020)

16. Krówczyński, M., Cecot, W.: A fast three-level upscaling for short fiber reinforced composites. Int. J. Multiscale Comput. Eng. **15**(1), 19–34 (2017)

17. Liu, P., et al.: Modelling and evaluation of aggregate morphology on asphalt compression behavior. Constr. Build. Mater. **133**, 196–208 (2017)

18. Mitra, K., Das, A., Basu, S.: Mechanical behavior of asphalt mix: an experimental and numerical study. Constr. Build. Mater. **27**(1), 545–552 (2012)

19. Mo, L., Huurman, M., Wu, S., Molenaar, A.: 2D and 3D meso-scale finite element models for ravelling analysis of porous asphalt concrete. Finite Elem. Anal. Des. **44**(4), 186–196 (2008)

20. Neumann, J., Simon, J.W., Mollenhauer, K., Reese, S.: A framework for 3D synthetic mesoscale models of hot mix asphalt for the finite element method. Constr. Build. Mater. **148**, 857–873 (2017)

21. Sawda, C.E., Fakhari-Tehrani, F., Absi, J., Allou, F., Petit, C.: Multiscale heterogeneous numerical simulation of asphalt mixture. Mater. Des. Process. Commun. **1**(3), e42 (2019)

22. Schüller, T., Jänicke, R., Steeb, H.: Nonlinear modeling and computational homogenization of asphalt concrete on the basis of XRCT scans. Constr. Build. Mater. **109**, 96–108 (2016)
23. Schüller, T., Manke, R., Jänicke, R., Radenberg, M., Steeb, H.: Multi-scale modelling of elastic/viscoelastic compounds. ZAMM J. Appl. Math. Mech. **93**, 126–137 (2013)
24. Tehrani, F.F., Absi, J., Allou, F., Petit, C.: Heterogeneous numerical modeling of asphalt concrete through use of a biphasic approach: porous matrix/inclusions. Comput. Mater. Sci. **69**, 186–196 (2013)
25. Temizer, I., Wriggers, P.: Homogenization in finite thermoelasticity. J. Mech. Phys. Solids **59**(2), 344–372 (2011)
26. Tu, Y., Li, J., Guan, C.: Heat transfer analysis of asphalt concrete pavement based on snow melting. In: 2010 International Conference on Electrical and Control Engineering, pp. 3795–3798 (2010)
27. Wang, L.: Mechanics of Asphalt: Microstructure and Micromechanics. The McGraw-Hill Companies, Inc. (2011)
28. Wimmer, J., Stier, B., Simon, J.W., Reese, S.: Computational homogenisation from a 3D finite element model of asphalt concrete - linear elastic computations. Finite Elem. Anal. Des. **110**, 43–57 (2016)
29. Woldekidan, M., Huurman, M., Pronk, A.: Linear and nonlinear viscoelastic analysis of bituminous mortar. Transp. Res. Rec. J. Transp. Res. Board **2370**(1), 53–62 (2013)
30. Xue-liang, Y., Bo-ying, L.: Coupled-field finite element analysis of thermal stress in asphalt pavement. J. Highw. Transp. Res. Dev. (Engl. Ed.) **2**(1), 1–6 (2007)
31. Yavuzturk, C., Ksaibati, K., Chiasson, A.: Assessment of temperature fluctuations in asphalt pavements due to thermal environmental conditions using a two-dimensional, transient finite-difference approach. J. Mater. Civ. Eng. **17**(4), 465 (2005)
32. You, T., Al-Rub, R., Darabi, M., Masad, E., Little, D.: Three-dimensional microstructural modeling of asphalt concrete using a unified viscoelastic-viscoplastic-viscodamage model. Constr. Build. Mater. **28**(1), 531–548 (2012)
33. Zbiciak, A.: Constitutive modelling and numerical simulation of dynamic behaviour of asphalt-concrete pavement. Eng. Trans. **56**(4), 311–324 (2008)
34. Ziaei-Rad, V., Nouri, N., Ziaei-Rad, S., Abtahi, M.: A numerical study on mechanical performance of asphalt mixture using a meso-scale finite element model. Finite Elem. Anal. Des. **57**, 81–91 (2012)

MaMiCo: Non-Local Means Filtering with Flexible Data-Flow for Coupling MD and CFD

Piet Jarmatz(✉)📖, Felix Maurer📖, and Philipp Neumann📖

Chair for High Performance Computing, Helmut Schmidt University,
Hamburg, Germany
jarmatz@hsu-hh.de

Abstract. When a molecular dynamics (MD) simulation and a computational fluid dynamics (CFD) solver are coupled together to create a multiscale, molecular-continuum flow simulation, thermal noise fluctuations from the particle system can be a critical issue, so that noise filtering methods are required. Noise filters are one option to significantly reduce these fluctuations.

We present a modified variant of the Non-Local Means (NLM) algorithm for MD data. Originally developed for image processing, we extend NLM to a space-time formulation and discuss its implementation details.

The space-time NLM algorithm is incorporated into the Macro-Micro-Coupling tool (MaMiCo), a C++ molecular-continuum coupling framework, together with a novel flexible filtering subsystem. The latter can be used to configure and efficiently execute arbitrary data-flow chains of simulation data analytics modules or noise filters at runtime on an HPC system, even including python functions. We employ a coupling to a GPU-based Lattice Boltzmann solver running a vortex street scenario to show the benefits of our approach. Our results demonstrate that NLM has an excellent signal-to-noise ratio gain and is a superior method for extraction of macroscopic flow information from noisy fluctuating particle ensemble data.

Keywords: Flow simulation · Non-Local means · LBM · GPU · Software design · Denoising · Data analytics · Transient · Two-way coupling · Molecular dynamics · HPC · Molecular-continuum

1 Introduction

Molecular dynamics (MD) simulations [16,17] are used in many applications within computational science, engineering or biochemistry. They can be coupled to a computational fluid dynamics (CFD) solver, creating a multi-scale molecular-continuum flow simulation [2,15]. On MD side, Brownian motion results in thermal fluctuations. Despite their importance and impact on molecular flow behavior, a direct incorporation of them into continuum-based simulations is not always desirable. It is rather the extraction of smooth flow data

© Springer Nature Switzerland AG 2021
M. Paszynski et al. (Eds.): ICCS 2021, LNCS 12746, pp. 576–589, 2021.
https://doi.org/10.1007/978-3-030-77977-1_46

(including smoothed, i.e. averaged, Brownian effects) that might be sufficient here, i.e. to improve stability and performance of the coupled simulation [8]. One way to obtain this smooth data is ensemble averaging using many MD instances [13]. However, since MD is the computationally most challenging part in molecular-continuum setups, running many MD simulations is expensive. Various filtering methods including anisotropic median diffusion [12] and proper orthogonal decomposition (POD) [8,9] have been proposed to handle molecular data. We have presented noise reduction using POD in 2019 [11] and introduced a noise reduction interface for the open source C++ coupling framework MaMiCo[1].

However, our latest research shows that POD filtering is rather sub-optimal in a molecular-continuum context: While POD is excellent for detecting temporal correlation in the input data, it does not fully exploit natural redundancy that exists in every single time step, because it views the MD data values as separate signal sources in the simulation space, without any information about their neighborhood relationships. Thus, in this paper we will propose a more advanced way to filter information from a particle system. Although the noise reduction interface we presented in our last paper [11] was already quite flexible, as it can be used to execute any noise filter or MD data analysis module, it could only run one filter at a time. However, it has been shown that in many cases a combination of more than one filter (e.g. POD+ wavelet transform) results in improved results, see e.g. [21]. If many filters are to be combined, they may be interdependent and a free parametrization and reordering even at runtime is desirable. From an HPC perspective, this can be very problematic, as many challenges arise in terms of parallelization and communication or memory access (e.g. avoiding unnecessary copy operations). While any such filter combination can easily be hard-coded, it is desirable to strive for a more generic and flexible solution, with the ability to execute many noise filters and data analysis modules arranged in any data-flow diagram including multiple data paths, configurable at runtime, but also with a strong focus on performance and scalability on HPC clusters.

In this paper, we present an approach in which we view CFD cells and cell-wise MD data as 'pixels', execute various operations or chains of operations on them, including the application of an image processing denoising method: Non-Local Means (NLM) [4] first presented by Buades et al. [3] in 2004 is one of the best algorithms in this area. The ability to detect fine structures and patterns in images, sustaining them instead of averaging them away like many other filters would, makes NLM perfect for MD data, where one wants to separate nano-scale flow features from thermal noise. NLM has been successfully applied for medical imaging [5], but to the authors' knowledge it was never used in CFD/MD so far. In this paper, we present a modified version of the NLM algorithm, optimized for filtering particle data. Due to the non-local nature of the NLM method, performance can be a critical issue. Coupé et al. [5] have presented approaches for fast implementation of NLM, we optimize and extend them

---

[1] https://github.com/HSU-HPC/MaMiCo.

for higher-dimensional data and execution on HPC clusters. We also extend the
C++ framework MaMiCo by new python bindings, our NLM implementation
and we introduce a new filtering system for configurable filter combinations, creating flexible data-flow filter chains, based on cell-wise operations. This system
also enables the reuse of existing filters in MaMiCo, for instance from the python
packages numpy and scipy, in an efficient and parallel way.

In Sect. 2 we explain our methodology for molecular-continuum coupling.
Section 3 introduces the original Non-Local Means idea, some optimizations of
it, and presents our new variant of the NLM algorithm. We also present a more
flexible novel data-flow filtering subsystem for MaMiCo in Sect. 4, specifying
details on software design and supported types of filters (Sect. 4.1). Section 4.2
presents a fast implementation approach for NLM. We analyze NLM and the
filtering subsystem more thoroughly by running a coupled molecular dynamics-
Lattice Boltzmann vortex street flow test scenario on a GPU that is described
in Sect. 5, giving simulation results, performance and signal-to-noise ratio measurements in Sect. 5.1. Section 6 summarizes our work, leading to perspectives
for potential future research topics.

2 Coupled Simulation

We consider transient, concurrent
coupling of MD and CFD, with nested
time stepping on MD side (ca. 50 MD
steps correspond to one CFD step).

On the molecular scale, our coupled flow simulation employs a simple Lennard-Jones particle system,
integrating the equations of motion
directly via explicit time stepping
in the NVT ensemble. It is implemented using the linked cells method
[7] and MaMiCo's built-in MD solver
SimpleMD, which is used only to
test the coupling algorithms and is
representative for more sophisticated
MD packages, such as LAMMPS [18]
or ls1 mardyn [16]; see [14] for an
overview of MD packages supported
in MaMiCo-based coupling.

The MD domain is covered by a

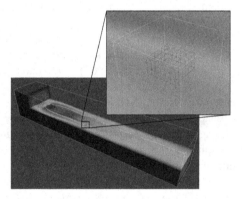

Fig. 1. Domain decomposition of coupled
simulation. Zoom-out: 3D vortex street
using 12.8 million Lattice Boltzmann cells.
Zoom-in: Yellow outline: Total coupling
region with ghost layer of macroscopic cells
and outer MD cells. Green wireframe: Inner
MD cells, where filtering is applied. (Color
figure online)

regular grid of cells which are used by MaMiCo for the coupling, called 'macroscopic cells'. Note that macroscopic cells neither have to match linked cells (they
could even be used with a non-cell-based MD code) nor do they have to correspond to the cells used by the CFD solver internally. These cells are rather the

main buffer data structure used for coupling and build the basis for all operations described in Sect. 4, where filters are always applied to multi-dimensional arrays of macroscopic cells.

On the continuum scale, similarly to the MD side MaMiCo can be coupled to many different solvers, but in the scope of this paper we use the lbmpy [1] package, which can automatically generate highly efficient GPU code for Lattice Boltzmann (LB) schemes. We employ a simple single-relaxation-time collision operator and a D3Q19 stencil.

Figure 1 shows an exemplary domain decomposition, where a small MD domain is placed (overlapping) inside a much larger CFD domain to establish molecular-continuum coupling [13,15]. The outer MD cell layers serve as a boundary condition to receive data from the continuum solver. We employ an advanced boundary forcing term [20] to model the thermodynamic pressure at the molecular scale at the continuum boundary and the USHER scheme [6] for particle insertion. In the inner MD cells, we sample cell-wise quantities such as density or average velocity out of the particle simulation instances, so that this data can be sent to the continuum solver. In the following, we will restrict considerations to these cells with regard to noise filtering.

3   Non-Local Means Algorithm

Buades et al. [3] proposed the Non-Local Means (NLM) algorithm after an extensive series of denoising experiments on natural images. It is inspired by various local smoothing methods and frequency domain filters, such as Gaussian convolution, anisotropic filtering, total variation minimization, Wiener filtering and wavelet thresholding. Already a very simple Gaussian filtering yields a powerful noise reduction by averaging away high-frequency noise, but it also removes high-frequency components of the original data; for instance sharp edges in an image are blurred. More advanced methods like an anisotropic filter can preserve large-scale geometrical structures like sharp edges, but all of these previously known methods still smooth out fine-scale textures. NLM on the contrary is constructed with the aim to preserve details and fine structures which behave like high-frequent noise in an image. This means that systematic patterns in particle data, for instance recurrent flow vortices or nano membrane diffusion structures will be retained by the filter instead of being averaged away.

For a noisy image $v = \{v(i) \mid i \in I\}$, NLM can be written as:

$$NLM[v](i) \; = \; \frac{1}{C_i} \sum_{j \in I} w(i,j) v(j) \tag{1}$$

We can see in Eq. (1) that the filtered result for the pixel at position i is always a weighted sum of all other pixel values $v(j)$ in the image. The weights $w(i,j)$ are defined for all pairs of pixel positions i and j; however, with the optimizations introduced in Sect. 4.2, $w(i,j)$ is sparse. The normalizing factor is chosen as

$$C_i \; = \; \sum_{j \in I} w(i,j) \tag{2}$$

so that the sum of weights is 1. Let N_i denote a subset of I that is a local neighborhood centered at position i. In [3] these neighborhoods in 2D images are often called *similarity windows* and have a size of, for instance, 7×7 pixels. On the other hand, [5] uses neighborhood windows in 3D images of size $3 \times 3 \times 3$ voxels. Here, we use the term *patch* for the small data array $v(N_i)$ in such a local neighborhood window N_i around i. Our patches of MD data will normally have a size of $3 \times 3 \times 3 \times 3$ cells, corresponding to space-time windows, simply because a size of 3 is the smallest reasonable choice for both, space and time.

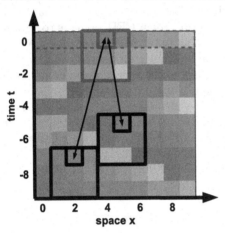

In the temporal dimension, we restrict the filtering to a time window size of T coupling cycles (for example $T = 10$ in Fig. 2). We typically require filter results for the present cycle at $t = 0$, with data from past cycles $t < 0$ available, but before MD data for future time steps at $t > 0$ can be computed. Thus, the patches which are centered on their cell in space have to be shifted in time, as illustrated in Fig. 2. Additionally, we introduce a temporal fade-out of weights (see Eq. (5)), so that recent data is more significant than older information.

Fig. 2. NLM principle, illustrated for noisy 1D wave propagation: To compute the filtered value of a cell i at $(x = 4, t = 0)$, its patch is compared with patches around every other cell j. Here, $(2, -7)$ is more similar and thus gets a higher weight, while $(5, -5)$ is less similar and has less influence on i.

To compare patches, we use a squared Euclidean distance (L2 norm), as recommended by [4]. The norm can be understood as a similarity metric used to measure the similarity of i and j by comparing their local neighborhood patches:

$$(i, j) = ((\boldsymbol{x}_i, 0), (\boldsymbol{x}_j, t)), t \in \{0, -1, -2, ..., -T + 1\} \qquad (3)$$

$$dist(i, j) = |\hat{v}(N_i) - \hat{v}(N_j)|^2 \qquad (4)$$

This distance can be used to define the weights as:

$$w(i, j) = \exp\left(-\frac{max(dist(i, j) - 2\sigma_N^2, 0)}{h^2}\right) \cdot \frac{1}{1 - t} \qquad (5)$$

Note that the expression $\exp(-\frac{d}{h^2})$ with $d \geq 0$ in Eq. (5) guarantees that all weights are in $[0, 1]$ and that h is a filtering parameter that controls the decay of the exponential function. The second filtering parameter σ_N that corresponds to the expected standard deviation of the noise, so that the weight of a patch with a distance smaller than $2\sigma_N^2$ is always 1, as described in [4].

Unlike previously described NLM versions, in our variant we introduce an additional pre-filtering function F (Eq. 6) into the similarity computation, inspired by regularization that is commonly used in a Perona-Malik filter. F can be any existing simple noise filter with a low gain of signal-to-noise (SNR) ratio, such as a Gaussian kernel. It is used to improve the quality of the similarity metric in high noise/low SNR scenarios.

$$\hat{v} = F[v] \tag{6}$$

Note that F is not restricted to a Gaussian kernel, but we allow any reasonable pre-filter. This has the advantage that either a computationally cheap F may be used or a filter with a higher SNR gain can be used, which improves the significance of the similarity metric. Especially, POD [8] can be applied here to exploit temporal correlation in v, and we show in Sect. 5.1 that this yields good results.

4   Implementation and Software Design

One of the most central concepts we developed with regard to creating a modular thus flexible filtering environment for MaMiCo is the notion of *filter sequences*. Let $\mathbb{F}$ be a set of filters operating on the same spatial domain. Then one can view a filter sequence S as a nonempty tuple or list $(f_0, .., f_n : f_i \in \mathbb{F})$ in which for all $i \geq 1$ the input of f_i shall be the output of f_{i-1}. The output of f_n is the sequence output of S. Similarly, the input of f_0 is the sequence input of S. The sequence input of all sequences must be defined before execution, it can be either MD data or another sequence.

In some cases, being restricted to unary functions as filters poses a problem. For example, our NLM algorithm asks for two sets of input (unfiltered/prefiltered). We thus implemented a generalization of the sequence concept, allowing for multiple in- and outputs: *filter junctions*. Filters that operate not in sequences but in junctions are called *junctors*. An exemplary filtering configuration using both junctors and filters is depicted in Fig. 4.

Sequences are managed by `FilteringService`, which also acts as an interface to the `MacroscopicCellService` (MCS). During MaMiCo startup, one MCS per MD instance gets initialized. It immediately constructs an instance of `FilteringService`. This service now interprets an XML configuration file, creating all `FilterSequences` specified. Every sequence is now instructed which cells will be the input for f_0 and where to write the output of f_n. Then, filter instantiation begins. Internally, each sequence stores two cell vectors V_1 and V_2. Let $I(f_i)$ denote the input and $O(f_i)$ the output of a filter f_i respectively. Then,

$$I(f_i) = \begin{cases} V_1 & \text{if } i \text{ is even} \\ V_2 & \text{if } i \text{ is odd} \end{cases} ; \quad O(f_i) = \begin{cases} V_2 & \text{if } i \text{ is even} \\ V_1 & \text{if } i \text{ is odd} \end{cases} .$$

This way, space required for cell vectors is constant in n.

4.1   Supported Types of Filters

For a D-dimensional scenario, any function applicable to both scalar and D-dimensional floating point in- and outputs can be utilized as a filter or junctor. This also means that read-only utilities can easily be wrapped in filters. For example, saving cell data to files is implemented as a `WriteToFile` filter, cf. Fig. 3. Such filters simply copy input cell data to corresponding output cells without prior modification.

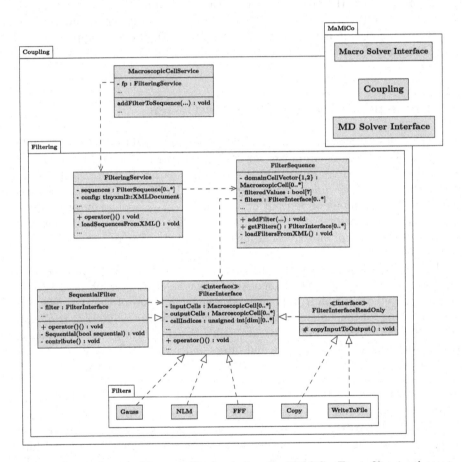

Fig. 3. Software design of the new filtering system for MaMiCo: Every filter implements the FilterInterface, they are combined together in FilterSequences and managed by the FilteringService.

Statically and Dynamically Linked Filters. As mentioned before, the main way of defining filters in a `FilterSequence` is via a configuration file. Naturally, the tree-like node structure of XML is very close to the structure of interlinked `FilterSequences`, cf. Fig. 4. Filters specified before runtime using this configuration file are called *statically linked*. These filters are written in C++, implementing the `FilterInterface` or expanded interfaces thereof. *Dynamically linked* filters on the other hand are linked at runtime by calling the method `addFilter()` of a `FilterSequence`. This method takes arbitrary `std::function` objects (of signatures fitting our definition of filters above) and constructs a `FilterFromFunction` (FFF) C++ object using that information. Since there are interfaces between `std::function` and other programming languages, this design omits the limitation of filters to be written in C++. For example, the pybind11 library [10] allows for python functions to be passed as `std::function` at runtime. Listing 1 gives an example as to how one could make use of SciPy[2] filters in MaMiCo.

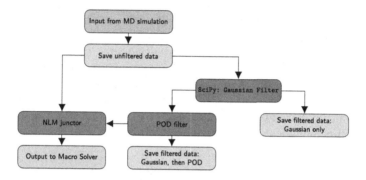

Fig. 4. Visualization of an exemplary *FilteringService* configuration diagram. Setups like this enable comparison between any number of filtering strategies with ease, e.g. between Non-Local Means, Gaussian and Gaussian followed by POD. Note that NLM is implemented as a junctor: it uses both, unfiltered data and POD output.

```
1   from scipy.ndimage import gaussian_filter
2
3   def gaussSigmaOne(data):
4       print("Applying gaussian filter. sigma_G = 1.")
5       return gaussian_filter(data, sigma = (1,1,1))
6
7   mcs = self.multiMDCellService.getMacroscopicCellService(0)
8
9   #Add scipy's gaussian filter in FilterSequence 'mySequence'
10  #at filter index 0, filtering scalar properties.
11  mcs.addFilterToSequence(filter_sequence = "mySequence", filter_index =
    ↪   0, scalar_filter_func = gaussSigmaOne, vector_filter_func = None)
```

Listing 1: Example showing how to use the MaMiCo python interface in order to add an arbitrary function as a filter at runtime. In this case, the `SciPy` package is utilized to apply a Gaussian filter to all scalar properties "mySequence" (defined via XML) operates on. With "`filter_index=0`" the filter is placed at f_0. `mcs` is the overarching `MacroscopicCellService`, cf. Fig. 3

---

[2] https://www.scipy.org/.

Sequentialized Filters in a Parallel Environment. One of MaMiCo's primary design goals is to support massively parallel solvers and thus be highly scalable. In particular, the MD simulation will usually be MPI-parallel. In that case, each MPI rank will have a unique instance of MCS and thus of FilteringService, managing filters entirely independent of other ranks. These filters will then only have a fraction of all cells in the global filtering domain, which can, depending on individual use cases, have either a negative or positive effect:

Some filter algorithms treat cells independently, in which case parallel filtering can imply performance gain. In other cases this separation is either suboptimal or entirely incompatible with the filter algorithm in use.

For the latter case we introduced the concept of *sequentialized filters*. If a filter is marked as *sequential*, that filter will not be applied in rank-wise independence. Instead, it has only one dedicated *processing rank*, while all other ranks are *contributors*. For application of a sequentialized filter, the FilteringService uses MPI_Gather and MPI_Bcast to manage the necessary communication steps. Although sequentialized filters represent only a special case within our methodology, they allow for simple incorporation of off-the-shelf filter implementations. e.g. a dynamically linked Gaussian filter can be used without having to manually manage MPI communication for neighbor information.

4.2 Non-Local Means: Optimized Implementation

Despite the name 'non-local', NLM is in practice not executed on the global domain, but the similarity comparison of patches is restricted to moving search windows. This means that j in Eq. (1) is not actually in I, but $\boldsymbol{x}_j$ is in a local neighborhood of constant spatial size $(2M+1)^3$, centered at i. Thus, since only a constant number of operations has to be performed in this local research zone, the computational complexity of the NLM implementation is asymptotically linear in the global number of domain cells.

To speed up the implementation, as described in [5], the computation of $w(i,j)$ can be skipped if $dist(i,j)$ is expected to be large, without the expensive evaluation of $|\hat{v}(N_i) - \hat{v}(N_j)|^2$. This is achieved by pre-computing and storing characteristic values of $\hat{v}(N_i)$ for every cell i, such as mean and standard deviation of this patch. Only if the relative differences of these values are small enough, i.e. inside configured bounds, then $dist(i,j)$ is computed, otherwise $w(i,j)$ is set to zero.

At the expense of extra memory consumption, the NLM execution can be further accelerated if patches are stored in a redundant, linearized way: This allows for a cache-efficient contiguous memory access and vectorization of the patch distance determination. Our implementation exploits this in a natural way – using the concept of *flowfield* and *patchfield* data structures. We define a *field* as a four-dimensional space-time array. Then, a flowfield is a field of quantities sampled from the particle system. A *patch* is a small flowfield together with characteristic values; and a patchfield is a field of patches. Our NLM implementation only constructs and accesses a patchfield, so that for the similarity computation,

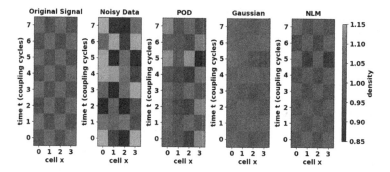

Fig. 5. Sound wave extraction test case: even with a very high level of noise in the input, our NLM implementation (using POD as pre-filter) can reconstruct the original pattern in the density data. NLM params used: $\sigma_N = 0$, $h^2 = 0.45$

it does not have to use a cell stencil or compute any neighbor indices. In our experiments we measured a speed-up of 20 % compared to direct neighbor cell accesses without linearized patch storage.

Since the filtering system of MaMiCo executes the filter only on the subdomain of inner macroscopic cells that belongs to this rank, the NLM implementation is MPI-parallel and can easily be scaled up to runs on large HPC-clusters. However, since we did not incorporate ghost layers for the patchfield initialization so far, the filtering quality at inter-rank boundaries can be slightly reduced due to missing neighbor values. For simplicity, we test NLM sequentially here.

4.3 NLM Test: Nano-Pattern (Sound Wave) Extraction

To test the filters under challenging conditions and demonstrate the ability of NLM to detect fine structures in MD data, we created a 3D flow scenario where ultra-high-frequent pressure wavefronts propagate at speed of sound (one cell per coupling cycle) in X direction through the domain. For simplicity, and to have an original true signal for comparison, we did not execute a real particle simulation here, but

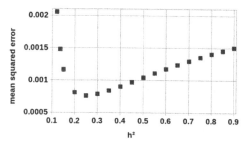

Fig. 6. NLM parametrization study: MSE against original signal (Fig. 5) over h^2

constructed synthetic pseudo-MD data by adding normally distributed artificial noise ($\sigma = 0.2$) to analytical oscillations (amplitude 0.05). Figure 5 shows the density values in four inner cells of the synthetic MD domain over time, and compares the noisy input with the respective filter outputs. Since we normalized the wavelength to two cell sizes, the original signal shows a checkerboard pattern in the $x - t$ plot. The filtering parameters N,k_{max},σ_G,T and F are exactly the same as in Sect. 5.1. While it can be observed that the Gaussian filter and POD [11] deliver blurry results, the new NLM approach captures the spatially varying pattern very well. Figure 6 shows the same test setup again, but evaluates the mean squared error (MSE) of the NLM output for various choices for the filtering parameter h^2. Each MSE value is computed over 64 cells and 50 coupling cycles here. We can identify an optimal value of $h^2 \approx 0.25$ for this scenario, and discover that too large values of h are less critical, while with too small values the filtering results fall off in quality much faster.

5   Simulation Results

Vortex Street Test Scenario. We chose a well-known vortex street setup for testing the filtering system and the NLM implementation in a transient one-way LB $\rightarrow$ MD coupled simulation, i.e. without sending data from MD to the macroscopic solver, so that we obtain reproducible results for validation and quantification of the denoising.

We use lbmpy [1] as macroscopic solver to set up the test case *3D-2Q* from [19] that models a flow around a cylinder with square cross-section, but we scale it down to a channel of size 663 nm × 108 nm × 108 nm. We keep the Reynolds number of $Re = 100$ and thus the flow properties constant, so that the flow is laminar and unsteady. More details defining the scenario precisely can be found in [19]. To validate the flow scenario, we measured the forces acting on the obstacle to compute drag and lift coefficients. Our experimentally determined values of $c_{D_{max}} = 4.39; c_{L_{max}} = 0.066$, as well as the Strouhal number we measured of $St = 0.37 \pm 0.03$ are in very good compliance with the results given in [19], which were obtained by multiple different research groups and using other numerical methods, such as finite element solvers. The LB simulation uses 780 × 127 × 127 cells and runs with a performance of 1937 MLUPS on a NVIDIA Tesla V100.

Using the novel MaMiCo python bindings, we couple lbmpy to a single MD instance placed at the position 332 × 80 × 54 nm in the vortex street, as shown in Fig. 1. We ensure that the dynamic viscosity is configured consistently between both solvers and for simplicity, we choose the same cell size for both MD and LB. Note that while the vortex street is not a common scenario in nanofluidics, it yields a multidimensional transient flow in the coupled simulation, which is excellent for studying the filtering methods on sufficiently complex particle data.

5.1   Vortex Street Filtering Results

We initialize the scenario described in Sect. 5 for only 2×10^4 LB time steps, so that there is no fully developed flow, but a start-up phase can be observed. At that point we enable coupling, sample and filter particle data for 1000 coupling cycles (with 100 MD time steps per cycle). The resulting values for the velocity in Y direction in one of the cells in the center of the inner MD domain are shown in Fig. 7. The main flow in X direction is not shown here, but instead only unsteady oscillating transversal fluid movements behind the obstacle (cf. Fig. 1).

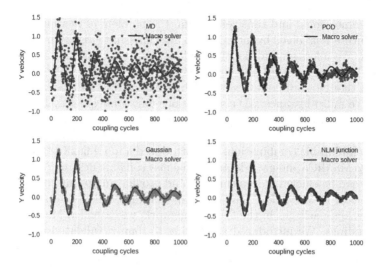

Fig. 7. Velocity data from particle system, compared to macroscopic flow, in only one single cell (center of MD domain). Red: raw MD (1 instance). Green: POD filtered, $N = 40$, $k_{max} = 2$. Yellow: Gaussian, $\sigma_G = (1, 1, 1)$. Purple: NLM, $T = 5$, F=POD. (Color figure online)

It can be seen in Fig. 7 that without filtering, there is a high level of thermal noise in the values sampled from MD. With all filtering methods, the results are much closer to the CFD values. We can quantify this precisely with the signal-to-noise ratio (as defined in [11]) in this cell: While the raw MD data has an SNR of −3.16 dB, this is improved to 6.06 dB with POD, 9.62 dB with the Gaussian filter and 10.57 dB with NLM. NLM is configured to use POD as pre-filter F. Compared to Sect. 4.3, the Gaussian performs better here, because in this example the flow has rather a large-scale structure instead of fine patterns, but NLM is still superior. Note that the time window size used is 40 coupling cycles for POD but only 5 for NLM, so that POD has 8 times more data available to use for filtering. Nevertheless, NLM achieves an increased SNR gain of 4.51 dB, corresponding to 2.82 times stronger noise reduction compared to POD.

The performance impact of the filtering system on the overall coupled simulation is minimal. For the configuration in Fig. 7 we measured the following average filter execution times per cycle, for all cells: POD 1.03 ms, Gaussian 0.33 ms, NLM 6.98 ms. This corresponds to less than 0.2 % of total simulation runtime spent in the filtering system in the worst case. For more details on performance and scaling tests on up to 512 cores of the coupled simulation including POD the reader is referred to [11].

6   Conclusions

We have discovered that the NLM algorithm from image processing is by design well suited to filter flow data from a MD system. When we generalized NLM to higher-dimensional and time series data, the problem of missing information from the future was solved by introduction of a time-shift (Fig. 2) in the definition of patches. We allow an additional pre-filtering step for an improved similarity metric in high noise setups and provided a formal definition of our new NLM variant. Although it operates on 4D data structures, it is efficient, especially compared to multi-instance MD computations, but some questions such as more elegant boundary handling and good parallelization strategies still have to be investigated. To be able to execute many data analysis or noise filtering modules in freely configurable combinations, we introduced a more flexible novel filtering system into MaMiCo and gave details on the new software design. We have shown how NLM can be efficiently implemented and provide an optimized implementation as junctor in MaMiCo. To allow for dynamically linked filters and reuse of existing python packages in the C++ framework, we extended MaMiCo by new python bindings. We introduced a coupling to lbmpy, validated a GPU-based vortex street test scenario, and used the vortex street flow to investigate NLM results and quantify the noise reduction, showing that NLM is superior to POD in a molecular-continuum context, at minimal performance expenses.

A prediction of the deviations between microscopic and continuum quantities is important for consistency in the coupling, i.e. to guarantee conservation of energy. This can easily be derived from statistical mechanics for raw MD data and ensemble averaging, but for filtering results such a generic error estimation tool constitutes a major challenge. Machine learning methods may be an approach to solve this. Other open questions for future work would be further investigation of possible limitations and drawbacks of the pattern extraction by NLM, as well as automatic dynamical selection of optimal filtering parameters to obtain a balanced ratio between and performance and accuracy.

Acknowledgments. Parts of this work have been supported through the HSU-IFF project "Resilience and Dynamic Noise Reduction at Exascale for Multiscale Simulation Coupling" and the dtec.bw project "MaST". We also want to thank the lbmpy developer Martin Bauer for fruitful discussions and support at using the software.

References

1. Bauer, M., Köstler, H., Rüde, U.: lbmpy: automatic code generation for efficient parallel lattice boltzmann methods. J. Comput. Sci. **49**, 101269 (2021)
2. Borg, M.K., Lockerby, D.A., Reese, J.M.: A hybrid molecular-continuum method for unsteady compressible multiscale flows. J. Fluid Mech. **768**, 388–414 (2015)
3. Buades, A., Coll, B., Morel, J.M.: On image denoising methods. CMLA. Preprint, vol. 5 (2004)
4. Buades, A., Coll, B., Morel, J.M.: Non-local means denoising. Image Process. OnLine **1**, 208–212 (2011)
5. Coupé, P., Yger, P., Barillot, C.: Fast non local means denoising for 3D MR images. In: Larsen, R., Nielsen, M., Sporring, J. (eds.) MICCAI 2006. LNCS, vol. 4191, pp. 33–40. Springer, Heidelberg (2006). https://doi.org/10.1007/11866763_5
6. Delgado-Buscalioni, R., Coveney, P.: USHER an algorithm for particle insertion in dense fluids. J. Chem. Phys. **119**(2), 978 (2003)

7. Gonnet, P.: A simple algorithm to accelerate the computation of non-bonded interactions in cell-based molecular dynamics simulations. J. Comput. Chem. **28**(2), 570–573 (2007)
8. Grinberg, L.: Proper orthogonal decomposition of atomistic flow simulations. J. Comput. Phys. **231**(16), 5542–5556 (2012)
9. Grinberg, L., Yakhot, A., Karniadakis, G.E.: Analyzing transient turbulence in a stenosed carotid artery by proper orthogonal decomposition. Ann. Biomed. Eng. **37**(11), 2200–2217 (2009)
10. Jakob, W., Rhinelander, J., Moldovan, D.: pybind11-seamless operability between C++ 11 and Python. https://github.com/pybind/pybind11 (2017)
11. Jarmatz, P., Neumann, P.: MaMiCo: parallel noise reduction for multi-instance molecular-continuum flow simulation. In: Rodrigues, J.M.F., et al. (eds.) ICCS 2019. LNCS, vol. 11539, pp. 451–464. Springer, Cham (2019). https://doi.org/10.1007/978-3-030-22747-0_34
12. Ling, H., Bovik, A.C.: Smoothing low-SNR molecular images via anisotropic median-diffusion. IEEE Trans. Med. Imaging **21**(4), 377–384 (2002)
13. Neumann, P., Bian, X.: Mamico: transient multi-instance molecular-continuum flow simulation on supercomputers. Comput. Phys. Commun. **220**, 390–402 (2017)
14. Neumann, P., Flohr, H., Arora, R., Jarmatz, P., Tchipev, N., Bungartz, H.J.: Mamico: software design for parallel molecular-continuum flow simulations. Comput. Phys. Commun. **200**, 324–335 (2016), ISSN 0010–4655
15. Nie, X., Chen, S., Robbins, M.O., et al.: A continuum and molecular dynamics hybrid method for micro-and nano-fluid flow. J. Fluid Mech. **500**, 55–64 (2004)
16. Niethammer, C., et al.: ls1 mardyn: the massively parallel molecular dynamics code for large systems. J. Chem. Theory Comput. **10**(10), 4455–4464 (2014)
17. Páll, S., Abraham, M.J., Kutzner, C., Hess, B., Lindahl, E.: Tackling exascale software challenges in molecular dynamics simulations with GROMACS. In: Markidis, S., Laure, E. (eds.) EASC 2014. LNCS, vol. 8759, pp. 3–27. Springer, Cham (2015). https://doi.org/10.1007/978-3-319-15976-8_1
18. Plimpton, S.: Fast parallel algorithms for short-range molecular dynamics. J. Comput. Phys. **117**(1), 1–19 (1995). ISSN 0021–9991
19. Schäfer, M., Turek, S., Durst, F., Krause, E., Rannacher, R.: Benchmark computations of laminar flow around a cylinder. In: Flow Simulation with High-Performance Computers II, pp. 547–566. Springer, Cham (1996). https://doi.org/10.1007/978-3-322-89849-4_39
20. Zhou, W., Luan, H., He, Y., Sun, J., Tao, W.: A study on boundary force model used in multiscale simulations with non-periodic boundary condition. Microfluidics Nanofluidics **16**(3), 1 (2014)
21. Zimoń, M., et al.: An evaluation of noise reduction algorithms for particle-based fluid simulations in multi-scale applications. J. Comput. Phys. **325**, 380–394 (2016)

Author Index

Printed in the United States
by Baker & Taylor Publisher Services